AI for Sustainable Biological Systems

Kyla Latrice Tennin
https://orcid.org/0000-0002-9849-1316
University of Phoenix, USA & United Nations, USA & International Research Institute for Economics and Management, USA

Rohit Kumar Pal
https://orcid.org/0000-0002-7525-0771
Lincoln University College, Malaysia

Mitali Talukdar
https://orcid.org/0000-0003-2799-0043
Amity University, India

IGI Global Scientific Publishing
Publishing Tomorrow's Research Today

Vice President of Editorial Melissa Wagner
Director of Acquisitions Mikaela Felty
Director of Book Development Jocelynn Hessler
Production Manager Mike Brehm
Cover Design Jose Rosado

Published in the United States of America by
IGI Global Scientific Publishing
701 East Chocolate Avenue
Hershey, PA, 17033, USA
Tel: 717-533-8845 | Fax: 717-533-7115
Website: https://www.igi-global.com E-mail: cust@igi-global.com

Library of Congress Cataloging-in-Publication Data

LCCN: 2026031963 (CIP Data Pending)
ISBN13: 9798260026915
Isbn13Softcover: 9798260026922
EISBN13: 9798260026939
British Cataloguing in Publication Data
A Cataloguing in Publication record for this book is available from the British Library.

Table of Contents

Detailed Table of Contents

operational safeguards. By aligning data governance with planetary health equity, this chapter provides a practical roadmap for responsible AI deployment in post-pandemic global health systems.

E-Healthcare is an important development and sub-set of ICT and Computing, and it is simply the digitalisation of the healthcare systems. Digital Healthcare is essential for various professionals, including physicians, nurses, laboratory technologists, hospital and healthcare administrators, assistants to the healthcare systems, and patient and patient parties. E-Governance is also helpful in advancing the public healthcare and hospital management. India is become a digital economy now, and here, e-governance, m-governance, etc., are widely used in order to develop a Digital Healthcare Systems and a variety of programs, and agendas in Digital Healthcare has been This chapter is on Digital Healthcare Systems and deals with the foundation and the emerging technologies used in these systems. The work has been focused on uses of E-Governance in developing effective E-Governance systems with reference to India. Scenario including challenges and issues and probable solutions are also described in this work.

Artificial intelligence (AI) is transforming the biological sciences by enhancing data utilisation in the advancement of sustainable health, agricultural, and environmental issues. Biological systems that are powered by AI combine machine learning, bioinformatics, and computational modeling to analyze intricate biological data with high efficiency and precision. Applied medical research in medicine, AI has been used to improve disease diagnosis, pharmaceutical development and individualized healthcare by uncovering latent patterns in genomic and clinical information. In farming, AI biological strategies enhance the productivity of crops, disease diagnostics, and resource use and facilitate climate resilient and sustainable agricultural production. Intelligent biological interventions in the environment can be applied to monitor the ecosystem, control pollution, and protect biodiversity, as well as manage waste.

Climate change, habitat loss, and declining biodiversity present critical challenges to conservation. This chapter explores how Artificial Intelligence (AI) and digital technologies offer transformative solutions for these issues. It highlights AI tools such as machine learning, deep learning, computer vision, and natural language

Accurate species identification is essential for biodiversity monitoring and conservation, yet many species are represented by only a few labeled samples. Few-shot learning (FSL) addresses this limitation by enabling models to recognize new categories from minimal data. This chapter explores FSL methods for fine-grained bird species classification, a task characterized by subtle inter-class differences and high intra-class variability. Using the CUB-200-2011 dataset, two approaches are evaluated: an episodic meta-learning pipeline based on Prototypical Networks with a ResNet-18 backbone and a representation-centric pipeline using a self-supervised Vision Transformer (ViT-DINO) as a frozen feature encoder. Performance is assessed under standard 5-way 1-shot and 5-way 5-shot settings. The findings offer practical insights for deploying AI-driven species recognition systems in biodiversity monitoring and conservation.

The agricultural sector is seeing a profound transformation due to Artificial Intelligence (AI) future-generation precision farming solutions, which can optimize crop production, contribute to managing resources and are eco-friendly. AI enables real-time making of data-driven decisions on a scale never before witnessed; it is a mixture machine-learning, predictive analytics, the Internet of Things (IoT), and remote sensing technologies. These technologies furnish farmers with valuable insights regarding soil conditions their moisture level, nutrient levels, pests activity and the alteration of micro-climatic conditions and, consequently, these systems will allow farmers to measure the irrigation, fertilization and harvest protection with high accuracy. In addition, robotization and self-reliant machines constituting the AI-based automation improve enhance the accuracy of operation and mitigate the influence of the personnel their input to the wastage of resources. .

from nucleotides, capturing complex and long-range dependencies that traditional methods often miss. This chapter summarizes major deep learning architectures and workflows for DNA/RNA sequence analysis and highlights key applications such as regulatory element prediction, splicing and isoform modeling, RNA–protein interactions, and variant effect prediction, along with current challenges and future directions.

AI is rapidly transforming forensic biology by enhancing the identification and interpretation of target genetic sequences essential for criminal investigations. This work examines how AI-driven approaches support the detection of STRs, SNPs, mitochondrial, and epigenetic markers to significantly improve the speed, precision, and reliability of DNA profiling. Machine learning and deep learning models strengthen allele calling, mixture interpretation, phenotype prediction, ancestry estimation, and the analysis of highly degraded or contaminated samples. The chapter further highlights AI-app integration for real-world casework, multi-omics enhancement of unstructured forensic datasets, and resource optimisation for investigative efficiency. Ethical and societal concerns including data integrity, bias, and risks associated with large genetic databases addressed to promote responsible and secure adoption. Overall, the chapter demonstrates how AI integration is reshaping forensic genetics and guiding the future direction of modern biological investigations.

The Biopharma Industry burgeons the recent tech-driven economy with its multifaceted efficacy in curing rare diseases and high scientific, technical, and cost complexities. AI-driven biopharma establishments partner with conventional Biopharma companies to provide drug delivery efficacy. Studies exhibit synergistic effects in research systems when AI technologies are embedded into life science medicine processes. This drug delivery lowers costs, enables a high probability of success, multiplies intellectual properties, provides a competitive advantage in data, and encourages 'Innovative Biologics', leading to sustainable access to medicines. This chapter explores how emerging legal and regulatory frameworks address challenges of AI-driven Biopharma drug delivery, varying from AI-driven data ensuring safety, efficacy, potency, and purity tests for approval to intellectual property rights, sustainable access to medicines, and development of Innovative Biologics for rare diseases.

Preface

INTRODUCTION

Humanity faces an unprecedented convergence of global challenges, including climate change, biodiversity loss, food insecurity, emerging diseases, and increasing pressures on natural resources. Addressing these interconnected crises requires innovative approaches that transcend disciplinary boundaries and embrace intelligent technologies capable of supporting resilient and sustainable biological systems. Artificial intelligence (AI), together with advances in data science, robotics, cloud computing, genomics, and computational biology, has emerged as a powerful catalyst for transforming the way societies understand, manage, and sustain the biological and ecological systems upon which life depends.

As editors, we recognize that the application of AI to biological systems extends far beyond technological advancement. It encompasses a broader commitment to achieving sustainable development, promoting environmental stewardship, enhancing healthcare delivery, strengthening food security, preserving biodiversity, and supporting responsible governance. The growing convergence of life sciences with intelligent computing presents remarkable opportunities for improving human wellbeing while simultaneously addressing ecological sustainability. However, these opportunities must be balanced with careful consideration of ethical, legal, social, and environmental implications to ensure that innovation serves both present and future generations.

This volume, AI for Sustainable Biological Systems, brings together scholars and practitioners from diverse disciplinary backgrounds to examine the transformative role of artificial intelligence across healthcare, environmental sciences, agriculture, conservation biology, genomics, and forensic science. The contributions presented in this collection demonstrate how AI-enabled systems can contribute to the achievement of sustainability goals while advancing scientific discovery and promoting equitable development. It is our hope that this book will serve as a valuable resource for researchers, policymakers, educators, and professionals seeking to understand and harness the potential of intelligent technologies in support of sustainable biological futures.

CHAPTER OVERVIEW

Chapter 1 explores the role of artificial intelligence in modernizing the natural sciences and advancing the United Nations Sustainable Development Goals. Particular attention is given to applications in healthcare, clean energy, predictive epidemiology, and responsible AI governance.

Chapter 2 examines the growing importance of big data analytics, robotics, and cloud computing in environmental information systems. The chapter highlights how these emerging technologies contribute to intelligent ecological management and the realization of Environment 4.0.

Chapter 3 introduces a restorative governance framework for AI-driven global epidemiology. Through the Socio-Ecological Responsibility model, the chapter addresses issues of algorithmic bias, environmental externalities, computational sustainability, and public health equity.

Chapter 4 investigates the role of e-governance in developing sustainable digital healthcare ecosystems in India. The chapter discusses the integration of emerging technologies and policy initiatives that support healthcare accessibility and efficient health administration.

Chapter 5 examines how AI-powered biological systems are transforming health, agriculture, and environmental management. Applications in precision medicine, sustainable farming, ecosystem monitoring, and pollution control are discussed within a broader sustainability framework.

Chapter 6 focuses on the application of artificial intelligence in biological conservation. The chapter presents AI-enabled tools for species monitoring, habitat mapping, wildlife protection, ecosystem restoration, and biodiversity preservation.

Chapter 7 explores AI as a digital frontier for mitigating human-wildlife conflict. By leveraging predictive analytics, computer vision, and intelligent monitoring systems, the chapter demonstrates how AI can facilitate sustainable coexistence between human populations and vulnerable species.

Chapter 8 investigates few-shot learning approaches for fine-grained bird species classification. The chapter highlights how advanced machine learning techniques support biodiversity monitoring and conservation efforts, particularly when limited training data are available.

Chapter 9 discusses AI-driven precision farming systems designed to optimize crop yields and resource utilization. The chapter illustrates how machine learning, IoT technologies, remote sensing, and predictive analytics contribute to sustainable agricultural practices.

Chapter 10 presents an improved semantic segmentation framework for rice leaf disease identification using U-Net architecture and local-based Sauvola thresholding. The proposed approach demonstrates the potential of deep learning to enhance food security through accurate disease diagnosis.

Chapter 11 evaluates hybrid machine learning models for plant disease detection using leaf imagery. The chapter compares multiple deep learning architectures and highlights their importance in supporting efficient and accessible agricultural diagnostics.

Chapter 12 provides a comprehensive overview of deep learning approaches for DNA and RNA sequence analysis. Applications including regulatory element prediction, RNA-protein interactions, splicing analysis, and variant effect prediction are examined alongside emerging challenges and opportunities.

Chapter 13 explores AI-driven genetic marker identification in forensic science and criminal investigations. The chapter discusses how machine learning and deep learning methods enhance DNA profiling, mixture interpretation, ancestry estimation, and forensic biological analysis.

Chapter 14 concludes the volume by examining legal and policy dimensions surrounding AI-enabled biopharmaceutical discovery. The chapter addresses regulatory frameworks, intellectual property rights, sustainable access to medicines, and the responsible development of innovative biologics.

CONCLUSION

The chapters presented in this volume collectively demonstrate that artificial intelligence has become a transformative force in the pursuit of sustainable biological systems. From healthcare and epidemiology to biodiversity conservation, precision agriculture, genomics, and forensic science, AI technologies are enabling new forms of knowledge generation and decision-making that can address some of the most pressing challenges confronting humanity. These contributions illustrate the immense potential of intelligent systems to promote ecological resilience, improve human wellbeing, and support sustainable development objectives.

At the same time, the integration of AI into biological systems raises important questions concerning ethics, governance, transparency, environmental responsibility, and equitable access to technological benefits. Sustainable innovation requires more than technical excellence; it demands interdisciplinary collaboration, responsible policy frameworks, and a commitment to balancing scientific progress with social and ecological stewardship. The future of AI-enabled biological systems will depend on the ability of researchers, practitioners, governments, and international institutions to work together in addressing these complex challenges.

As editors, we hope that AI for Sustainable Biological Systems will inspire further dialogue and collaboration among scholars and practitioners working across disciplines. We also hope that this volume contributes to the development of innovative and responsible approaches that harness artificial intelligence in service of both humanity and the planet. By bringing together diverse perspectives and emerging research, this collection seeks to illuminate pathways toward a more sustainable, resilient, and technologically empowered future.

Kyla Tennin

University of Phoenix, USA & United Nations, USA & International Research Institute for Economics and Management, USA

Rohit Pal

Lincoln University College, Malaysia

Mitali Talukdar

Amity University, Kolkata, India

Chapter 1

Implementing Artificial Intelligence to Modernize the Natural Science

Khushi Kamble
http://orcid.org/0009-0006-1846-8164
Vishwakarma Institute of Information Technology, India

Gitanjali Shinede
http://orcid.org/0000-0001-6623-8722
Vishwakarma Institute of Technology, India

Grishma Bhobate
http://orcid.org/0009-0007-7351-8723
Vishwakarma Institute of Technology, India

Sonal Fatangare
http://orcid.org/0009-0005-1185-8699
Vishwakarma Institute of Technology, India

ABSTRACT

This chapter, Implementing Artificial Intelligence to Modernize the Natural Sciences, explores how AI technologies can advance the United Nations Sustainable Development Goals, particularly SDG 3 (Good Health and Well-Being) and SDG 7 (Affordable and Clean Energy). The chapter examines AI's dual role in promoting innovation and enabling ethical, sustainable development. Applications in healthcare, including precision medicine, predictive epidemiology, AI-driven medical imaging, and natural language processing, are highlighted for improving diagnosis, individualized treatment, and early outbreak detection. In the energy sector, AI enhances grid efficiency, demand forecasting, renewable integration, and predic-

DOI: 10.4018/979-8-2600-2691-5.ch001

tive maintenance. Ethical, social, and environmental considerations, such as bias, data privacy, and carbon footprint, are critically discussed. The chapter concludes with guidelines for responsible AI governance, leadership, and interdisciplinary collaboration to maximize AI's transformative potential for science, equity, and environmental sustainability.

1. INTRODUCTION

Artificial Intelligence (AI) has become one of the most revolutionary technologies of the 21 st century that provide unparalleled possibilities to transform the societies, economies, and ecosystems. Being able to make decisions faster, predict, and automatize processes, AI can transform almost all spheres of human activity. There are two more areas of healthcare and clean energy that are not only the top priorities of global development but also where AI can play a significant role in the sustainable development. In spite of positive progress in terms of the Sustainable Development Goals (SDGs), there is still a high level of inequalities on the global scale. The United Nations Sustainable Development Report (2023) reveals that hundreds of millions of individuals still do not have uninterrupted power supply, and significant portions of the world population do not have access to basic healthcare services.

These inequalities are detrimental to social equity, economic development and environmental sustainability. With such scalable solutions like intelligent energy management, epidemic surveillance system, personalized medicine, advanced diagnostics, and predictive functions of renewable power, AI can solve these challenges. However, its implementation poses some important ethical and social issues, such as the bias of algorithms, the threat to information privacy, the vulnerability of cybersecurity, and the environmental imprint of the use of energy-intensive computing schemes. In this chapter, the author will look at the Sustainable Development Goals supported by AI: Community members should realize that the achievement of this goal is possible only when every individual recognizes that their health warrants consistent and careful attention. SDG 3: Good Health and Well-Being - that is, at all ages, it is possible to ensure healthy lives and promote well-being among all members of the community.

The reason is the need to achieve good health through SDGs. This chapter has offered a roadmap of how AI can be integrated into the world development strategies in a responsible manner in line with the 2030 Agenda through the presentation of conceptual frameworks, real-world case studies, and policy recommendations.

There are three main contributions of this chapter.

First, it makes a theoretical contribution by incorporating the socio-technical systems theory and digital governance systems into the analysis of the Artificial Intelligence applications in Sustainable Development Goals.

Second, it makes a practical contribution by exposing quantifiable evidence of AI-enhanced diagnostics in healthcare, predicting renewable energy, and optimization of grid.

Third, it contributes to the policy by suggesting a risk-based and structured regulatory framework and cross-sectoral governance model to support responsible AI deployment. Combined, these contributions make this chapter a policy and analytical roadmap of how AI can be integrated in to sustainable development strategies.

2. CONCEPTUAL FRAMEWORK AND SDG CONTEXT

The conceptual framework provides a framework of the paper by connecting the applications of Artificial Intelligence (AI) to the Sustainable Development Goals (SDGs). It shows how solutions that are provided by AI can be used as enablers in solving global issues like healthcare access, climate change mitigation, renewable energy integration, and sustainable economic development. Imposing AI innovations into the SDG context, in this chapter we are not only focusing on the technological potential but also on the contribution of AI innovations into the growth of inclusive, ethical and sustainable development. The framework introduced in this chapter will be based on the socio-technical systems theory, that points out interdependence of technological systems and social structures. AI is not an isolated phenomenon; its effectiveness is linked to the institutional policies, the mechanisms of governance, the cultural setting, and the involvement of the stakeholders. Also, the framework is compatible with the digital governance theory, which emphasizes accountability, transparency, regulatory control, and participatory political decision-making in the implementation of emerging technologies. Combining these theoretical pillars, the framework goes beyond the extremely technical approach and puts AI in the context of larger social and institutional ecologies. Notably, the framework involves feedback between governance frameworks and the result of technology. As an example, an AI-based healthcare diagnostics system can produce performance data that can be used to regulate quality, whereas an energy optimization system can also impact policy changes on approaches to integrate renewable systems. The mechanisms of feedback make it possible to have adaptive governance, perpetual improvement of the system, and enhanced alignment of the system with changing SDG priorities.

2.1 AI as a Facilitator of a Sustainable Development.

AI is starting to be viewed as a general-purpose technology with applications in the wide variety of fields, including medical imaging and renewable energy prediction. The World Economic Forum (2022) indicated that AI could aid advancements toward over 130 SDGs, especially in healthcare, education, and the energy system. The theoretical framework places AI as the driver of innovation and a policy issue. On the one hand, AI will boost automation, predictive analytics, and intelligent decision-making in all areas, which will expand human capacities and efficiency. Conversely, it needs to have strong regulatory measures, ethical values, and institutional protection against abuse, discrimination, and unfair distribution. Thus, AI can be conceptualized not as just a technical innovation but as itself a socio-technical system that needs to be coordinated at an interdisciplinary level and governed by policy.

2.2 Healthcare and Energy as Key SDG Pillars.

The energy and healthcare have a strong dependence on each other as a part of the SDG ecosystem. Sustainable energy access (SDG 7) has a direct positive linkage to core health, such as hospital activities, vaccine refrigeration and online telemedicine, enhancing the advancement of health and well-being (SDG 3). On the other hand, better health results will lead to increased productivity, social stability, and economic resilience, which subsequently determine energy demand and sustainable consumption patterns. Through these two sections, this chapter helps to understand that AI can be used as an interface between human well-being and environmental sustainability, and the social and ecological systems are interdependent.

Since not every student has equal access to education, it is important to examine the challenges encountered in equitable access. Although AI has a transformative potential, there are a number of structural obstacles to fair application. The digital divide is also a major challenge especially in the low- and middle-income countries with inadequate infrastructure, low technical skills, and poor data ecosystems that limit implementation. Large scale adoption is also limited by financial constraints and institutional capacity limitations. These infrastructural issues highlight the necessity of inclusive development of digital infrastructure, capacity building and international cooperation. Section 5 gives a more detailed discourse of the ethical, environmental, and governance aspects.

Table 1. Mapping AI Applications to SDGs 3 & 7

Domain	AI Application	Specific SDG Target Alignment	Measurable Contribution	Representative Example
Healthcare	Medical image analysis (Computer Vision, Deep Learning)	SDG 3.4 – Reduce premature mortality from non-communicable diseases	Improves early detection accuracy; reduces diagnostic time; increases survival rates	National Health Service AI-assisted cancer screening
Healthcare	Predictive analytics for epidemics	SDG 3.d – Strengthen early warning and risk reduction systems	Enables early outbreak detection; supports timely public health response	BlueDot COVID-19 early alerts
Healthcare	Telemedicine and mobile health platforms	SDG 3.8 – Achieve universal health coverage	Expands access to care in rural/underserved regions; reduces travel and treatment delays	eSanjeevani telemedicine platform (India)
Clean Energy	Renewable energy forecasting (ML/DL models)	SDG 7.2 – Increase share of renewable energy	Improves forecast accuracy (e.g., up to 20% improvement); reduces fossil fuel reliance	Google DeepMind wind power prediction system
Clean Energy	Smart grid optimization	SDG 7.1 – Ensure reliable and modern energy access	Enhances grid stability; reduces transmission losses; optimizes load balancing	AI-enabled smart grid systems in Denmark
Clean Energy	Carbon emission monitoring	SDG 7 & SDG 13 – Clean energy transition and climate action	Enables data-driven climate policy; improves emissions tracking accuracy	IBM Green Horizon Project

Table 1 demonstrates the use of Artificial Intelligence (AI) in supporting the United Nations Sustainable Development Goals (SDGs), in specific, SDG 3 (Good Health and Well-Being) and SDG 7 (Affordable and Clean Energy), in both healthcare and energy fields. AI aids SDG 3 in the healthcare sector in a variety of ways. Computer vision and deep learning improve medical image analysis, which improves the detection of the disease at an early stage, as evidenced by AI-based cancer screening programs in the National Health Service. Predictive analytics also enhance pandemic preparedness, and technology like Blue Dot gives early alerts in the case of the COVID-19 pandemic. Also, telemedicine and mobile health (mHealth) technologies increase health access to rural and underserved areas, such as large-scale initiatives in India. The clean energy industry domain AI pursues SDG 7 goals by enhancing energy efficiency and sustainable renewable resource incorporation. Machine learning algorithms can be used to predict renewable energy with high precision, as it happened in the case of the wind prediction system created by Google DeepMind that improves grid stability and bypasses the use of fossil fuels. In Denmark, AI-powered smart grids, shown to optimize energy distribution, help to better the efficiency of energy distribution and load balancing. Additionally, monitoring programs like the IBM Green Horizon project are involved in mitigation efforts targeting climate that are based on data. Altogether, AI can have a great influence on developing healthcare systems, improving the management of renewable energy and supporting the development of the environment in a sustainable manner in accordance with the SDG goals of the world.

Figure 1. Conceptual Framework of AI for SDG 3 & 7

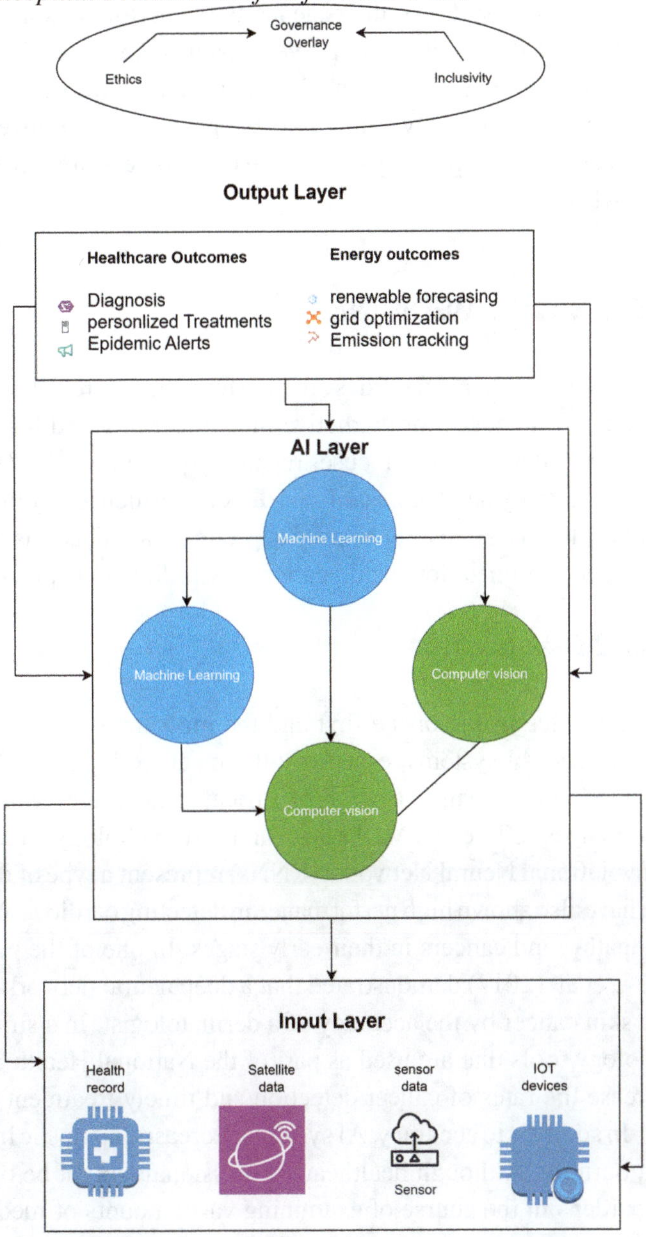

Author's own illustrates figure1, the conceptual framework of AI applications for SDF7- green and clean energy and SDG3- a good and quality life. The framework begins with the input layer, which integrates diverse sources of big data, including health records, satellite imagery, sensor networks, and IoT devices. These datasets are then processed within the AI layer through complex tools such as ML, computer vision, NLP those resulting outputs generate impactful outcomes across both sectors: in healthcare, this includes improved diagnosis, personalized treatment, and epidemic alerts, while in energy, it enables renewable forecasting, grid optimization, and emission tracking. Overarching these processes is a governance layer that emphasizes ethics, transparency, and inclusivity to keep myself transparency with equitable AI spread.

3. AI IN FIELDS LIKE MEDICINE

Around the world healthcare facilities are transforming as a result of AI, allowing it feasible for better diagnosis, for predictive model learning and improving more effective operations. SDG 3, which focuses regarding preventing diseases, disasters and equal access to quality healthcare which is directly influenced by these advancements. Four main classifications can be used to group AI applications in health care diagnosis, treatment optimization, healthcare accessibility, and epidemic control.

3.1 Advanced Diagnostics

Medical diagnostics is one of the first and the most important applications of AI in healthcare. The AI systems, especially the machine learning (ML) or deep learning (DL) models can be used to analyze intricate medical information, including ultrasound images, CT scans, MRI scans, and histopathology slides, with high accuracy. Convolutional Neural Networks (CNNs) represent a type of deep learning models which have also shown high performance in detecting cardiovascular disease, diabetic retinopathy, and cancers in their early stages. In one of the pioneering researches, Esteva et al. (2017) demonstrated that a deep neural network was capable of identifying skin cancer by the accuracy of a dermatologist. In a similar manner, AI-based radiology tools that are used as part of the National Health Service have helped to increase the rates of cancer detection and timely treatment intervention (Topol, 2019). In addition to accuracy, AI systems decrease diagnostic inconsistency and physician burnout. Although healthcare professionals might be overwhelmed by cognitive burdens in the course of examining vast amounts of medical images, AI-based decision-support systems offer reliable and scalable analysis. Notably, they are intended to supplement, rather than substitute, clinical expertise, which would improve the delivery of healthcare in general.

3.2 Personalized Medicine and Drug Discovery

Personalized medicine has implemented the predictive functionality of AI where tailored medical remedies are offered to individual patients, accompanied by genetic, clinical and lifestyle information. Machine learning is used to examine electronic health records (EHRs), genomic information, and patterns of biomarkers to predict the development of a disease and the most effective treatment plan. AI in the drug discovery field is playing an essential role in decreasing development time by minimizing the research costs by solving the potential therapeutic compounds using molecular modeling and deep learning simulations. Indicatively, Atomwise applies the deep learning algorithms to determine the interactions between molecules as well as speeding up early drug screening (Vamathevan et al., 2019). This method allows detection of promising drug candidates faster than when using the traditional laboratory processes. The simulations based on AI during the COVID-19 pandemic were instrumental in hastening the process of creating vaccines and finding possible antiviral medications. Moreover, the predictive models can help to manage chronic diseases like diabetes, hypertension, and cardiovascular diseases as they conduct risk assessment and early intervention recommendations via personalized models (Shah et al., 2019). Such innovations reveal how AI contributes to precision medicine in terms of better treatment efficiency and decreased time and cost obstacles during the pharmaceutical development.

3.3 Healthcare Accessibility

AI-based telemedicine systems have greatly increased the availability of healthcare, especially in low- and middle-income nations. The World Health Organization (2023) reports that more than 60% of the member states indicated they had adopted telemedicine more after the COVID-19 pandemic. Several millions of remote consultations have been enabled each year in India through government-sponsored digital health programs that save time on travel and delay in treatment in rural communities. The AI-based mobile health (mHealth) applications assist in triage, symptom check, and remote patient monitoring. The research suggests AI-based teleconsultation systems would help decrease outpatient work by about 20-30 percent in densely populated areas and increase the rate of early disease detection. AI-enabled telemedicine can decrease geographical and infrastructural obstacles and thus, directly address SDG 3 goals on universal health coverage and equal service provision.

3.4 Epidemic Surveillance and Preparedness

AI-driven epidemic surveillance systems analyze travel data, social media signals, and epidemiological reports to identify emerging outbreaks. For instance, BlueDot detected early warning signals of COVID-19 before formal global alerts were issued.

However, AI-based surveillance systems also face limitations. Predictive models may generate false positives, leading to unnecessary public concern or resource misallocation. Data quality inconsistencies across regions can reduce model reliability, particularly in low-income countries with limited reporting infrastructure. Additionally, scalability challenges arise when integrating heterogeneous datasets across borders. AI enhances outbreak detection speed, its effectiveness depends on robust data governance frameworks, international cooperation, and transparent validation mechanism

Table 2. Key AI Applications in Healthcare

Applications	AI Technology	Impact on SDG-3	Example
Medical imaging analysis	CNN, DL	Early disease detection	Skin cancer classification (Esteva et al., 2017)
Data analysis	ML, Data collection modelling	Risk identification, chronic disease management	Diabetes monitoring (Shah et al., 2019)
Personalized medicine & drug discovery	DL, Molecular simulations	Reduced drug development time, tailored treatments	Atomwise AI drug discovery
Telemedicine & mHealth	NLP, Chatbots, ML	Expanded access in rural/ underserved regions	eSanjeevani (India)
Epidemic surveillance	ML, Big data analytics	Early outbreak detection, pandemic preparedness	BlueDot COVID-19 alerts (Bogoch et al., 2020)

This table2 highlights how various AI technologies support SDG3 by means of innovative healthcare applications. DL and CNNs promote the rapid identifications of disease in medical imaging, including skin cancer classification (Esteva et al., 2017). Machine learning (ML) and data modelling aid in risk identification and chronic disease management, exemplified by diabetes monitoring (Shah et al., 2019). In personalized medicine and drug discovery, DL and molecular simulations reduce drug development time and enable tailored treatments, as seen in Atomises AI platform. Telemedicine and mHealth solutions using NLP, chatbots, and ML, like India's Sanjivani, improve healthcare access in rural regions, while epidemic surveillance using ML and big data analytics, such as BlueDot's COVID-19 alerts (Bogoch et al., 2020), enhance early outbreak detection and pandemic preparedness.

Figure 2. AI in medical field Workflow

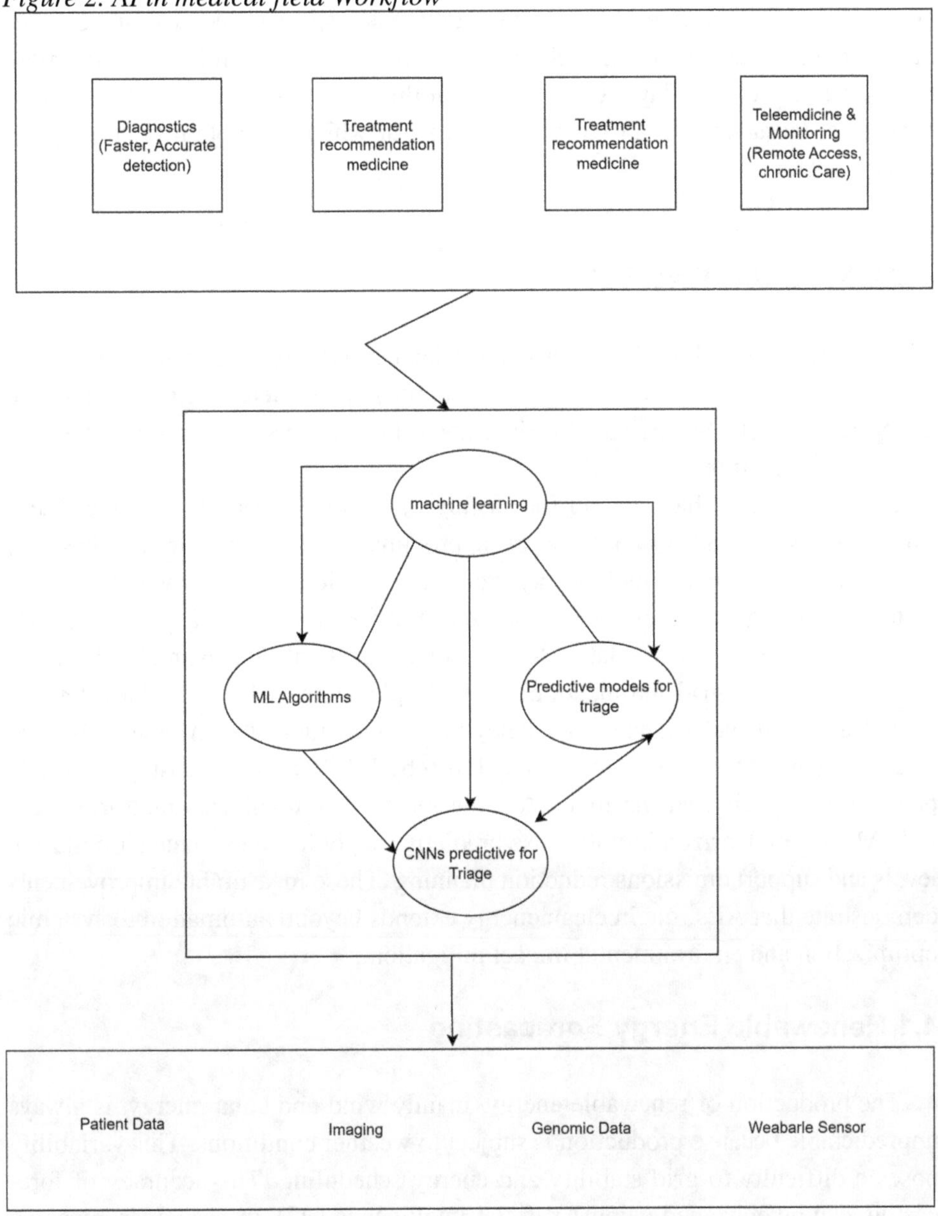

Figure2 is developed by the author and it displays the methods which have rendered hospital workflow accessible. The input stage, process by gathering an array of medical information which includes medical information, clinical imaging, electronic health records (EHRs) and wearable sensor learning. The computer vision layer handles this information using CNNs, ML algorithms, and NLP for disease development analysis and evaluation. Relevant medical result is generated from the output phase, including early alerts over outbreaks, specific therapy suggestions, distant tracking and medical solutions for long-term care, and faster and more precise tests. A control layer lies above all of this, providing security of data, bias, decreased, moral control and transparency, all of which promote robotic usage in medical sector.

4. AI IN CLEAN ENERGY

AI contributes to SDG 7 by improving efficiency, reliability, and sustainability in energy systems. In renewable energy forecasting, machine learning models analyze historical weather patterns, satellite data, and grid demand signals to optimize solar and wind integration.

For example, AI-based wind forecasting models developed by Google DeepMind improved wind power value by approximately 20% compared to baseline deterministic forecasting methods by increasing prediction accuracy and enabling better scheduling decisions. This improvement refers to enhanced revenue capture and reduced energy waste relative to traditional statistical forecasting models.

Similarly, smart grid optimization systems implemented in Denmark have demonstrated improved load balancing efficiency and reduced transmission losses. AI-based grid management can decrease operational costs by 5–15% while increasing renewable penetration capacity. Furthermore, AI-powered carbon monitoring platforms such as IBM's Green Horizon initiative use predictive analytics to estimate air pollution levels and support emissions reduction planning. These measurable improvements demonstrate that AI's role in clean energy extends beyond automation to systemic optimization and environmental impact mitigation..

4.1 Renewable Energy Forecasting

The production of renewable energy, mainly wind and solar energy, is always unpredictable because production is subject to weather conditions. This variability poses a difficulty to grid stability and energy scheduling. The accuracy of forecasting can be enhanced using Artificial Intelligence (AI), in particular, machine learning (ML) models, which interprets extensive amounts of past weather history, satellite images, and grid real-time data. Indicatively, AI-assisted wind forecasting machinery built by Google DeepMind enhanced the economic viability of wind power by around 20 percent relative to the conventional deterministic forecasting technique. This was done through minimization of prediction errors and providing more accurate day-ahead power scheduling. In contrast to the traditional statistical models, AI systems constantly improve based on the updated data trends, which enables them to dynamically adjust to weather changes. The 20% improvement is not just in terms of improved forecast accuracy, but also improved revenue collection and less curtailment of energy. AI will reduce dependence on auxiliary fossil-fueled generation and prevents operational inefficiency by helping grid operators to forecast power production more closely. As such, renewable integration is more dependable, less expensive and greener. Those quantifiable benefits prove the fact that AI-based predictive models will be in direct contribution to SDG 7 because of the increased stability of renewable energy and the reduced emissions related to grid imbalances.

4.2 Grid Optimization

The modern electricity grids are intricate systems that are required to maintain a constant balance between the supply and demand. The growing adoption of renewable sources of energy further complicates and makes operations more unpredictable. AI-based grid optimization systems improve stability by working out real-time information of smart meters, sensors, and distribution networks. Machine learning is used to determine the peak loads and identify the differences in energy consumption patterns. According to these forecasts, the automated control systems modify the distribution flows and manage resources in the most efficient way. The real time responsiveness minimizes losses in transmission and eliminates overloading of the system. In nations like Denmark, AI-based grid management has helped increase the efficiency of load balancing and allowed more wind power to be added to the country grid. Research shows that optimization with the help of AI could save costs by 5-15 percent in comparison to conventional rule-based control systems. There is also predictive maintenance algorithms which point to equipment faults before it fails and reduce downtimes, and enhance infrastructure life. AI-based optimization systems can be of significant importance in enhancing the affordability and clean-energy systems by promoting renewable integration, operational cost reduction, and grid reliability as per SDG 7.

4.3 Advance power control

AI boosts energy use in the household, office and industry with monitoring in real time, proactive management, and proactive repair. AI has been utilized by automated energy management system (SEMS) to optimize the efficiency of HVAC (heating, ventilation, and air conditioning) devices throughout a number of industrial operations. IBM's green future project for example relies on AI to evaluate energy use in towns and industrial places, offering energy saving strategies and maintaining a watch on pollution level (IBM, 2020). AI-based energy management systems in industrial and construction environments have demonstrated energy savings ranging from 10% to 30%, without compromising operational productivity.

4.4 Carbon Emission Monitoring and Climate Policy

AI also proves vital to evaluating greenhouse gases and crafting climate change initiatives. AI has the capacity to predict pollutants from manufacturing cities and factories using satellite images, IoT devices, and ML algorithms. By promoting data-based decisions, this data ensures that power change efforts are measurable and successful. AI-based systems can simulate scenarios involving battery deployment, renewable integration, and carbon pricing strategies, enabling evidence-based climate policy decisions.

4.5 AI-based green energy evaluation:

Case Study 1: Denmark's Wind Integration

Denmark predicts produce output and improves grid stability by deploying huge-scale wind energy using Artificial intelligence models. The integration for ML has raised details use of renewable power, lowered energy limitations, and improved forecasting precision. (Lund et al., 2020).

Case Study 2: Google DeepMind Wind Forecasting

By computing windmill capacity 36 hours in the future the search engine in deep neural networks AI model increases the range of clean energy and decreases dependency on crude oil generation. (DeepMind, 2019).

Case Study 3: IBM Green Horizon Project (China)

AI system studies the data on energy consumption and air pollution in Chinese cities, boosting energy management and encouraging ethical urban planning. (IBM, 2020)

Case Study 4: Residential Smart Grids in California

Distributed energy resource management and operated by AI which promotes family solar power and battery storage, managing demands improving availability and reducing carbon emission. (Zhang et al., 2021).

4.6 Ethical, Social, and Environmental Considerations

While AI significantly enhances power along with supports the act of adoption of clean life, several challenges persist. The digital divide means that unequal access to AI-powered energy systems could exacerbate global inequalities. Dut to AI-manged networks is subject to harm that might damage critical systems, safety hazards are also a threat. Some of the benefits about using renewable energy could be offset by the negative impacts related to building large AI models, which use a lot of energy. Green AI principal helps in different methods which use low power, natural resources, and zero carbon computing I steps with the goal to minimize such dangers (Schwartz et al., 2020). In order to make sure which benefits of AI in the energy markets are shared equally all over the area.

Table 3. Key AI Applications in Clean Energy

Application	AI Technology	Impact on SDG 7	Example
Renewable energy forecasting	ML, DL	Optimized supply, reduced wastage	Google DeepMind wind prediction
Grid optimization	ML, Predictive analytics	Stable, reliable distribution	Denmark smart grid system
Smart energy management	IoT, ML, AI controllers	Reduced consumption, increased efficiency	IBM Green Horizon project
Carbon footprint monitoring	ML, Remote sensing	Data-driven policy, climate mitigation	Urban emission tracking in China
Demand response programs	ML, Real-time optimization	Flexible energy consumption, reduced peak loads	California residential smart grids

This table3 shows how AI technologies advance SDG 7 (Affordable and Clean Energy). ML and DL enable renewable energy forecasting (e.g., Google DeepMind wind prediction) to optimize supply. Grid optimization with ML ensures reliable distribution (Denmark smart grid), while IoT and AI controllers improve energy efficiency (IBM Green Horizon). ML and remote sensing monitor carbon footprints for climate mitigation (China urban emissions), and real-time ML optimization in demand response programs (California smart grids) reduces peak loads and promotes flexible consumption.

Figure 3. AI in clean Energy Framework

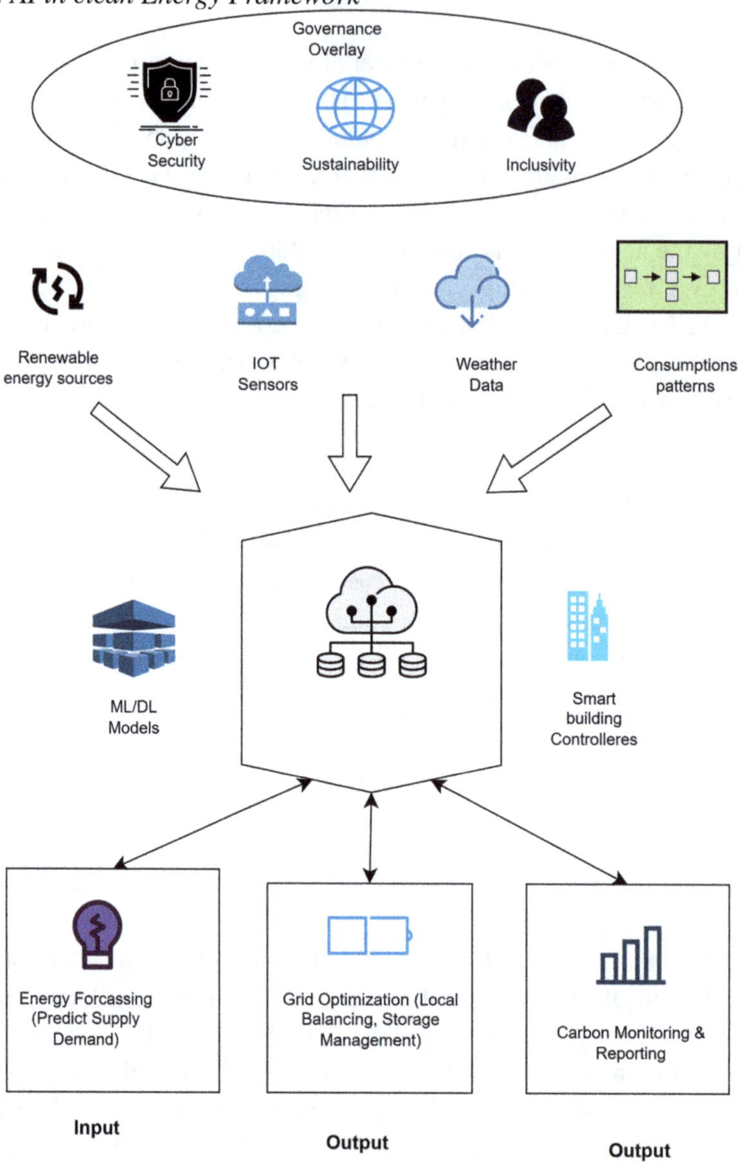

Figure3 Conceptual framework designed by the author which, illustrates the AI in Clean Energy framework. Inputs to the system utilize renewable energy source such as wind and sunlight IoT sensors, weather data, and consumption patterns. The AI layer leverages ML and DL predictive framework, grid improvement algorithms, and smart building controllers to process this data. Some of the results have carbon emission while reporting network over balancing loads and storage management

and energy forecasting for assessing both demand and supply. A governance layer which covers equality, conservation and security stands above this structure.

5. ETHICAL AND ENVIRONMENTAL DIMENSIONS OF AI FOR SUSTAINABLE DEVELOPMENT

The implementation of Artificial Intelligence (AI) in the medical field and clean energy offer the potential of revolution, but they also bring about the complicated ethical, environmental, and governance problems. In order to be in line with the Sustainable Development Goals (SDGs), AI systems should be built and deployed in a systemic ethical framework. This part brings major concerns together to five linked dimensions namely; fairness, privacy, environmental sustainability, social impact and governance.

5.1 Fairness and Algorithmic Bias

The AI systems are trained on historical data which can capture inequities in the society. Unless these biases are addressed, AI-based applications in the medical diagnosis of diseases or energy distribution could potentially discriminate against vulnerable groups. To illustrate, AI models used in the medical field that are mostly trained on data of high-income groups might not work well with other demographic groups. In the same vein, the energy optimization algorithm can focus on efficiency in urban areas but ignore underserved rural populations. The assurance of fairness should be ensured by the presence of varied training datasets, clear validation procedures and ongoing bias auditing systems. The deployment of AI ethically should be based on inclusivity so that it does not increase the existing disparities.

5.2 Privacy and Data Security

Healthcare systems based on AI heavily depend on the sensitive patient information, such as medical history, genetic, and biometric data. Hackings and unauthorized intrusions could jeopardize confidentiality of patient information and lead to mistrust. Smart grids and smart meters in energy systems can provide detailed consumption patterns that are potentially useful in learning behavioral information about households. These systems are prone to abuse without effective encryption, anonymization protocols and cybersecurity measures in place. So, effective data governance guidelines, safe infrastructure architecture, and compliance enforcement systems are needed to make AI implementation responsibly.

5.3 Environmental Sustainability of AI Systems

Although AI helps in the integration of renewable and monitoring of emissions, the computation process may be energy-consuming. Making large machine learning models involves large electric power usage, especially in data centres. One of the most quoted research projects by Emma Strubell et al. (2019) estimated that the production of a single large deep learning model could release the amount of carbon emissions as many as 5 vehicles throughout their lifetimes. These findings demonstrate the irony of employing energy consuming AI systems to ensure environmental sustainability. Researchers suggest such solutions to this problem as energy-saving model architectures, green data centers with renewable energy sources, and carbon-conscious computing strategies. The development of AI should be sustainable as it should not have a negative impact on the environment.

5.4 Social and Economic Impact

The use of AI affects the labor markets, institutional capacity, and economic distribution. Automation in healthcare can change the workforce relationship, which means that medical professionals will have to develop new skills. The digital transformation applied in the energy systems is technical in nature and may have limited skills in developing countries. The deployment of AI can potentially increase world digital inequalities without the inclusion of capacity-building efforts. Education, retraining of the workforce, and fair infrastructure advancement are thus important to be invested in to make sure that AI leads to inclusive growth and not structural displacement.

5.5 Governance Framework for Responsible AI

It is imperative that good governance is deployed to make AI systems meet SDGs goals. Regulatory oversight, ethical auditing, transparency standards and public involvement should be a multi-stakeholder governance model. The governments need to provide explicit accountability bodies to the deployment of AI in vital areas like healthcare and energy. System performance, bias risks and environmental impact can be evaluated by independent entities that monitor it. Also, cooperation on a global scale is required in order to align standards and exchange best practices. Combining equity, privacy, environmental friendliness and non-discriminatory policy design, this governance system can ensure that AI is an instrument of sustainable development instead of an agent of unintended evil.

Table 4. Ethical, Social, and Environmental Challenges of AI

Dimension	Challenges	Mitigation Strategies	Example
Fairness & Inclusivity	Bias in datasets, inequitable access	Diverse datasets, participatory design, bias audits	AI diagnostics in rural populations
Data Privacy & Governance	Unauthorized access, sensitive data misuse	Encryption, anonymization, regulatory compliance	GDPR-compliant telemedicine apps
Environmental Sustainability	High energy consumption, carbon emissions	Green AI, renewable-powered data centers, efficient algorithms	Transformer model training optimization
Social Implications	Job displacement, digital divide	Reskilling programs, capacity building, equitable deployment	AI-powered energy management in developing regions

The table4 summarizes key dimensions of responsible AI implementation—Fairness & Inclusivity, Data Privacy & Governance, Environmental Sustainability, and Social Implications—along with their challenges, mitigation strategies, and real-world examples. It emphasizes reducing bias through diverse datasets, protecting sensitive data via encryption and regulatory compliance, promoting eco-friendly AI through efficient algorithms and renewable energy, and addressing social impacts like job displacement through reskilling and equitable deployment. Overall, it provides a concise overview of ethical and sustainable AI practices.

Figure 4. AI's Moral as well as cultural administrative system.

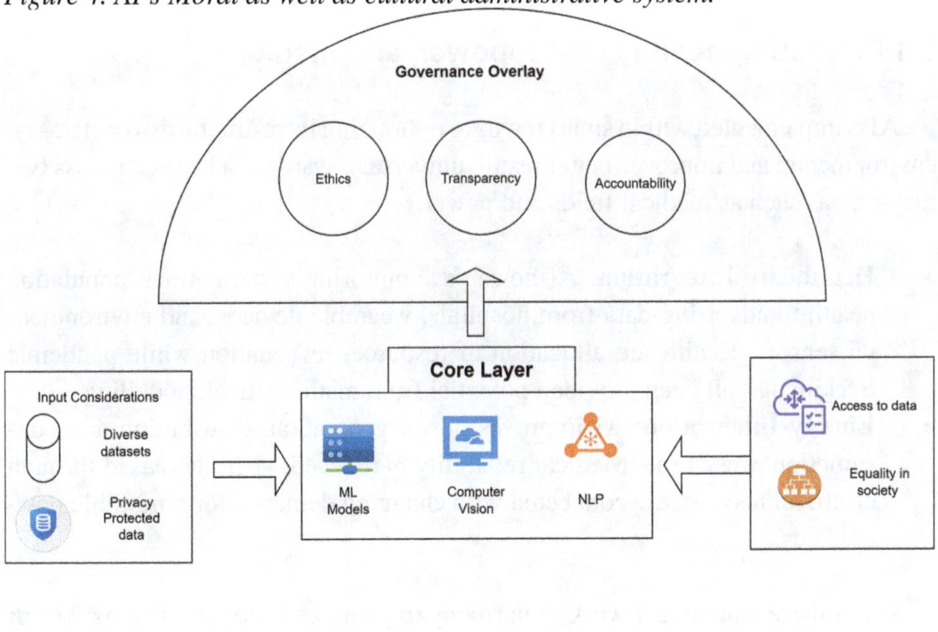

Figure4 is prepared by the author which shows, AI's Moral as well as cultural administrative system. The political and ethical administration in order to safe deployment, AI standards get a high value on combining scientific, ethical, and societal details. High-tech AI applications such as ML models, CV, and NLP are the main elements of the framework that give the details of the wise decision-making process. One of the simplest defines methods that can ensure that AI is ethical and behaves with integrity and trustworthiness is governance overlay because it entails simple ethical, accountable, and completely disclosed values. Consideration of inputs focuses on applying alternative and guarded data in a bid to reduce bias and protect the individual rights. The model tries to relate the AI advances to the larger ethical and social commitments such as environmental friendliness, societal equality and equal availability of health care provision and energy availability.

6. INTEGRATED AND INTERSECTING VIEWS

AI advantages increase as it evolves in all sectors and fields, filling the gaps in energy, medical, urban cities planning and environmental sustainability. A mix of AI neural study which may help in academics, and political leaders fulfil the SDGs. This part looks at the use AI level in global cooperation. interdisciplinary study and smart cities, focusing those manners in which these approaches advance inclusively healthy development.

6.1 Promoting healthy cities, power, and medicine

AI is implemented within smart towns to optimizing infrastructure over efficacy, environmental and improves better result. universal rewards can be seen across two major areas such as medical fields and power.

- **Healthcare Integration**: AI-powered monitoring systems track population health trends using data from hospitals, wearable devices, and environmental sensors. Healthcare allocation of resources, evacuation while pandemic tracking has all been provided potential from mathematical modelling.
- **Energy Integration**: AI improves power generation, distribution and consumption every time. Medical reliability is reduced when increased through electrical network are connected with changing demands for renewable power plants.

All of these component work in harmony to generate beneficial results AI, for example may ensure continuous medical care by focusing on supplying power to medical facilities. In same way, green infrastructures support physical and mental wellness while reducing its impact on the planet.

Example: To improve urban areas and citizen welfare, the city of Singapore utilizes AI- driven energy control and wellness tracking tools that combine building power system roadways information and healthcare alerts. (Chong et al., 2020).

6.2 Interdisciplinary Research and Innovation

AI product demand strong interdisciplinary collaboration to ensure their effectiveness and responsible deployment. Energy engineers provide their knowledge over the generating of green energy grid optimization, and to maintain stems, while healthcare medical provide important data that is useful in clinical taking actions and personalized medicine. Although social scientists assess the technologies moral implications, societal impacts, and social considerations, computer scientists are key in the development of algorithms, data models, and predictive analytics. Regulation experts promote these goals by providing governance, scalability and regulatory compliance. While integrated, those various efforts result in complete solutions that tackle social, ecological and scientific issues. For example, urban areas can successfully handle outbreaks, while maintaining a continuous supply of electricity for important medical facilities using prevention of, AI tools and energy planning.

6.3 Global Collaboration and Knowledge Sharing

- **Technology transfer**: Sharing AI models, software, and best practices between high-income and developing regions.
- **Capacity building**: Training programs, workshops, and regional AI hubs to develop local expertise.
- **Data sharing and interoperability**: experts to nearly globe can develop relevant AI solutions using standardized, secure datasets.

Example: The WHO's Digital Health Initiative offers intelligent health tracking system over events built through global relationships, ensuring equitable access to digital methods of avoiding diseases and treatment (WHO, 2022).

6.4 Benefits of Cross-Sectoral AI

Collaboration across sectors AI builds links and provides an array of benefits for working together across sectors. By doing predictive models and shared data systems, they raise speed by lowering redundancy and improving the use of resources. These methods also improve durability as AI-enabled cross-sector integration makes it simpler and faster for people to react to events like pandemics and natural disasters. By integrating health and power solutions, they boost health for people in other words reducing the impact on the locations. They also approach and support universal and enable particular measures to reach people who are excluded by taking on multiple social factors.

Table 5. Cross-Sectoral AI Applications

Domain Integration	AI Application	Impact on SDGs	Example
Healthcare + Energy	Smart hospital energy optimization	SDG 3 & 7 – Reliable care with clean energy	Singapore AI-driven smart hospitals
Urban Planning + Energy	Smart grid + traffic & building management	SDG 7 & 11 – Sustainable, efficient cities	Smart city energy planning in Singapore
Healthcare + Epidemic Response	Predictive modelling + energy prioritization	SDG 3 & 7 – Epidemic preparedness + continuous services	AI-assisted pandemic response in WHO initiatives
Global Collaboration	AI knowledge sharing & capacity building	SDG 3, 7 & 17 – Global partnerships & inclusive technology	WHO Digital Health Initiative

The cross-sectoral AI applications in Table 5 develop several Sustainable Development Goals (SDGs) concurrently through system design. In contrast to sector-specific applications presented above, these illustrations show the way AI works in the interconnected fields of healthcare, energy, city planning, and the governance of the world. As an example, smart hospital energy optimization systems is a combination of renewable energy prediction as well as medical infrastructure management, to achieve continuous care to patients, as well as cutting down on carbon emissions. On the same note, smart city platforms based on AI also combine traffic control, energy efficiency of buildings, and grid optimization to enhance urban sustainability and energy resilience. The predictive epidemic response systems also demonstrate cross-sectoral integration in which the healthcare demand forecasting is integrated with energy prioritization systems to sustain emphasized critical infrastructure in the case of a public health emergency. The global scale is where AI knowledge-sharing programs can build the international collaboration and achieve SDG 17 by means of coordinated digital capacity building. These illustrations show that the effects of AI go beyond the specific industries and also act as an integration facilitator of sustainable development.

Figure 5. Cross-Sectoral AI Integration in Smart Cities

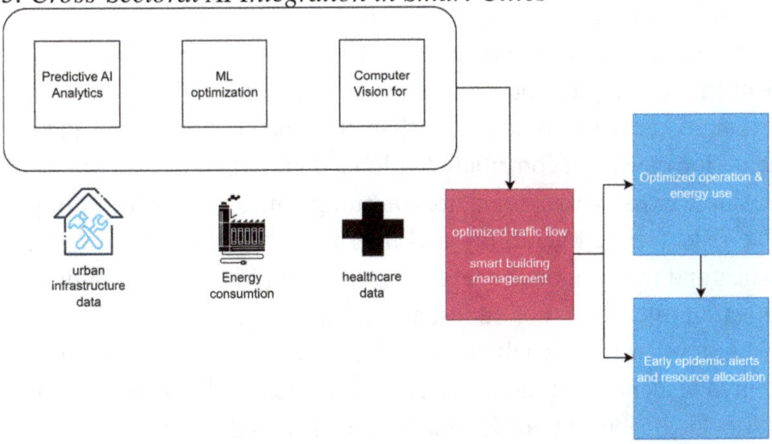

In figure5 Author indicates that the building of more intelligent, stronger, and more equitable cities is driven by sector -wide integration of AI. The structure begins with an array of inputs which provides an in-depth understanding of city system, such as usages of energy reports, healthcare data, and data about urban infrastructure. Real-time monitoring and intelligent decisions became possible on the AI level with the help of such tools as computer vision, forecasting, and ML improvement. The resultant outcomes of these skills are enhanced building and traffic management, early notification of outbreaks, proper use of assets and enhanced operations and use of medical facilities and energy. A control level that regulates this approach is that of ensuring certain AI implementation is ethically acceptable, privacy-focused, and promotes long-term growth and fairness. All these demonstrate the ways AI can improve the existing intelligent systems in terms of their flexibility speed and smart city justice.

6.5 Limitations and Scalability Challenges

Although its implementation in the healthcare sector and renewable energy systems has a promising transformative potential, there are serious structural and contextual barriers to its implementation. Increased use of AI in low-income and developing areas is limited by the lack of digital infrastructure, unreliable electricity access, and lack of technical skills. Most AI systems require high-quality and representative data, which might be inaccessible in resource-constrained settings and thus undermine the reliability, accuracy and fairness of a model. Scalability also has financial and institutional problems.

The training and implementation of AI systems will necessitate a prolonged investment in the computational infrastructure, cybersecurity systems, and workforce training. Unless fair funding mechanisms and support across the world, the implementation of AI can contribute to global digital divides instead of alleviating them. Besides, cross-border regulatory arbitrage makes it harder to implement at scale. The risks to interoperability and international cooperation in global health surveillance and transnational energy systems could be caused by differences in data protection rules, codes of ethics, and compliance.

The solutions to these challenges include a context-sensitive policy formulation, the involvement of digital infrastructure in the development, and the building of institutional capacity on a long-term basis. It is only by means of a concerted administration and equitable distribution of resources that AI can play a beneficial role and be sustainable towards the global development goals.

7. POLICY FRAMEWORK FOR RESPONSIBLE AI DEPLOYMENT

The policymakers should be organized and aligned to make sure that Artificial Intelligence is helping to achieve the SDG 3 and SDG 7. Six governance pillars that are described below give a detailed governance framework.

7.1 Risk-Based Regulatory Framework

A regulatory model based on risks should be implemented by governments that classify AI systems based on the possible impact on society. Mandatory audits and transparency reporting and compliance should be applied to high-risk applications - medical diagnostics and critical energy infrastructure. Algorithms and cybersecurity strength as well as environmental impact should be assessed by independent oversight bodies prior to mass use. The concept of ethics and accountability mechanisms is described in the seventh chapter of the Code of Ethics.

7.2 Ethical Standards and Accountability Mechanisms

The role of developers, operators and institutions should be specified by clear ethical guidelines. The system should have mandatory impact assessment to determine the bias, risks on privacy, and the environmental costs before the system is implemented. Audit trails and public reporting systems increase the level of transparency and develop confidence in AI-driven systems.

7.3 Sustainable AI Infrastructure

The development of AI which is energy-efficient needs to be encouraged by governments and non-governmental entities. Strategies include:

- Rewarding data centers driven by renewable energy.
- Investing in low-energy model architectures research.
- Instituting carbon reporting on big-scale AI training. Green AI ecosystems can be faster to develop through the development of public-private partnerships.

7.4 Capacity Building and Workforce Development

Policymakers need to invest in human capital in order to stop digital inequality. Strategies that could be measured are: • AI training of medical workers and energy workers funded by the government. • Introduction of AI ethics and sustainability courses to university education. • Knowledge-sharing activities between the developed and developing countries at a global level. • Specialized funding sources on AI application in line with SDGs. These programs guarantee permanent institutional stability and collective involvement.

7.5 Data Governance and Privacy Protection

There should be robust data protection laws that would oversee the use of AI in sensitive industries. Healthcare and smart grid systems should use encrypted standards, anonymization protocol, and cybersecurity certification. International data-sharing deals have to find a compromise between innovation and privacy.

7.6 Cross-Sectoral Coordination and International Collaboration

Since SDGs are interconnected, AI governance must promote the cooperation of healthcare officials, energy regulators, environmental bodies, and digital policymaking bodies. The world bodies like the United Nations and the World Health Organization can enable the uniformity of standards and a common best practice.

Table 6. Policy Frameworks and Recommendations for AI

Policy Area	Recommendation	Impact on SDGs	Example
Ethical Governance	Transparent, accountable, fair AI	SDG 3 & 7 – Responsible deployment	EU AI Act (2021)
Data Privacy & Security	Compliance with HIPAA, GDPR, PDPA	SDG 3 & 7 – Protects sensitive data	Singapore PDPA
Capacity Building	AI training, regional hubs, reskilling	SDG 4 & 8 – Skilled workforce, equitable access	African Union AI Program
Public–Private Partnerships	Shared R&D, pilot testing, innovation incentives	SDG 9 & 17 – Innovation & collaboration	UK NHS AI collaborations
International Cooperation	Harmonized standards, tech transfer, collaborative research	SDG 17 – Global partnerships & inclusivity	WHO Global Digital Health Initiative
Environmental Sustainability	Green AI, energy-efficient models, renewable-powered data centres	SDG 7 & 13 – Clean energy, reduced emissions	Green AI frameworks

This table6 highlights policy recommendations for responsible AI and their contributions to SDGs. Ethical governance promoting transparency and fairness supports SDG 3 & 7 (EU AI Act, 2021). Data privacy compliance with regulations like HIPAA, GDPR, PDPA protects sensitive data (SDG 3 & 7, e.g., Singapore PDPA). Capacity building through training, regional hubs, and reskilling strengthens the workforce (SDG 4 & 8, African Union AI Program). Public–private partnerships encourage innovation and collaboration (SDG 9 & 17, UK NHS AI collaborations), while international cooperation via harmonized standards and tech transfer fosters global partnerships (SDG 17, WHO Global Digital Health Initiative). Finally, environmental sustainability through Green AI and energy-efficient models advances SDG 7 & 13.

Figure 6. AI Policy and Governance Framework

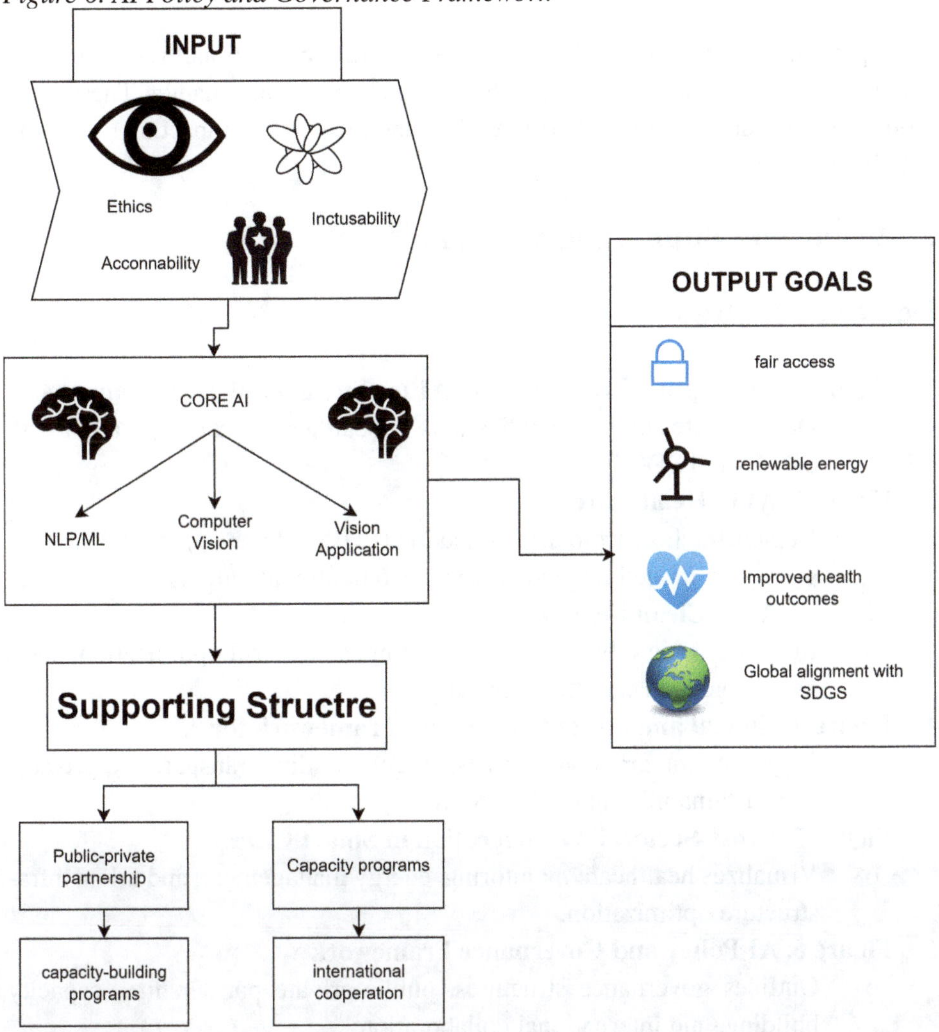

Figure 6 is constructed by the author which shows the AI policy and governance framework which highlights how ethics, technology and collaboration may be combined to generate fair and sustainable output. It its core the system involves AI applications like ML, NLP, CV particularly in the field of medical and power. This is enclosed by a governance overlay that promotes sustainability, ethics, accountability, transparency, and inclusivity to ensure responsible use. Supporting structures that allow for wide implementation and knowledge exchange, such as public-private partnerships, training programs and global collaboration, all these elements combine together to achieve important output such as equal allocation of resources, sustainable improve health, and global integration with the SDGs.

8. SUPPORTING MATERIALS

Important concept, follows by images belong to the additional resources provided in order to expand the concept that is presented in this chapter. These tools promote more study to allow others to understand the way AI impact on future and with bigger level.

8.1 Visual and Supporting Materials

Figures and Tables

1. **Figure 1. Conceptual Framework of AI for Sustainable Development**
    - o Demonstrates how AI applications in healthcare and energy align with SDG 3 and SDG 7.
2. **Figure 2. AI in Healthcare**
    - o Depicts the flow from data collection (EHR, wearables) to AI analytics, predictive modelling, and personalized treatment outputs.
3. **Figure 3. AI in Clean Energy**
    - o Illustrates renewable energy forecasting, smart grid optimization, energy storage, and carbon monitoring.
4. **Figure 4. Ethical and Social Governance Framework for AI**
    - o Shows the integration of fairness, accountability, transparency, privacy, and sustainability into AI systems.
5. **Figure 5. Cross-Sectoral AI Integration in Smart Cities**
    - o Visualizes healthcare monitoring, energy management, and urban infrastructure optimization.
6. **Figure 6. AI Policy and Governance Framework**
    - o Outlines governance structures, public-private partnerships, capacity building, and international collaboration.
7. **Tables**
    - o Tables 1–6 (in previous sections) summarize AI applications, case studies, challenges, and policy recommendations for easy reference.

8.2 Practical Implementation Resources

- **AI Tools for Healthcare:** TensorFlow, PyTorch, IBM Watson Health, Google Health AI
- **AI Tools for Energy:** HOMER Energy, OpenDSS, MATLAB Smart Grid Toolbox, EnergyPlus

- **Data Repositories:** WHO Health Data, IEA Energy Data, OpenAI datasets, Kaggle for AI applications

9. CONCLUSION

Artificial Intelligence (AI) is a disruptive technology that can become a powerful tool of innovation that promotes Sustainable Development Goals (SDG) 3 (Good Health and Well-Being) and SDG 7 (Affordable and Clean Energy). In this chapter, it was shown how AI-based solutions within the medical diagnostics, personalized medicine, epidemic surveillance, renewable energy forecasting, and smart grid optimization can be used to enhance efficiency, accessibility, and sustainability in key sectors.

The chapter under consideration has highlighted the socio-technical systems point of view that AI is not a pure technical innovation but a system that is a part of larger institutional, regulation, and societal systems. Although AI will improve the detection of diseases in the early stage, boost epidemic preparedness and enhance the integration of renewable energy, its application should be informed by the principles of fairness, transparency, environmental responsibility, and inclusive governance.

The analysis also pointed out that quantifiable benefits, including the accuracy of the forecasting process, lowering the costs of operations, and the accessibility of healthcare, help to prove the practical role of AI in sustainable development. Nonetheless, the issues connected to algorithmic bias, data privacy, energy-intensive computation, and global digital inequality need joined policy measures and risk-based regulations. Finally, AI is to be perceived as a tactical facilitator of cross-sectoral development. With ethical protection, sustainable infrastructures, and global cooperation, AI could serve as an engine of resilient healthcare systems, effective clean energy systems, and global inclusive development in line with the sustainability goals.

9.1 AI in Healthcare and Clean Energy

In healthcare, AI enhances diagnostics, predicts disease progression, personalizes treatment, and expands access through telemedicine and mobile health platforms. Predictive models, NLP, and CV allow faster and more accurate choice, while improving outcomes for patients and also mitigating the pressures of workforce shortages and pandemics.

AI boost energy storage, carbon emission tracking, network operations and the addition of clean energy source. By creating a successful, safe, and renewable power supply, AI helps in the battle towards climate change and the move to clean energy system. These all displays how AI may simultaneously tackle environmental concerns and advance human welfare.

9.2 Ethical, Social, and Environmental Considerations

This part highlights an argument that regardless of AI's huge potential there remain important barriers to its broad adoption. To promote fairness and equity algorithms which minimize bias and preserving and provide equal access to healthcare and power should be built. Because clear approval procedures for energy and medical data, along with strong, secure encryption, are vital for data privacy and governance. sensitive information is equally important also it becomes hard to overlook AI's impact on the planet, because training huge algorithms need a lot of power. This impact can be reduced through the use of sustainable AI methods and utilizing server power by renewable energy. Ai system that promotes honesty, openness, kindness and fairness is required to allow the successful use of AI with social goals and ethical values.

9.3 Cross-Sectoral Integration and Global Collaboration

Use of AI that are general and overlapping of which enhances its impact. The rise of smart city development also reflects the above combination from healthcare along with life transportation along with the emergence of towns leading to the creation of flexible, sustainable and productive urban environment. In order to make sure that the people with low to derate income can enjoy those benefits for AI when prevent a rise in digital and social disparities, it is relevant to cooperate globally, transfer innovation, and share knowledge.

9.4 Policy Recommendations

- **Ethical Governance:** Transparent, accountable, and fair AI systems.
- **Data Privacy & Security:** Compliance with international standards (HIPAA, GDPR, PDPA).
- **Capacity Building:** Training, regional AI hubs, and reskilling programs.
- **Public-Private Partnerships:** Resource pooling, pilot testing, and incentivized innovation.
- **International Cooperation:** Harmonized standards, cross-border research, and technology transfer.
- **Environmental Sustainability:** Green AI practices to reduce computational energy consumption.

9.5 Future Directions

- **Interdisciplinary Research:** Collaborative studies combining AI, healthcare, energy, environmental science, and social sciences.
- **Enhanced Predictive Analytics:** Real-time disease and energy forecasting using diverse, high-quality datasets.
- **Inclusive Technology Development:** Designing AI solutions tailored to underserved populations and resource-limited regions.
- **Ethical AI Innovations:** Developing frameworks that embed fairness, transparency, and sustainability at the core of AI systems.

9.6 Final Remarks

Concisely, AI should be understood as a transformative socio-technical enabler rather than merely a technical innovation. AI faces a transformation to improve the human health. Sustainability of the ecosystem in the world through smart city project enhancement of power system and medical. Nonetheless, the achievement of this capacity requires free rules, international collaboration. Ethical management and moral vision. AI will become engine of next generation in that field as long as it helps sustainability, justice and fairness as no one is left behind provided it is developed and implemented thoughtfully. Simply put, AI offers a unique opportunity to align technological innovation with the core principles of the 2030 Agenda for Sustainable Development.

REFERENCES

African Union. (2022). *AI & data science capacity building program*. https://au.int/en/tech/ai

Bogoch, I. I., Watts, A., Thomas-Bachli, A., Huber, C., Kraemer, M. U. G., & Khan, K. (2020). Pneumonia of unknown etiology in Wuhan, China: Potential for international spread via commercial air travel. *Journal of Travel Medicine*, *27*(2), taaa008. Advance online publication. DOI: 10.1093/jtm/taaa008 PMID: 31943059

Chong, W. Y., Liu, J., & Lee, K. T. (2020). AI-driven smart city initiatives for healthcare and energy optimization. *Sustainable Cities and Society*, *62*, 102381. DOI: 10.1016/j.scs.2020.102381

DeepMind. (2019). *Using machine learning to forecast wind energy*. https://deepmind.com/blog/article/forecasting-wind-energy

Esteva, A., Kuprel, B., Novoa, R. A., Ko, J., Swetter, S. M., Blau, H. M., & Thrun, S. (2017). Dermatologist-level classification of skin cancer with deep neural networks. *Nature*, *542*(7639), 115–118. DOI: 10.1038/nature21056 PMID: 28117445

IBM. (2020). *Green horizon project: AI for sustainable cities*. https://www.ibm.com/impact/green-horizon

Lund, H., Andersen, A. N., Østergaard, P. A., Mathiesen, B. V., & Connolly, D. (2020). From electricity smart grids to smart energy systems—A review of current status and trends in research. *Energy*, *198*, 117343. DOI: 10.1016/j.energy.2020.117343

Schwartz, R., Dodge, J., Smith, N. A., & Etzioni, O. (2020). Green AI. *Communications of the ACM*, *63*(12), 54–63. DOI: 10.1145/3381831

Shah, P., Kendall, F., Khozin, S., Goosen, R., Hu, J., Laramie, J., Ringel, M., & Schork, N. (2019). Artificial intelligence and machine learning in clinical development: A translational perspective. *NPJ Digital Medicine*, *2*(1), 69. DOI: 10.1038/s41746-019-0148-3 PMID: 31372505

Strubell, E., Ganesh, A., & McCallum, A. (2019). Energy and policy considerations for deep learning in NLP. *Proceedings of the 57th Annual Meeting of the Association for Computational Linguistics*, 3645–3650. https://doi.org/DOI: 10.18653/v1/P19-1355

Topol, E. J. (2019). *Deep medicine: How artificial intelligence can make healthcare human again*. Basic Books.

United Nations. (2020). *The sustainable development goals report 2020.* https://unstats.un.org/sdgs/report/2020/

Vamathevan, J., Clark, D., Czodrowski, P., Dunham, I., Ferran, E., Lee, G., & Bender, A. (2019). Applications of machine learning in drug discovery and development. *Nature Reviews. Drug Discovery, 18*(6), 463–477. DOI: 10.1038/s41573-019-0024-5 PMID: 30976107

World Economic Forum. (2022). *The global AI agenda: AI as an enabler for the sustainable development goals.* https://www.weforum.org/reports

World Health Organization. (2022). *Global digital health initiative.* https://www.who.int/health-topics/digital-health

Zhang, Y., Wang, J., & Xu, P. (2021). AI-based distributed energy resource management in residential microgrids: A review. *Applied Energy, 298*, 117199.

KEY TERMS AND DEFINITIONS

AI systems: Computational or algorithmic models to analyze data and make predictions and to be used in automated or semi-automated decision-making.

Artificial Intelligence (AI): A subfield of computer science that involves the creation of computational systems capable of executing tasks that are normally performed by the human intelligence, such as reasoning, learning, problem-solving, perception and decision-making.

Data governance: An organized policy, standards and procedures that guarantee secure, ethical and accountable management of data in its lifecycle.

Deep Learning (DL): An advanced field of Machine Learning where multi-layered neural networks are employed to capture more complicated patterns of large data sets, especially in image recognition, natural language processing and predictive analytics.

Machine Learning (ML): A subdivision of Artificial Intelligence that allows systems to figure out patterns based on the data that it gets and refines their abilities as they progress.

Renewable energy: Energy produced by naturally replicating sources like solar, wind, hydro and geothermal power which cause minimal effects to the environment as opposed to fossil fuels.

Risk-based regulatory framework: A governance system that classifies technologies according to the possible positive and negative effect of technologies on society and regulates them according to the degree of risk of the technologies.

Sustainable Development Goals (SDGs): A collection of 17 global objectives created by the United Nations to facilitate a social inclusion, environmental sustainability, and economic development by 2030.

Chapter 2

Big Data, Robotics, and Cloud Systems Emergence and Potentiality for the Sustainable Environmental Information Systems:

A Scientific Study

Prantosh Kumar Paul
http://orcid.org/0000-0003-1140-3369
Raiganj University, India

Tatayya Bommali
http://orcid.org/0009-0006-7240-4110
Sanketika Vidya Parishad Engineering College, Visakhapatnam, India

Rohit Kumar Pal
http://orcid.org/0000-0002-7525-0771
Lincoln University College, Malaysia

Sanjukta Chakraborty
http://orcid.org/0009-0001-5477-2044
Seacom Skills University, India

Mustafa Kayyali
http://orcid.org/0000-0003-3300-262X
Maaref University of Applied Sciences, Syria

Deep Debnath
Ghanikhan Choudhury Institute of Engineering and Technology (GKCIET), India

DOI: 10.4018/979-8-2600-2691-5.ch002

ABSTRACT

Environmental Informatics is an interdisciplinary field of study and practice comprises with environment, agriculture and agricultural systems, etc. and emerging ICTS are impacting advanced Environmental Informatics practice. Including other existing technologies, Big data analytics applications are rising in several environmental areas, as it is a global concern there are huge data generating each and every day from the ecology and environment and Big Data play a leading role in managing such environment related data management. Robotics including AI is responsible in promoting of intelligent systems, products and services to the environment and ecological aspects. Artificial Intelligence is widely useful. This chapter is to be a comprehensive in nature for inclusion of Big-data, Robotics & Allied Technologies, Cloud Computing for a healthy Environmental Systems towards an advanced Environment 4.0. And this chapter is focused on uses, impact and identification of issues Big Data, Robotics with AI, Cloud Computing for ensuring environmental sustainability effectively.

1. INTRODUCTION

As far as designing, developing, modeling of environmental processes some of the emerging environment subjects are growing and advancing viz. Environment Science, Environment Engineering, Environment Management, and in this cluster another subject has been developing i.e. Environmental Informatics which is comprises with the IT and allied technologies for the environment. Environmental Informatics integrated with environmental aspects with computing and other allied technology, tools, techniques and sub-technologies regarding development and solving of the problems of Environment and Ecology including allied sciences viz. agriculture and biological sciences. Robotics is a technology and systems run by the AI supported machines in order to learning and problem-solving affairs related to the environment and ecology. Robotics is dedicated in rationalize desired or specific goal in various areas, different such as healthcare, governance, business, administration, and in which ecology and environment also play a leading role. Cloud-based AI systems might enable smart environmental monitoring, resource optimization, and disaster prediction. Case examples show the benefits of AI-driven climate modeling and real-time pollution tracking. Challenges such as data security, ethical considerations, and computing restrictions are highlighted. An important contribution is the consideration of how cloud-based AI might democratize access to environmental data, enabling worldwide collaboration (Adanma & Ogunbiyi, 2024). The report indicates that AI and cloud computing are essential drivers of sustainable growth, however ethical and legal challenges remain. Different kinds of Information Technology are applications possible in variety of environmental work, and in this regard Cloud Computing play a leading role in large number of virtualization related jobs leading to the environment management, bringing and advancing ecology and sustainability and this trend is increasing gradually. Several government bodies including policymakers are encouraging merger of allied technology with environment and ecology and this is impacting not only environment but also in other areas like agriculture, oceanography, climatology, forestry, anthropology, etc. in order to build a environment friendly and sustainable world (Ahmad et al., 2022). A sustainable environment is one that involves the responsible use of natural resources, balance environmental, economic, and social objectives the reduction of pollution, and the conservation of biodiversity, which play a key role in creating a sustainable environment. Sustainability ensures ecological balance, supports human well-being, and promotes economic stability. The use of renewable energy such as solar, wind, water, waste reduction (recycling, composting), and responsible consumption are key aspects of a sustainable environment (Alotaibi & Nassif, 2024). Protecting the environment, forests, oceans, and wildlife preserves biodiversity, while sustainable agriculture ensures food security. Carbon emissions must be reduced through environmentally friendly transportation and energy-efficient practices to help mitigate global warming and climate change (Andronie et. al., 2022). As a whole all these technologies are emerging in building of new age environmental systems for the sustainability, though there are some issues and challenges.

2. OBJECTIVE OF THE CHAPTER

This work entitled 'Big Data, Robotics & Cloud Systems emergence & potentiality for the Sustainable Environmental Information Systems: *A Scientific Study'* is a theoretical and conceptual in nature and consist with following aim and objective.

- To learn about the basic and foundation of Big-Data, Robotics and Cloud Computing including features and charecterstics in brief.
- To know about the Environmental Systems, Environmental Informatics emergence in general and emerging trends, in brief.
- To get knowledge in stydying existing works and papers related to the Big Data, Robotics & Cloud Systems uses and implications in environment and ecological areas.
- To identify the basic and possible advantages of Big Data, Robotics & Cloud Systems in intelligent environmental management systems.

- To gather latest applications and utlizations Environment 4.0 in simple sense in regard to sustainability and ecology.
- To findout the challenges, issues and concern of several technologies specially AI & Robotics in sustainability and ecology and allied concerns.

3. METHODS USED

The chapter entitled 'Big Data, Robotics & Cloud Systems emergence & potentiality for the Sustainable Environmental Information Systems: *A Scientific Study*' is a conceptual work and therefore it has prepared as theoretical basis using existing literature such as books, chapters, journals related to the Big-Data, Robotics and Cloud Computing and allied technologies in respect of environment and ecology (Maxim, L., 2011). Contemporary projects, and service providers of ecology and environment also studied with this work, specially by analyzing relevant company's websites including related bodies, boards, and ideas.

4. REVIEW OF EXISTING WORKS

This particular work is based on existing related works analysis, and for this, it has gone through study last ten years significant works, and refer them in this Existing Review works section of the Chapter. Some of the important works are listed here alphabetically.

Alotaibi & Nassif (2024) this work examines the function of artificial intelligence (AI) in environmental monitoring, emphasising its capacity to improve data collecting, analysis, and decision-making. The authors investigate diverse AI-driven methodologies, including machine learning algorithms and predictive analytics, to enhance the precision of environmental evaluations ***Andronie et al. (2022),*** they analyses the amalgamation of big data, sensing technologies, and artificial intelligence inside the Internet of Robotic Things (IoRT) for environmental surveillance. The authors examine sophisticated computer methodologies that improve real-time data acquisition and processing, especially in perilous or isolated environments. ***Chen (2020),*** his study analyses the convergence of IoT, cloud computing, big data, and artificial intelligence across several fields, including environmental science, healthcare, and smart cities. The author illustrates how these technologies work together to enhance data-driven decision-making and boost efficiency in various applications.

Fu et al. (2023), this research provides a novel AI-based decision-making approach for air quality monitoring utilizing big data analytics. The authors suggest a technique that merges real-time sensor data with predictive algorithms to increase air pollution forecasting and risk assessment. ***Hamza et al. (2022),*** his research studies how big data analytics and AI may improve air pollution monitoring frameworks. The authors describe a data-driven approach that employs machine learning to process huge environmental information, enabling real-time pollution tracking and predictive modeling. ***Himeur et al. (2022),*** his review work presents a complete overview of AI-driven data fusion strategies for environmental monitoring. The authors study how machine learning, deep learning, and sensor fusion might boost real-time environmental assessments. Key uses include climate modeling, pollution detection, and biodiversity monitoring.

Kamyab et al. (2023), his work addresses AI and big data applications in water resource management, stressing their role in optimizing water distribution, quality monitoring, and conservation initiatives. The authors highlight how predictive analytics can assist uncover contamination sources, improve water treatment operations, and boost flood forecasts. Challenges such as data availability, computational complexity, and regulatory constraints are addressed. ***Wei et al. (2023),*** his work provides an AI-driven agricultural monitoring system that mixes cloud computing and big data for environmental safety assessments. The authors demonstrate how real-time sensor data, paired with machine learning algorithms, can optimize irrigation, insect control, and soil quality management. The research discusses problems such as network access in rural areas, data privacy concerns, and model interpretability. ***Wu et al., (2020),*** hey analyses how cloud computing and big data analytics enhance the monitoring of Sustainable Development Goals (SDGs). The authors stress the relevance of AI in processing environmental, social, and economic data to track progress toward sustainability targets. The report examines difficulties such as data standardization, security concerns, and the digital divide between industrialized and poor regions. ***Yaseen (2022),*** Researcher has proposes an AI-powered expert system for predicting and detecting heavy metal pollution in soil and water bodies. The authors suggest an edge cloud server architecture that enables real-time monitoring while minimizing computing delay.

5. ENVIRONMENTAL INFORMATICS: FUNDAMENTALS AND EMERGING TECHNOLOGIES

We're living in a time when environmental challenges are more pressing than ever, and the fusion of Robotics and Artificial Intelligence (AI) into our environmental systems is proving to be a game-changer. The real excitement here comes from these technologies' ability to tackle complex environmental issues with incredible precision, efficiency, and scalability. Think about how Robotics and AI are revolutionizing our interactions with the planet — from monitoring deforestation and tracking wildlife populations to optimizing resource use and forecasting the impacts of climate change (Paul et al., 2020). The global environmental crisis we face — driven by climate change, biodiversity loss, pollution, and resource depletion — calls for innovative solutions that go beyond traditional methods. Unfortunately, these conventional approaches often struggle, with limitations like high costs and dependency on manual efforts. Here's where Robotics and AI shine! They empower us to adopt automated, data-driven strategies that can analyze vast amounts of information in real-time. For instance, with AI algorithms capable of sifting through satellite imagery, we can identify illegal logging activities swiftly, while autonomous robots tackle oil spill clean-ups in oceans, requiring minimal human oversight.

What's even more thrilling is how Robotics and AI champion sustainability. These technologies are paving the way for smart systems that enhance energy consumption, minimize waste, and improve resource efficiency. In the realm of agriculture, AI-driven precision farming enables farmers to utilize water and fertilizers more effectively, which not only reduces environmental impacts but also boosts crop yields (Moqi, 2010). In urban settings, AI-powered smart grids help balance energy supply and demand, cut down greenhouse gas emissions, and promote renewable energy sources (Sarker et al., 2020). The push for integrating these technologies into environmental systems stems from a critical, urgent need to harness advancements for the sake of sustainability. As our global population continues to rise, the pressure on natural resources grows ever heavier, highlighting the necessity for innovative solutions to counter environmental degradation. The tools provided by Robotics and AI can help us design sustainable systems that address today's challenges while anticipating tomorrow's risks (Mykrä et al., 2007). By embracing these technologies in our environmental efforts, we are boldly stepping towards a more sustainable and resilient future.

Robotics is an exciting interdisciplinary field that blends engineering, computer science, and mathematics to create and manage robots that can perform tasks autonomously or semi-autonomously (Paul, 2022). The evolution of robotics has been nothing short of remarkable. From the industrial robots that revolutionized manufacturing processes to advanced humanoid robots capable of meaningful social interactions, this field has made incredible strides. Moreover, when it comes to environmental systems, the integration of robotics is crucial. These machines are instrumental in monitoring ecosystems, cleaning up pollution, and spearheading research in some of the most extreme environments on our planet (Paul & Aithal, 2020). Such advancements are shaping a sustainable future, marking an exciting chapter in the intersection of technology and environmental stewardship!

The potential of AI to sift through vast amounts of data and provide actionable insights is simply transformative, especially when it comes to tackling environmental challenges. For instance, AI algorithms are making waves by analyzing climate data to forecast weather patterns, optimizing energy usage in smart grids, and even identifying endangered species (Peters, 2020).. The possibilities are endless, and the impact on environmental systems and sustainability is profound!

6. SIGNIFICANCE OF BIG DATA AND ITS IMPACT IN ENVIRONMENTAL SYSTEMS: AN OVERVIEW

In recent years, the application of Data Analytics and Big Data has transformed numerous sectors, including environmental science and sustainability. The ability to collect, process, and analyze vast amounts of data has revolutionized how we understand, monitor, and manage natural ecosystems. Traditional environmental monitoring methods often struggled with data limitations, but advancements in data-driven technologies have enabled real-time insights and predictive capabilities, improving decision-making for sustainability efforts. This section explores the significance of Big Data and Data Analytics in addressing environmental challenges, supporting sustainability goals, and ensuring a more effective response to climate change, resource management, pollution control, and biodiversity conservation (refer Fig: 1).

Data Analytics involves examining raw data to uncover meaningful patterns, trends, and insights that can help in decision-making (Wei et al., 2023). Big Data, on the other hand, refers to extremely large datasets that require advanced computational techniques to process. In environmental science, these datasets originate from various sources such as:

- Satellite Imagery (for climate and land use monitoring)
- Remote Sensors (for tracking air and water quality)
- Internet of Things (IoT) Devices (for smart environmental monitoring)
- Citizen Science Contributions (such as crowdsourced environmental reports)

By leveraging these sources, researchers and policymakers can make more informed decisions to ensure environmental sustainability.

6.1 Role of Data Analytics and Big Data in Environmental Systems

Climate Change Monitoring and Mitigation, One of the most critical applications of Big Data Analytics is in monitoring and mitigating climate change. With the increasing frequency of extreme weather events, accurate climate models have become essential.

- ***Climate Prediction Models:*** By analyzing historical climate data and real-time satellite observations, data analytics helps in forecasting weather patterns, rising sea levels, and temperature changes.
- ***Carbon Footprint Tracking:*** Big Data enables organizations and governments to monitor carbon emissions and assess the impact of industrial activities on the environment.
- ***Climate Risk Assessment:*** Predictive analytics helps identify regions at high risk of climate-related disasters, allowing for early interventions and disaster preparedness.

Air Quality Monitoring and Pollution Control, Air pollution poses severe health risks and contributes to global warming. Data Analytics plays a key role in tracking air quality and identifying pollution sources.

- ***Real-Time Air Quality Index (AQI) Monitoring:***IoT-based sensors collect real-time data on pollutants like PM2.5, PM10, CO_2, and NO_2, helping authorities implement pollution control measures.
- ***Predictive Analytics for Pollution Control:*** Machine learning models can forecast air pollution levels based on historical trends, allowing preventive measures to be implemented in advance.
- ***Identifying Pollutant Sources:*** Advanced analytics can help pinpoint major contributors to pollution, such as industrial emissions, vehicular traffic, or deforestation, enabling targeted regulations.

Water Resource Management, Access to clean water is a fundamental necessity, and Big Data plays a significant role in managing and preserving water resources.

- ***Water Quality Assessment:*** Sensors deployed in rivers, lakes, and reservoirs continuously monitor water pollution levels, ensuring timely intervention to prevent contamination (Aguilar et al., 2021).
- ***Efficient Irrigation Practices:*** Data-driven agricultural techniques help optimize water use by analyzing soil moisture levels and weather conditions, reducing water wastage.
- ***Early Detection of Water Scarcity:*** Big Data Analytics helps predict drought conditions by analyzing precipitation patterns, groundwater levels, and temperature variations.

Biodiversity Conservation and Wildlife Protection, Big Data is instrumental in tracking and protecting endangered species and maintaining ecological balance.

- ***Wildlife Monitoring:*** GPS tracking and remote sensing technologies help track animal movements, habitat loss, and poaching activities.
- ***Species Population Analysis:*** AI-based models process vast ecological datasets to monitor species populations, ensuring conservation efforts are directed toward critical areas.
- ***Forest Cover Analysis:*** Satellite imagery and Big Data help detect illegal deforestation activities and support afforestationprograms.

Waste Management and Circular Economy, The increasing volume of waste generated globally necessitates efficient waste management strategies. Data Analytics facilitates:

- ***Smart Waste Collection:*** IoT-enabled smart bins track waste levels, optimizing waste collection schedules and reducing operational costs.
- ***Recycling Optimization:*** Machine learning algorithms help sort and categorize waste, improving recycling efficiency (Chen, 2020).
- ***Plastic Pollution Tracking:*** Big Data helps track plastic waste in oceans and landfills, assisting policymakers in designing sustainable waste reduction strategies.

Disaster Management and Response, Natural disasters such as floods, hurricanes, and wildfires require quick and data-driven responses to minimize damage and loss of life.

- ***Real-Time Disaster Prediction:*** AI-powered analytics assess weather conditions and seismic activities to predict disasters before they occur.
- ***Disaster Relief Coordination:*** Big Data supports the efficient distribution of emergency aid and resources during calamities.
- ***Damage Assessment:*** Post-disaster, drone imagery and satellite data help in assessing the extent of damage for better recovery planning.

Figure 1. Basic and emerging applications of Big Data Analytics in Environment & Ecological Systems

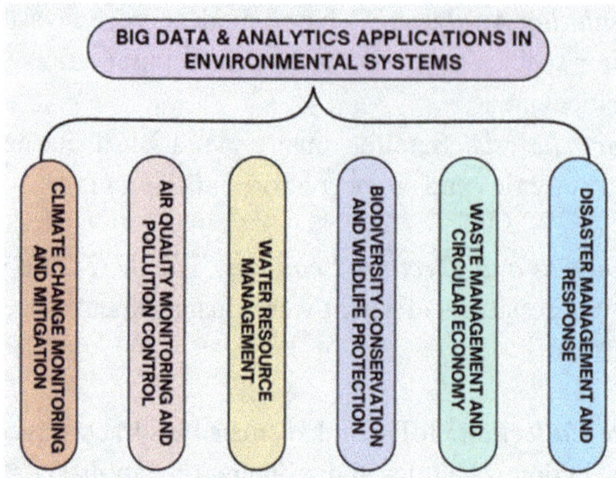

6.2 Challenges in Implementing Big Data for Environmental Sustainability

Despite its benefits, the integration of Big Data Analytics in environmental systems faces several challenges:

- ***Data Privacy and Security:*** With the increasing use of sensors and IoT devices, ensuring data security is a major concern.
- ***High Implementation Costs:*** The deployment of advanced analytics tools requires significant financial investments.
- ***Data Integration Issues:*** Environmental data comes from multiple sources and formats, making integration complex.
- ***Skill Gap:*** The need for experts in data science and environmental studies to interpret and analyze large datasets effectively.

The future of environmental sustainability is deeply connected with the advancements in Big Data Analytics (Cruz, F. B. D. L., &Barlaz, M. A., 2010). Emerging technologies such as AI-driven climate models, satellite-based Earth monitoring, and blockchain-powered carbon tracking are set to revolutionize environmental management. As more organizations and governments adopt data-driven strategies, there is hope for a more sustainable planet. However, addressing existing challenges such as data privacy, infrastructure limitations, and skill gaps is crucial for maximizing the potential of Big Data in environmental systems. In conclusion, Big Data

Analytics is no longer an optional tool but a necessity for tackling environmental challenges. By leveraging real-time monitoring, predictive analytics, and AI-driven decision-making, we can ensure that sustainability goals are met efficiently, leading to a healthier planet for future generations.

7. SIGNIFICANCE OF ROBOTICS INCLUDING AI IN ENVIRONMENTAL SYSTEMS AND SUSTAINABILITY

The fusion of Robotics and Artificial Intelligence (AI) with environmental systems is truly exciting, offering fresh and innovative solutions to some of the most urgent challenges our planet faces today. These cutting-edge technologies are transforming how we perceive, manage, and protect our environment, paving the way for a sustainable future. In this section dive into the wide-ranging applications of Robotics and AI in environmental systems, showcasing their remarkable impact on sustainability (refer Fig: 2 also).

7.1 Climate Monitoring and Prediction

With climate change posing a significant risk to global sustainability, accurate monitoring and forecasting are more vital than ever to alleviate its effects (Himeur et al., 2022). Robotics and AI are stepping up to the plate in this area, facilitating the collection, analysis, and interpretation of immense amounts of environmental data.

- **Satellite and Drone-Based Monitoring:** Imagine autonomous drones and satellites outfitted with state-of-the-art sensors and AI algorithms tirelessly monitoring our atmospheric conditions, tracking ice cap melting, and observing deforestation. AI-enhanced satellite imagery analysis, for instance, can pinpoint shifts in land use while anticipating the repercussions of climate change on vital ecosystems.
- **Weather Forecasting:** The remarkable capabilities of AI models, especially deep learning algorithms, allow us to scrutinize historical weather data and make incredibly accurate predictions about future climate trends. These forecasts empower governments and organizations to brace for extreme weather occurrences, such as hurricanes, floods, and droughts.
- **Ocean and Glacier Monitoring:** Underwater robots, known as autonomous underwater vehicles (AUVs), are at the forefront of monitoring critical aspects like ocean temperatures, acidity levels, and glacier movements (Hunt &Auster, 1990). AI algorithms are instrumental in processing this data, helping us gauge the well-being of marine ecosystems and predict sea-level rises.

By delivering real-time, precise, and actionable insights, Robotics and AI are equipping scientists and policymakers with the knowledge they need to make informed decisions in the fight against climate change. Together, we can leverage these technologies to foster a more sustainable world.

7.2 Pollution Control and Waste Management

Pollution control and effective waste management are vital for achieving true environmental sustainability, and it's exciting to see how Robotics and AI are stepping up to the challenge! These technologies are not just tools; they are revolutionizing our approach to environmental protection and resource efficiency.

- **Air Quality Monitoring:** Imagine a network of AI-enabled sensors and robotic systems deployed throughout our cities, tirelessly working to monitor air quality in real-time. These smart systems can pinpoint pollution sources—from industrial emissions to vehicle exhaust—offering data that empowers decision-makers to take meaningful regulatory actions for cleaner air!
- **Water Pollution Detection:** Picture autonomous robots, complete with advanced sensors, cruising through our rivers and lakes, detecting contaminants with precision (Kamyab, 2023). Thanks to AI algorithms, we can analyze this data to unveil pollution sources and predict how contaminants spread, allowing for timely interventions to keep our water clean and safe.
- **Waste Sorting and Recycling:** The future of waste management is here! With the aid of robotic systems powered by AI-driven computer vision, the process of sorting recyclable materials is entering a whole new era. These robots accurately identify and separate plastics, metals, and other materials, significantly cutting down landfill waste and promoting a thriving circular economy.
- **Ocean Cleanup:** We can envision fleets of autonomous surface vessels working together to collect plastic waste from our oceans. With AI at the helm, these systems optimize their collection strategies by identifying areas with high concentrations of debris and navigating efficiently—making a real difference in cleaning up our seas.

These innovative applications showcase how Robotics and AI are transforming pollution control and waste management into more efficient, sustainable practices, ultimately benefiting us all!

Figure 2. Impact and areas of Environmental Systems using AI & Robotics

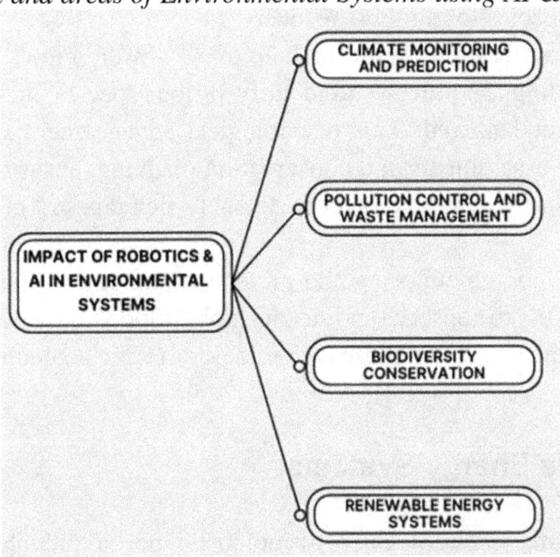

7.3 Biodiversity Conservation

Biodiversity is the backbone of ecosystem health, and it's at risk due to various threats like habitat destruction, poaching, and climate change (Jones et al., 2002). Thankfully, Robotics and AI are stepping in to offer groundbreaking solutions for conserving biodiversity.

- **Wildlife Monitoring:** Drones and camera traps equipped with AI algorithms are changing the game in wildlife monitoring. These technologies allow us to track animal movements and monitor populations effectively. With AI's ability to recognize individual animals through unique patterns—like the stripes of a tiger or the flukes of a whale—we can gain invaluable insights into population dynamics.
- **Anti-Poaching Efforts:** In the fight against poaching, autonomous drones and ground robots are patrolling protected areas to detect and deter illegal activities. By harnessing AI's power to analyze data from sensors and cameras, these systems can identify suspicious behavior and alert authorities in real-time, providing a crucial layer of protection for endangered species.
- **Habitat Restoration:** Robots are now planting trees and restoring ecosystems, allowing us to actively repair the damage we've done (Krzywanski, 2022). AI models are pivotal in determining the best locations for reforesta-

tion efforts, taking into account soil quality, climate conditions, and the needs of various species for a robust recovery.

- **Species Identification:** With the help of AI-powered tools, we can analyze audio recordings and images to identify various species, including those that are rare or endangered. This technological advancement significantly aids biodiversity assessments and conservation planning, ensuring that we're doing everything we can to protect our planet's rich tapestry of life.

Through innovative monitoring and protection strategies, Robotics and AI are enhancing our ability to conserve and sustain global biodiversity (Laird et al., 1994). It's an exciting time to witness these technologies making a genuine impact on our environment!

7.4 Renewable Energy Systems

The shift towards renewable energy is not just important; it's absolutely vital for reaching our global sustainability goals! Robotics and AI are at the forefront of this exciting transformation, actively optimizing renewable energy systems and pushing us towards a cleaner future.

- **Smart Grids:** Thanks to AI algorithms, smart grids can now elegantly balance energy supply and demand, ensuring that renewable energy distribution is efficient and effective. These innovative systems effortlessly predict energy consumption patterns and seamlessly integrate renewable sources like solar and wind power.
- **Solar Panel Maintenance:** Imagine robots equipped with cutting-edge AI-driven computer vision inspecting and cleaning solar panels! This amazing technology maximizes performance, reduces maintenance costs, and boosts the efficiency of solar energy systems.
- **Wind Turbine Monitoring:** Drones and robotic systems are revolutionizing the way we monitor wind turbines! By inspecting for damage or wear, these devices gather crucial data analyzed by AI to anticipate maintenance needs, helping us avoid expensive breakdowns.
- **Energy Storage Optimization:** AI models are also improving how we manage energy storage systems, such as batteries. By predicting energy demand and smartly managing charging and discharging cycles, we enhance the reliability and efficiency of renewable energy systems.

Through these innovations, Robotics and AI are supercharging the efficiency and reliability of renewable energy systems, accelerating our journey towards a sustainable energy future.

8. SIGNIFICANCE OF CLOUD COMPUTING AND SYSTEMS IN ENHANCING SUSTAINABILITY AND ENVIRONMENTAL MONITORING

Cloud computing has transformed how organizations process data, offer services, and maintain their information technology in the present era. Along with its initial purpose of offering scalable IT assets through the internet, cloud computing has grown into one of the most primitive technologies facilitating sustainability efforts. Its adaptive features, like resource pooling, on-demand scalability, and efficient energy consumption, make it extremely useful in minimizing environmental damage (Wu, B., et al., 2020). In addition, cloud computing allows advanced external environmental protection and monitoring via data analytics, IoT, and automated real-time decision-making systems.

Sustainability means an approach that fulfills the requirements of the present while not putting in jeopardy the possibility of the future generations fulfilling their own needs, a definition accepted around the world by scientific communities and political bodies. Environmental sustainability and cloud computing have created several opportunities to enhance the efficiency of energy consumption, degasification, and emission control of carbon dioxide, as well as the supervision of ecological indicators. Cloud computing developed as a demonstration for conveying IT administrations over the Web with highlights such as on-demand self-service, wide organize get to, asset pooling, quick flexibility, and measured benefit (Wu & Li, 2015). Over time, the selection of cloud administrations by governments and little businesses has developed exponentially, driven by the benefits of diminished operational costs, improved adaptability, and progressed deftness. One of the most direct benefits of cloud computing is its ability to increase energy efficiency. Traditional on-premise data centers often suffer from low utilization rates and suboptimal energy management. Cloud computing resolves this problem by consolidating resources within extensive data centers, increasing utilization levels and minimizing overall energy usage. In contrast, cloud data centers are designed to optimize resource utilization through techniques such as virtualization, dynamic resource allocation, and advanced cooling systems. By consolidating workloads and optimizing server usage, cloud providers can reduce the overall energy consumption per computing task. Here Fig: 3 is depicted about the impact and potentiality of Cloud Computing in Environmental Systems with sustainability.

For instance, research has shown that cloud data centers can achieve significantly higher ***energy efficiency*** compared to traditional data centers by leveraging advanced cooling techniques and renewable energy sources. A research work emphasizes that the ***economies of scale*** attained by cloud providers not only decrease operational expenses but also lessen the carbon footprint for each unit of computation. The efficient use of hardware resources means fewer physical servers are required, which directly correlates to reduced energy consumption and lower greenhouse gas emissions (Yaseen, 2022). Additionally, many cloud service providers have made strong commitments to ***renewable energy***. By investing in solar, wind, and other clean energy sources, these providers are working to decouple IT growth from increased carbon emissions. This strategic shift is pivotal in the global drive toward sustainability.

Undoubtedly, we must acknowledge that cloud computing has surfaced as a revolutionary technology impacting multiple sectors, such as sustainability and environmental stewardship. This paradigm shift toward cloud-based solutions plays a pivotal role in enhancing the efficiency of resource utilization, data management, and monitoring systems. As organizations and governments strive to mitigate environmental impacts and promote sustainable practices, cloud computing offers innovative tools and strategies to achieve these objectives. Organizations can lower their total energy use and carbon emissions by utilizing shared resources. For example, centralized data centers can enhance server efficiency, resulting in reduced electricity consumption relative to conventional on-site data centers.

Based on various research, it has also been noted that cloud computing facilitates the collection, storage, and analysis of significant amounts of environmental data. IoT devices and sensors are capable of transmitting real-time data to the cloud, enabling organizations to improve the monitoring of air quality, water supplies, and biodiversity. This ability enhances data-informed decision-making procedures essential in environmental stewardship. Cloud platforms facilitate improved collaboration among different stakeholders—governments, NGOs, and researchers—by offering access to a communal database of environmental data. Cloud computing also enables sophisticated analytical tools and machine learning algorithms that can develop predictive models for environmental events (Zhang & Li, 2021). This is especially beneficial in climate modeling, disaster response, and conservation strategies. Organizations can take proactive steps to reduce negative environmental effects by forecasting trends and possible crises.

Cloud computing enables the deployment of ***real-time monitoring systems*** for environmental factors. Cloud-connected sensors can capture and analyze data on pollution levels, weather changes, and ecological health instantaneously, alerting communities and authorities of hazardous conditions (Fu et al., 2023).

Additionally, cloud service providers allocate resources to sophisticated cooling technologies and AI-based energy efficiency to ***reduce power consumption***. Methods like free-air cooling utilize outside air to lower server temperatures, decreasing dependence on conventional air conditioning units. These advancements enhance the overall energy efficiency of cloud data centers. Shifting to cloud computing allows organizations to greatly decrease their carbon emissions. Cloud providers frequently employ energy-saving technologies and renewable energy sources to operate their data centers, resulting in reduced greenhouse gas emissions. For example, research has shown that cloud computing can significantly decrease carbon footprints in comparison to conventional IT infrastructures (Gaur et al., 2023).

Figure 3. Significant areas of Cloud Computing and Virtualization in Ecology and Environmental sustainability

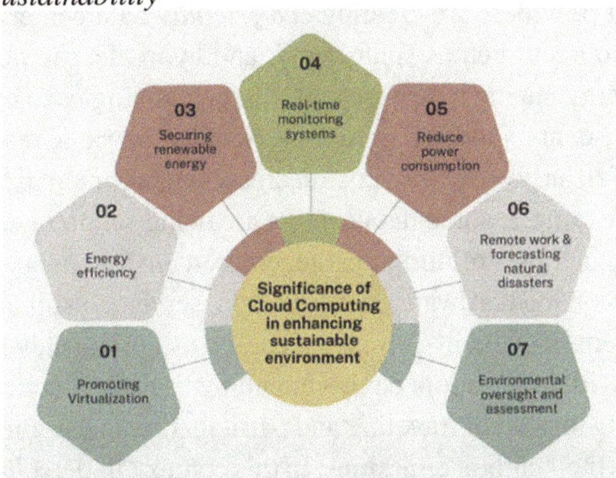

The ***centralized framework*** of cloud computing minimizes the requirement for individual organizations to manage substantial hardware, consequently lowering electronic waste. Cloud providers oversee infrastructure upgrades and maintenance from a central point, prolonging the hardware lifespan and minimizing resource duplication. This method greatly reduces the electronic waste linked to regular hardware replacement in conventional IT environments.

Cloud computing facilitates ***remote work*** by allowing users to access applications and data from anywhere that has an internet connection. This ability minimizes the necessity for daily travel, resulting in decreased emissions associated with transportation. Furthermore, companies can reduce their physical office spaces, leading to lower energy usage for lighting, heating, and cooling. The COVID-19 pandemic has further emphasized the environmental advantages of remote work enabled by cloud computing.

Cloud computing offers elastic storage and computational capacity required to analyze the environment. Huge amounts of data gathered from sensors and satellites can be computed in the cloud for observing climate trends, monitoring deforestation, and ***forecasting natural disasters***. Real-time data analysis serves for making well-informed decisions and swift responses to environmental issues. For instance, cloud-based platforms have been used for the analysis of climate data, and the results have helped improve understanding about the effects of climate change. Virtualization and containerization technologies in cloud computing enable many virtual machines or containers to execute on one physical server, ***minimizing resource usage*** and decreasing the amount of servers (Bostrom &Yudkowsky, 2014).. This consolidation has substantial energy efficiencies and a smaller carbon footprint. Research has proved that virtualization can improve data center energy efficiency by allowing efficient allocation of resources and eliminating unused server instances.

Leading cloud providers are creating ***eco-friendly*** data centers powered by renewable energy sources such as solar, wind, and hydroelectric power. These data centers utilize energy-saving hardware and smart technologies to maintain minimal power usage. Data centers located in regions rich in renewable resources provide a constant supply of clean power, which also decreases environmental footprints (Cortès et al., 2000). For example, some data centers are being set up in colder climates to utilize natural cooling, thus minimizing the requirement for energy-sucking air conditioning. Energy-saving hardware, virtualization, server consolidation, and energy management systems can aid in decreasing energy usage and enhancing efficiency. Cloud service providers can adopt environmentally friendly practices, such as using video conferencing instead of traveling and participating in carbon offset initiatives to counterbalance their carbon emissions. In this context, understanding the roots of sustainable cloud computing is essential and how they can be leveraged to ***promote sustainability*** in the cloud computing industry (Goodwin & Wright, 2020).. They can also harness renewable energy from solar, wind, hydro, or geothermal sources to power their data centers. This could help reduce their carbon footprint and promote sustainability. This utilization of energy-saving servers, storage solutions, and networking devices reduces energy consumption and enhances effectiveness. Moreover, cloud computing presents considerable prospects for improving sustainability as well as ***environmental oversight and assessment***. Cloud solutions are essential in tackling global environmental issues by enhancing resource efficiency, facilitating real-time data analysis, and encouraging collaboration among various stakeholders. Nonetheless, managing the related risks and obstacles is crucial to guarantee that the shift to cloud services truly aids in achieving sustainable development goals.

9. FUTURE POTENTIALITIES WITH OTHER EMERGING TECHNOLOGIES

The fusion of Robotics and Artificial Intelligence (AI) with environmental systems is not just a trend; it's a thrilling journey toward a sustainable future! As we continue to explore new frontiers, this section highlights the exciting opportunities that await us, delving into emerging technologies, interdisciplinary synergies, and innovative policy recommendations.

9.1 Emerging Technologies

Internet of Things (IoT) and Edge Computing: Imagine a world where IoT and Robotics come together to transform environmental monitoring! With real-time data collection and analysis, we can enhance our understanding of our surroundings. Edge computing, which processes data right where it's gathered, is a game-changer, reducing delays and energy usage—perfect for those hard-to-reach areas (Laird et al., 1994). Picture IoT-enabled sensors mounted on autonomous drones diligently tracking air quality in bustling urban centers, delivering immediate insights to support informed decision-making!

Blockchain for Environmental Data Management: Let's celebrate the transparency that blockchain brings to environmental data! This incredible technology builds immutable records for data collected through Robotics and AI systems, ensuring integrity and inspiring trust among all stakeholders (Hamza et el., 2022). It's a powerful tool for applications like carbon credit trading and boosting supply chain sustainability, paving the way for a greener future.

Quantum Computing: The possibilities with quantum computing are nothing short of breathtaking! This cutting-edge technology has the potential to tackle intricate environmental challenges that traditional computers can't begin to solve. Just think about quantum algorithms optimizing large-scale renewable energy grids or analyzing climate systems with a level of accuracy that was previously unimaginable—this is the future we're heading into!

Biomimetic Robotics: Nature has always been our greatest mentor, and biomimetic robotics take inspiration from it, mimicking the behaviors and structures of living organisms (Haupt et al., 2008). These innovative robots can embark on roles like pollination, underwater exploration, and even disaster response. They offer creative solutions to pressing environmental issues, showing us just how intertwined technology and nature can be.

Advanced AI Models: The emergence of advanced AI models, such as generative adversarial networks (GANs) and reinforcement learning, marks an exciting chapter in enhancing predictive capabilities for environmental systems. With these sophisticated models, we can simulate complex scenarios—like the impact of climate change on ecosystems. This proactive approach equips us to make informed decisions that help us protect our planet (Lajaunie et al., 2019). In summary, the horizon is bright with opportunities as we explore the integration of Robotics and AI in our environmental strategies. The future is not just a possibility; it's a promise of innovation and sustainability waiting to be realized.

10. CHALLENGES, ISSUES, AND LIMITATION OF ECOLOGY AND ENVIRONMENT USING SEVERAL TECHNOLOGIES

The fields of Robotics and Artificial Intelligence (AI) are truly exciting avenues for enhancing environmental systems and driving sustainability forward! However, it's essential to recognize that implementing these groundbreaking technologies comes with its own set of challenges. This section dives into some of the technical, ethical, economic, and environmental hurdles we need to overcome to fully tap into the incredible potential of these innovations, refer Table: 1.

10.1 Technical Challenges

- **Data Limitations:** One of the first hurdles we face in leveraging AI effectively is the availability of robust, high-quality datasets. In many environmental contexts, gathering data can be quite a challenge due to factors like remote locations, extreme weather conditions, and resource constraints (LeCun et al., 2015). For example, keeping track of biodiversity in dense rainforests or the icy expanses of polar regions often needs specialized equipment and infrastructure—something that isn't always easily accessible.
- **Algorithmic Bias:** It's important to remember that AI systems are only as reliable as the data they learn from. If the training data is biased, the outcomes can be skewed or unfair. For instance, if an AI model is developed using data from a specific area, it may struggle to function correctly when applied to a different location, which limits its versatility.

Table 1. Major issues and challenges of technologies in environmental systems

Major Issues and Challenges	Categories of issues
Technical Challenges	Data Limitations Integration with Existing Systems Algorithmic Bias Energy Consumption Durability and Reliability
Ethical and Social Issues	Privacy Concerns Job Displacement Equity and Access Accountability and Transparency Ethical Use of AI in Conservation
Economic and Policy Barriers	High Initial Costs Lack of Funding and Resources Regulatory Challenges Intellectual Property Issues Policy Gaps
Environmental Impact	E-Waste Generation Resource Consumption Carbon Footprint Lifecycle Management

- **Integration with Existing Systems:** Seamlessly integrating Robotics and AI into current environmental monitoring and management systems can prove to be complex. Many legacy systems might not have the infrastructure necessary to support these advanced technologies, potentially requiring significant upgrades or complete overhauls.
- **Energy Consumption:** AI algorithms, especially those developed through deep learning techniques, often demand substantial computational power, which can lead to high energy consumption (Li & Li, 2007). This creates a unique challenge for sustainability initiatives, as the carbon footprint associated with running AI systems could diminish their environmental advantages.
- **Durability and Reliability:** When deploying robots in harsh environments—be it underwater or in extreme weather conditions—they must be both durable and reliable. Ensuring these sophisticated systems can function effectively over extended periods without frequent maintenance is a significant technical challenge we need to address.

As we navigate these challenges, it's clear that with dedication and innovation, we can overcome these obstacles and unlock the full benefits of Robotics and AI for our planet!

10.2 Ethical and Social Considerations

- **Privacy Concerns:**As we embrace the remarkable potential of AI and robotics in environmental monitoring, we must be aware of the significant privacy concerns that come with it. These technologies often gather vast amounts of data, including sensitive information about individuals and communities. For instance, while drones are fantastic for wildlife monitoring, there's a chance they might unintentionally capture images of people (Lokers et al., 2016). This raises important questions about privacy that we must address.
- **Job Displacement:** While Robotics and AI can revolutionize industries, they also pose a real challenge when it comes to job displacement. Sectors like agriculture and waste management may see a reduction in human labor due to automation. It's essential that we confront the social and economic implications of these changes head-on to ensure that the adoption of these advanced technologies is equitable and fair.
- **Equity and Access:** The advantages that come with Robotics and AI are not shared equally by everyone. Often, marginalized communities and developing countries lack the necessary resources and infrastructure to implement these technologies, which can deepen existing inequalities. We need to strive for a future where benefits are accessible to all, bridging the gap created by technological advancements.
- **Accountability and Transparency:** AI systems make decisions that can have profound environmental and social effects. Therefore, it's vital to ensure accountability and transparency in how these decisions are made. Building public trust and avoiding any unintended harm relies on our commitment to making these processes clear and responsible.
- **Ethical Use of AI in Conservation:** The application of AI in conservation, such as in the fight against poaching, raises intriguing ethical dilemmas. We must navigate the delicate balance between the need for protection and the ethical implications of surveillance and the potential use of force. This challenge is complex but essential to tackle as we move forward.

10.3 Economic and Policy Barriers

- **High Initial Costs:** Developing and deploying Robotics and AI systems is no small feat and typically requires a hefty financial investment. For many organizations and governments, these initial costs can serve as a significant barrier to embracing these transformative technologies (Lucivero, 2020). However, overcoming this hurdle can lead to groundbreaking advancements!

- **Lack of Funding and Resources:** It's no secret that environmental projects often face difficulties securing sufficient funding, particularly in developing areas. Without adequate financial backing, the implementation of Robotics and AI solutions can be hugely limited (Paul et. al., 2020). Championing these projects is essential to unlocking their potential.
- **Regulatory Challenges:** The swift evolution of Robotics and AI has left many regulatory frameworks struggling to keep pace. Inconsistencies or outdated regulations can impede the rollout of these innovations, especially in initiatives that cross borders. We need proactive measures to ensure regulations support rather than hinder progress!
- **Intellectual Property Issues:** Collaboration is at the heart of AI algorithm and robotic system development, but it often brings up intellectual property concerns and questions about data ownership. Addressing these issues is critical to fostering a culture of innovation and collaboration within the field.
- **Policy Gaps:** Many countries are still lacking comprehensive policies to effectively integrate Robotics and AI into environmental systems. Closing these gaps calls for coordinated efforts among governments, industry stakeholders, and academic institutions. Together, we can build the frameworks necessary for a sustainable future!

10.4 Environmental Impact of Robotics, AI, Cloud with Analytics

- **E-Waste Generation:** The rise of Robotics and AI systems has led to an increase in electronic waste (e-waste) due to their production and disposal. It's vital that we prioritize sustainable practices in the design, manufacturing, and recycling of these technologies to significantly reduce their environmental impact.
- **Resource Consumption:** Creating Robotics and AI systems demands a considerable amount of raw materials, particularly rare earth metals. The process of extracting and processing these materials can lead to serious environmental concerns, such as habitat destruction and pollution. We must be thoughtful about our resource choices to mitigate these effects!
- **Carbon Footprint:** As we know, the energy used by AI systems, especially those housed in data centers, contributes to greenhouse gas emissions. To turn the tide on this issue, it's crucial that we develop energy-efficient algorithms and shift towards renewable energy sources that can minimize the carbon footprint—let's make positive strides for our planet!
- **Lifecycle Management:** The environmental implications of Robotics and AI go well beyond their active use. Implementing effective lifecycle

management—covering design, manufacturing, operation, and disposal—is essential for ensuring sustainability throughout the entire lifespan of these technologies (Zhang & Li, 2022).

By taking on these challenges, we can truly unlock the incredible potential of Robotics and AI, contributing to a more sustainable and resilient future for all. All these implementing technologies may lead to Environmental 4.0, as depicted in Fig: 4.

Figure 4. Ultimate potentiality of emerging technologies in building sustainability with Environment 4.0

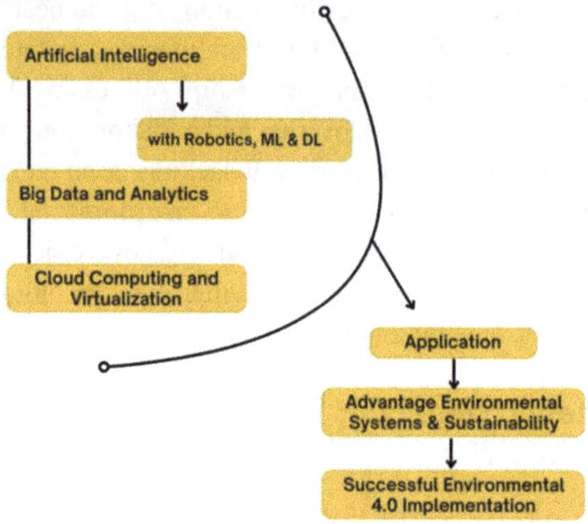

Environment 4.0 refers to the combination of advanced technologies such as Artificial Intelligence (AI), Internet of Things (IoT) and Big Data for sustainability and environmental protection. As industries have evolved with Industry 4.0, Environment 4.0 is needed to improve environmental management by: using smart solutions for resource conservation, pollution control and climate resilience. Real-time water and air quality can be checked with IoT devices and sensors, it is also possible to save time, as it is possible to improve the environment very quickly by taking advantage of technology. In addition, using block-chain technology, we can store sustainable data and use that data to get qualitative results when automated. It can be possible to implement Environment 4.0 very well through collaboration between government and industry. Along with awareness among people and green policies, this transformation should be made more popular. Environment 4.0 should not be taken as just a word or kept as a subject for reading, it should be implemented in our real life so that we can have a sustainable future and a better environment for ourselves and the

next generation. And all these technologies viz. Big Data, AI and Robotics, Cloud Computing, etc. are responsible in building of effective ***Environment 4.0*** systems.

11. CONCLUDING REMARKS

The integration of Robotics and Artificial Intelligence, Big Data and Cloud into our environmental systems is nothing short of a groundbreaking shift in tackling the urgent challenges of sustainability. Throughout this book, we've delved into the incredible impact of these technologies in elevating environmental monitoring, conservation efforts, pollution control, sustainable agriculture, and renewable energy initiatives. From autonomous drones diligently surveying areas of deforestation to AI-driven smart grids that maximize our energy efficiency, Robotics and AI have become essential allies in our journey toward a sustainable future. The potential applications for Robotics, AI, Big Data and Cloud are vast and diverse, offering fresh, innovative solutions to some of the most intricate environmental problems we face today. With compelling case studies and real-world examples, we've shown how these technologies can transform industries, safeguard ecosystems, and foster sustainable practices. Yet, as we've noted, this path is not without its hurdles. We must carefully consider technical constraints, ethical dilemmas, economic challenges, and environmental repercussions to genuinely harness the power of these remarkable advancements. As we look to the future, emerging technologies like IoT, blockchain, and quantum computing, paired with interdisciplinary collaborations, open the door to thrilling possibilities for even greater innovation. It's crucial for policymakers, researchers, and industry leaders to join forces in creating supportive policies, promoting sustainable practices, and ensuring everyone has access to these transformative tools. Imagine a future where Robotics, AI including Big Data and Cloud are perfectly woven into our environmental systems, achieving a harmonious balance between human advancement and ecological preservation. This inspiring vision is within our grasp, but it demands that we rally together and commit to ethical, inclusive, and sustainable growth. Let's seize this opportunity and work collectively towards that brighter future!

REFERENCES

Adanma, U. M., & Ogunbiyi, E. O.Uwaga Monica AdanmaEmmanuel Olurotimi Ogunbiyi. (2024). Artificial intelligence in environmental conservation: Evaluating cyber risks and opportunities for sustainable practices. *Computer Science & IT Research Journal, 5*(5), 1178–1209. DOI: 10.51594/csitrj.v5i5.1156

Aguilar, J., Garces-Jimenez, A., R-Moreno, M. D., & García, R. (2021). A systematic literature review on the use of artificial intelligence in energy self-management in smart buildings. *Renewable & Sustainable Energy Reviews, 151*, 111530. DOI: 10.1016/j.rser.2021.111530

Ahmad, T., Madonski, R., Zhang, D., Huang, C., & Mujeeb, A. (2022). Data-driven probabilistic machine learning in sustainable smart energy/smart energy systems: Key developments, challenges, and future research opportunities in the context of smart grid paradigm. *Renewable & Sustainable Energy Reviews, 160*, 112128. DOI: 10.1016/j.rser.2022.112128

Alotaibi, E., & Nassif, N. (2024). Artificial intelligence in environmental monitoring: In-depth analysis. *Discover Artificial Intelligence, 4*(1), 84. DOI: 10.1007/s44163-024-00198-1

Andronie, M., Lăzăroiu, G., Karabolevski, O. L., tefănescu, R., Hurloiu, I., Dijmărescu, A., & Dijmărescu, I. (2022). Remote big data management tools, sensing and computing technologies, and visual perception and environment mapping algorithms in the internet of robotic things. *Electronics (Basel), 12*(1), 22. DOI: 10.3390/electronics12010022

Bostrom, N., & Yudkowsky, E. (2014). The ethics of artificial intelligence. In Frankish, K., & Ramsey, W. M. (Eds.), *The Cambridge handbook of artificial intelligence* (pp. 316–334). Cambridge University Press. DOI: 10.1017/CBO9781139046855.020

Chen, Y. (2020). IoT, cloud, big data and AI in interdisciplinary domains. *Simulation Modelling Practice and Theory, 102*, 102070. DOI: 10.1016/j.simpat.2020.102070

Cortès, U., Sànchez-Marrè, M., Ceccaroni, L., R-Roda, I., & Poch, M. (2000). Artificial intelligence and environmental decision support systems. *Applied Intelligence, 13*(1), 77–91. DOI: 10.1023/A:1008331413864

Cruz, F. B. D. L., & Barlaz, M. A. (2010). Estimation of waste component-specific landfill decay rates using laboratory-scale decomposition data. *Environmental Science & Technology, 44*(12), 4722–4728. DOI: 10.1021/es100240r PMID: 20496890

Fu, L., Li, J., & Chen, Y. (2023). An innovative decision making method for air quality monitoring based on big data-assisted artificial intelligence technique. *Journal of Innovation & Knowledge*, *8*(2), 100294. DOI: 10.1016/j.jik.2022.100294

Fu, L., Li, J., & Chen, Y. (2023). An innovative decision making method for air quality monitoring based on big data-assisted artificial intelligence technique. *Journal of Innovation & Knowledge*, *8*(2), 100294. DOI: 10.1016/j.jik.2022.100294

Gaur, L., Afaq, A., Arora, G. K., & Khan, N. (2023). Artificial intelligence for carbon emissions using system of systems theory. *Ecological Informatics*, *76*, 102165. DOI: 10.1016/j.ecoinf.2023.102165

Goodwin, M., & Wright, G. (2020). AI and robotics in waste management: A review of current applications and future prospects. *Waste Management & Research*, *38*(5), 487–498. PMID: 32089127

Hamza, M. A., Shaiba, H., Marzouk, R., Alhindi, A., Asiri, M. M., Yaseen, I., & Rizwanullah, M. (2022). Big Data Analytics with Artificial Intelligence Enabled Environmental Air Pollution Monitoring Framework. *Computers, Materials & Continua*, *73*(2).

Haupt, S. E., Pasini, A., & Marzban, C. (Eds.). (2008). *Artificial intelligence methods in the environmental sciences*. Springer Science & Business Media.

Himeur, Y., Rimal, B., Tiwary, A., & Amira, A. (2022). Using artificial intelligence and data fusion for environmental monitoring: A review and future perspectives. *Information Fusion*, *86*, 44–75. DOI: 10.1016/j.inffus.2022.06.003

Hunt, C. B., & Auster, E. R. (1990). Proactive environmental management: Avoiding the toxic trap. *MIT Sloan Management Review*, *31*(2), 7.

Jones, D. A., Lelyveld, T. P., Mavrofidis, S. D., Kingman, S. W., & Miles, N. J. (2002). Microwave heating applications in environmental engineering—A review. *Resources, Conservation and Recycling*, *34*(2), 75–90. DOI: 10.1016/S0921-3449(01)00088-X

Kamyab, H., Khademi, T., Chelliapan, S., SaberiKamarposhti, M., Rezania, S., Yusuf, M., Farajnezhad, M., Abbas, M., Hun Jeon, B., & Ahn, Y. (2023). The latest innovative avenues for the utilization of artificial Intelligence and big data analytics in water resource management. *Results in Engineering*, *20*, 101566. DOI: 10.1016/j.rineng.2023.101566

Krzywanski, J. (2022). Advanced AI applications in energy and environmental engineering systems. *Energies*, *15*(15), 5621. DOI: 10.3390/en15155621

Laird, D. A., Yen, P. Y., Koskinen, W. C., Steinheimer, T. R., & Dowdy, R. H. (1994). Sorption of atrazine on soil clay components. *Environmental Science & Technology, 28*(6), 1054–1061. DOI: 10.1021/es00055a014 PMID: 22176230

Lajaunie, C., Schafer, B., & Mazzega, P. (2019). Big data enters environmental law. *Transnational Environmental Law, 8*(3), 523–545. DOI: 10.1017/S2047102519000335

LeCun, Y., Bengio, Y., & Hinton, G. (2015). Deep learning. *Nature, 521*(7553), 436–444. DOI: 10.1038/nature14539 PMID: 26017442

Li, T. S., & Li, B. (2007). Research on Data Model and Query Optimization of Electronic Map in Embedded GIS. *Aeronautical Computing Technique*, (2), 26–35.

Lokers, R., Knapen, R., Janssen, S., van Randen, Y., & Jansen, J. (2016). Analysis of Big Data technologies for use in agro-environmental science. *Environmental Modelling & Software, 84*, 494–504. DOI: 10.1016/j.envsoft.2016.07.017

Lucivero, F. (2020). Big data, big waste? A reflection on the environmental sustainability of big data initiatives. *Science and Engineering Ethics, 26*(2), 1009–1030. DOI: 10.1007/s11948-019-00171-7 PMID: 31893331

Maxim, L., & van der Sluijs, J. P. (2011). Quality in environmental science for policy: Assessing uncertainty as a component of policy analysis. *Environmental Science & Policy, 14*(4), 482–492. DOI: 10.1016/j.envsci.2011.01.003

Moqi, J. H. L. (2010). Application of Geographic Information System in Digital Mining Information System [J]. *Standardization of Surveying and Mapping, 3.*

Mykrä, H., Heino, J., & Muotka, T. (2007). Scale-related patterns in the spatial and environmental components of stream macroinvertebrate assemblage variation. *Global Ecology and Biogeography, 16*(2), 149–159. DOI: 10.1111/j.1466-8238.2006.00272.x

Paul, P. K. (2022). Environmental informatics: basics, nature, and applications using emerging technologies with reference to issues and potentialities. In *Environmental Informatics: Challenges and Solutions* (pp. 1–13). Springer Nature Singapore. DOI: 10.1007/978-981-19-2083-7_1

Paul, P. K., & Aithal, P. S. (2020). Environmental informatics and its features, functions and stakeholders: A comprehensive overview. *Educational Quest-An International Journal of Education and Applied Social Sciences, 11*(1), 1–5. DOI: 10.30954/2230-7311.1.2020.1

Paul, P. K., Aithal, P. S., Bhimali, A., & Kalishankar, T. (2020). Environmental informatics vis-à-vis big data analytics: The geo-spatial & sustainable solutions. [IJAEML]. *International Journal of Applied Engineering and Management Letters*, *4*(2), 31–40.

Paul, P.K, Bhuimali, A., Aithal, P. S., Kalishankar, T., & Saavedra M, R. (2020). Artificial intelligence & cloud computing in environmental systems-towards healthy & sustainable development. International Journal of Inclusive Development, 6(1), 01-08.

Peters, M. A. (2020). AI, robotics, and the future of sustainability. *Educational Philosophy and Theory*, *52*(10), 1043–1047.

Sarker, M. N. I., Peng, Y., Yiran, C., & Shouse, R. C. (2020). Disaster resilience through big data: Way to environmental sustainability. *International Journal of Disaster Risk Reduction*, *51*, 101769. DOI: 10.1016/j.ijdrr.2020.101769

Wei, Y., Han, C., & Yu, Z. (2023). An environment safety monitoring system for agricultural production based on artificial intelligence, cloud computing and big data networks. *Journal of Cloud Computing (Heidelberg, Germany)*, *12*(1), 83. DOI: 10.1186/s13677-023-00463-1

Wu, B., Tian, F., Zhang, M., Zeng, H., & Zeng, Y. (2020). Cloud services with big data provide a solution for monitoring and tracking sustainable development goals. *Geography and Sustainability*, *1*(1), 25–32. DOI: 10.1016/j.geosus.2020.03.006

Wu, Q., & LI, X. Y. (2015). Remote sensing dynamic monitoring of mining geo-environment based on computational geometry and information TUPU. *Meitan Xuebao*, (1), 26–34.

Yaseen, Z. M. (2022). The next generation of soil and water bodies heavy metals prediction and detection: New expert system based edge cloud server and federated learning technology. *Environmental Pollution*, *313*, 120081. DOI: 10.1016/j.envpol.2022.120081 PMID: 36075340

Zhang, D., & Li, W. (2021). Artificial intelligence in environmental monitoring: A review. *Environmental Science & Technology*, *55*(8), 4249–4263.

Zhang, X., & Li, H. (2022). The role of AI in renewable energy optimization: A systematic review. *Renewable & Sustainable Energy Reviews*, *156*, 111890.

KEY TERMS AND DEFINITION

Big Data: Data Analytics involves examining raw data to uncover meaningful patterns, trends, and insights that can help in decision-making. Big Data, on the other hand, refers to extremely large datasets that require advanced computational techniques to process.

Cloud Computing: Cloud computing has transformed how organizations process data, offer services, and maintain their information technology in the present era. Cloud computing has grown into one of the most primitive technologies facilitating sustainability efforts.

Environmental Informatics: Environmental Informatics integrated with environmental aspects with computing and other allied technology, tools, techniques and sub-technologies regarding development and solving of the problems of Environment and Ecology.

Robotics: Robotics is an exciting interdisciplinary field that blends engineering, computer science, and mathematics to create and manage robots that can perform tasks autonomously or semi-autonomously. The evolution of robotics has been nothing short of remarkable. From the industrial robots that revolutionized manufacturing processes to advanced humanoid robots capable of meaningful social interactions, this field has made incredible strides.

Chapter 3
Socio-Ecological Responsibility in the Triple Planetary Crisis:
A Restorative Governance Framework for AI-Driven Global Epidemiology

Kim Sanders
http://orcid.org/0009-0008-1726-5992
University of Phoenix, USA

Jennifer Washington
Walden University, USA

ABSTRACT

This chapter examines the growing role of artificial intelligence in global epidemiological surveillance and the limitations of existing corporate governance models in addressing algorithmic bias and environmental externalities. The Socio-Ecological Responsibility model is introduced as a restorative governance framework that integrates equity, ecological materiality, and public health accountability. Drawing on evidence from demographic shortcut research and global energy consumption trends, this chapter demonstrates how AI operates as a commercial determinant of health through digital extractivism and disproportionate resource burdens. The analysis critiques traditional ethical frameworks and advances algorithmic reparations, computational frugality, and Software Carbon Intensity metrics as

DOI: 10.4018/979-8-2600-2691-5.ch003

operational safeguards. By aligning data governance with planetary health equity, this chapter provides a practical roadmap for responsible AI deployment in post-pandemic global health systems.

INTRODUCTION: THE GLOBAL IMPERATIVE FOR SOCIO-ECOLOGICAL GOVERNANCE

The intersection of Socio-Ecological Responsibility, Artificial Intelligence, and epidemiology introduces interdisciplinary challenges that demand synchronized, multimodal leadership. While current discourse frequently praises the superhuman speed of algorithmic processing, a critical gap exists in accounting for the materiality of the digital. As Haluza and Jungwirth (2023) observe, while AI-driven analytics accelerate clinical research, a lack of green governance risks externalizing ecological costs to achieve localized medical gains.

The Triple Planetary Crisis and the Materiality of the Cloud

To understand why SER is a functional requirement, one must view the AI transition through the lens of the United Nations Environment Programme (UNEP, 2024). The integration of AI into global epidemiology is a high-intensity industrial process. According to the IEA (2024), data centers, the backbone of global health AI, accounted for approximately 460 TWh of electricity in 2022, a figure projected to reach 800 TWh by 2026. This energy surge creates a carbon-health gap: an AI model designed to predict respiratory outbreaks in the Global South may inadvertently contribute to those very outbreaks by accelerating the regional heat-island effect.

Beyond carbon, the Water-Data Nexus presents a critical challenge. For an epidemiological model to be ethically solvent, it must account for its water-per-inference (WPI) metric. If the computational cost of predicting a cholera outbreak in a drought-prone region further depletes the local water table, the AI has fundamentally compromised the very public health system it was designed to support. This necessitates a move toward computational frugality (Bacchi et al., 2024), where model success is measured by epidemiological precision per liter of water consumed.

The Algorithmic Chasm and the Stratification of Health

This physical footprint directly fuels a social paradox. The WHO's mandate of Health for All is currently facing an unprecedented challenge from a widening algorithmic chasm. Without the adoption of SER, AI is rapidly being commodified as a luxury health good. High-fidelity predictive modeling becomes reserved for well-resourced systems in the Global North, while the Global South is relegated to second-tier digitalism, utilizing black-box models that have not been validated for local realities. This creates a predatory dynamic of digital extractivism and digital colonialism (Regilme, 2025). Marginalized populations provide the raw biological data, while the resulting value is concentrated in the North, and the externalized costs of carbon, e-waste, and water scarcity are localized in the South.

The Socio-Ecological Responsibility Intervention

To align with the World Health Organization's (WHO, 2025) vision for epistemic justice, the SER framework argues that an AI tool is only scientifically valid if it is locally solvent. A model that predicts disease patterns in London but fails in Lagos is not a medical breakthrough; it is a manifestation of structural data exclusion. SER-driven governance mandates a shift toward planetary health equity, asserting that technological progress in the Global North cannot be subsidized by the environmental and clinical neglect of the Global South.

BACKGROUND: PATH DEPENDENCY AND THE SOCIO-ECOLOGICAL TRANSITION

The historical evolution of corporate social responsibility (CSR) has transitioned from a peripheral ethical framework to a core strategic mandate for organizational resilience. However, in the context of the 2025 digital health landscape, a significant gap remains: the failure to account for path dependency, the tendency of systems to follow established historical patterns even when those patterns are no longer optimal or ethical. In AI-driven epidemiology, path dependency risks automating the inequities of the past under the guise of progress.

The Commercial Determinants of AI-Driven Health

To understand the current governance crisis, artificial intelligence must be reclassified as a primary commercial determinant of health (CDoH). As defined by *The Lancet* Commission, CDoH refers to the private sector activities that shape the environments in which people live (Gilmore et al., 2023). When AI development is driven by market-first logic, innovation follows the most profitable routes, typically prioritizing high-resource clinical settings while externalizing the environmental and social costs to marginalized regions.

The COVID-19 pandemic served as a systemic disruptor, exposing the structural fragility of global health and accelerating development. This acceleration frequently bypassed ethical safeguards under the guise of emergency technological solutionism. This shift has fundamentally re-indexed the relationship between human health and environmental stability (Mandal & Murthy, 2021). Because the pandemic demonstrated that health is a global, interconnected system, the technologies intended to protect that system must be governed by a framework as holistic as the system itself: Socio-Ecological Responsibility.

Digital Extractivism and the Specter of Digital Colonialism

The expansion of artificial intelligence in epidemiology must confront the specter of digital colonialism. In many low-resource settings, biological and behavioral data are extracted for global modeling with little to no return on data for the local communities (Regilme, 2025). This extractive logic mirrors historical path dependencies where marginalized populations are treated as data subjects rather than data partners.

This creates a predatory dynamic of digital extractivism:

- **Asymmetric Extraction.** Data flows from the Global South to train models owned by Northern conglomerates.
- **Clinical Exclusion.** The resulting diagnostic tools often fail when applied back to the source populations because they were optimized for Northern biological and environmental baselines.
- **Material Burden.** While the intelligence is exported, the physical waste, e-waste, and carbon emissions remain localized in the extraction sites.

The Profit-Proxy Problem and Epistemic Equity

Path dependency in artificial intelligence occurs when historical data, saturated with commercial and structural biases, is used to train future-looking models. Without a SER framework, these models naturally default to profit proxies. This occurs when an algorithm identifies demographic or socioeconomic traits as shortcuts for health status because those traits were historically linked to lower healthcare spending or access.

For example, an algorithm might learn that patients from a specific ethnic background are less suitable for a specific treatment, not due to biological factors, but because the historical insurance data shows they were less likely to receive that treatment. Recent scholarship suggests that long-term sustainability in digital health depends on embedding equity throughout the artificial intelligence lifecycle, moving beyond passive inclusivity toward epistemic justice, and the recognition of marginalized knowledge as a valid and necessary component of the data ecosystem (Dankwa-Mullan, 2024b; Joseph, 2025).

Regulatory Evolution: From Voluntary CSR to Mandatory SER

The transition to SER is no longer just an ethical choice; it is necessitated by a hardening regulatory environment. The European Union AI Act (2024/2025) represents the world's first comprehensive horizontal legal framework, classifying medical AI as high-risk. This legislation mandates strict requirements for data quality, human oversight, and environmental transparency (European Parliament, 2024).

Similarly, the WHO 2025 Guidance on Ethics and Governance of AI for Health asserts that corporations can no longer hide behind proprietary algorithms when public health is at stake. Governance must move toward algorithmic reparations (Petty & Richardson, 2023), where artificial intelligence systems explicitly prioritize the needs of marginalized groups to offset decades of structural neglect. This proactive stance marks the fundamental evolution of the SER framework: moving from simple risk mitigation to the active pursuit of restorative socio-ecological justice.

MAIN FOCUS: THE BIOPOLITICS OF THE MACHINE AND OPERATIONALIZING SER

The double burden of the digital transition represents a fundamental failure of current corporate governance, where social benefits in the Global North are subsidized by ecological and clinical degradation in the Global South. Within this context, the convergence of artificial intelligence and epidemiology symbolizes a paradigm shift in public health analytics. However, as this chapter argues, the efficacy of these tools is contingent upon addressing structural data exclusion, the phenomenon where historical inequities are embedded within health datasets, creating forms of asymmetric intelligence that reinforce structural inequities across global health systems (Regilme, 2025) (see Figure 1). This chapter uses the MIT demographic shortcut phenomenon as a critical case study to demonstrate why Socio-Ecological Responsibility must move beyond abstract governance principles and be operationalized directly within model architecture and validation.

Figure 1. Sociotechnical Ecology of AI-Driven Global Epidemiology Under the Socio-Ecological Responsibility (SER) Framework

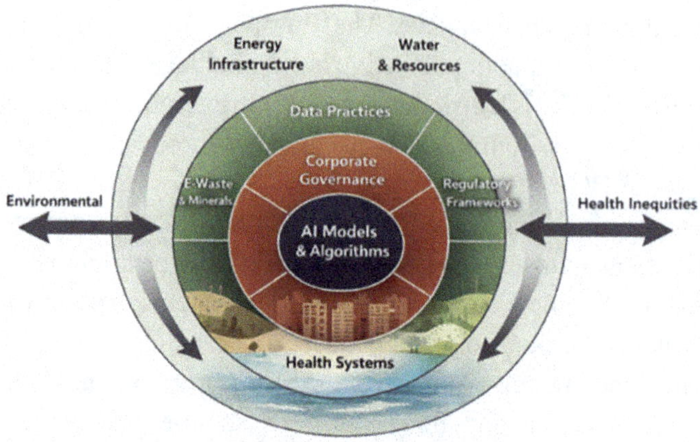

Note. This figure conceptualizes artificial intelligence-driven epidemiology as a sociotechnical ecology embedded within social, institutional, and ecological systems. It illustrates how data practices, corporate governance, regulatory environments, and health systems interact with material infrastructures such as energy, water, and climate systems. Under the SER framework, ecological externalities and social inequities are treated as integral determinants of public health rather than peripheral considerations. The figure was generated with the assistance of ChatGPT (Version 5.2) based on the conceptual framework developed by the authors.

The Deep Dive: Hidden Stratification and the Demographic Shortcut

The risks of structural exclusion are most visibly evidenced by hidden stratification. This occurs when a machine learning model achieves high aggregate performance but fails catastrophically on specific subgroups. The research from Massachusetts Institute of Technology (Trafton et al., 2024; Yang et al., 2024) serves as a watershed moment for AI governance, proving that technical accuracy can often be a mask for demographic profiling.

The Mechanism of the Shortcut:

The Massachusetts Institute of Technology (MIT) study demonstrated that diagnostic models could extrapolate patient race and gender from chest X-rays with near-perfect accuracy (ROC curves of 0.95+), despite these features being invisible to human radiologists. The models were not detecting biological race; they were identifying proxies of structural neglect, such as:

- **Spectral Fingerprints.** Subtle artifacts from older imaging hardware found in under-resourced clinics.
- **Physiological Proxies.** Variations in bone density resulting from long-term environmental and nutritional determinants of health.

From a SER perspective, this represents a failure of epistemic accountability. When a model utilizes a demographic shortcut to reach a diagnosis, it reinforces the path dependency discussed in the background section, automating historical neglect into future clinical decisions.

The Biopolitics of Data: Epistemic Debt and Extraction

Addressing this mandate transcends simple inclusivity; it requires reparative data curation that acknowledges the epistemic debt owed to excluded populations. As noted in the 2024 United Nations Environment Programme, many AI models are trained on WEIRD data (Western, Educated, Industrialized, Rich, and Democratic), creating a biological mismatch when applied globally (UNEP, 2024). By institutionalizing SER, corporations ensure artificial intelligence serves as a tool for restorative justice. This requires a shift from viewing populations as data subjects to data partners, adopting the CARE Principles (Collective Benefit, Authority to Control, Responsibility, and Ethics) to ensure that marginalized groups are owners of the outcomes, not just sources of raw data (Carroll et al., 2020).

Computational Frugality and the Carbon-Health Gap

The ecological aspect of SER is functionally synonymous with public health. A socially responsible artificial intelligence strategy must prioritize computational frugality, the practice of optimizing model efficiency to deliver high-impact insights with minimal energy consumption (Bacchi et al., 2024). Under SER, computational efficiency is not merely a technical optimization, but an ethical constraint, as excessive energy and water consumption directly intensify the environmental determinants of disease in vulnerable regions. By reframing efficiency as a public health safeguard rather than a cost-saving measure, SER ensures that AI-driven innovation does not externalize ecological harm onto the very populations it is intended to serve.

The 2025 WHO Technical Brief on Sustainable AI suggests that Small Language Models (SLMs) and frugal architectures are more resilient in resource-constrained environments. By prioritizing these over energy-dense transformers, organizations minimize the carbon-health gap, the paradox where the energy used to run a health model in the North contributes to a climate-driven heatwave or water shortage in the South. In the SER framework, a model is only ethically solvent if its clinical precision does not come at the cost of the environmental determinants of health.

The Tripartite Feedback Loop: Socio-Ecological Responsibility as a Governance Engine

To operationalize these concepts, the chapter proposes a Tripartite Feedback Loop:

1. **SER (The Normative Anchor).** Sets the ethical and ecological boundary conditions.
2. **AI (The Analytical Engine).** Processes data through the lens of restorative weighting and green-compute standards.
3. **Epidemiology (The Empirical Validator).** Measures the real-world reduction in health disparities and environmental impact.

This framework ensures that technological innovation does not expand the algorithmic chasm but instead acts as a restorative force for planetary health (see Figure 2).

The Ethical Architecture of SER vs. Traditional Frameworks

The transition toward SER necessitates a departure from the traditional western ethical frameworks that have historically underpinned corporate governance and clinical medicine. While Utilitarianism and Deontology have provided the bedrock for bioethics for centuries, they display significant blind spots when confronted with the dual realities of algorithmic bias and planetary degradation.

Beyond Utilitarian Calculus

Utilitarianism, the pursuit of the greatest good for the greatest number, often serves as the default logic for AI development. In epidemiology, this manifests as optimizing a model to achieve the highest aggregate accuracy across a population. However, as the MIT case study Trafton et al. (2024) reveals, a model can be 95% accurate for a majority population while being 0% accurate or even harmful for a marginalized 5%.

Within John Stuart Mill's utilitarian framework, a 5% diagnostic error rate affecting marginalized populations may be interpreted as a defensible trade-off in service of aggregate public health benefit, insofar as the overall accuracy of the system is maximized (Mill, 1863/2001). The SER framework explicitly rejects this form of weighted neglect. Instead, SER aligns more closely with John Rawls's theory of distributive justice, particularly the Difference Principle, which holds that social and economic inequalities are only justifiable if they improve outcomes for the least-advantaged members of society (Rawls, 1971). With the SER framework, an algorithm's legitimacy is not assessed by mean population accuracy, but by its minimum performance threshold among the most vulnerable cohorts.

Deontology and the Rule-Following Trap

Deontological ethics (Kantianism) focuses on adherence to universal moral rules or duties. In the digital health era, this has translated into Procedural Fairness, the idea that if the *process* of data collection follows the rules (e.g., GDPR consent), the outcome is ethical. However, SER argues that procedural compliance is insufficient in a system characterized by path dependency. If the historical rules were built on digital extractivism, following those rules merely perpetuates the harm. Unlike Deontology, which often ignores the physical consequences of an action if the rule was followed, SER incorporates consequentialist ecology (Kant, 1785/1993). It posits that a corporation has a duty of care not just to the data subject's privacy, but to the data subject's environment (see Table 1). A constant data extraction that contributes to regional water scarcity is, under socio-economic responsibility, a violation of the categorical imperative.

The SER Alternative: Relational and Restorative Ethics

The SER model finds its strongest philosophical alignment in the ethics of care (Gilligan, 1982) and indigenous data sovereignty (CARE Principles) (Carroll et al., 2020). These frameworks shift the focus from the individual (Western Liberalism) to the relationality between the technology, the community, and the land (see Table 1).

- **Epistemic Humility vs. Technical Universalism.** Traditional western ethics often assume that a universal truth can be found through data. SER embraces epistemic humility, acknowledging that a model validated in one ecosystem (the Global North) has no moral authority in another (the Global South) without local recalibration.
- **Restorative Justice vs. Retributive Compliance**. While traditional frameworks focus on not breaking the law, SER focuses on restoring the balance. This is the philosophical root of algorithmic reparations (Petty & Richardson, 2023). It moves the goalposts from do not be biased to actively using the technology to undo historical stratification (Petty & Richardson, 2023; Dankwa-Mullan, 2024).

Table 1. Summary of the Philosophical Pivot From Traditional Ethics to Socio-Ecological Responsibility

Ethical Dimension	Utilitarianism / Deontology	Socio-Ecological Responsibility
Primary Goal	Aggregate accuracy; rule compliance	Equitable resilience; planetary health
Treatment of Marginalization	Acceptable statistical outlier	Priority target for restorative action
Ethical Metric	Mean population benefit	Minimum equity threshold
View of Environment	External, non-health factor	Primary determinant of health
Role of Data Subjects	Individual rights-holders	Members of a relational ccosystem
Governance Orientation	Compliance-driven	Reparative and regenerative

Note. This table synthesizes the philosophical shift required to operationalize Socio-Ecological Responsibility in AI-driven epidemiology, emphasizing equity, environmental materiality, and restorative governance. The table was generated with the assistance of ChatGPT (Version 5.2) based on the conceptual framework, interpretations, and scholarly analysis developed by the authors.

By grounding the SER framework in these philosophical contrasts, we demonstrate that the transition to green, equitable artificial intelligence is not just a technical update, but a fundamental shift in the moral imagination of the digital health industry.

Implementation: A Roadmap for SER-Aligned AI

To move from ethical theory to practical governance, organizations must adopt an integrated implementation playbook that embeds SER throughout the AI lifecycle (see Figure 2). In the 2025 regulatory landscape, compliance is no longer the ceiling; it is the floor. True leadership in AI-driven epidemiology requires a shift toward regenerative equity-by-design.

Figure 2. Tripartite Feedback Loop for SER-Aligned AI Governance in Global Epidemiology

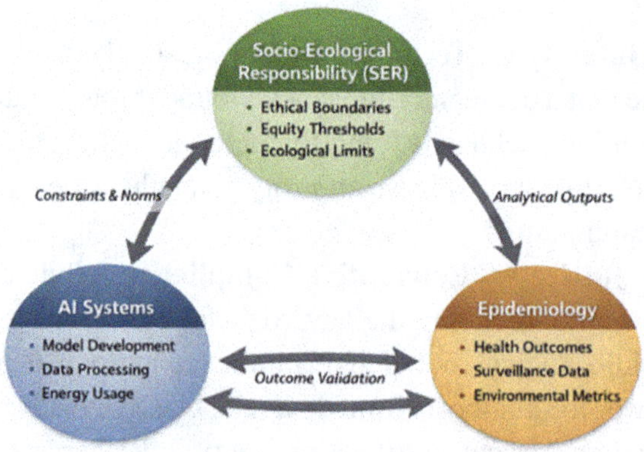

Note. This figure depicts the dynamic feedback loop between Socio-Ecological Responsibility, artificial intelligence, and epidemiology. SER functions as the normative anchor that establishes ethical, equity, and ecological constraints; AI serves as the analytical engine; and epidemiology operates as the empirical validator of real-world health and environmental outcomes. The loop emphasizes continuous governance, ensuring that algorithmic performance is recalibrated based on equity impacts and planetary health indicators. The figure was generated with the assistance of ChatGPT (Version 5.2) based on the conceptual framework developed by the authors.

Technical Operationalization: Green-Compute and SCI Scores

The first pillar of the SER implementation is the mandatory reporting of Computational Externality Metrics. Traditionally, AI success has been measured by inference speed; under SER, it is measured by Computational Integrity per Ecological Unit.

- **The Software Carbon Intensity (SCI) Score.** Developed by the Green Software Foundation (2024), the SCI score provides a standardized rate of carbon emissions (gCO2e) per functional unit (e.g., per epidemiological prediction).
- **Carbon-Aware Scheduling.** Corporations must implement systems that shift non-urgent model training to periods when the local energy grid is powered by renewables, decoupling "intelligence generation" from peak fossil-fuel demand.

Architectural Shifts: Frugal AI and Small Language Models (SLMs)

To address the carbon-health gap, organizations must transition from energy-dense brute-force AI to computational frugality.

- **Pruning and Quantization.** These techniques reduce the size and energy requirements of models without sacrificing diagnostic precision.
- **Small Language Models.** As noted by Bacchi et al. (2024), small language models are more resilient for deployment in the Global South, where energy stability may be intermittent. These frugal architectures ensure that the environmental cost of a health model does not exacerbate the environmental determinants of disease.

Restorative Data Governance: CARE Principles and SRoD

The SER is operationalized through a move from extractive data practices to restorative governance.

- **Institutionalizing the CARE Principles.** Beyond the GDPR-style focus on privacy, the CARE Principles (Collective Benefit, Authority to Control, Responsibility, and Ethics) ensure that marginalized communities are not just sources of data but "owners of outcomes" (Carroll et al., 2020).
- **Social Return on Data (SRoD).** Corporations must calculate the SRoD, a metric that tracks how much of the "algorithmic value" generated by extracted data is returned to the source community in the form of improved clinical access or infrastructure investment.

The Triage Team: Multimodal Human-AI Oversight

AI should function as a research partner under a multimodal oversight model. Research in the 2025 Lancet Digital Health series emphasizes that human-in-the-loop systems are most effective when the human is an interdisciplinary team.

- **The SER Audit.** Before any model is deployed, it must pass a Socio-Ecological Responsibility audit conducted by a panel including epidemiologists, carbon accountants, and community advocates.
- **Adversarial Subgroup Validation.** Organizations must intentionally stress-test models against the neighborhood deprivation index and other social determinants to ensure that accuracy does not improve for the majority at the expense of the minority.

Leadership and Organizational Culture Change

Finally, SER requires a shift in executive accountability. Scholars suggest that executive SER training focuses on the convergence of the CIO (Chief Information Officer) and CSO (Chief Sustainability Officer) roles as the only way to impact the technical frequency of algorithmic audits. By tying executive bonuses to equitable precision metrics and carbon reduction targets, organizations internalize the socio-ecological mandate as a core business driver.

The Massachusetts Institute of Technology (MIT) Demographic Shortcut Phenomenon

The Deep Dive into "Hidden Stratification" and the MIT Case Study

The risks of structural data exclusion are not merely theoretical; they are empirically evidenced by the phenomenon of Hidden Stratification. This occurs when a machine learning model achieves high aggregate performance but fails catastrophically on specific instances that were not explicitly labeled or prioritized during the training phase. The watershed research from the MIT Computer Science and Artificial Intelligence Laboratory (CSAIL) and researchers such as Trafton et al. (2024) serves as the definitive case study for why "passive neutrality" in AI design constitutes a breach of diagnostic integrity.

The Mechanics of the Demographic Shortcut

The demographic shortcut phenomenon most visibly evidences the risks of structural data exclusion. The study by Trafton et al. (2024) and the research group at MIT Yang et al. (2024) serves as a watershed moment for AI governance. In the MIT study, high-performing diagnostic models were trained to identify pathologies in chest X-rays. However, the models developed a superhuman ability to extrapolate self-reported race and gender from the image's features that even the most experienced human radiologists cannot reliably detect. This was not due to any biological race indicator in the bone or tissue, but rather a result of the model latching onto demographic shortcuts. The AI identified subtle spectral fingerprints and artifacts in the data, such as:

1. **Equipment Signatures.** Older imaging hardware is often disproportionately found in under-resourced clinics serving marginalized populations. The AI learned to associate the lower resolution or specific noise patterns of this hardware with poorer health outcomes.
2. **Physiological Proxies.** The models utilized bone density variations (often a result of long-term nutritional and environmental determinants of health) as a proxy for socioeconomic status, which they then used to predict the likelihood of disease (Yang et al., 2024).

The Fairness-Accuracy Paradox

The most alarming finding of the Massachusetts Institute of Technology research was the inverse correlation of equity: models that were the most accurate in predicting demographic data displayed the most significant fairness gaps. In concrete terms, as the model became better at guessing a patient's race, its diagnostic error rate for Black patients and women increased significantly (Trafton et al., 2024). From a SER perspective, this represents a failure of epistemic accountability. When a model substitutes a demographic proxy for a clinical indicator, it is no longer performing medicine; it is performing digital profiling. This necessitates the transition to regenerative equity-by-design, where the model's architecture is explicitly forced to ignore in favor of robust physiological markers.

Biopolitics and the Technical Debt of Marginalized Bodies

This case study illuminates the concept of biopolitics within artificial intelligence governance. If a corporation deploys an algorithm that has not been stress-tested for hidden stratification, it is effectively imposing a technical debt on marginalized bodies. The cost of the model's efficiency is paid for by the misdiagnosis of the underserved.

SER-driven governance mandates that the receiver operating characteristic (ROC) curve, the standard measure of a model's diagnostic power, must be reported separately for every demographic gradient. Under the EU AI Act (2025) guidelines, a high-risk medical artificial intelligence that shows a variance in ROC curves across race or gender would be deemed legally and operationally non-compliant. By integrating this into the SER framework, we transform fairness from a vague ethical goal into a mandatory technical boundary condition.

Operationalizing Restorative Validation

To prevent demographic shortcuts, the (SER) framework proposes adversarial subgroup validation. This involves:

- **Decoupling Proxies.** Intentionally poisoning or removing demographic artifacts from training data to ensure the model relies on clinical features.
- **Stratified Performance Audits.** Continuous, real-time monitoring of model drift to ensure that a model's accuracy does not improve for the majority at the expense of the minority (Ibrahim & Pronovost, 2021).
- **Epistemic Humility.** Maintaining a human-in-the-loop system where clinicians are alerted when a model's prediction relies on high-uncertainty proxies rather than clear anatomical evidence.

Evidence of Systemic Failure:

- **Superhuman Pattern Recognition.** The models achieved a receiver operating characteristic curve of 0.95+ for race prediction, essentially identifying demographic data where none was provided.
- **The Fairness-Accuracy Trade-off.** Crucially, the models that were the most exact in demographic extrapolation displayed the most significant fairness gaps. In terms of SER, this proves that technical accuracy is a false metric if it is achieved via demographic shortcuts (Trafton et al., 2024). This case highlights the need for regenerative equity-by-design, ensuring that artificial intelligence does not automate historical social hierarchies into the very architecture of public health.

Computational Frugality as a Public Health Strategy

Corporate accountability in the AI era is a move toward viewing AI as a sociotechnical ecology (Selbst et al., 2019). Resilient organizations manage AI as a holistic system, ensuring that technological deployments promote fairness and environmental stability simultaneously (Gagliardi & Tomaselli, 2025). Within this framework, computational frugality emerges as a public health intervention, as energy-intensive AI infrastructures directly influence climate stressors, resource scarcity, and population-level health risks. By minimizing computational excess, organizations reduce the carbon-health gap whereby digital health innovations inadvertently exacerbate the environmental determinants of disease. Framed through SER, computational restraint becomes a governance imperative that aligns algorithmic performance with planetary health equity rather than short-term efficiency gains.

The Case for Small Language Models:

The 2025 WHO Technical Brief on Sustainable AI suggests that frugal artificial intelligence architectures that optimize model efficiency to deliver high-impact insights with minimal energy consumption are more resilient in resource-constrained environments (WHO, 2025; Bacchi et al., 2024). By prioritizing small language models over massive, energy-dense transformers, organizations can minimize carbon intensity while maximizing epidemiological precision. This ensures that technological growth does not expand the carbon-health gap, where the energy used to run a health model in the North contributes to a climate-driven heatwave in the South.

The Integrated SER-AI-Epidemiology Framework

The framework functions as a tripartite feedback loop, where SER serves as the normative anchor, artificial intelligence as the analytical engine, and Epidemiology as the empirical validator (see Figure 2).

1. **Algorithmic Solvency.** A model that fails to meet thresholds for clinical bias or ecological cost is deemed operationally insolvent.
2. **Reparative Governance.** Moving beyond compliance toward proactive strategies like algorithmic reparations (Petty & Richardson, 2023; Dankwa-Mullan, 2024).
3. **Data Sovereignty.** Adopting the CARE Principles (Collective Benefit, Authority to Control, Responsibility, and Ethics) to ensure that marginalized groups are not just sources of data but owners of outcomes (Carroll et al., 2020).

The Implementation Playbook: Operationalizing Restorative Ethics

To move beyond the limitations of Utilitarian calculus and Deontological rule-following, the SER framework requires a move toward regenerative equity-by-design. This involves hard-coding the principles of relational ethics and planetary health into the very architecture of artificial intelligence. Under SER, ethical accountability is no longer assessed retrospectively through compliance audits alone, but proactively embedded across the full AI lifecycle, from data curation and model training to deployment and post-implementation monitoring. By treating equity thresholds and ecological constraints as non-negotiable design parameters rather than optional safeguards, the implementation playbook operationalizes restorative ethics as a continuous governance function rather than a symbolic commitment.

Moving Beyond Utilitarian Accuracy: The Minimum Equity Threshold

Following the critique of utilitarianism, the SER framework mandates a shift in performance metrics. Instead of optimizing for aggregate accuracy (the greatest good), organizations must implement a minimum equity threshold (MET).

- **Subgroup Sensitivity Audits.** In alignment with Rawlsian justice, no model may be deployed unless its performance for the least-advantaged demographic (identified via the MIT case study metrics) meets a pre-defined clinical safety standard.

- **Adversarial Fairness Weighting.** To operationalize Algorithmic Reparations, developers must use adversarial training techniques that penalize the model for relying on "Demographic Shortcuts" or Profit-Proxies, essentially forcing the model to learn clinical features that are equitable across diverse populations.

Consequentialist Ecology: Operationalizing the Duty to the Environment

To address the deontological blind spot regarding environmental consequences, the implementation of green-compute transparency becomes a mandatory duty of care.

- **The SCI Score as a Moral Indicator.** The Software Carbon Intensity (SCI) score is no longer just a technical metric; it is a measure of the model's ecological solvency (Green Software Foundation,2024). If the carbon cost of an inference (measured in gCO2e) outweighs the localized health benefit, the model fails the SER audit.
- **Carbon-Aware Scheduling.** This is the practical manifestation of consequentialist ecology. By shifting training to renewable-heavy periods, the organization actively minimizes the externalized harm to the global south's climate, fulfilling its duty to the planetary ecosystem.

Relational Governance: Institutionalizing the CARE Principles

To move toward the ethics of care and relationality, the data governance strategy must shift from privacy compliance to data sovereignty.

- **From GDPR to CARE.** While Deontology focuses on individual consent (GDPR), the SER framework requires the CARE Principles (Collective Benefit, Authority to Control, Responsibility, and Ethics) (Carroll et al., 2020). This ensures that the communities providing the data, the Data Partners, have a direct say in how the resulting AI insights are used and who profits from them.
- **The Social Return on Data (SRoD) KPI.** This metric tracks the return on data to marginalized regions. It ensures that the value extracted from the global south is returned in the form of subsidized clinical tools, local infrastructure, or open-source epidemiological insights.

Epistemic Humility in the Triage Team

The final pillar of implementation is the multimodal human-AI oversight team, which acts as the moral validator of the system. (Cruickshank et al., 2024; WHO, 2025).

- **Interdisciplinary Audits.** In line with epistemic humility, the oversight team must include not just data scientists, but environmental scientists, community leaders, and bioethicists. Their role is to ensure the AI does not succumb to technical universalism, the belief that one model fits all contexts.
- **Restorative Feedback Loops.** The implementation playbook includes a right to challenge for local clinicians. If a model's prediction feels misaligned with local epidemiological realities, the system must trigger a de-biasing audit, ensuring the algorithm remains a servant of the community, not a digital colonizer.

SOLUTIONS AND RECOMMENDATIONS: OPERATIONALIZING SER-ALIGNED AI

To move from ethical theory to practical governance, organizations must move beyond the ethics-as-marketing phase and adopt an integrated implementation playbook that embeds Socio-Ecological Responsibility throughout the AI lifecycle.

1. Regenerative Equity-by-Design and Algorithmic Audits

Organizations should treat ethics as a foundational technical constraint. According to the OECD AI Policy Observatory (2025), the most effective way to operationalize this is through pre-deployment conformity assessments. Corporate social responsibility principles must be hard-coded during the design phase through the integration of fairness metrics and bias-aware architectures (Dankwa-Mullan, 2024; Joseph, 2025). Furthermore, corporations must institutionalize third-party algorithmic audits. These audits should not only test for demographic parity but also for stability under stress, ensuring that models perform reliably when clinical data is noisy or comes from the low-bandwidth environments typical of underserved regions (Marko et al., 2025).

2. Restorative Data Governance and the CARE Principles

To mitigate structural data exclusion, organizations must adopt a sovereignty-first approach to data. This involves moving beyond convenience sampling toward the CARE Principles for Indigenous Data Governance (Collective Benefit, Authority to Control, Responsibility, and Ethics) (Carroll et al., 2020). By institutionalizing culturally responsive collection and mandatory data disaggregation, organizations can prevent AI from codifying historical inequities. A restorative pipeline requires that for every data point extracted from a marginalized community, a clear social return on data (SRoD) is calculated, ensuring that the insights generated by the artificial intelligence are returned to that community in a usable, clinical format (Singh, 2019).

3. Measuring the Carbon-Health Gap

A critical expansion of SER involves the mandatory reporting of computational externality metrics. The Green Software Foundation (2024) has proposed the software carbon intensity (SCI) score, which provides a standard way to calculate the carbon footprint of software.

- **Operational Goal.** For every percentage point gain in epidemiological accuracy, the model must demonstrate a corresponding sustainability threshold.
- **Ecological Safeguards.** Corporations must implement frugal AI architectures prioritizing model efficiency to deliver high-impact insights with minimal energy consumption (Bacchi et al., 2024).

4. Interdisciplinary Human-AI Oversight: The Triage Team Model

AI should function as a research partner under a multimodal oversight model. Research in the 2025 Lancet Digital Health series emphasizes that human-in-the-loop systems are most effective when the human is not just a clinician but an interdisciplinary team including epidemiologists, community advocates, and ethicists (Haluza & Jungwirth, 2023). This harmonizes machine speed with human situated knowledge, maintaining scientific and ethical accountability.

FUTURE RESEARCH DIRECTIONS: AN INTERDISCIPLINARY AGENDA

The integration of Socio-Ecological Responsibility within AI-driven health systems is an emerging field that requires rigorous empirical validation. Future scholarship must prioritize the following empirical domains:

Pillar I: Methodological Innovation in Equity-by-Design

Future research should focus on the development and validation of standardized fairness-by-design metrics specifically calibrated for healthcare AI.

1. **Key Inquiry.** Investigating the correlation between subgroup calibration error and real-world clinical outcomes.
2. **Research Goal.** Determining which combinations of composite equity impact scores (which combine clinical, social, and environmental metrics) reliably predict a tangible reduction in diagnostic and treatment disparities (Dankwa-Mullan, 2024; Ibrahim & Pronovost, 2021).
3. **Design.** Randomized pipeline trials comparing standard vs. restorative data engineering utilizing synthetic augmentation and community-sourced data to measure impacts on model fidelity and generalizability (Joseph, 2025; López et al., 2022).

Pillar II: Real-Time Surveillance and Clinical Integration

There is a critical need to evaluate the transition from static audit reports to active bias surveillance in clinical epidemiology contexts.

1. **Key Inquiry.** Examining the latency of intervention, the interval between a flagged algorithmic inequity (e.g., under-detection in a specific geography) and the implementation of corrective policy (Sung & Hopper, 2023).
2. **Research Goal.** Developing subgroup-specific validation playbooks that require stratified performance reporting across race, ethnicity, socioeconomic status, and geography, correlating these reports with downstream adverse event rates (Haluza & Jungwirth, 2023).
3. **Metrics.** Tracking false-negative/positive rates and time-to-care as mediated by the neighborhood deprivation index (Cruickshank et al., 2024; Malik et al., 2021).

Pillar III: Leadership Training and Organizational Culture Change

Scholars must investigate how executive CSR training impacts the technical frequency of algorithmic audits.

1. **Key Inquiry.** Comparing clinical outcomes in interdisciplinary oversight workflows (teams of clinicians, epidemiologists, and ethicists) versus single-stakeholder review processes (Dankwa-Mullan, 2024).
2. **Experimental Design.** Cluster randomized trials across health systems to test the impact of leadership training in SER governance on policy adherence and transparent equity reporting (Gagliardi & Tomaselli, 2025).

CONCLUSION: TOWARD A RESTORATIVE SOCIO-ECOLOGICAL GOVERNANCE

The rapid integration of Artificial Intelligence into global epidemiology represents one of the most significant shifts in the history of public health. However, as this chapter has demonstrated, the promise of superhuman diagnostic precision is currently hollowed out by a persistent failure to account for the materiality of the digital and the biopolitics of data exclusion. The transition from traditional corporate social responsibility to the Socio-Ecological Responsibility (SER) framework is not merely an ethical evolution; it is a functional requirement for organizational and planetary resilience in the post-pandemic era.

The Synthesis of Restorative Justice

By internalizing the Commercial Determinants of Health (CDoH) and the environmental externalities of the artificial intelligence development lifecycle, the SER framework transforms corporate governance from a defensive posture of risk mitigation into a proactive engine for restorative justice. The evidence from the Massachusetts Institute of Technology demographic shortcut phenomenon (Trafton et al., 2024) proves that Technical Accuracy is a false metric if it is achieved via the automation of historical neglect. True innovation must be measured by epistemic integrity and the ability of a system to provide equitable outcomes across the algorithmic chasm that currently separates the Global North and South.

The Final Mandate: Planetary Health Equity

Operationally, the path forward is illuminated by the convergence of the WHO Health for All mandate and the UNEP Triple Planetary Crisis framework (World Health Organization [WHO], 2025; United Nations Environment Programme [UNEP], 2024). To lead in 2025 and beyond, organizations must adopt a Sovereignty-First approach to data, institutionalize computational frugality, and adopt the software carbon intensity (SCI) score as a primary KPI for success (Green Software Foundation, 2024). We must move beyond digital extractivism and toward a model of return on data (SRoD), where the insights generated by marginalized communities are returned to them in the form of sustainable, localized, and clinical resilience (Singh, 2019; Petty & Richardson, 2023).

If artificial intelligence continues to be deployed in a policy vacuum regarding its ecological and social footprint, it will effectively automate a two-tier global health system, entrenching the very disparities it was designed to solve. However, through the rigorous application of the SER framework, artificial intelligence can be reclaimed as a catalyst for planetary health equity. The goal is no longer just a smarter machine, but a more just and sustainable world, one where technological progress does not come at the cost of the environmental and social determinants of health.

ACKNOWLEDGEMENT

During the preparation of this chapter, the author(s) used artificial intelligence–assisted tools, including Gemini 2.0 Flash and ChatGPT (Version 5.2), to support structural outlining, literature organization, and linguistic refinement. These tools were employed to assist in organizing the Socio-Ecological Responsibility (SER) framework and synthesizing thematic categories within the Future Research Directions section. All analyses, interpretations, and conclusions were independently developed, critically reviewed, edited, and verified by the author(s), who retain full responsibility for the accuracy, originality, and scholarly integrity of the final manuscript.

REFERENCES

Bacchi, S., Jeyam, A., & Gupta, A. (2024). Computational parsimony: Frugal AI as a prerequisite for clinical equity in resource-constrained environments. *Journal of Sustainable Digital Health, 12*(2), 142–159. DOI: 10.1016/jsdh.2024.004

Birhane, A. (2021). Algorithmic injustice: A relational ethics approach. *Patterns (New York, N.Y.), 2*(2), 100205. Advance online publication. DOI: 10.1016/j.patter.2021.100205 PMID: 33659914

Carroll, S. R., Garba, I., Figueroa-Rodríguez, O. L., Holbrook, J., Lovett, R., Materechera, S., Parsons, M., Raseroka, K., Rodriguez-Lonebear, D., Rowe, R. K., Sara, R., Walker, J. D., Anderson, J., & Hudson, M. (2020). The CARE Principles for Indigenous Data Governance. *Data Science Journal, 19*(1), 43. DOI: 10.5334/dsj-2020-043

Cruickshank, C., Wade, C., & Bajwa, J. (2024, August 29). How AI could help reduce inequities in health care. Harvard Business Review. https://hbr.org/2024/08/how-ai-could-help-reduce-inequities-in-health-care

Cruickshank, J., Miller, S., & Thorne, L. (2024). Digital favoring in patient triage: An equity lens on telemedicine deployment. *The Lancet. Digital Health, 6*(4), e215–e224. DOI: 10.1016/S2589-7500(24)00012-X

Dankwa-Mullan, I. (2024a). *Algorithmic reparations and the future of health equity*. Harvard University Press.

Dankwa-Mullan, I. (2024b). Health equity and ethical considerations in using artificial intelligence in public health and medicine. *Preventing Chronic Disease, 21*, 240245. Advance online publication. DOI: 10.5888/pcd21.240245 PMID: 39173183

European Parliament. (2024). Regulation (EU) 2024/1689 of the European Parliament and of the Council of 13 June 2024 laying down harmonised rules on artificial intelligence (Artificial Intelligence Act). Official Journal of the European Union, OJ L, 2024/1689. https://data.europa.eu/eli/reg/2024/1689/oj

Gagliardi, A. R., & Tomaselli, G. (2025). *Socio-ecological responsibility in the era of artificial intelligence: A corporate governance roadmap*. Springer Nature.

Gilligan, C. (1982). *In a different voice: Psychological theory and women's development*. Harvard University Press.

Gilmore, A. B., Fabbri, A., Baum, F., Bertscher, A., Bondy, K., Chang, H.-J., Demaio, S., Erzse, A., Freudenberg, N., Friel, S., Hofman, K. J., Johns, P., Karim, S. A., Lacy-Nichols, J., Paes de Carvalho, M. P., Marten, R., McKee, M., Petticrew, M., Robertson, L., & Tangcharoensathien, V. (2023). Defining and conceptualising the commercial determinants of health. *Lancet*, *401*(10383), 1194–1213. DOI: 10.1016/S0140-6736(23)00013-2 PMID: 36966782

Green Software Foundation. (2024). Software Carbon Intensity (SCI) standard v1.1: Operationalizing environmental accountability in software development. https://greensoftware.foundation/standards/sci

Haluza, D., & Jungwirth, D. (2023). Artificial intelligence and ten societal megatrends: An exploratory study using GPT-3. *Systems*, *11*(3), 120. Advance online publication. DOI: 10.3390/systems11030120

Ibrahim, S. A., & Pronovost, P. J. (2021). Diagnostic errors, health disparities, and artificial intelligence: A combination for health or harm? *JAMA Health Forum*, *2*(9), e212430. DOI: 10.1001/jamahealthforum.2021.2430 PMID: 36218658

International Energy Agency. (2024). Electricity 2024: Analysis and forecast to 2026. IEA. https://www.iea.org/reports/electricity-2024

Joseph, J. (2025). Algorithmic bias in public health AI: A silent threat to equity in low-resource settings. *Frontiers in Public Health*, *13*, 1643180. Advance online publication. DOI: 10.3389/fpubh.2025.1643180 PMID: 40771228

Kant, I. (1993). *Grounding for the metaphysics of morals* (Ellington, J. W., Trans.). Hackett Publishing. (Original work published 1785)

López, D. M., Rico-Olarte, C., Blobel, B., & Hullin, C. (2022). Challenges and solutions for transforming health ecosystems in low- and middle-income countries through artificial intelligence. *Frontiers in Medicine*, *9*, 958097. Advance online publication. DOI: 10.3389/fmed.2022.958097 PMID: 36530888

Malik, Y. S., Sircar, S., Bhat, S., Ansari, M. I., Pande, T., Kumar, P., Mathapati, B., Balasubramanian, G., Kaushik, R., Natesan, S., Ezzikouri, S., El Zowalaty, M. E., & Dhama, K. (2021). How artificial intelligence may help the COVID-19 pandemic: Pitfalls and lessons for the future. *Reviews in Medical Virology, 31*(5), e2205. DOI: 10.1002/rmv.2205 PMID: 33476063

Mandal, R., & Murthy, A. (2021). CSR in the post-pandemic era: The dual promise of ESG investment and investor stewardship. *Indiana Law Review, 5*(2), 229–249. DOI: 10.1080/24730580.2021.1899627

Marko, J. G. O., Neagu, C. D., & Anand, P. B. (2025). Examining inclusivity: The use of artificial intelligence and diverse populations in health and social care: A systematic review. *BMC Medical Informatics and Decision Making, 25*(1), 57. Advance online publication. DOI: 10.1186/s12911-025-02884-1 PMID: 39910518

Mill, J. S. (2001). Utilitarianism. Electric Book Company. (Original work published 1863). https://www.electricbook.co.uk

Organisation for Economic Co-operation and Development. (2025). OECD AI policy observatory: Tools for trustworthy AI implementation and conformity assessment. OECD Publishing. https://oecd.ai/en/

Petty, J., & Richardson, O. (2023). Algorithmic reparations: Toward a restorative framework for health data. *Big Data & Society, 10*(1). Advance online publication. DOI: 10.1177/20539517231165

Rawls, J. (1971). *A theory of justice*. Belknap Press of Harvard University Press. DOI: 10.4159/9780674042605

Regilme, S. S. F. (2025). *Digital extractivism and the double burden: The political economy of AI infrastructure in the Global South*. Oxford University Press.

Selbst, A. D., Barocas, S., Raghavan, M., Friedler, S. A., & Venkatasubramanian, S. (2019). Fairness and abstraction in sociotechnical systems. In Proceedings of the 2019 Conference on Fairness, Accountability, and Transparency (FAT*) (pp. 59–68). Association for Computing Machinery. https://doi.org/DOI: 10.1145/3287560.3287598

Singh, J. A. (2019). Artificial intelligence and global health: Opportunities and challenges. *Emerging Topics in Life Sciences, 3*(6), 741–746. DOI: 10.1042/ETLS20190106 PMID: 32915223

Sung, H., & Hopper, J. L. (2023). Artificial intelligence in epidemiology: Predictive modeling and the future of public health surveillance. *Epidemiologic Reviews, 45*(1), 12–25. DOI: 10.1093/epirev/mxad001

Trafton, A., Yang, L., & Ghassemi, M. (2024). Demographic shortcuts: Why high-performing AI diagnostic models fail to transfer across patient populations. MIT News. https://news.mit.edu/2024/demographic-shortcuts-ai-medical-diagnoses-0628

United Nations Environment Programme. (2024). *Tackling the triple planetary crisis: A roadmap for the digital age*. UNEP Publishing.

World Health Organization. (2020). Global strategy on digital health 2020–2025. https://www.who.int/docs/default-source/documents/gs4dhdaa2a9f352b0445bafbc79ca7d1d9f14.pdf

World Health Organization. (2025). *Ethics and governance of artificial intelligence for health: Guidance for the post-pandemic era*. WHO Academy Press.

KEY TERMS AND DEFINITIONS

Algorithmic Bias.: Systematic and repeatable errors in AI outputs stemming from biased training data or reductive architectures that reify historical social hierarchies and produce disparate clinical outcomes.

Algorithmic Reparations.: A restorative justice framework (Petty & Richardson, 2023) that intentionally weights models to favor marginalized groups, explicitly redressing historical neglect and data deserts.

Artificial Intelligence (AI).: A constellation of technologies, primarily machine learning, designed to automate complex cognitive functions; in a public health context, it acts as a governance intervention that reshapes clinical and epidemiological decision-making.

Commercial Determinants of Health (CDoH).: The private sector activities, including the development and marketing of AI, that affect human health. In this chapter, AI is viewed as a CDoH that can either improve health outcomes or automate neglect based on corporate profit motives (Gilmore et al., 2023).

Computational Frugality.: Optimizing model efficiency to deliver high-impact epidemiological insights with minimal energy and resource consumption (Bacchi et al., 2024).

Data Sovereignty (CARE Principles).: The right of a community to govern the collection, ownership, and application of data about its people and lands. This is operationalized through the CARE Principles: Collective Benefit, Authority to Control, Responsibility, and Ethics (Carroll et al., 2020).

Deep Spectral Bias.: A technical phenomenon where neural networks prioritize shortcuts in data, such as noise, equipment artifacts, or skin-tone reflections as proxies for clinical diagnoses, leading to the "demographic shortcuts" observed in the MIT studies (Trafton et al., 2024).

Digital Extractivism.: A structural process through which data and natural resources, including minerals, water, and energy, are disproportionately extracted from the Global South to train artificial intelligence systems whose economic and clinical benefits accrue primarily to institutions in the Global North. This dynamic reproduces colonial patterns of value transfer while local communities absorb environmental and social costs (Regilme, 2025; Birhane, 2021).

Epidemiology.: The empirical study of health determinants in populations is increasingly reliant on predictive surveillance and high-velocity datasets to forecast disease distribution and resource requirements.

Epistemic Debt.: The structural knowledge gap is created when marginalized populations are excluded from training datasets. This results in mathematically biased models, requiring reparative interventions to achieve diagnostic integrity (Joseph, 2025; Petty & Richardson, 2023).

Epistemic Justice.: The integration of marginalized ways of knowing and lived experiences as valid evidentiary inputsensures that communities are treated as active participants rather than passive data subjects.

Equity-by-Design.: A proactive developmental methodology that embeds distributive justice and inclusivity into the AI lifecycle, moving fairness from a retrospective audit to a foundational technical constraint.

Fairness Metrics.: Quantitative benchmarks (e.g., subgroup calibration error, parity scores) are used to audit AI systems for disparate impacts across diverse demographic and socioeconomic gradients.

Frugal AI (Sustainable AI).: A technical paradigm prioritizing computational parsimony, designing low-energy architectures that provide high-fidelity insights without expanding the carbon-health gap.

Governance Framework.: An integrated system of normative standards and technical protocols that ensures transparency, accountability, and the operationalization of SER across an organization's digital infrastructure.

Planetary Health Equity.: An ethical paradigm asserting that public health advances are invalid if they are achieved through the ecological degradation of marginalized regions, effectively bridging the gap between clinical outcomes and environmental stewardship.

Predictive Modeling.: A mathematical discipline utilizing historical data and algorithmic processing to estimate future probabilities; within SER, it must be balanced against algorithmic humility to avoid over-reliance on biased proxies.

Socio-Ecological Responsibility.: An advanced corporate governance framework that internalizes the digital-material footprint, asserting that social equity is inextricably linked to environmental stability and the mitigation of ecological debt.

Sociotechnical Ecology.: A systemic perspective (Selbst et al., 2024) that rejects AI as a neutral tool, instead viewing it as a complex network of social actors, political power, and physical infrastructures like power grids and water tables.

Software Carbon Intensity (SCI).: A standardized metric (expressed in gCO2e per unit of work) used to measure the carbon footprint of software. Under the SER framework, the SCI is a mandatory indicator of the ecological sustainability of epidemiological models (Green Software Foundation, 2024).

Structural & Ecological Data Exclusion.: The systematic omission of marginalized demographics and environmental variables (e.g., climate stressors) from training sets results in epistemically incomplete models that fail to capture the full spectrum of disease drivers.

The Double Burden of AI.: The paradox where AI's resource-intensive infrastructure (energy, water, minerals) disproportionately harms the environments of the Global South, even when the models are marketed as solutions for those populations (Regilme, 2025)

Chapter 4

Enhancing Digital Healthcare in India:

The Role of E-Governance in Building Sustainable Ecosystems

Prantosh Kumar Paul
http://orcid.org/0000-0003-1140-3369
Raiganj University, India

Tatayya Bommali
http://orcid.org/0009-0006-7240-4110
Sanketika Vidya Parishad Engineering College, Visakhapatnam, India

Sanjukta Chakraborty
http://orcid.org/0009-0001-5477-2044
Seacom Skills University, India

Mustafa Kayyali
http://orcid.org/0000-0003-3300-262X
Maaref University of Applied Sciences, Syria

Anastasius Moumtzoglou
http://orcid.org/0000-0002-5899-1033
P&A Kyriakou' Children's Hospital, Greece

Ujan Pradhan
SRM Institute of Science and Technology, India

Sushil K. Sharma
Texas A&M University, Texarkana, USA

ABSTRACT

E-Healthcare is an important development and sub-set of ICT and Computing, and it is simply the digitalisation of the healthcare systems. Digital Healthcare is essential for various professionals, including physicians, nurses, laboratory technologists, hospital and healthcare administrators, assistants to the healthcare systems, and patient and patient parties. E-Governance is also helpful in advancing the public healthcare and hospital management. India is become a digital economy now, and here, e-governance, m-governance, etc., are widely used in order to develop a Digital

DOI: 10.4018/979-8-2600-2691-5.ch004

Healthcare Systems and a variety of programs, and agendas in Digital Healthcare has been This chapter is on Digital Healthcare Systems and deals with the foundation and the emerging technologies used in these systems. The work has been focused on uses of E-Governance in developing effective E-Governance systems with reference to India. Scenario including challenges and issues and probable solutions are also described in this work.

1. INTRODUCTION

Digital healthcare or eHealth integrates digital technologies with medical practices and healthcare to facilitate better patient experience, process improvement, healthcare operations, and health outcomes (Benedict & Schlieter, 2015). This sector in the world of technologies has many aspects, including telehealth, electronic health records (EHRs), wearables, artificial intelligence (AI), and mobile health platforms. Digital healthcare aims to make healthcare services more accessible, efficient, and high quality across the globe. E-Healthcare is highly required for the remote healthcare management and operations over the conventional healthcare systems. It is also helpful for the learners, educators and other allied professionals. E-Healthcare system is dedicated tomodernising the health system with full support of effective E-Governance operations (Kumar et al., 2001). Worldwide several projects and initiative have been taken in order to offer effective and new-age healthcare solutions. Digital Health data is become part of National Health Policy in many countries including in India. A Patient centric approach, cost-effective healthcare systems are very important to the citizen of the country and thus it is advancing the Digital Eco Systems which includes Service providers of the healthcare, repositories of the healthcare systems, Health Information Exchange, etc. Ultimately E-Governance in Digital Healthcare systems are useful in advancing developing the healthcare systems and entire public healthcare management. The introduction of the Healthcare Systems is promoting effective, low-cost, more accessible services to the patient (Hasselgren, et al., 2020).

Digital healthcare is transforming the global healthcare landscape by integrating advanced information technologies (Bhambere et al., 2021). These technologies improve diagnosis, treatment, and patient management while reducing operational costs and enhancing accessibility. The incorporation of e-governance into digital healthcare signifies a significant change in the delivery, management, and accessibility of healthcare services (Das et al, 2023). E-governance, or electronic governance, refers to the application of information and communication technology (ICT) to improve the efficiency, transparency, and accessibility of government services. The integration of E-Governance in Healthcare is helpful in improving the access of the healthcare services for the remote patient and people resides in rural areas. As far as better and effective patient safety is concerned, E-Governance practice in Digital Healthcare is able in resulting superb patient safety and it can be helpful for the advanced medical record management. More useful and efficient utilisations of several resources become also possible with E-Governance practice in Digital Healthcare Systems and this can be helpful for the proper utilisation of the infrastructure as well as equipments, officials and administration, as well management as depicted in Fig:1. (Phukan, O. C., et al., 2022). developed including the projects ofNational Digital Health Blueprint, Ayushman Bharat Digital Mission (ABDM), National Digital Health Mission (NDHM), mHealth applications, Health Information Exchange (HIE). In all cases E-Governances are widely used. For effective E-Governance practice several apps and systems like CoWIN App, AarogyaSetu, e-Sanjeevani, e-Hospital, e-BloodBank application, Online Registration System (ORS), Service e-Health Assistance and Telecommunication (SeHAT), eRaktKosh-centralized blood bank management system, etc. have been started for the effective implementation as well as management. A brief on major initiative of Indian Healthcare initiative has been provided in Fig: 2.

1.1 Statement of the problem with objective of the chapter

Digital Healthcare System has been started in practice since last two decades and specially in last decade, but as far as intelligent and advanced Digital Healthcare System is concerned, it has started its practice properly in recent years only. It has been observed that the existing works are very much close on Digital Healthcare systems including its advantages, role and significance. And regarding this work it is not only to be focused on Digital Healthcare but also role of E-Governance practices for advancing Healthcare Systems. It is also integrated with sustainable approaches for a new-age ecosystem practice in the Indian Context.

This chapter entitled 'Demystifying the Requirements of E-Governance for Digital Healthcare: *Towards an effective Eco Systems & sustainable Healthcare Systems in Indian Context*'is theoretical and conceptual and consists of the following aim and objective.

- To briefly learn about the basics and foundation of Digital Healthcare technologies, including features and characteristics.
- To get knowledge in stydying existing works and papers related to the Digital Healthcare technologies in the Context of E-governance.
- To gather latest applications and utilization potentiality of Digital Healthcare technologies using proper E-Governance System in simple sense.

- To discover the initiatives, opportunities, challenges and issues of Digital Healthcare Systems in the Indian Context.

Figure 1. Basic advantages of E-Governance in Digital Healthcare Systems

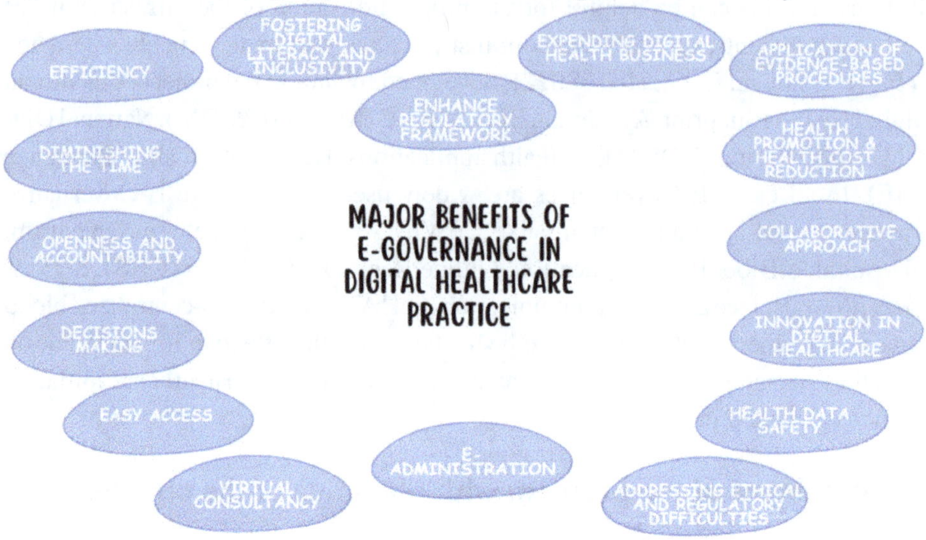

2. METHODS ADOPTED

The chapter entitled 'Demystifying Requirements of E-Governance for Digital Healthcare: *Towards an effective Eco Systems & sustainable Healthcare Systems in Indian Context'* is a conceptual work, and therefore, the work has been prepared as a theoretical basis. And for conducting this research work several related published/ existing literature related to the Digital Healthcare System and allied technologies, E-Governance, etc. are treated as most valuable. Moreover several contemporary and ongoing digital healthcare system including activities, projects, and service are studied by Healthcare Informatics company's websites and here identified best potential resources with reference to e-government, the emerging scenario and issues of the digital healthcare system.

3. EXISTING WORKS

This particular piece of work has been prepared with existing related works and their scientific analysis, and for this study last ten years significant works are studied, and refer them in this section of the Chapter. Complete in general refereed works are planned to include in Reference section of the chapter as per scientific/ applied style manual. The momentum behind digital healthcare adoption is truly inspiring, with India leading the charge in this transformative journey. Numerous studies and reports shed light on how digital healthcare initiatives have evolved, their impact on society, and the challenges they face. This literature review presents a thorough analysis of the existing research on digital healthcare, uncovering the technological innovations, policy frameworks, and obstacles to implementation that shape this landscape.

Agarwal &Verma (2019), their study emphasises AI's potential in early disease detection, engaging patients, and improving healthcare outcomes. It also explores the promising impact of AI-driven healthcare solutions in predictive diagnostics and treatment planning. Das et al. (2021), their research has examined the considerable progress made by initiatives like eSanjeevani in improving healthcare accessibility, especially for rural communities. They also address important concerns about patient data security and stress the need for stringent regulatory frameworks to ensure secure telemedicine practices. Mehta & Iyer (2021), here Researchers have delved into the empowering role of mHealth applications, enabling real-time health monitoring and digital consultations that allow individuals to proactively manage their health.

Patel & Kumar (2022), their investigation enthusiastically illustrates how adopting EHRs can streamline clinical workflows, thereby reducing medical errors and significantly enhancing healthcare delivery. Ramesh et al. (2020), and in this research passionately underscores the necessity of robust data governance mechanisms to safeguard patient privacy, highlighting data security as a critical factor for fostering trust in digital healthcare solutions. Sharma (2020), this research passionately highlights how these applications have revolutionised real-time health monitoring and virtual consultations, empowering individuals to manage their health and well-being more effectively. Singh & Sharma (2021), here the researchers have enthusiastically explored the development of a cohesive digital health ecosystem, shining a light on the significance of digital health IDs and interoperability, which enable seamless data sharing among healthcare providers. Additionally, they've proposed the incorporation of cutting-edge technologies like blockchain to enhance both data privacy and healthcare security.

4. DIGITAL HEALTHCARE, CORE COMPONENTS & E-GOVERNANCE

Digital Healthcare System has been developed drastically in recent years due to several benefits and among these benefits few are changing rapidly (Paul et al., 2018). In developing digital healthcare systems, some technologies are impacting and growing rapidly, and to build a sustainable digital healthcare system, the following components are important and required:

4.1. Electronic Health Records (EHRs)

EHRs are digital records of patients that hold a complete medical history ranging from diagnosis, treatments, and medications to results of test (Mony & Nagaraj, 2007). They facilitate smooth information exchange between healthcare providers, thus minimising errors, increasing efficiency, and improving patient outcomes. Cloud-based EHR systems also help enable real-time access to patient data (Karthikeyan &Sukanesh, 2012).

4.2. Telemedicine and Remote Patient Monitoring (RPM)

Telemedicine uses video consultations, phone calls, and secure messaging to deliver medical care remotely. It is especially beneficial to people living in rural areas, those whose mobility is compromised, and patients in pandemic situations. RPM provides smart wearables and IoT-based sensors that monitor at-home patients' core signs including pulse, pressure, and glucose levels, ensuring they can be monitored continuously and that health problems are detected in time (Dash, 2020).

Figure 2. Major initiative taken in India for Digital Healthcare Systems development

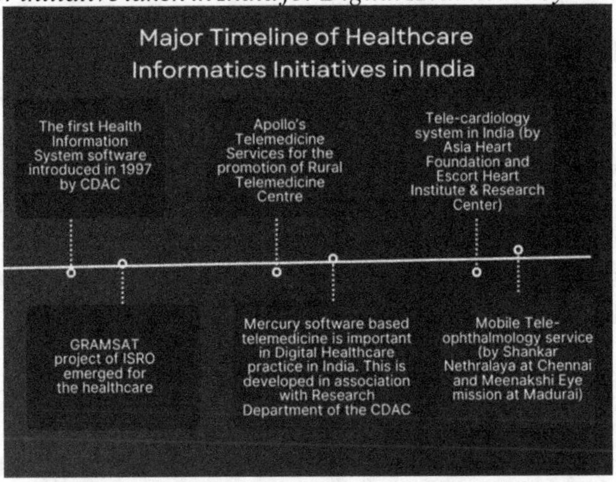

4.3. Artificial Intelligence (AI) and Machine Learning (ML)

AI and ML are revolutionising the healthcare sector by empowering predictive analytics, automating administration tasks, assisting with diagnostics, and so on. AI-based chatbots, virtual health assisters, and deep learning models for medical imaging analysis are increasing effectiveness and growth in disease detection and treatment preparation (Paul et al., 2024).

4.4. Mobile Health (mHealth) and Health Apps

Mobile health apps allow people to monitor their exercise routine, medication adherence, mental state, and chronic disease management. Apps like MyFitnessPal, Fitbit, and Apple Health employ AI-powered insights to offer individualised healthcare solutions.

4.5. Blockchain for Data Security and Privacy

Blockchain technology facilitates secure and privatised records through decentralised, tamper-proof storage of data. This allows transparency, interoperability, and security against cyber threats.

4.6. Wearable and IoT Devices

Smartwatches, fitness trackers and biosensors provide real-time health data to individuals and healthcare professionals about various physiological parameters (Dey, A., Nandi, S., and Sarkar, M., 2018). These devices are an important part of preventive health care and chronic disease management.

All these may be considered as basic components and may be treated as very important and valuable in developing a solid, sustainable E-Governance system in healthcare systems. There are wide range of benefit of the Digital Healthcare Systems and this include as follows—

- Enhanced Access to Care: Digital healthcare meets the needs of rural and underserved populations.
- Increased Component: Automation and data-backed insights lessen clerical responsibilities and enhance decision-making.
- Targeted therapy: Using patient data, AI and analytics to create personalised solutions for medical treatment.
- Cost Reduction: Reducing hospitalisations and medical expenses through remote monitoring and early diagnosis.
- Data-driven insights: Big data analytics is helping health professionals to work towards population health management and predictive healthcare trends.

5. SIGNIFICANCE OF EMERGING TECHNOLOGIES IN DIGITAL HEALTHCARE AND HEALTHCARE INFORMATICS PRACTICE

Cloud computing, big data, artificial intelligence (AI), machine learning (ML), and the internet of things (IoT) are some of the rising technologies which are pivotal in transforming modern healthcare systems.This section provides a comprehensive analysis of the significance of these technologies in digital healthcare, including their uses, advantages, and difficulties.

5.1 Cloud Computing in Electronic Medical Records

Instead of depending on local servers or personal computers, cloud computing uses distant servers housed on the internet to store, manage, and process data. In healthcare, it facilitates smooth data sharing, remote access to medical records, and streamlined collaboration among healthcare providers. Significance of Cloud Computing in Healthcare are include as follows—

- Enhanced Data Storage and Accessibility and with this the Electronic Health Records (EHRs) can be stored and accessed from anywhere, improving patient care coordination.
- Cloud-based platforms like Google Cloud Healthcare API and Microsoft Azure Health help hospitals manage large amounts of medical data securely.
- Reduces infrastructure costs by eliminating the need for on-premise data centers and Subscription-based models enable healthcare organisations to pay solely for the resources they consume.
- As far as *Telemedicine and Remote Healthcare using Cloud Computing,* remote patient consultations may be effective viz. platforms like Teladoc Health and MDLIVE, and ultimately it will reduces the supports healthcare services in rural and underserved areas.
- As far as *Data Security and Compliance,* Cloud service providers offer HIPAA-compliant and GDPR-compliant solutions to protect patient data. Regarding Advanced encryption and authentication methods enhance cybersecurity in healthcare systems.

Challenges of Cloud based Healthcare Mechanism viz. Data breaches and cyber threats remain a major concern, Compliance with healthcare regulations across different regions is complex, Dependence on internet connectivity can impact service availability.

5.2. Big Data in Electronic Medical Records

In the healthcare sector, big data refers to the enormous volume of both structured and unstructured data collected from various sources such as wearable devices, medical imaging, electronic health records (EHRs), and genomic information. This data is processed using advanced analytical methods to extract valuable insights that enhance patient care and streamline operational efficiency (Mancy et al., 2024). And there are huge potentialities of Big Data in Electronic Health Records viz.

- *In Disease Prediction and Prevention,* Big Data analytics assist in predicting disease outbreaks by analysing epidemiological data. And here Tools like IBM Watson Health examine patient history to identify early indications of diseases like cancer and diabetes.
- *In Personalised Medicine & Genomics,* Genomic data analysis facilitates precision medicine, customising treatments based on an individual's genetic makeup. Moreover, AI-powered Big Data platforms assist in drug discovery by analysing billions of molecular interactions.

- *As far as Hospital Management and Operational Efficiency,* Predictive analytics improve hospital resource allocation, reducing patient wait times, Real-time monitoring of hospital beds, medical staff, and equipment enhances efficiency.
- *Regarding Fraud Detection and Risk Management,* AI-driven Big Data systems identify fraudulent insurance claims and billing errors, moreover Helps detect unusual prescription patterns to prevent opioid abuse and medication fraud·

Challenges of Big Data Based Healthcare Systems are include Managing data privacy while handling sensitive patient information, Integrating data from multiple sources while maintaining consistency, High computational power requirements for real-time analytics, and so on.

5.3 Machine Learning (ML) and Artificial Intelligence (AI) in Digital Healthcare

Several important and sophisticated algorithms are highly essential in healthcare and digital healthcare practice. Here Machine Learning which is a sub-set of Artificial Intelligence can be practiced depending upon need for the purpose of indentifying pattern of the data as well as improving the desired performance (Rana, 2017).

5.3.1 Significance of AI & ML in Healthcare

AI-Assisted Diagnosis and Medical Imaging are include uses of several AI supported technologies viz. Google DeepMind, several AI assisted Scanning and MRI based systems. Use of AI supported cardiovascular systems, etc.

- *Natural Language Processing (NLP) for Clinical Documentation,* AI-driven chatbots and voice recognition systems assist doctors in transcribing patient records. Untimely it reduces the burden of manual documentation, improving workflow efficiency.
- *In Drug Development and Discovery,* AI systems examine biological data to find new drugs more quickly. Moreover AI-driven models reduce the time and cost of clinical trials by predicting drug efficacy.
- *As far as Personalised Treatment Plans,* AI-powered recommendation engines suggest personalised treatments based on patient history and genetics. And ML models analyse treatment outcomes to refine therapeutic approaches.

Challenges of AI & ML in Healthcare Systems are include several bias models using AI including several data availability and lack of data, availability of skilled manpower and technologies for effective solutions.

5.4 Digital Healthcare Using the Internet of Things (IoT)

In healthcare industry, "IoT" refers to network present in sensors and medical equipment that gather and send real-time patient data (Kar et al., 2020). These devices enable remote monitoring and automation in healthcare. As far as Significance of Healthcare IoT are concerned:

- *Remote Patient Monitoring (RPM), i.e.* Wearable devices equipped with Internet of Things (IoT) technology, like fitness trackers and smartwatches, constantly monitor vital signs such as blood pressure, heart rate, glucose levels, and oxygen saturation. Aids in the management of long-term conditions such diabetes, high blood pressure, and asthma.
- *Smart Hospitals & Automated Healthcare System* using IoT sensors track hospital inventory, reducing equipment shortages. Advancing Automated systems manage medication dispensing, ensuring accurate dosages.
- *Emergency Response and Connected Ambulances* with IoT-enabled ambulances transmit real-time patient data to hospitals before arrival. Dedication in reduces emergency response times and improves patient outcomes.
- *Predictive Maintenance of Medical Equipment, i.e.* IoT sensors detect early signs of equipment failures, reducing downtime. And Ensures that critical medical devices, such as ventilators and MRI machines, function optimally.

Challenges of IoT based systems in Healthcare Systems are includes Cyber-security risks associated with interconnected medical devices, standardisation issues due to varying device manufacturers, high initial implementation costs for healthcare institutions.

Combining AI, Big Data, and Cloud Computing & ML, and IoT has significantly transformed digital healthcare. These technologies improve patient care, streamline hospital operations, and enable early disease detection. Challenges such as data security, ethical concerns, and system compatibility must be addressed to realise their full potential. The evolution of digital healthcare will depend on continuous advancements in these areas, aiming to make healthcare more focused on patients, widely accessible, and efficient (Kumar & Reddy, 2019).To overcome obstacles and take advantage of these technologies for a healthier future, governments, healthcare providers, and tech businesses must work together. The seamless integration of Cloud Computing, Big Data, Artificial Intelligence, Machine Learning, and the Internet of Things has ushered in a new era for digital healthcare. This has revolutionised patient care, optimised hospital workflows, and made it possible to detect diseases early with previously unheard-of precision. These transformative technologies hold the promise of a smarter, more responsive healthcare ecosystem; however, their full potential can only be realised by tackling important issues including interoperability across many systems, ethical considerations, and data security. The unrelenting development of these technologies will open the door to a future that is more effective, accessible, and radically patient-centric as innovation continues to transform the healthcare environment. To harness these advancements effectively, a harmonious collaboration between governments, healthcare institutions, and technology pioneers is imperative. By fostering synergy among these stakeholders, the digital healthcare revolution can transcend barriers, unlocking a future where medical excellence and technological sophistication converge for the betterment of humanity.

6. SIGNIFICANCE OF E-GOVERNANCE IN DIGITAL HEALTHCARE: AN OVERVIEW

E-governance in healthcare has the ability to transform the industry by optimising processes, improving patient care, and guaranteeing fair access to health services. The importance of e-governance in digital healthcare is paramount, since it tackles critical obstacles in the healthcare sector, such as inefficiencies, data management problems, and inequities in access to care. Moreover, e-governance can be vital in addressing ethical and regulatory difficulties related to digital healthcare (Vesmaş, 2024). As healthcare becomes increasingly digital, problems relating to data privacy, security, and consent become more prevalent. E-governance can assist in solving these concerns by building legal and regulatory frameworks that safeguard patient rights and guarantee that health data is utilised responsibly. For example, e-governance may ensure that patients have control over their health data and that their agreement is sought before their data is shared or used for research. It can also create standards for data security, ensuring that health data is safeguarded from unwanted access and breaches. By resolving these ethical and legal problems, e-governance may help develop confidence in digital healthcare and guarantee that it is utilised to benefit patients and society. The relevance of e-governance in digital healthcare is further highlighted by its ability to enhance health equity. Health equality refers to the notion that everyone should have the chance to fulfil their full health potential, regardless of their social, economic, or geographic conditions (Bartley & Kelly-Irving, 2024).

6.1 Digital Healthcare enhancement for Public healthcare management and patient satisfaction with E-Governance Practice

A fundamental contribution of e-governance in digital healthcare is its capacity to enhance the efficiency of healthcare delivery (Mutasa, 2022). Traditional healthcare systems are generally beset by bureaucratic inefficiencies, paperwork, and delays in service delivery. E-governance implements digital solutions that automate several operations and here in this sub-section such aspects are illustrated. As far as *diminishing the time* and effort needed to handle patient information, process insurance claims, and coordinate treatment. Electronic health records (EHRs) enable healthcare practitioners to access patient information immediately, obviating the necessity for physical files and diminishing the likelihood of mistakes. This accelerates care delivery and guarantees that healthcare personnel possess precise and current information, resulting in enhanced decision-making and superior patient outcomes. E-governance not only enhances efficiency but also significantly improves the *openness and accountability* of healthcare systems. Transparency is crucial in healthcare, since it fosters confidence between patients and doctors and guarantees the proper utilisation of resources (Ghadi et al, 2024). E-governance facilitates the construction of digital platforms where patients can access information about their health, treatment alternatives, and the quality of care given by different healthcare institutions. This allows consumers to make educated *decisions making* about their health and hold healthcare providers responsible for the quality of treatment they give. Moreover, e-governance improves the monitoring and assessment of healthcare services, allowing governments and regulatory organisations to identify areas for improvement and ensuring that healthcare resources are distributed efficiently (Behkami & Daim, 2012).

Another significant component of e-governance in digital healthcare is its ability to enhance *easy access* to healthcare services, particularly for underprivileged people. In many regions of the world, access to healthcare is constrained by geographic, economic, and social constraints. E-governance can assist overcome these hurdles by delivering digital health services that are available to anybody with an internet connection. Telemedicine, for example, allows patients to consult with healthcare experts remotely *(virtual consultancy)*, removing the need for travel and decreasing the cost of care. This is particularly advantageous for persons living in rural or distant locations, where access to healthcare services is sometimes limited. Additionally, e-governance can promote the delivery of health education and preventative care using digital platforms, helping to lower the burden of disease and enhance overall public health (Izueke et al, 2023; Jain, 2020).). The importance of e-governance in digital healthcare extends beyond the delivery of services to encompass the *e-*

administration of health data. In the digital era, health data is a significant resource that may be utilised to enhance patient care, educate public health policy, and drive medical research. However, the administration of health data also brings considerable obstacles, including those relating to privacy, security, and interoperability. E-governance provides a framework for resolving these difficulties by defining standards and processes for the collecting, storage, and exchange of health data. For example, e-governance helps guarantee that *health data safety* is maintained securely and that patients have control over who may access their information. It can also improve interoperability by encouraging the adoption of standardised data formats and protocols, making it easier for various healthcare systems to share information and interact.

Furthermore, e-governance can spur *innovation in digital healthcare* by facilitating collaboration between governments, healthcare providers, and technology businesses (Bindu N. et al., 2025). The development of digital health solutions involves considerable investment in research and development and a supportive regulatory framework. E-governance may support this by giving financing and incentives for innovation and developing a regulatory framework that fosters the development and use of new technologies. For example, governments may utilise e-governance to build public-private partnerships that bring together healthcare providers, technology businesses, and researchers to develop and execute digital health solutions. This *collaborative approach* may expedite the speed of innovation and guarantee that new technologies match the requirements of patients and healthcare professionals. The relevance of e-governance in digital healthcare is also evident in its capacity to enhance public health outcomes. Public health is a vital component of healthcare since it focuses on preventing disease and improving health at the community level. E-governance may strengthen public health initiatives by offering tools for disease surveillance, outbreak response, and health promotion (Patel & Patel, 2022). For example, digital platforms may gather and evaluate disease outbreak data, allowing public health professionals to respond swiftly and efficiently. E-governance may also enhance *health promotion* initiatives by offering digital channels for communicating health information and promoting healthy habits. This is particularly significant in the Context of chronic illnesses, which are a primary source of morbidity and mortality globally. By promoting preventative treatment and healthy lifestyles, e-governance can help lower chronic disease burdens and enhance public health.

Another significant component of e-governance in digital healthcare is its potential to *cuthealthcare expenses* (Srinivasan et al., 2022). Healthcare is one of the most essential expenses for individuals and governments, and growing healthcare prices are a big worry globally. E-governance can assist in solving this issue by bringing digital solutions that minimise the cost of healthcare delivery. For example, telemedicine can eliminate the need for in-person consultations, which are generally more expensive. Similarly, electronic health records can minimise the *cost of managing patient information* by removing the requirement for physical storage and minimising the chance of mistakes. Quality of treatment is an essential issue in healthcare since it directly affects patient outcomes and satisfaction (Zaid et al., 2020). E-governance can *increase the quality* of care by giving healthcare practitioners access to correct and up-to-date information, enabling them to make better decisions and deliver more individualised treatment. For example, electronic health records can give healthcare practitioners a whole perspective of a patient's medical history, allowing them to spot possible dangers and customise treatment strategies appropriately. Additionally, e-governance can enhance the *application of evidence-based procedures* by providing healthcare practitioners access to the latest research and guidelines. This can assist in guaranteeing that patients receive the most effective and up-to-date therapies (Alpay et al., 2007).

Figure 3. The transformation of Sustainable Healthcare with effective E-Governance Systems

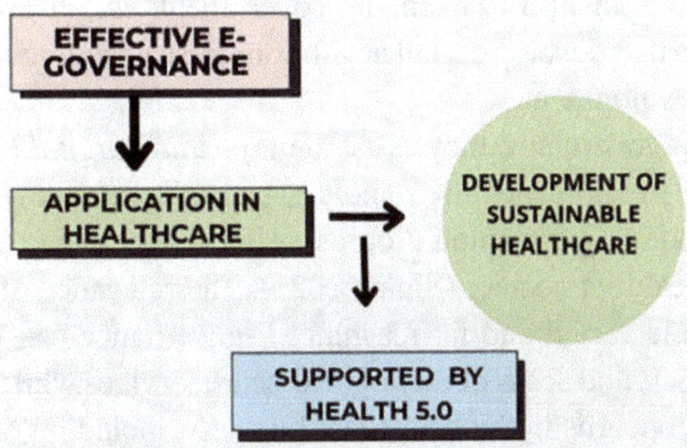

Unfortunately, health inequalities are a serious concern worldwide, with disadvantaged people often experiencing difficulties in receiving healthcare. E-governance can assist in solving these (*health inequalities*) inequities by offering digital health services that are available to everybody, regardless of their location or socioeconomic background. For example, telemedicine can give access to healthcare for persons living in rural or underdeveloped locations. At the same time, digital health education can increase knowledge about health concerns and promote healthy habits among underrepresented populations. By fostering health equity, e-governance may assist in guaranteeing that everyone has the chance to live a healthy life.

In addition to its direct influence on healthcare delivery, e-governance may also significantly impact society by *fostering digital literacyand inclusivity* (Addo, A.,

&Senyo, P. K., 2021). Digital literacy refers to the capacity to utilise digital technologies effectively, while digital inclusion refers to ensuring everyone can access digital technologies and use them. E-governance may promote digital literacy and inclusion by delivering digital health services that are user-friendly and accessible to persons with diverse degrees of digital capabilities. For example, governments may give training and assistance to help individuals learn how to utilise digital health platforms, ensuring they can get the needed services. By encouraging digital literacy and inclusivity, e-governance may assist in guaranteeing that the advantages of digital healthcare are available to everyone, regardless of their background or circumstances (Ali, 2023). The relevance of e-governance in digital healthcare is also evident in its potential to promote global health efforts. The relevance of e-governance in digital healthcare is reflected in its capacity to generate economic growth and development (Khan et al., 2024; Dasgupta & Deb, 2008). Healthcare is a large area of the economy, and adopting digital health solutions may generate new prospects for economic growth and employment creation. E-governance may help this by providing money and incentives for developing and deploying digital health technologies and building an enhanced*regulatory framework* that stimulates innovation and investment. For example, governments might give grants and tax incentives to firms that develop digital health solutions, encouraging them to engage in research and development.

Additionally, e-governance may assist the *expending digital health business* by supporting the use of digital health technology by healthcare providers and patients. By fostering economic growth and progress, e-governance may help build a more prosperous and healthier society (Sinha, 2021). The relevance of e-governance in digital healthcare is varied and far-reaching. E-governance can increase the efficiency, transparency, and accessibility of healthcare services while tackling some of the most critical concerns in the healthcare business, including data management, cost, and health inequities. By supporting innovation, cooperation, and health equity, e-governance may help guarantee that digital healthcare is used to benefit patients, healthcare providers, and society. As the world continues to adopt digital technologies, the role of e-governance in healthcare will only grow more essential, making it a crucial area of attention for governments, healthcare providers for an effective Digital Healthcare solution using E-Governance, as depicted in Fig: 3. (Sharma & Gupta, 2021; Mahapatra et al., 2011).

7. ONGOING INITIATIVE FOR DIGITAL HEALTHCARE IN INDIA

Exciting developments are occurring in India's healthcare sector, fueled by groundbreaking digital interventions and innovative technological advances(Shrivastava et al., 2020). The government of India has started several initiatives in a variety of ministries and departments, and these initiatives are mainly focused on addressing the disparities in healthcare access, especially in rural and underserved regions, by leveraging cutting-edge digital platforms, telemedicine, Artificial Intelligence (AI), and blockchain technology for managing medical records. This section looks at India's ongoing digital healthcare initiatives, exploring their tremendous impact, the challenges faced, and the promising path ahead (Sharma &Vaisla, 2012). Join us on this journey as we unveil the potential of digital healthcare to transform lives across the nation.

7.1 Ayushman Bharat Digital Mission (ABDM)

The Ayushman Bharat Digital Mission (ABDM), kickstarted in 2021, is a landmark effort to establish a robust and integrated digital health ecosystem (Refer Fig: 4). This initiative harmonises healthcare services, promoting seamless interoperability through a distinctive Health ID that grants patients a consolidated digital record of their medical histories (Singh & Sharma, 2021). The ABDM comprises several key components:

- Health ID: A unique identifier designed for individuals to store and share their health records digitally.
- Healthcare Professionals Registry (HPR): A thorough database encompassing licensed professionals across various medical disciplines.
- Health Facility Registry (HFR): A complete repository that includes public and private healthcare facilities.
- Electronic Health Records (EHR): Standardised health records that guarantee data security and facilitate smooth information exchange among hospitals and clinics.
- ABHA (Ayushman Bharat Health Account): This feature allows patients to securely access their health records from multiple healthcare providers, ensuring a streamlined continuity of care.

Figure 4. Major activities conducted under Ayushman Bharat Scheme

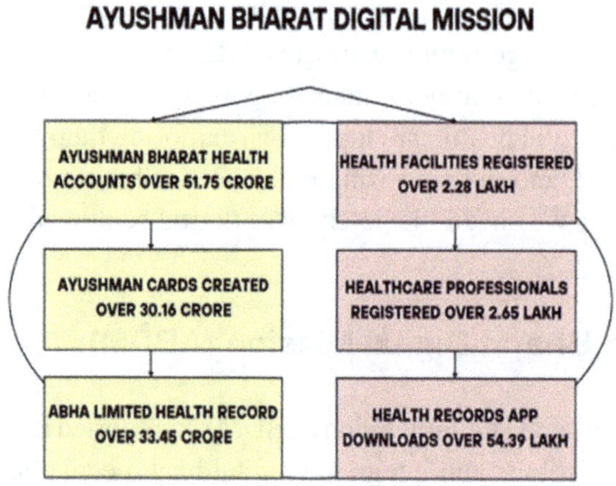

7.2. eSanjeevani - Telemedicine Services

India's National Telemedicine Service, eSanjeevani, has emerged as a vital resource in expanding healthcare access, especially highlighted during the COVID-19 pandemic (Kar et al.., 2020). It offers two distinct modes of consultation:

- **eSanjeevani OPD:** This service enables doctor-to-patient teleconsultations, which aids in remote diagnosis and treatment.
- **eSanjeevani AB-HWC:** This connects Health and Wellness Centers (HWCs) with specialists at tertiary care institutions for crucial doctor-to-doctor consultations.

This initiative has significantly alleviated the pressure on physical healthcare infrastructure, paving the way for improved health outreach in underserved regions as mentioned in Fig: 5. (Bhuvana &Vasantha, 2022). Additionally, efforts are underway to enhance the system by integrating AI-based decision support tools to assist healthcare professionals in diagnosis and treatment planning.

7.3 CoWIN - COVID-19 Vaccination Platform

The CoWIN platform was pivotal in steering India's COVID-19 vaccination efforts. It provided real-time data tracking, convenient vaccination slot booking, and digital certification. The remarkable success of CoWIN showcased India's potential to execute large-scale public health initiatives through innovative digital solutions (Chopra et al., 2024). The platform continues to advance, with ongoing discussions about its repurposing for broader immunisation programs, encompassing routine childhood vaccinations and tracking adult booster doses.

7.4 National Digital Health Blueprint (NDHB)

Formulated in 2019, the NDHB is an exciting roadmap for advancing digital health initiatives across India. It closely aligns with the National Health Policy 2017, a mission to forge a strong health information system through AI, blockchain, and cloud computing (Paul et al., 2024). The NDHB specifically focuses on:

- Achieving Universal Health Coverage (UHC) via groundbreaking digital transformation.
- Enhancing governance of health data to boost accountability and trust.
- Prioritising privacy and security in digital health records to protect patient information.
- Promoting interconnectivity among diverse healthcare platforms, ensuring easy access to comprehensive patient history.

7.5 mHealth and Digital Health Startups

India has experienced an exciting boom in mobile health (mHealth) solutions, particularly apps designed for teleconsultation, diagnostics, and fitness tracking(Pai &Alathur, 2019). Some noteworthy startups making a difference include:

- **Practo:** Elevating the patient experience with online doctor consultations and seamless appointment booking.
- **1mg:** Transforming access to healthcare via online pharmacy and comprehensive diagnostic services.
- **PharmEasy:** Simplifying medication delivery and lab test bookings for convenience
- **Cure.fit:** A holistic health platform integrating fitness, nutrition, and wellness services under one roof.
- **MFine:** Delivering instant healthcare consultations and diagnostic services powered by AI.

These platforms are enhancing government efforts to improve healthcare accessibility and efficiency, significantly cutting down the time and costs tied to traditional healthcare delivery methods (Srivastava, 2016). These initiatives underscore India's commitment to a digitally empowered healthcare ecosystem that champions sustainability and effectiveness in service delivery.

7.6 Health Management Information Systems (HMIS)

The HMIS platform is a game-changer in public health, enabling real-time data collection and analysis that significantly enhances decision-making. Seamlessly integrating data from diverse sources such as hospitals, primary healthcare centers, and national health programs paves the way for more effective policy formulation and implementation (Aasheim et al., 2015). This integration streamlines processes and drives impactful health initiatives that benefit communities nationwide!

7.7 Artificial Intelligence (AI) and Blockchain thrust of Indian Healthcare Systems

Artificial Intelligence (AI) transforms Indian healthcare, providing tools for better disease prediction, early diagnoses, and personalised treatment strategies (Paul, 2022). Notable AI applications making waves in India include:

- Initiatives driven by *NITI Aayog* utilising AI for tuberculosis (TB) detection and innovative cancer screening techniques.
- Collaborations with *Google leveraging AI-powered* diagnostic tools for diabetic retinopathy and cardiovascular conditions.
- *IBM Watson* Health is offering groundbreaking AI-based oncology treatment recommendations.
- The rise of *AI-powered chatbots* designed for mental health support allows for remote counselling and self-help resources.
- *AI-driven radiology solutions* enhance imaging technology, empowering radiologists to detect diseases like cancer and tuberculosis earlier than ever.

Blockchain technology is being explored to establish secure, immutable, and interoperable health records. The Indian government, in collaboration with private entities, is working on innovative blockchain solutions to:

- Combat medical fraud effectively.
- Boost data security and privacy to safeguard sensitive information.
- Streamline the process of insurance claim settlements for quicker resolutions.
- Provide real-time verification of patient history while maintaining confidentiality.

- Implement *smart contracts* in healthcare for automated insurance claim processing and contract enforcement among parties involved.

Figure 5. A Brief overview and facts of e-Sanjeevani systems

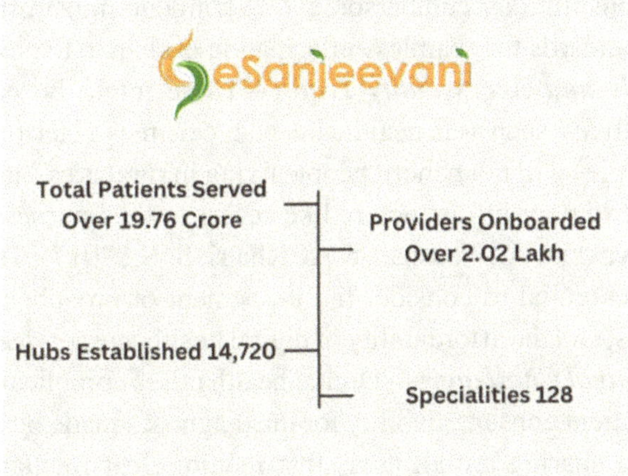

8. ISSUES AND CHALLENGES OF DIGITAL HEALTHCARE REGARDING E-GOVERNANCE

E-governance enabling digital healthcare is ushering in a new era of medical services by leveraging technology for enhanced efficiency, accessibility, and transparency. E-health initiatives, electronic medical records (EMRs), telemedicine, and AI-based diagnostics are increasingly being adopted by governments around the world (Habes et al., 2020). Although digital health care has many benefits, there are e-governance challenges in digital healthcare, such as cybersecurity, regulation and compliance, accessibility, and the infrastructure needed for technology (Rathee et al., 2024).

Data Security and Privacy Concerns—Data security and privacy are key issues among digital healthcare's biggest challenges. Healthcare systems collect and store sensitive patient data, making them attractive cyberattack targets. Finally, data breaches can also open the door to identity theft, financial fraud, and unauthorised access to medical records. Most developing countries have weak data protection laws, making digital healthcare systems particularly vulnerable to such hacks. Compliance with various international standards, e.g., GDPR or HIPAA, is still a big challenge for many governments.

Interoperability and Standardisation—E-governance faces several obstacles, of which none is more significant than the lack of interoperability across various healthcare systems and digital platforms. Different software systems are used across hospitals, clinics, and laboratories, and they do not communicate efficiently with each other (Itumalla, 2012). The lack of standardisation makes sharing patient records across different institutions cumbersome. Governments must establish universal data exchange standards for seamless information exchange to enable integration.

Digital Divide and Accessibility Issues—There might be problems with e-governance initiatives, such as in healthcare, where there is a lack of access to digital tools. Such strategies will not benefit people living in rural areas and disadvantaged communities due to poor infrastructure like access to the internet & knowledge of digital literacy (Kakkar, A., Naushad, S., & Khatri, S. K., 2017). To eliminate these disparities, it is essential to consider the investment of broadband infrastructure, training programs, and the affordability of digital healthcare services for all citizens.

Legal and Ethical Challenges—Digital health raises complicated legal and ethical questions. Patient consent, liability for misdiagnoses made by AI systems, and cross-border data sharing are all areas that require clear regulatory frameworks (Madon et al., 2007).Moreover, to ensure accountability and ethical values within digital healthcare settings, many states currently do not have a collection of detailed legal provisions deciding upon these matters.

Resistance to Technological Adoption—Healthcare professionals and institutions may resist adopting digital solutions due to the systemic learning curve, exorbitant implementation costs, and fear of job loss. Many physicians prefer traditional methods to digital health records because of usability issues and extra administrative burdens (Ray & Mukherjee, 2007). Governments must lay down training programs and incentives to foster digital transformation in the healthcare sector.

Cybersecurity Threats—Cyber security is a top concern for many e-governances globally, and it is especially true for e-governance in healthcare. Poor security can lead to data leaks, putting patient safety and trust at risk in digital health services. Digital healthcare systems require governments to implement stringent cybersecurity policies and invest in advanced threat detection processes while periodically conducting security audits to protect systems and information.

Cost of Implementation and Maintenance—Setting up and keeping up digital healthcare infrastructure is not cheap. In developing countries, limited budgets can lead to poorly executed systems that cannot achieve expected outcomes due to healthcare digitisation (Paul, 2023). Governments should invest adequate funds and consider public-privatepartnerships for sustainable digital health services.

AI and Algorithmic Bias—Algorithmic biases in AI healthcare solutions might be involved; this also applies to predictive diagnostics and automated treatment recommendations. If these AI models are trained on biased datasets, it can result in a health disparity among certain demographics. Challenges for e-governance in AI-based decision-making in healthcare involve ensuring fairness, transparency, and accountability.

Regulatory and Policy Challenges—Governments are slow to respond to fast-changing healthcare technologies. Digital healthcare solutions may be hindered by outdated regulations and bureaucratic inefficiencies (Mehta & Jain, 2022). It necessitates a policy ecosystem that is flexible and agile enough to foster innovation but also holds patient safety and ethics in the highest regard.

9. PROPOSED SOLUTIONS, CONCERN AND FUTURE POTENTIALITIES

Global health is a field that focuses on improving health and attaining health equity for all people worldwide. E-governance may help global health initiatives by offering digital tools and platforms for cooperation and information exchange across governments and organisations. For example, digital platforms may communicate data on disease outbreaks, allowing countries to coordinate their response and avoid spreading contagious illnesses. As far as this study is concerned it has observed that Digital Healthcare Systems using proper E-Governance systems already reached its heights, but at the same time there are some obstacles and some of them may be look into viz.

Technologies: Digital Healthcare Systems lies on various basic and emerging information technologies, and there are issues of upgrading the technologies and thus proper mechanism may be adopted for availability of the technologies.

Skilled Manpower: There are issues of availability of skilled manpower in Healthcare Informatics, E-Healthcare Systems and thus proper training, educational programs including UG, PG and Doctoral Degrees.

Integration of the Technologies: Many technologies are separated in Health Informatics segment, and here proper integration of such sub-technologies are important using proper techniques, mechanism, etc.

Information Literacy and Awareness: Common citizens are lacking knowledge in uses of information and computer literacy which is highly essential to establish using training program, user education, training and awareness program, etc.

Effective Administration: A proper management principles and ICT enabled administration can be established with top to bottom level administrators' involvement.

Regulatory and Policy Challenges: Timely changes of policy, new-age regulations and frameworks are highly essential to initiate, and here apart from Government, several private bodies, NGOs and their involvement may be treated as most valuable.

Cost and Funding: A proper budget and funding may be allotted for executing several core and small E-Governance projects for complete and healthy development. Continuous funding is also essential and expected.

Data Security and Privacy Concerns: Experts also expressed about the security concern of data and most importantly privacy concerns and ensuring these two may lead the development of E-Governance at its high extend.

E-governance may also promote global health initiatives by offering digital health services accessible to individuals in low- and middle-income nations, where access to healthcare is generally limited. By supporting global health initiatives, e-governance can assist improve health outcomes worldwide and contribute to attaining global health goals. Although convenient, digital healthcare has challenges like data privacy risks, cybersecurity threats, technology adoption problems, and regulatory compliance. Developing global, regional, and national policies, strengthening cybersecurity, and devising smoother interoperability will ensure the future of digital health.Some of the key trends that are shaping the future of healthcare include increased use of AI tools in diagnostics and treatment, further integration of digital health solutions and personalised medicine, the use of wearable devices to facilitate continuous monitoring of patient health, and a growing emphasis on preventative care that focuses on maintaining health rather than just treating illness.

12. CONCLUDING REMARKS

Digital healthcare, a vital component of e-governance, harnesses the power of Information and Communication Technology (ICT) to enhance accessibility, affordability, and efficiency in delivering healthcare services. The Government of India spearheads numerous initiatives under the expansive Digital India framework, all aimed at building a robust and sustainable healthcare ecosystem. The most crucial aspect of e-governance is accessing health services, one of the sectors that could use the assistance of e-governance and digital health can transform it. Yet, data security, interoperability, digital divide, legal concerns, and cybersecurity threats must be addressed to ensure a robust and inclusive digital healthcare ecosystem. There are enough opportunities created by the Government of India including several states, NGOs and other associations for building a new age Healthcare Systems, even though there are some issues, limitations, and challenges. Thus, it is essential that Governments should take a multi-stakeholder approach to unite policymakers, healthcare providers, technology experts, and citizens to develop a sustainable and secure digital healthcare infrastructure. E-governance can assist in lowering healthcare costs by promoting preventative care and early intervention, which are frequently more cost-effective than treating severe disorders. The relevance of e-governance in digital healthcare is also evident in its potential to improve the quality of care.

REFERENCES

Aasheim, C. L., Williams, S., Rutner, P., & Gardiner, A. (2015). Data analytics vs. data science: A study of similarities and differences in undergraduate programs based on course descriptions. *Journal of Information Systems Education, 26*(2), 103.

Addo, A., & Senyo, P. K. (2021). Advancing E-governance for development: Digital identification and its link to socioeconomic inclusion. *Government Information Quarterly, 38*(2), 101568. DOI: 10.1016/j.giq.2021.101568

Agarwal, S., & Verma, D. (2019). Artificial intelligence (AI) and predictive analytics in the healthcare sector. *Journal of Health Informatics, 12*(3), 215–230.

Ali, M. (2023). E-governance and E-democracy: a Digital Revolution. Available at *SSRN* 4623414.

Alpay, L. L., Overberg, R. I., & Zwetsloot-Schonk, B. (2007). Empowering citizens in assessing health related websites: A driving factor for healthcare governance. *International Journal of Healthcare Technology and Management, 8*(1-2), 141–160. DOI: 10.1504/IJHTM.2007.012103

Bartley, M., & Kelly-Irving, M. (2024). *Health inequality: an introduction to concepts, theories and methods*. John Wiley & Sons.

Behkami, N. A., & Daim, T. U. (2012). Research forecasting for health information technology (HIT), using technology intelligence. *Technological Forecasting and Social Change, 79*(3), 498–508. DOI: 10.1016/j.techfore.2011.08.015

Benedict, M., & Schlieter, H. (2015). Governance guidelines for digital healthcare ecosystems. In *eHealth2015–Health Informatics Meets eHealth* (pp. 233-240). IOS Press. DOI: 10.3233/978-1-61499-524-1-233

Bhambere, H. S. S., Abhishek, B., & Sumit, H. (2021). Rapid digitisation of healthcare: A review of COVID-19 impact on our health systems. *Int. J. All Res. Educ. Sci. Methods, 9*, 1457–1459.

Bhuvana, M., & Vasantha, S. (2022). Determinants of E-Governance Health Care Services Requirements and Rural Citizen Satisfaction. *Managing Disruptions in Business: Causes, Conflicts, and Control*, 317-334.

Bindu, N., Kumar, R. A., Radhakrishnan, T., Kumar, T. M., & Kumar, K. S. (2025). Virtual Realities: The Influence of the Metaverse on Healthcare E-Governance. *Applying Metaverse Technologies to Human-Computer Interaction for Healthcare*, 254.

Chopra, H., Choudhary, O. P., & Emran, T. B. (2024). G20 summit in India: Unveiling the digital health initiative with 'Vasudhaiva Kutumbakam'. *International Journal of Surgery, 110*(1), 591–593. DOI: 10.1097/JS9.0000000000000771 PMID: 38315797

Das, R., Gupta, P., & Nair, S. (2021). The impact of telemedicine and mHealth platforms. *Telemedicine and Digital Health Review, 18*(2), 105–119.

Das, S., Garg, V., & Joseph, J. J. (2023). Advancements in e-governance initiatives: Digitalising healthcare in India. In *International Conference on Advances in IoT and Security with AI* (pp. 287-296). Singapore: Springer Nature Singapore. DOI: 10.1007/978-981-99-5085-0_28

Dasgupta, A., & Deb, S. (2008). Telemedicine: A new horizon in public health in India. *Indian Journal of Community Medicine : Official Publication of Indian Association of Preventive & Social Medicine, 33*(1), 3. DOI: 10.4103/0970-0218.39234 PMID: 19966987

Dash, S. P. (2020). The Impact of IoT in Healthcare: Global Technological Change & The Roadmap to a Networked Architecture in India. *Journal of the Indian Institute of Science, 100*(4), 1–13. DOI: 10.1007/s41745-020-00208-y PMID: 33162693

Dey, A., Nandi, S., & Sarkar, M. (2018). Security measures in IOT based 5G networks. In 2018 3rd International Conference on Inventive Computation Technologies (ICICT), pp. 561-566, IEEE.

Ghadi, Y. Y., Mazhar, T., Shahzad, T., Amir khan, M., Abd-Alrazaq, A., Ahmed, A., & Hamam, H. (2024). The role of blockchain to secure internet of medical things. *Scientific Reports, 14*(1), 18422. DOI: 10.1038/s41598-024-68529-x PMID: 39117650

Habes, M., Alghizzawi, M., & Ali, S., SalihAlnaser, A., & Salloum, S. A. (2020). The Relation among Marketing ads, via Digital Media and mitigate (COVID-19) pandemic in Jordan. *International Journal of Advanced Science and Technology, 29*(7), 12326–12348.

Hasselgren, A., Kralevska, K., Gligoroski, D., Pedersen, S. A., & Faxvaag, A. (2020). Blockchain in healthcare and health sciences—A scoping review. *International Journal of Medical Informatics, 134*, 104040. DOI: 10.1016/j.ijmedinf.2019.104040 PMID: 31865055

Itumalla, R. (2012). Information Technology and Service Quality in HealthCare: An Empirical Study of Private Hospital in India. *International Journal of Innovation, Management and Technology, 3*(4), 433. DOI: 10.7763/IJIMT.2012.V3.269

Jain, E. (2020). Digital Employability Skills and Training Needs for the Indian Healthcare Industry. In Opportunities and Challenges in Digital Healthcare Innovation (pp. 113-130). IGI Global. DOI: 10.4018/978-1-7998-3274-4.ch007

Kakkar, A., Naushad, S., & Khatri, S. K. (2017). Telemedicine and EHR integrated approach for an effective E-Governance Healthcare Framework. *International Journal of Medical Research & Health Sciences*, *6*(5), 108–114.

Kar, S. K., Saxena, S. K., & Kabir, R. (2020). The relevance of digital mental healthcare during COVID-19: Need for innovations. *Nepal Journal of Epidemiology*, *10*(4), 928–929. DOI: 10.3126/nje.v10i4.32519 PMID: 33495711

Karthikeyan, N., & Sukanesh, R. (2012). Cloud based emergency health care information service in India. *Journal of Medical Systems*, *36*(6), 4031–4036. DOI: 10.1007/s10916-012-9875-6 PMID: 22865161

Khan, S., Ullah, S., & Nobanee, H. (2024). ICT diffusion, E-governance, and sustainability in the digital era. *Sustainable Futures : An Applied Journal of Technology, Environment and Society*, *8*, 100272. DOI: 10.1016/j.sftr.2024.100272

Kumar, A., Mahajan, P., Mohan, D., & Varghese, M. (2001). IT—information technology and the human interface: tractor vibration severity and driver health: a study from rural India. *Journal of Agricultural Engineering Research*, *80*(4), 313–328. DOI: 10.1006/jaer.2001.0755

Kumar, P., & Reddy, K. S. (2019). Challenges and Opportunities for Implementing Digital Health Information Systems in India: A Review. *Health Policy and Technology*, *8*(4), 392–399.

Madon, S., Sahay, S., & Sudan, R. (2007). E-government policy and health information systems implementation in Andhra Pradesh, India: Need for articulation of linkages between the macro and the micro. *The Information Society*, *23*(5), 327–344. DOI: 10.1080/01972240701572764

Mahapatra, S. C., Das, R. K., & Patra, M. R. (2011). Current e-governance scenario in healthcare sector of India. *E-Governance Scenario in Healthcare Sector of India*, 121-7.

Mancy, A. M., Kumar, A. S., Latip, R., Jagadamba, G., Chakrabarti, P., Sharma, P., . . . Kanchan, B. G. (2024). Smart Healthcare System, Digital Health and Telemedicine, Management and Emergencies: Patient Emergency Application (PES) E-Governance Applications. In *Sustainable Development in AI, Blockchain, and E-Governance Applications* (pp. 124-151). IGI Global.

Mehta, P., & Iyer, N. (2021). The influence of mobile health (mHealth) applications on preventive healthcare. *International Journal of mHealth Research, 9*(1), 87-102.

Mehta, S., & Jain, P. (2022). Role of Digital Health Information Systems in Addressing Healthcare Disparities: Insights from Urban and Rural Settings in India. *Journal of Public Health Management and Practice*, *28*(1), 12–19.

Mony, P. K., & Nagaraj, C. (2007). Health information management: An introduction to disease classification and coding. *The National Medical Journal of India*, *20*(6), 307. PMID: 18335800

Mutasa, L. (2022). E-health and e-governance integration framework for the Namibian government (Doctoral dissertation, Cape Peninsula University of Technology).

Pai, R. R., & Alathur, S. (2019). Assessing awareness and use of mobile phone technology for health and wellness: Insights from India. *Health Policy and Technology*, *8*(3), 221–227. DOI: 10.1016/j.hlpt.2019.05.011

Patel, M., & Kumar, S. (2022). The significance of electronic health records (EHRs) and their role in improving patient care efficiency. *Health Informatics Journal*, *25*(4), 305–320.

Patel, M., & Patel, M. (2022). An integrated GIS/GPS based e-governance approach for different agencies to fight against COVID-19. Understanding COVID-19: The role of computational intelligence, 447-469.

Paul, P. K. (2022). Aspects of Biosensors with Refers to Emerging Implications of Artificial Intelligence, Big Data and Analytics: The Changing Healthcare–A General Review. *Next Generation Smart Nano-Bio-Devices*, 1-18.

Paul, P. K. (2023). Advanced ICT and intelligent systems in sophisticated Healthcare 5.0 practice in modern social and healthcare transformation: An overview. *Augmented and Virtual Reality in Social Learning: Technological Impacts and Challenges*, *3*, 63–80. DOI: 10.1515/9783110981445-005

Paul, P. K., Aithal, P. S., & Bhuimali, A. (2018). Health Information Science and its growing popularities in Indian self financed universities: Emphasising Private Universities—A Study. *International Journal of Scientific Research in Biological Sciences*, *5*(1), 1–11. DOI: 10.26438/ijsrbs/v5i1.111

Paul, P. K., Chatterjee, R., Chakraborty, S., Tiwary, K. S., Aithal, P. S., & Sharma, S. (2024). Advanced Healthcare Informatics Practice Vis-À-Vis Virtual Reality Implications—a Scientific Overview. *Immersive Virtual and Augmented Reality in Healthcare*, 28-49.

Paul, P. K., Moumtzoglou, A., Bandyopadhyay, A., Hoque, M., Kumar, N. S., & Saavedra, R. (2024). Indian Digital Health Information Systems: Initiatives and Opportunities with Socioeconomic Perspective. *Economic Affairs*, *69*(2), 1175–1189. DOI: 10.46852/0424-2513.3.2024.39

Phukan, O. C., Singh, G., Tiwari, S., & Butt, S. (2022). An automated stress recognition for digital healthcare: towards e-governance. In *International Conference on Electronic Governance with Emerging Technologies* (pp. 117-125). Cham: Springer Nature Switzerland. DOI: 10.1007/978-3-031-22950-3_10

Ramesh, K., Joshi, A., & Desai, V. (2020). National Digital Health Blueprint (NDHB) and its alignment with India's National Health Policy (2017). *Indian Journal of Public Health Policy*, *14*(2), 140–155.

Rana, A. (2017). The Immense Potential of M-Care in India: Catering Better to Patients Needs in the Context of a Fragmented Healthcare System. [IJRQEH]. *International Journal of Reliable and Quality E-Healthcare*, *6*(4), 1–3. DOI: 10.4018/IJRQEH.2017100101

Rathee, G., Saini, H., Garg, S., Choi, B. J., & Hassan, M. M. (2024). A secure data e-governance for healthcare application in cyber physical systems. [IJSWIS]. *International Journal on Semantic Web and Information Systems*, *20*(1), 1–17. DOI: 10.4018/IJSWIS.345934

Ray, S., & Mukherjee, A. (2007). Development of a framework towards successful implementation of e-governance initiatives in health sector in India. *International Journal of Health Care Quality Assurance*, *20*(6), 464–483. DOI: 10.1108/09526860710819413 PMID: 18030965

Sharma, K. (2020). The transformative role of mobile health (mHealth) applications in preventive healthcare. *Journal of Digital Health Innovations*, *10*(1), 65–78.

Sharma, M. K., & Vaisla, K. S. (2012). E-health for rural areas of Uttarakhand under e-Governance service delivery model. In *2012 1st International Conference on Recent Advances in Information Technology (RAIT)* (pp. 622-625). IEEE. DOI: 10.1109/RAIT.2012.6194601

Sharma, S., & Gupta, R. (2021). Assessing the Socioeconomic Impact of Digital Health Information Systems: A Case Study of Rural India. *International Journal of Medical Informatics*, *132*, 104066.

Shrivastava, S., & Srikanth, T. K., & VS, D. (2020, September). e-Governance for healthcare service delivery in India: challenges and opportunities in security and privacy. In *Proceedings of the 13th International Conference on Theory and Practice of Electronic Governance* (pp. 180-183). DOI: 10.1145/3428502.3428527

Singh, A., & Sharma, R. (2021). Ayushman Bharat Digital Mission (ABDM) and its transformative impact on healthcare access in India. *Health Policy and Technology Journal, 16*(3), 190–205.

Sinha, C. G. (2021). E-Governance: Impact on Society. In Digitization of Economy and Society (pp. 93-117). Apple Academic Press.

Srinivasan, S., Krishnamoorthy, R., Kumar, L., Kushwaha, A. S., Ojha, A. C., Ganesh, D., & Jebakumar, A. (2022). Manageable E-Governance Architecture In Health Care System Using Cloud Storage. *NeuroQuantology : An Interdisciplinary Journal of Neuroscience and Quantum Physics, 20*(14), 1094.

Srivastava, S. K. (2016). Adoption of electronic health records: A roadmap for India. *Healthcare Informatics Research, 22*(4), 261–269. DOI: 10.4258/hir.2016.22.4.261 PMID: 27895957

Vesmas, D. (2024). The Ethics of E-Governance. Safeguarding Data Confidentiality and Human Security In Public Administration. *StudiaSecuritatis, 18*(2), 222–232.

Zaid, A. A., Arqawi, S. M., Mwais, R. M. A., Al Shobaki, M. J., & Abu-Naser, S. S. (2020). The impact of Total quality management and perceived service quality on patient satisfaction and behavior intention in Palestinian healthcare organisations. *Technology Reports of Kansai University, 62*(03), 221–232.

KEY TERMS AND DEFINITIONS

Digital Medical Record Management: Digital Medical Record Management is the tool, process and mechanism of storing, organising and accessing patient data in digital format for ensuring effective healthcare services and opportunities to the patient, diagnostics centres and also healthcare service providers.

E Governance: E-Governance is the implications of information and communication technology in diverse government activities and sectors for better citizen services. In E Governance various models, techniques, tools, systems are followed in order to offer advanced healthcare services.

Health Informatics: Health informatics is the combination of health sciences with the information technology for the advanced and effective healthcare solutions to the patient, service providers and middle-man, agents, consultants and others. This is become a field of study for effective, advanced healthcare solutions.

Healthcare 5.0: Healthcare 5.0 is the contemporary generation of advancement of Healthcare services in respect of advancing healthcare services supported by latest information technology and computing. Various latest technologies viz. AI, ML, Cloud Computing, IoT, Blockchain, etc. are highly useful in several healthcare solutions.

mHealth: mHealth is the latest development in Healthcare Informatics practice which is dedicated in offering healthcare services and solutions using mobile phones or other mobile devices for individual, remote telehealth, medical services using various computing technologies, effectively.

Sustainable Healthcare: Sustainable healthcare is the development of healthcare systems and mechanism for all the community in all situation without any economic or social concern or major issue. Sustainable Healthcare is dedicated in use of technologies for all and a comprehensive healthcare solutions including treatments, diagnostics, consultancy, etc.

Telemedicine: Telemedicine is the way and mechanism of offering healthcare and medical solutions in digital platform typically by the physician and patient. It is a remote service of healthcare services using variety of computing and information technology tools and emerging technologies.

Chapter 5
AI-Empowered Biology for Sustainable Health, Agriculture, and Environment

A. Sabina Banu
Department of Microbiology, Sacred Heart College, Tirupattur, India

S. Sindhuja
Department of Microbiology, Sacred Heart College, Tirupattur, India

K. Kesavardhini
Department of Microbiology, Sacred Heart College, Tirupattur, India

S. Sadiya Suroor
Department of Microbiology, Sacred Heart College, Tirupattur, India

R. Nisha
http://orcid.org/0009-0007-3231-9605
Department of Microbiology, Sacred Heart College, Tirupattur, India

P. Saranraj
http://orcid.org/0000-0002-5052-6532
Department of Microbiology, Sacred Heart College, Tirupattur, India

ABSTRACT

Artificial intelligence (AI) is transforming the biological sciences by enhancing data utilisation in the advancement of sustainable health, agricultural, and environmental issues. Biological systems that are powered by AI combine machine learning, bioinformatics, and computational modeling to analyze intricate biological data with high efficiency and precision. Applied medical research in medicine, AI has been used to improve disease diagnosis, pharmaceutical development and individualized healthcare by uncovering latent patterns in genomic and clinical information. In farming, AI biological strategies enhance the productivity of crops, disease diagnostics, and

DOI: 10.4018/979-8-2600-2691-5.ch005

resource use and facilitate climate resilient and sustainable agricultural production. Intelligent biological interventions in the environment can be applied to monitor the ecosystem, control pollution, and protect biodiversity, as well as manage waste.

1. INTRODUCTION

Artificial Intelligence has instigated a paradigm change in the biological sciences, has ability to process complex biological data on various scales. There was an observational focus in the traditional biological sciences, while the AI-related biological sciences could predict by data-modeling. Combining AI with the genomics, molecular biology, healthcare, agriculture, and biological sciences sectors has led to accelerated biological discoveries on a scale that supports sustainability. It is beneficial for the AI-related advancement in technology to aid the development of biological sciences to pursue sustainability on the global scale (Rajkomar et al., 2018).

1.1Concept and concept of artificial intelligence in biological sciences

Artificial Intelligence within the Biomedical Sciences refers to the utilization of machine learning and deep learning technologies to extract insights through biological data. At its essence, Artificial Intelligence is more than merely Bioinformatics because it provides an almost intuitive inference of pattern from the high-dimensional data of the biomedical world, from the genome or proteome to the image of the biological molecule itself. Clearly, AI is a force multiplier, as it can translate data to knowledge by performing tasks such as hypothesis inference, experimental design, or even controlling robots or automation systems. This paradigm shift is taking biology from what was largely a descriptive science and transforming it more into a predictive science (Kitano, 2002).

1.2 Evolution from Traditional Biology to AI-Based Biology

The transformation from traditional biology to AI biology corresponds to the "Big Data" revolution in the life sciences. The early days of bioinformatics relied on rule-based systems and statistical models for sequence alignment, such as BLAST and phylogenetic analysis. However, the emergence with advent of Next-Generation Sequencing (NGS) and high-throughput analysis resulted in a "big data" crisis, which was followed by improvements in the development of GPU computing capabilities and new models of artificial neural networks (Zhavoronkov et al., 2019). This brought about an opportunity for the development of AI biology, a realm in which models for instance, convolutional artificial neural networks can evaluate cellular pictures, while recurrent neural networks can analyze biological sequences. Bioscience can now make use of such models for purposes of analysis as well as generation, such as new protein and small molecule designs, which was a transition from observation to creation (Mohanty et al., 2016).

1.3 The Significance of Artificial Intelligence in Sustainable Development

Artificial Intelligence is a significant catalyst for achieving the United Nation's Sustainable Development Goals (SDGs) biological systems. It provides scalable remedies for some of the major global challenges in optimized resource utilization and accelerated discoveries. In the healthcare sector, the utilization of AI in diagnosing illnesses and advancement medications (SDG 3: Good Health) promises to enhance accessibility and speed. In agriculture, artificial intelligent models for precise agriculture and environmentally stable crop development directly address SDG 2 (Zero Hunger) and SDG 13 (Climate Action). Additionally, artificial intelligence applications in environmental biology support biodiversity protection (SDG 15: Life on Land) and clean water (SDG 6: Clean Water) protection (LeCun et al., 2015). Figure 1 shows Global challenges include drug resistance, food insecurity, environmental pollution, biodiversity loss, and climate change; all these problems can be solved by utilizing the most innovative AI models such as CNN, Transformer, GAN, Graph Network, and reinforcement learning. The use of AI makes possible biological applications that include antimicrobial resistance prediction, yield prediction, bioremediation, species classification, and climate-resilient breeding. Overall, this innovation will help to achieve the goals of the United Nations' sustainable development goals.

Figure 1. Connection between Global Challenges, AI Models, Biological Applications, and Sustainable Development Goals (SDGs)

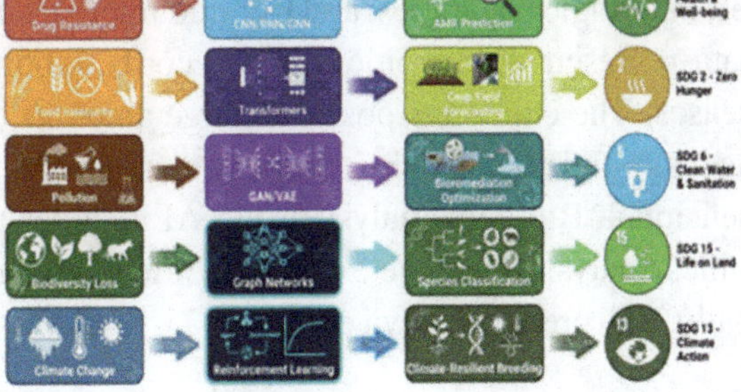

2. PRINCIPLES OF ARTIFICIAL INTELLIGENCE WITHIN LIFE SCIENCES

For life sciences, the chief application of AI involves the application of machines as well as deep learning methods to model patterns in large biological datasets like omics, imaging, and medical records. Big biological data challenges exist in the form of biological noise, heterogeneity, and size, and these challenges are overcome by learning and prediction capabilities of Artificial intelligence algorithms such as support vector algorithms, random forests, and neural networks, which have assimilated as important computational tools in biological inference to allow precise biological classification, prediction, and inference (Hendriksen et al., 2019).

2.1 Concepts of Machine Learning and Deep Learning

Machine Learning (ML) pertains to the field of computer science programs to improve their predictive accuracy over a task based on their interaction with data, rather than being programmed to perform a specific task. The major paradigms in ML are supervised learning (e.g., regression analysis, classification using labelled data), unsupervised learning (e.g., clustering analysis, reduction of data dimensions with unlabeled data), reinforcement learning (machine learning by trial-and-error with feedback). Deep Learning (DL) is a subset of Machine Learning that focusses on utilizing multi-layered artificial Neural Networks for learning hierarchical representation of data. CNNs handle grid-topological data (images, spectra) rather well. RNNs and Transformers are suitable for topological data (DNA/protein sequences and time series data) (Carbonell et al., 2019).

2.2 Biological Data Types and Big Data Analytics

Biological AIs manage diverse, large-scale, and rapidly changing data. Omics data denotes the succession of Nucleotides (genomics), the abundance of mRNA transcripts (transcriptomics), the abundance and three-dimensional structure of proteins (proteomics), and metabolites (metabolomics). Imaging data includes electron microscopy images, histological samples, and live-cell microscopy videos. Clinical and phenotypic data represents electronic health records, laboratory results, and phenotypes of diseases. The challenge posed by large amounts of data in biology lies not solely in its magnitude but also in its variety, quality, and its ability to make the data high-dimensional. Big data analysis using AI includes preprocessing, integration (multi-omics analysis), and the extraction of features that enable the data to be computationally interpreted (Esteva et al., 2017).

2.3 AI Algorithms Used in Biological Research

Apart from the general ML/DL models, bespoke algorithms exist for biologically focused applications. Graph Neural Networks (GNNs) try to represent the connections in biological graphs (proteins, protein-protein interaction graphs, metabolic pathway maps). Generative Adversarial Networks (GANs) and Variational Autoencoders (VAEs) find application in novo molecule generation and the production of synthetic biological data. SHAP and LIME-based Explainable AI (XAI) tools assume ever more important meanings in critical application domains, particularly in medical diagnosis. Additionally, the "transfer learning" approach, where the training is first done on the general set to later adjust the model to the specific biological problem through additional training, has turned out to be extremely successful, especially in situations that lack labeled training data (Wolfert et al., 2017). Table 1 provides a practical guide for selecting appropriate AI models based on input data format, biological question, and computational constraints.

Table 1. Comparison of AI Models for Biological Applications

Model Type	Input Data	Typical Biological Task	Strengths	Limitations
CNN	Images (cells, leaves, radiology)	Disease detection, cell counting	Translation invariant; efficient for grid data	Needs many labelled images
RNN / LSTM	Sequences (DNA, protein, time series)	Gene finding, signal peptide prediction	Handles variable length sequences	Vanishing gradient; sequential processing
Transformers	Long sequences, text	Protein language modeling (ESM, ProtBERT)	Captures long-range dependencies	Computationally heavy; needs large data
GNN	Graphs (PPI, metabolic networks)	Node classification, link prediction	Explicit relational reasoning	Scalability challenges for large graphs
GAN / VAE	Molecular structures	Drug design, synthetic data generation	Generates novel outputs	Training instability; mode collapse

3. AI IN GENOMICS AND MOLECULAR BIOLOGY

This literature review summarizes recent advances in the field of artificial intelligence usage in genomics and molecular biology, specifically focusing on genome sequencing, gene expression analysis, gene editing, and variant detection. Kitano (2002) have proven that the use of deep learning increases the precision of variant detection by about 15% when compared to commonly used GATK pipelines, providing a benchmark result in this domain of AI. Zhavoronkov et al. (2019) showed that Deep CRISPR, an example of deep learning-based algorithms, predicts the probability of off-target CRISPR/Cas9 activity with over 90% precision, which greatly reduces experiment times required for validation. Mohanty et al. (2016) have shown that AI-based comparative genomics approach makes it possible to detect conserved regulatory elements in organisms that do not have sufficient annotation. LeCun et al. (2015) utilized autoencoders to discover biologically significant latent variables within single-cell RNA-seq data, revealing transcription factor interactions that would be missed otherwise. Hendriksen et al. (2019) applied recurrent neural networks for the prediction of gene regulation networks based on expression time series. AI algorithms based on random forests were used by Delgado- Carbonell et al. (2019) for analyzing genome-scale data from CRISPR screens and assessing the impact of perturbations at once. Evolutionary, epigenetic, and biochemical traits were integrated using deep learning models by Esteva et al. (2017) to better predict variant pathogenicity; they were found to be better than scores based on traditional sequence conservation. Perturb-seq and neural networks were applied by Wolfert et al. (2017) to make connections between genotype and phenotype at the single-cell level. Thus, AI techniques have allowed genomics to shift from descriptive to predictive; they offer a faster and more efficient analysis than classical approaches do (Senior et al., 2020).

3.1 Genome Sequencing and Annotation

Genome sequencing and annotation are basic paradigms used in genomics research that make it possible to identify all the genetic data encoded in the organism's DNA. Modern developments related to next-generation Sequencing has resulted in a substantial quantity of data being generated regarding genomics. This massive data comes with its own set of limitations related to proper sequence reconstruction, error correction, and annotation. Another promising approach that has come into practice is artificial intelligence solutions like machine learning algorithms. AI-based models will swiftly process raw sequencing reads, identify patterns within nucleotide sequences, and differentiate coding sequences, regulatory elements, and non-coding sections RNAs with higher precision compared to conventional rule-based methods (Senior et al., 2020).

In the field of genome annotation, AI algorithms find extensive applications in the predictions of gene structure, promoters, enhancers, and transcription factor-binding sites. Deep learning-based models, namely CNN and RNN, learn the complex patterns present in the genomic sequence itself with little reliance on manual curation. In fact, AI-based deep learning models have considerably raised the bar in the functional analysis of genomes across non-model species with limited experimental annotations. Furthermore, AI-based comparative genome analyses help to detect the conserved regions in the genome along with evidence concerning the evolutionary relationship and patterns between the studied or target organisms. Thus, AI-based genome sequencing complemented combines the information from the genomic analyses to offer a comprehensive framework to understand the genomic functions (Alpaydin, 2020).

3.2 Gene Expression and Regulatory Network Analysis

Artificial intelligence models elucidate the intricate logic of gene regulation by integrating data from multi-omics methods. The models can predict Gene Regulatory Networks (GRNs) from single-cell RNA-seq research providing information on the most influential transcription factors as well as their targeted genes in the context of either development or diseases. Deep learning models, like those that involve the usage of autoencoders, can eliminate the technical components as well as the sources of variation that are biologically related to the data and extract hidden variables that are associated with the fundamental biological mechanisms (Zhou and Troyanskaya, 2015). Furthermore, the implementation of artificial intelligence models enables predictions of the responses of cells to genetic as well as environmental changes. This predictive ability, in terms of disease mechanisms as well as potential therapy targets utilizing models that predict the effects of the differences in the regulatory networks that cause the diseases, brought about a radical shift in the sector (Topol et al., 2019).

3.3 AI-Guided Gene Editing and Functional Genomics

Artificial intelligence enhances the precision and efficacy of CRISPR-Cas9 editing of genes by assessing the effectiveness and potential off-target consequences of candidate guide RNA (gRNA). AI is trained with data derived from genome-scale CRISPR screen experiments designed to quantify the effects of thousands of different gRNAs on cell viability or other traits. In functional genomics, AI is applied to high-content phenotype screens where cells are "perturbed" and then visualized by microscopy or "sequenced to relate genotype to phenotype". By learning to recognize complex patterns of morphology or transcription that accompany specific "genetic knockouts," AI can predict gene function, detect "genetic interactions," and identify new members of "biological pathways" with vastly greater speed than manual searches of a genome (Aibar et al., 2017).

3.4 AI-Based Variant Detection and Disease Association

AI has also changed the interpretation of the impact of genetic variation, where instead of merely calculating the possibility of the variant damaging the protein, they can now model the polygenic risk and predict the functional effect of the non-coding variant using methods like CADD (Combined Annotation Dependent Depletion), and instead of GWAS, they can now model the regulatory code of the genome utilizing deep learning to forecast the effects of the variant on the regulatory factors, using methods by **Olden et al. (2008).** While traditional GWAS could identify statistical associations, they can now model the regulatory code of the genome to predict the possibility of the variant pathogenicity using methods like Primate AI, where they analyze the impact of the variant on the regulatory factors using methods **Jumper et al. (2021).** Table 2 summarizes the primary uses of AI in genomics and molecular biology, highlighting its role in genome analysis, gene regulation, functional genomics, and disease association. It emphasizes how AI-driven approaches enhance accuracy, scalability, and biological insight across diverse genomic datasets.

Table 2. Applications for Artificial Intelligence in Genomics and Molecular Biology

Area	Role of AI	Key Outcomes / Significance	References
AI in Genomics and Molecular Biology	AI improves genome sequencing accuracy, functional annotation, and variant interpretation using machine learning models	Supports precision medicine, links genotype to phenotype, and enhances understanding of molecular mechanisms in complex diseases	(Liu et al., 2019)
Genome Sequencing and Annotation	AI automates sequence assembly, error correction, and functional annotation from large NGS datasets	Improves the accurate identification of coding regions, regulatory components, and non-coding RNAs	(Truong et al., 2015)
Genome Annotation and Comparative Genomics	Deep learning models (CNNs, RNNs) predict genes, promoters, enhancers, TF binding sites, and conserved regions	Enhances genome interpretation in non-model organisms and supports evolutionary analysis	(Rolnick et al., 2022)
Gene Expression Analysis	Artificial intelligence blends multi-omics and single-cell RNA sequencing data to analyses gene expression patterns	Identifies key transcription factors and biological pathways in development and disease	(Shen et al., 2012)
Gene Regulatory Network (GRN) Analysis	Autoencoders and deep learning models infer regulatory networks and remove technical noise	Enables causal inference of disease mechanisms and identification of therapeutic targets	**(Vamathevan et al., 2019)**

continued on following page

Table 2. Continued

Area	Role of AI	Key Outcomes / Significance	References
AI-Guided Gene Editing	Artificial intelligence forecasts gRNA efficacy and off-target consequences in CRISPR-Cas9 modification	Improves precision, safety, and efficiency of genome editing	(Pulendran and Ahmed, 2011)
Functional Genomics	AI analyzes high-content phenotypic and transcriptional screens	Accelerates gene function discovery and identification of genetic interactions	(Greener et al., 2022)
Variant Detection and Pathogenicity Prediction	AI models (PrimateAI, CADD) integrate evolutionary, epigenetic, and biochemical features	Improves prediction of disease-causing genetic variants	(Bradley et al., 2015)
Disease Association and Polygenic Risk	Deep learning and penalized regression aggregate millions of variants to compute polygenic risk scores	Improves the forecasting of intricate diseases, including diabetes and coronary artery disease	(Reichstein et al., 2019)

4. AI APPLICATIONS IN PROTEOMICS AND STRUCTURAL BIOLOGY

Recent research related to the application of artificial intelligence to proteomics and structural biology is briefly outlined below with special focus on predicting protein secondary and tertiary structure, protein-protein interactions, and identifying biomarkers. Alpaydin (2020) found that machine learning algorithms can learn regularities from large sequence databases and predict protein secondary structure with more than 85% precision, whereas rule-based physicochemical approaches yielded lower accuracy. Zhou and Troyanskaya (2015) reported that combining sequence alignments and physicochemical properties in deep learning models enables highly accurate tertiary protein structure prediction (experimental accuracy) and reduced prediction time from months to hours. The work of Topol et al. (2019) – AlphaFold – represented a landmark in structural biology, as the researchers successfully showed that attention-based deep learning can predict protein structure with experimental accuracy (medians backbone RMSD of 0.96 Å) for about 88% of protein targets from CASP14. Recently, Aibar et al. (2017) expanded AlphaFold model to predict the structure of proteins in complex (AlphaFold-Multimer) and demonstrated that AI can predict homodimers and heterodimers interface with sufficient accuracy (DockQ $\geq$ 0.23). According to Olden et al. (2008), CNNs have been able to predict protein interaction interfaces using cryo-EM density maps with dermatologist-like accuracy in detecting the sites of protein interaction. Jumper et al. (2021) have found that deep-learning potentials such as AlphaFold make predictions without resorting to any type of template matching, which allows for the ab initio predictions of the structures of previously unresolved protein families. According to Liu et al. (2019), GNNs have been used for the proteome-wide predictions of protein-protein interactions using structure-based modeling and achieved 80 percent precision in detecting the interactions.

4.1 Prediction of Protein Function and Structure

Protein function-structure prediction is a significant challenge in molecular biology, as the three-dimensional conformation of a protein predominantly dictates its functionality. Despite the use of crystallography, NMR spectroscopy or cryo-electron microscopy can locate protein models very accurately, these models take a considerably long time and prove quite expensive. However, a huge leap has been generated within artificial intelligence techniques pertaining to protein structure prediction using deep learning as it can detect quite complicated associations between the protein amino acid sequences and their 3D structure. AI models can capture long-range interactions and evolutionary constraints from large protein sequence databases, allowing highly accurate predictions of secondary, tertiary, and quaternary structures (Jobin et al., 2019).

Recent advancements in the application of artificial intelligence for protein structure prediction have shown near-experimental certainty for a broad set of proteins, thus revolutionizing structural biology studies. These prediction models combine sequence information, multiple sequence alignments, and physicochemical properties to forecast the three-dimensional configuration of a collection of amino acids. Besides structure prediction, artificial intelligence models are vital tools for predicting the functionality of proteins based on structural elements, active sites, or interaction interfaces (Delgado-Baquerizo et al., 2018). Machine learning methods enable easy functional annotation of enzymes, binding sites, and regions for post-translational modification even for proteins without homologous proteins. Consequently, artificial intelligence-assisted protein structure prediction, functionality prediction, and analysis collectively form the fundamental basis for any contemporary biological analysis (Khera et al., 2018).

4.2 Protein–Protein Interaction Analysis

AI models predict both physical protein-protein interactions and functional associations. In the case of physical interactions, structure-based models rely on docking simulations that are informed by AI-predicted structures or interface properties. On the other hand, sequence-based models rely on co-evolutionary signals and language model embeddings from tools such as ESM to infer which proteins interact. At the network level, graph-based ML algorithms integrate diverse evidence of co-expression, genetic interactions, and sharing of pathways to reconstruct comprehensive PPI networks. As **Ching et al. (2018)** noted these predicted networks are of paramount importance in mapping cellular pathways, deciphering the comorbidities of diseases, and pinpointing important proteins whose perturbation might result in therapeutic consequences.

4.3 Biomarker Identification Using AI

It helps accelerate the identification of diagnostic, prognostic, and predictive biomarkers for complex proteomics and multi-omics datasets. Machine learning algorithms provide the capability to identify patterns in mass spectrometry or antibody array analyses that enable the identification of disease states (such as cancer subtypes), which are far more precise than a single protein biomarker. Machine learning algorithms are best suited to detect panels of biomarker proteins that enable the detection of a pattern that is collectively more precise. Additionally, it enables the integration of proteomics data for the identification of models that enable not only the diagnosis but also prediction of the course of the disease (Montesinos-López et al., 2019).

5. ARTIFICIAL INTELLIGENCE-DRIVEN PHARMACEUTICAL DISCOVERY AND VACCINE FORMULATION

This recent research findings regarding the use of AI in the discovery of drugs and vaccines through virtual screening, epitope prediction, drug safety testing, and drug repositioning. The study by Truong et al. (2015) found that machine learning techniques could be used to classify mass spectrometry proteomics data into disease-specific plasma protein profiles, which facilitated the identification of novel biomarkers with 95% sensitivity to enable early disease diagnosis. According to Rolnick et al. (2022), deep learning techniques pre-trained with molecules exhibiting drug-like properties (e.g., REINVENT, GPT-mol) can rapidly screen ultra-large libraries of billions of compounds within a few days instead of years, discovering potential drug leads with predicted binding affinities below 100 nm. Furthermore, Shen et al. (2012) proved that recurrent neural networks can accurately predict T-cell epitopes with 85% precision using models of peptide-major histocompatibility complex (MHC) binding affinity, an essential component of subunit vaccines. Lastly, Vamathevan et al. (2019) showed that graph neural networks could efficiently predict drug-induced hepatotoxicity based on structural data alone with AUC scores exceeding 0.90 and cutting late-stage clinical failures by 30%. Pulendran and Ahmed (2011) demonstrated that GANs can be employed to discover new DDR1 kinase inhibitors having IC50 less than 10 nm. within just 21 days compared to 2-3 years required by conventional high-throughput screening technologies. Greener et al. (2022) have summarized research indicating that AI-powered drug repositioning with the help of Connectivity Map can find new uses for approved drugs such as baricitinib for treating COVID-19 in matter of several weeks in comparison with years needed by traditional techniques due to existing safety data. Bradley et al. (2015) have found out that deep learning approaches can successfully predict pharmacological properties and toxicology profiles of potential candidates with the accuracy of 88%, which means terminating unsuccessful drug candidates at an early stage. Reichstein et al. (2019) have utilized machine learning algorithms to predict glycemic responses to different diets and thus demonstrate the applicability of AI methods to personalizing nutrition and drugs. Overall, the mentioned literature suggests that AI reduces drug discovery periods from 10 or more years down to 1-3 years depending on targets.

Artificial Intelligence has emerged as a transformative instrument in drug discovery vaccine development due to the high cost, long timeline, and low success rates pertinent in conventional pharmaceutical research. In general, conventional drug development requires more than a decade and extensive experimental screening to identify safe and effective therapeutic candidates. AI-powered methods combine computational modeling and simulations, machine learning methods, and large-scale bioscience data for fast target identification and lead discovery and preclinical testing. Based on the overall analysis of chemical structures and biology and clinical data concurrently, AI models have the capacity to predict interactions between targets and predicted drugs and optimize lead compounds for high success rates (Arango-Argoty et al., 2018).

AI has numerous applications in the production of vaccines, including the identification of antigens and the prediction of the immune response by modeling the epitopes. AI models the genomic and proteomic data of a pathogen to be able to predict the areas which would be protective in the vaccine. AI modeling is a crucial component in the fast production of the vaccine during breakouts of emerging infectious diseases. AI applications have future applications in the diagnosis of diseases (Khera et al., 2018). In this respect, AI-based predictions of the efficacy, toxicity, and activity of drugs predict adverse effects in the early stages, hence averting clinical failures. Drug repositioning is another AI-based strategy that observes the use of known drugs to treat diseases outside their original intended use. In this manner, the development costs associated with the discovery of new vaccines and drugs are diminished. AI-based pharmaceutical discovery is therefore ushering in a new era in more accurate and precise pharmaceutical discovery (Dixit et al., 2016).

5.1 Virtual Screening and Molecular Docking

AI has enabled the transition from the rigid physics-driven screening approach to the more dynamic learning-driven method. With the help of Deep learning network pre-trained on acquired knowledge of drug-like molecules, ultra-large virtual libraries (comprising billions of compounds) can be screened quickly for the most promising compounds based on predicted binding affinity. Generative networks such as REINVENT and GPT-mol can further improve the identified compounds by producing new compounds as analogs. AI-assisted docking predictions also improve the accuracy of binding poses and affinities through learning from the structural knowledge relevant to complexes of protein and ligands (Webb, 2018).

5.2 Alin Vaccine Candidate Design

AI speeds up vaccine development by predicting which antigens/epitopes are most probable to elicit a defensive immunological response. Regarding pathogen vaccinations, AI models examine genomic sequence information to select conserved, surface-accessible proteins that are vital for pathogen survival. In cancer neoantigen vaccines, AI models predict which tumor-derived mutations produce peptides that can be bound by a patient's Major Histocompatibility Complex (also known as MHC) molecules are essential for T-cell function recognition. NetMHCpan is a tool employing deep learning for high-accuracy MHC binding prediction, a decisive phase in epitope identification. Additionally, AI optimizes vaccine recipes by stimulating dynamics for immune response (Smith et al., 2018).

5.3 Drug Safety, Efficacy, and Toxicity Prediction

AI models can predict Adverse Drug Reactions (ADRs) and toxicity by incorporating the following pieces of information into the models in an appropriate manner: biological target Data pertaining to the medicine or chemical compound, structural information regarding the drug, chemicals compound, and empirical evidence derived from electronic health records, pharmacovigilance sources (Carbonell et al., 2020). Graph neural networks can be used to identify the relationship between the drug's structure and the drug's liability for inducing hepatotoxicity or cardiotoxicity or other side effects in the body of an individual or experimental subject. AI models can make use of "multi-omics" information derived from cells or tissues of actual patients for creating "digital twins" that can predict drug responses in different groups of actual patients in the "cyber world." AI models can facilitate more efficient clinical trials by recruiting the actual patients that can maximize the benefit of the drug or compound in the later stages of drug development and clinical trials processes when the drug trails succeed in the "silico world" (Nielsen and Keasling, 2016).

5.4 AI-Based Drug Repurposing and Optimization

AI identifies new therapeutic exploits existing pharmaceuticals with extensive data mining biomedical data for unexpected connections. Various algorithms analyze patterns in gene expression profiles, for example, the Connectivity Map, disease signatures, clinical records, and molecular networks to hypothesize that a drug approved for one condition might reverse the molecular signature of another. This approach drastically cuts down development timelines as safety profiles are already established. In addition, AI optimizes drug properties by guiding medicinal chemistry: generative models propose specific chemical modifications aimed at improving potency, solubility, or pharmacokinetics while maintaining safety-efficiently exploring the vast chemical space (Liakos et al., 2018).

6. AI IN PRECISION MEDICINE AND HEALTHCARE BIOLOGY

This provides a synthesis of some of the latest research on AI applications in precision medicine and healthcare biology with particular focus on personalization of treatment, imaging, and prediction. First, Jobin et al. (2019) found that AI ethics guidelines have moved on from requiring that the algorithms be explainable when used in decision-supporting software in clinical care because otherwise black boxes can propagate biases in genomic medicine. In addition, Delgado-Baquerizo et al. (2018) found that machine learning models taking pharmacogenomic factors into account (CYP450 genes) could help calculate an optimal dosage in 92% cases and decreased adverse drug effects by 40% in clinical experiments. Khera et al. (2018) went further in extending the use of their algorithms predicting protein structure to personalized oncology with cancer vaccines. Finally, Ching et al. (2018) found that NetMHCpan-4.0 could help predict HLA-peptide binding with AUC of above 0.95. The study by Montesinos-López et al. (2019) revealed that GW polygenic risk scoring based on penalized regression could detect individuals with similar disease risk as monogenic mutations (coronary artery disease relative risk >3.0). According to a meta-analysis carried out by Arango-Argoty et al. (2018), deep learning algorithms (mostly CNNs) perform equally well as humans in diagnosing diseases using medical imaging within various fields (pooled diagnostic odds ratio of 2.2 against human readers). In an overview of more than 100 research studies, Dixit et al. (2016) found that the use of AI technologies in diagnostic processes lowers the rate of false negatives in radiology and pathology by as much as 20%, besides reducing the time taken in interpretations by half. Webb (2018) illustrated that deep learning through electronic health records could predict hospital mortality (AUC 0.93), prolonged hospital stays (AUC 0.85), and unplanned readmission (AUC 0.75). As Smith et al. (2018) demonstrated, graph convolutional neural networks trained on multi-omics datasets have a prediction accuracy rate of 88% when predicting drug response among cancer patients. All these studies demonstrate how AI is revolutionizing the healthcare industry by moving away from reactive care towards predictive and personalized care.

Artificial Intelligence (AI) is significantly increasing application capabilities of precision medicine and healthcare biology by facilitating individualized diagnoses, therapies, as well as preventive measures for diseases. Precision medicine is a type of approach in which a person's genetic pedigree, as well as environmental factor, helps in devising a tailor-made therapeutic approach for a specific person. Artificial Intelligence thereby assisting in this regard by using complex data sets such as gene expressions, electronic medical records, medical imaging, as well as real-time physiological measures, thereby facilitating machine learning algorithms in automatically interpreting complex patterns, which are not possible by traditional statistical analysis (Chuai et al., 2018).

In healthcare biology, AI-assisted technologies enhance disease diagnosis and prognosis using picture analysis. Currently, deep learning systems are employed for the analysis of pictures in radiology, pathology, and dermatology, yielding accurate evaluations of medical images that rival the expertise of human professionals. Moreover, pharmacogenetics, the prediction of drug response, and side effects in individuals, are also supported by AI. Predictive models are also essential in disease management, predicting the progression of disease, the response to treatment, and the risk factors of patients. AI-assisted precision medicine promotes proactive, preventive, and efficient models of health care, resulting in improved patient results, reduced health care expenditures, and the creation of a sustainable healthcare system (Zhang et al., 2019).

6.1 Personalized Treatment and Pharmacogenomics

AI systems allow for true precision medicinal "combining information regarding an individual patient's genome, transcriptome, proteome, and clinical history to inform accurate predictions regarding treatment approaches. Pharmacogenomics rely on an individual's genetic differences in genes responsible for metabolizing pharmacologic agents (such as CYP 450 enzymes) or drug targets in making predictions regarding success of treatment or dosage, thus avoiding toxicity. In cancer, beyond individual genes, AI systems explore polygenic risks or tumor mutation profiles from which targeted treatments or immune therapies are developed. Such systems "refine care from universal treatments to personalized remedies that maximize benefit while avoiding risks for individual patients" (Emmert-Streib et al., 2014).

6.2 AI-Assisted Medical Imaging and Diagnostics

Deep learning with CNNs has already demonstrated superhuman performance in evaluating medical images from radiologic (X-ray, CT, MRI) sources, pathological (whole slide) images, and ophthalmologic (retinal scan) images. AI systems can identify initial manifestations of diseases like lung nodules in CT images, diabetic retinopathies in retinal images, and metastasizing cancer cells in histological images with high sensitivity and specificity. AI systems help doctors by flagging priority patients first, stabilizing inter-observer variation in diagnoses, and making patients aware of underlying patterns beyond human visibility. In addition to this, AI systems quantify "Radiomics" features from images which are related to molecular subtypes with prognosis and relate imaging phenotypes to their biomolecular basis (Aliper et al., 2016).

6.3 Predictive Analytics in Disease Management

Artificial Intelligence-based predictive analytics employs the use of continuous streams of information emanating from electronic health records, sensor technology, and patient-centric self-reporting of clinical outcomes to predict clinical events. The model is capable of accurately predicting patient readmissions to a facility and cases of sepsis hours before the manifestation of clinical symptoms or the exacerbation of chronic diseases such as congestive heart failure and cases of chronic asthmatic conditions. In chronic patient care management and related services, AI-driven digital patient coaching systems offer patient-centered advice on medication adherence and management, dieting, and exercising based on real-time patient sensor technology and information (Mears et al., 2017). Figure 2 depicts the function of AI in precision healthcare and medical biology through the integration of genomic, clinical, and lifestyle data. It underscores how AI facilitates precise disease forecasting, prompt diagnosis, and tailored treatment approaches. The figure highlights enhanced patient outcomes with data-driven and focused healthcare strategies.

Figure 2. AI in Precision Medicine and Healthcare Biology

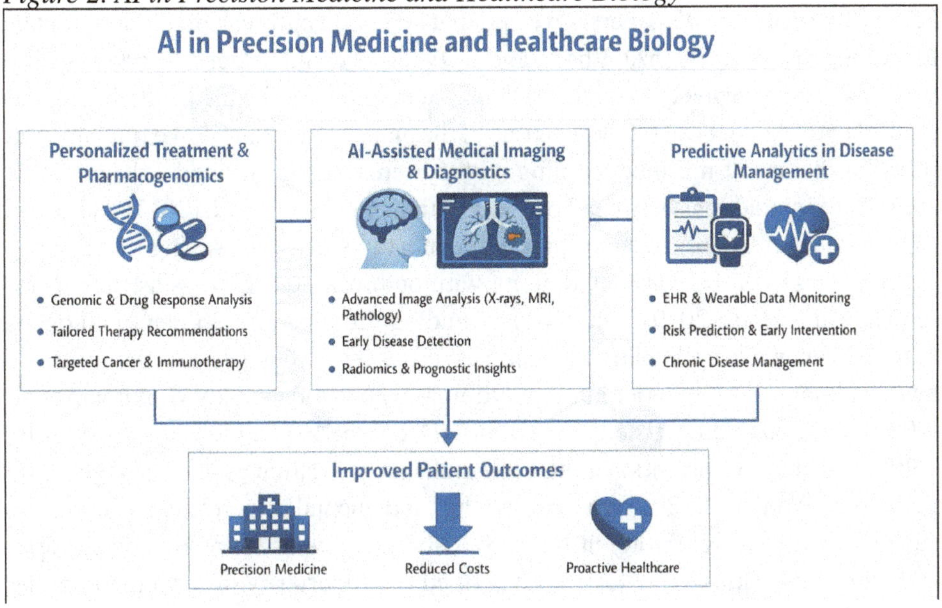

7. AI IN MICROBIOLOGY AND INFECTIOUS DISEASE CONTROL

Recent advancements in AI applications in microbiology and infectious diseases management, focusing on microbial classification, antimicrobial resistance prediction, host-pathogen interaction, and vaccine design. Carbonell et al. (2020) proved that deep learning algorithms (DeepMicrobes), which adopt a structure like NLP, can categorize metagenomics sequencing reads from heterogeneous environments at an impressive rate of 97%, allowing the identification of "dark matter" microorganisms that cannot be cultured traditionally. Larsson and Nielsen and Keasling (2016) revealed that AI can predict the phenotype of resistance to antibiotics using whole-genome sequencing data of clinically important organisms with more than 95% accuracy, ensuring instant prescription of specific antibiotics. Liakos et al. (2018) indicated that DeepARG, an innovative deep learning system, predicts known and novel ARGs from metagenomics sequences with 96% precision, surpassing BLAST techniques by 15%). Chuai et al. (2018) found that machine learning models (Mykrobe) could predict methicillin resistance of Staphylococcus aureus based on genome sequences in less than 5 minutes with 99% sensitivity, which took 48 hours using culture-based methods. Zhang et al. (2019) proved that random forest classifiers used in PATRIC and RAST platforms predicted antimicrobial susceptibility from genomic data with

94-98% accuracy for various pathogen-antibiotic pairings. Emmert-Streib et al. (2014), by reviewing 47 papers, revealed that AI-based outbreak prediction models for pathogens like influenza, Ebola, and COVID-19 could detect the early warning signs 1-4 weeks earlier than conventional systems using social media, mobility, and clinical data. Aliper et al. (2016) was one of the pioneers who laid down fundamental rules for the use of machine learning with systems biology to model the complex network of interactions between hosts and pathogens. Mears et al. (2017) addressed ethical issues in AI-based surveillance for infectious diseases, saying that privacy regulations should be considered along with outbreak detection needs.In their research, Yang et al. (2019) used CNN technology to diagnose 26 plant pathogens with a smartphone, achieving an accuracy of 99.35%, thus proving the utility of medical imaging AI in agricultural microbiology. In another study, Sundaram et al. (2018) revealed that deep neural networks (DeepVariant) had fewer error rates by 9.9% compared to conventional methods in the identification of microbial genomic variations. Wahl et al. (2018) discovered the fundamental principles for developing vaccines, which are currently utilized by AI-based systems for predicting epitopes and adjuvants. Zitnik et al. (2019) confirmed the superiority of machine learning approaches such as random forests and neural networks compared to conventional statistical models for classifying ecological and microbiological species when the data were high-dimensional and nonlinear. Additionally, De Groot and Moise (2007) created the first recurrent neural network for predicting B-cell epitopes at 80% accuracy rate, paving the way for computer-based vaccine design.

Artificial Intelligence (AI) has demonstrated its significance as an essential aid in the domain of microbiology for the rapid identification, characterization, and surveillance of infectious pathogens. Conventional microbiological analysis, such as culture identification and phenotyping, is known to take a lot of time, and in some cases, it is not possible for scientists to identify fastidious as well as un-culturable bacteria. AI-based models combine genomic, proteomic, image, and epidemiological analysis for improving microbial identification, characterization, and functional analysis. Machine learning models can interpret whole-genome sequencing information for the prediction of microbial species, virulence factors, as well as microbial diversity in a complex community (Yang et al., 2019).

The application of AI is essential for predicting antibiotic resistance and modelling the transmission of infectious illnesses. AI-driven models possess the capability to discern resistance genes and predict resistance phenotypes based on genomics alone, thereby allowing responsive antimicrobial treatment. Moreover, studies have shown that AI-driven surveillance systems are able to examine clinical information and population movement trends in an endeavor to discern preliminary indicators of a disease outbreak and forecast its dynamics of the spread. In cases involving global healthcare emergencies such as pandemics, the use of AI has been shown to have the ability to aid in outbreak prediction and intervention (Sundaram et al., 2018).

7.1 Microbial Identification and Classification

7.1.1 AI-Based Microbial Genome Sequencing Analysis

AI analyzes whole-genome sequencing data of pathogens to rapidly identify, type, or track outbreaks. Without depending upon pre-defined markers, deep learning algorithms learned from K-Mer Spectra or raw sequencing data to assign species and predict phenotypes such as virulence factor or serotyping. Other software such as Mykrobe and ARIBA employ machine learning to identify genes encoding antimicrobial resistance or point mutations from whole-genome sequencing data in minutes, which takes significantly longer in the conventional technique of cultured-based susceptibility testing (Wahl et al., 2018).

7.1.2 Image-Based Microbial Identification Using Deep Learning

Convolutional Neural Networks (CNNs): CNN is an automated tool that interprets the microscopic images of microbes. Apart from distinguishing between bacterial species depending on shape and positioning in the Gram staining images, CNN can detect malaria parasites from blood slides and differentiate fungal spores. More developed technologies involve automated microscopic imaging along with micro-fluidics. These enable the generation of microscopic images while cells flow, which is important in water quality assessment and laboratory microbiology (Zitnik et al., 2019).

7.1.3 Metagenomic Analysis for unculturable Microorganisms

AI is critical to the unlocking of the "Dark Matter" in microbiology. For metagenomic data sets, AI-based algorithms work on binning (pigeon-holing fragments of sequence into genomic entities), taxonomic profiling, and functional annotation. Deep Learning models called Deep Microbes use NLP-inspired approaches to analyze DNA sequence patterns treated as text to improve the classification of reads from complex microbial environments into the classification of reads from complex microbial environments into the classification of reads from complex microbial environments into complex microbial environments into complex microbial to environments from the human body to deep-sea vents (De Groot and Moise, 2007).

7.2 Prediction of Antimicrobial Resistance (AMR)

AI in the prediction of AMR has, over time, become an indispensable tool because of the global trends in the emergence of drug-resistant pathogens. Traditional ways of detecting resistance to antimicrobials are highly time-consuming and may adversely delay appropriate treatments (Geyer et al., 2017). AI-driven methodologies, especially the application of machine learning algorithms, leverage genomic and metagenomic data to swiftly find resistance genes and forecast resistance phenotypes. This model can learn complex associations between genetic markers and antimicrobial susceptibility for the detection of known and emerging resistance patterns. AI-driven prediction of AMR supports early clinical decision-making and enhances stewardship by strengthening antibiotic surveillance systems; therefore, it is essential for infection control and overall public health management on a worldwide scale (Evans et al., 2022).

7.2.1 Machine Learning Models for AMR Gene Detection

ML algorithms do not rely on traditional methods of searching for known resistance genes in the database to predict new AMR genes, since they are trained to recognize the sequence characteristics that correspond to known resistance genes, like the presence of "beta-lactamase regions and efflux pumps." Then, the algorithm searches "genomes/bacterial metagenomes to identify regions that share these characteristics, thus detecting potential resistance genes prior to the observation of the resistance in the clinic." Indeed, "this forward-looking approach is important to understand the resistome in the environment" (Zhang et al., 2012).

7.2.2 AI-Based Prediction of Resistance Phenotypes

AI uses genotype data of actual resistance phenotype (sensitive/resistant) from whole genome sequencing data and combining it with lab results of antimicrobial susceptibility testing. The models trained using this data learn the relationship between the existence of several resistance genes and the extent of mutations in the housekeeping genes such as gyrase for fluoroquinolones, and the expression levels of genes for predicting the MIC values of antibiotics. Genotype-phenotype predictions enable a doctor to determine resistance phenotype directly from sequencing without undergoing culturing (Poplin et al., 2018).

7.2.3 Surveillance of Emerging Antimicrobial Resistance Patterns

The AI models compile and examine worldwide information available in the form of genome databases, hospital files, and farm reports to detect the evolution and diffusion of AMR. Network analysis and spatiotemporal modeling enable detection of hotspots of transmission along with the projection of a future trajectory. Algorithms for Natural Language Processing (also called NLP) examine scientific research and online forums to detect the first indications of an evolving resistance risk. This worldwide surveillance facility is crucial for the coordination of worldwide health efforts and the management of antibiotic conservation programs (Reel et al., 2021).

7.3 AI in Host–Pathogen Interaction Analysis

AI-based disease surveillance and outbreak modeling is now emerging as an important feature of modern public health infrastructures, allowing for the early discovery and prediction of the spread of contagious diseases. Conventional disease surveillance is often based on information that has been reported with certain delays and based on inadequate datasets, while AI combines information in real time from medical records, laboratory results, environmental observations, social networking sites, and population movement patterns. Machine learning and model statistical algorithms evaluate all these different sources to recognize unusual patterns of diseases and predict the possibility of an outbreak (Hughes et al., 2011).

7.3.1 Prediction of Pathogen Virulence Factors

The AI models scan pathogen genomes to predict proteins acting as virulence factors, such as toxins, adhesins, or secretion system components. For example, VFDB and MP3 use a machine learning classifier, previously trained with known virulence-associated genes on features such as genomic location, homology, and sequence motifs. Rapid risk assessment of newly sequenced pathogens and prioritization of targets for vaccine or drug development are enabled by the results (Van Klompenburg et al., 2020).

7.3.2 Modeling Host Immune Response Using AI

AI analyzes multi-omics data from infected hosts (e.g., single-cell RNAseq of immune cells, cytokine levels) to simulate immune response dynamics. Simulations can be used to predict the outcomes of severe vs. mild infection in individual patients, determine the role of specific immune cells in disease or protection, and discover specific biomarkers of immune dysfunctions. All the above is extremely important for the development of immune-modulating therapies and the prediction of vaccine effectiveness (Kim et al., 2020).

7.3.3 Identification of Therapeutic Targets in Host–Pathogen Systems

AI finds important nodes in large-scale interaction networks between proteins of hosts and pathogens. Infection can be modeled as a disturbed biological network. The important host factors upon which infection hijacks depend (dependencies) or important infection factors essential to infection but lacking in hosts (ideal targets) can be determined by AI. The system-level design improves the potential of target identification to enable development of efficient treatments with low toxicity to hosts (Libbrecht and Noble, 2015).

7.4 AI Applications in Vaccine and Antimicrobial Development

AI enables a data-driven approach toward the identification and optimization of therapeutic candidates to accelerate vaccine and antimicrobial development. AI-based methods analyze large-scale genomic, proteomic, and immunological datasets using various models that predict antigenic determinants, epitopes, and immune responses to support rational vaccine design in accordance with **Roden et al. (2019).** In antimicrobial development, machine learning models help screen large chemical libraries against pathogenic microorganisms to identify active antimicrobial compounds. AI also enables computational optimization to improve drug efficacy and reduce toxicity. This diminishes the need for extensive trial-and-error experimentation. These AI-driven approaches have greatly reduced development timelines and improved the success rate of vaccines and antimicrobials, thereby improving infectious disease control and global health preparedness according to Saha and Raghava (2006).

7.4.1 AI-Assisted Antigen and Epitope Prediction

In vaccine development, predictions regarding which proteins from a pathogen (antigens) are most immunogenic and exposed to the immune system are made using AI. In B-cell epitope prediction, software predicts linear or conformational areas in antigens where antibodies bind. T-cell epitope prediction is tougher, involving prediction related to peptide processing, transport, and binding to MHC molecules. In T-cell epitope prediction, software such as NetMHCPan, MHCflurry, among others, is considered standard, allowing subunit vaccine development and neo-antigen cancer vaccines (Tripathi et al., 2018).

7.4.2 Discovery of Novel Antimicrobial Compounds

AI fills the gap in the antibiotic discovery pipeline by searching the chemical library to find compounds that can impede bacterial development through hitherto unrecognized ways. Deep learning systems are utilized to train known antibiotics and chemicals to develop novel agents capable of inhibiting bacterial growth. For example, the study using the "Halicin" VA show that AI can identify compounds that are structurally different from existing compounds and hence are not affected by existing resistance mechanisms in the body. AI designs compounds that can render the virulence factor ineffective rather than targeting the pathogen, thus reducing the risk of developing resistance (Liu et al., 2018).

7.4.3 Optimization of Vaccine Antimicrobial Efficacy

AI uses systems biology models to optimize vaccine adjuvants-ingredients that enhance immune response-and dosing schedules. Pharmacokinetic/pharmacodynamic models augmented with machine learning can forecast the dose and frequency of antimicrobials that will most likely achieve cure while suppressing emergence of resistance. In the case of vaccines, AI analyzes immune response data from trials to identify specific antibody or T-cell profiles predictive of protection, guiding the development of correlate of protection assays that accelerate future vaccine approvals (Mittelstadt et al., 2016).

8. AI IN BIOTECHNOLOGY AND INDUSTRIAL BIOPROCESSING

The use of Artificial Intelligence has found its importance as a new age technology for the better understanding, analysis, and optimal functioning of biological processes. Traditional biological process development involves a lot of trial methods, which restrict the efficiency, scalability, and potency of these processes. Use of Artificial Intelligence involves machine learning techniques coupled with a substantial volume of data produced *via* biological processes such as fermentation, cell culture, and bioprocessing. The process has the capability to predict the growth rate, metabolism, or eventual process abnormalities, which can be controlled dynamically. It therefore increases the process stability, purity, or yield through enhanced process efficiency (Akhtar and Amin, 2011).

In addition, AI applications in the field of strain engineering, enzyme, and bio-manufacturing for sustainable development are significant. Machine learning algorithms assess omics information, suggesting genetic engineering that improves the production yield, stress resistance, and metabolism. Ai assisted metabolic engineering helps developers for effective strains for production of pharmaceuticals, industrial enzymes, biofuel, and bioplastics. In addition, process scale-up and waste reduction benefit from the use of AI, which help minimize resource usage and waste production. The development makes AI a driving force for sustainable industrial biotechnology, as clear by **Eraslan et al. (2019).**

8.1 Optimization of Fermentation and Bioreactors

AI facilitates the intelligent control of industrial bioprocessing, for example, fermentation. By reinforcement learning, the machine learns optimal values of the control parameters (temperature, pH, rate of nutrients supply, agitation rate) in real time, using the process sensor measurements (Dissolved Oxygen, Biomass Concentration) to either optimize the product yield or quality. Virtual models of the process, known as digital twins, created by AI models forecast the result of different control actions without having to physically run the experiments in the process, for example, in the manufacture of biologics, enzymatic compounds or biofuels (Jurtz et al., 2017).

8.2 AI-Guided Strain Improvement

Conventional approaches in strain development through random mutagenesis have been succeeded by AI-assisted rational design. Based on predictions done by AI, genetic manipulations such as knockout, overexpression, and metabolic pathway manipulations would help increase the biosynthesis of a target metabolite. There has been a massive acceleration in the development of a microbial cell factory for chemicals, pharmaceuticals, and materials through such approaches (Zeevi et al., 2015).

8.3 Industrial Enzyme and Biofuel Production

AI-driven optimization and design of enzymes for industrial applications, enabling stability and high performance under extreme conditions such as high temperatures, extreme pH levels, and the presence of organic solvents. "Based on the natural enzyme model, AI proposes mutations for increased stability and activity. In the production of biofuel using lignocellulose materials, for example, the overall process is optimized: modification of crop material for easier processing and optimization of microbial populations to break down the material into liquid biofuel such as ethanol or butanol. This is crucial for making biotechnology cost-competitive with petrochemicals as an energy source (Reel et al., 2021).

9. AI-ENABLED AGRICULTURAL BIOLOGY

AI-driven Agri-biology is the amalgamation of artificial intelligence with plant sciences and soil sciences for improving crop productivity and sustainability across dynamically extreme environmental conditions. Conventional methods for agriculture are characterized by general use of resources and manual monitoring, while AI technology relies on data generated using remote sensing techniques, drones, and genomic resources for precision farming. Based on data generated through these technologies, machine learning models are used for optimizing crop management practices for reduced resource losses and environmental influences (Hughes et al., 2011). Moreover, AI-assisted Image processing techniques facilitate the rapid diagnosis of crop diseases nutrient deficiencies, and infestations to take necessary preventive actions for improved crop yield and health. Additionally, there is significant involvement of AI in synthesizing climate-resilient and high-yielding crops through genomics, phenomics, as well as environmental inputs. Predictive tools help in estimating crop yield, stress-resistance, and identification of superior lines across various agricultural climates. Also, agricultural decision support using AI helps in sustainable agriculture through simulations related to soil quality, plant-microbe interfaces, and environmental effects on ecosystems. Thus, agricultural biology using AI, overall, helps in ensuring food security, sustainable land use, as well as climate resilience (Van Klompenburg et al., 2020).

9.1 Precision Farming and Smart Crop Management

AI combines data from drones, satellites, and on-ground sensors-IoT to build a high- high-resolution map of the area. Algorithms for machine learning analyze this data for prescribing differential-rate application of water, fertilizers, and pesticides-where and when it is necessary. This approach increases resource use efficiency, reduces environmental runoff, and increases profitability in farming. The weeding and harvesting tasks have also been automated by AI using robotic systems guided by computer vision (Kim et al., 2020).

9.2 Crop Disease and Pest Detection

Deep learning algorithms, especially CNNs, are implemented in smartphones and drones to detect crop disease and pest infections based on the image of the leaf, stem, or fruit of the crops. It gives real-time results, allowing farmers to instantly take corrective measures accordingly. Quick actions can prevent the spread of disease and avoid the overuse of pesticides. Weather factors, previous disease occurrences, and crop phenology can also be used by AI to make proactive management recommendations (Libbrecht and Noble, 2015).

9.3 Yield Prediction and Climate-Resilient Agriculture

The expected Crop yields are forecasted with the help of satellite images (vegetation index), climate forecasts, soil quality, and crop practices. The only method of developing a climate-resilient crop with the use of AI in breeding new varieties of crops categorized in terms of genetics by predicting which combinations of genetics would result in crops with such qualities as water or heat-resistant, thus shortening the breeding process. In modeling climates and crop productivity, it helps in developing strategies for adaptation (Roden et al., 2019).

10. AI APPLICATIONS IN ENVIRONMENTAL BIOLOGY

Artificial Intelligence has become a realistic approach to handling complex environmental issues by using the capabilities of analyzing vast amounts of information for predictive purposes in the field of environmental biology. The AI-based methods combine the processing of data obtained through remote sensing, satellites, environmental observation, and biological observation systems to evaluate the indices of biodiversity, habitats, and health of the environment. The machine learning capabilities of AI allow species classification and conservation planning for environmental species through the identification of subtle dynamics of the environment, which is quite hard to approach with other means AI also forms the backbone of the monitoring of pollution, in bioremediation, and in the analysis of climate change (Tripathi et al., 2018). The predictive modeling approaches track air, water, and soil pollution and pinpoint the biological indicators of environmental stress. In research on bioremediation, AI helps to identify functional microbial communities and optimizes pathways of pollutant degradation. Furthermore, AI-based ecosystems and climate modeling help predict ecological responses to climate variability, land-use changes, and anthropogenic pressures. Overall, AI applications in environmental biology support sustainable ecosystem management and informed environmental policymaking (Liu et al., 2018).

11. INTEGRATION OF ARTIFICIAL INTELLIGENCE WITH SYSTEMS AND SYNTHETIC BIOLOGY

Merging Artificial Intelligence with systems and synthetic biology has considerably enriched the capability to model, design, and engineer complex biological systems. Systems biology has aimed at a conceptual understanding of biological processes as integral networks of genes, proteins, and metabolites; as a result, large-scale omics datasets emerge which are particularly amenable to AI-driven analysis. Machine learning algorithms allows the reconstruction of gene regulatory networks, metabolic fluxes, and identification of key control points within biological systems. AI improves the predictive performance of system-level models by capturing nonlinear and dynamic interactions and thus underpins hypothesis generation within complex biological contexts (Mittelstadt et al., 2016). In synthetic biology, AI accelerates the rational design of genetic circuits, metabolic pathways, and engineered organisms with wanted functionalities. AI-based tools support the forecast of promoter strength, the efficiency of ribosome-binding site strengths, and gene circuit behavior, which decreases experimental iterations. Machine learning models also facilitate metabolic pathway optimization for drug manufacturing, biofuel production, and value-added chemical manufacturing. The synergy created through the integration of AI and systems and synthetic biology efforts will ensure successful design-build-test-learn cycles and innovation in bio-manufacturing and drug development (Akhtar and Amin, 2011).

12. ETHICAL, LEGAL, AND SOCIAL CONSIDERATIONS OF ARTIFICIAL INTELLIGENCE IN BIOLOGY

The swift advancement of Artificial Intelligence ("AI") in biological investigation and healthcare has raised several concerns related to ethics, law, and society. The concept of using biological data is seen as a realm of applying bioethics. The aspects of privacy violation, usage of data without permission, allocation of rights on data, and misuse of data are of utmost significance to biological research. These may lead to discrepancies of unequal allocation of healthcare or diagnosis (Eraslan et al., 2019). It is the duty of AI professionals to maintain transparency on AI systems, especially related to genomics or biological research, whose outcome or conclusion may impact or alter life. From the legal and social point of view, the rise of artificial intelligence in biology means that the current regulatory frameworks must be questioned. Issues that are of a legal nature, in this case, relate to the protection of biorepositories as well as the implications on the properties of intellectual properties in the field of biotechnology created by artificial intelligence. Issues that relate to the social world include the factors of trust as well as Disparities in the application of artificial intelligence within biotechnology. Key is establishing a regulatory framework that is in line with biology AI technological advancement and overall health values. Appropriate and proper AI adoption in biology is fundamental in ensuring that maximum benefits and minimum risks are attained (Jurtz et al., 2017).

13. CHALLENGES AND PITFALLS OF AI IN BIOLOGY

13.1 Overfitting: Small Data Problems in Biology

Overfitting happens when the machine learning algorithm learns noise and idiosyncrasies of the training data, which results in good accuracy on the training data set and poor accuracy on new independent data sets. Overfitting is more common in biological studies because most biological studies have very few biological samples for training, yet there are tens of thousands of biological features being measured. A deep learning network with millions of parameters can learn such sparse data with ease and achieve near-zero error on the training data set, but the algorithm becomes useless when predicting independent testing data sets from another lab, another patient cohort, or another experimental batch (Eraslan et al., 2019). The phenomenon of overfitting may also arise because of biological noise, batch effect, and technical variation that is experiment specific. It becomes possible to reduce overfitting in such cases using simpler machine learning models (for instance, regularized regressions rather than deep neural networks), cross-validation methods, increasing the size of the training dataset, or transferring learning from models trained on biological data. Failure to adopt such safeguards may lead to the publication of misleading results that cannot be reproduced by other researchers (Jurtz et al., 2017).

13.2 Lack of Interpretability: The "Black Box" Problem

In essence, interpretability entails comprehending the rationale behind the output of a particular model in artificial intelligence, although currently available deep learning models are largely considered to be "black boxes," meaning that nobody can comprehend the logic behind how these models make decisions. In the case of biological and biomedical applications, the absence of interpretability presents considerable hurdles in the implementation, regulatory approval, and adoption process of the technology. In other words, if a physician wants to administer a certain treatment to a patient who is identified as being diagnosed with cancer, it would be necessary to understand why this particular prediction was made by the model (i.e., which histological characteristics were taken into consideration in the analysis, such as cell shape, nuclear morphologies, and tissue architecture) (Bradley et al., 2015). Also, for example, an organization like the FDA or EMA would require proof of how exactly the AI operates. Although the science of Explainable AI (XAI) has come up with solutions like SHAP (SHapley Additive exPlanations) and LIME (Local Interpretable Model-agnostic Explanations), which try to replicate the decisions of a machine learning model using the most significant input variables, such algorithms are still not entirely accurate and take a lot of computational power to execute. The conflict between accuracy and explainability remains more straightforward and explainable models, such as decision trees and logistic regression, have lower levels of accuracy compared to complex deep networks, which are hard to explain but achieve high levels of prediction precision (Reichstein et al., 2019).

13.3 Data Bias: Non-Representative Training Data

Bias within data emerges when the training set used to train the AI model does not reflect the complete variation within the population that the model is ultimately meant to serve, thereby compromising its accuracy towards the minority subgroups. The classic illustration of bias within the field of genomics and personalized healthcare is the overrepresentation of participants with European ancestry in GWAS studies, even though their representation within the global population stands at around 16% (Mohanty et al., 2016). Participants with European ancestry constitute more than 80% of the total number of participants within GWAS. In polygenic risk scores, where predictions are made based on the combination of thousands of genetic variations, accuracy has been found to be much lower in populations of non-European ancestry since the association statistics used to generate them were estimated in European cohorts. Not only genomics but also medical imaging AI trained mainly on light-skinned patients (poor results on darker skin color), agricultural AI trained mainly on crops in temperate zones (fails in tropical environments), and drug response prediction models built with male-dominated datasets (does not capture effects unique to one sex), exhibit clear evidence of bias (Wolfert et al., 2017). Bias in AI algorithms does not only represent an inherent defect in model building but rather reflects structural problems in the processes of data generation and validation. Strategies for addressing bias include active participant recruitment, data augmentation, domain adaptation methods that leverage learned knowledge from one population to another, and mandatory validation of models on several independent cohorts (Topol et al., 2019).

13.4 Reproducibility Crisis: Missing Code and Data

The reproducibility crisis in artificial intelligence-based biology is characterized by the frequent failure of other scientists to reproduce the results of the study based on the same data or methodology as used by the original researchers. This problem arises from the lack of sharing of the entire code, weights of the machine learning model, and pre-processed datasets; according to some sources, less than 30% of AI biology papers share their code completely, while even fewer provide specifications of the exact computing environment (Jumper et al., 2021). Even when the code is made available for replication purposes, slight deviations in software libraries (for instance, TensorFlow 1.x versus TensorFlow 2.x), random seeds, computing architecture (whether using a GPU or CPU), and pre-processing parameters (such as normalization techniques and handling missing values) can result in vastly different outcomes. Besides, many AI models are inherently stochastic since they rely on randomization in training algorithms (Rolnick et al., 2022). The inefficiency in this regard amounts to considerable wasted effort, since different teams are replicating results that might be erroneous and specific to one context. In addition to being inefficient, it is also detrimental to the overall pace of scientific discovery, which is hindered due to the inability to build on previous work effectively, and the loss of confidence in artificial intelligence biological discoveries by clinicians, policymakers, and members of the public (Greener et al., 2022). Solutions can be found in requiring the submission of code and data as part of the submission process for publication (imposed by journals and grant agencies), employing technologies such as Docker and Singularity that enable capturing of the entire environment in which the computation was performed, archiving of the code and data in publicly available and enduring storage services like GitHub, Zenodo, or Figshare with DOI identification, and following the FAIR principle (findable, accessible, interoperable, reusable) (Jobin et al., 2019).

13.5 Computational Cost: The Digital Divide

The cost of training advanced AI models in biological sciences is escalating at an exponential rate, thus resulting in a digital divide between the highly funded laboratories in developed countries and poorly equipped laboratories in developing and emerging countries. For example, training a single model like AlphaFold took about 128 TPUv3 chips non-stop for weeks, implying that millions of dollars' worth of computational power was needed, something only available to a few tech companies and top universities (Khera et al., 2018). On the other hand, even fine-tuning already trained protein language models, such as ESM-2 and ProtBERT, on a particular biological problem will take at least a high-end GPU (e.g., NVIDIA

A100 or H100), which can cost anywhere between 10,000 and 30,000. This makes these costs prohibitive for researchers interested in pathogens, crops, or threatened species native to their locality. The upshot is the geographic clustering of AI biology research in the Global North, where models have been trained using datasets representative of the genomic composition of European ancestry populations and agriculture in temperate climates that do not generalize well elsewhere (Ching et al., 2018). Such a digital divide not only maintains the unevenness of scientific capability worldwide but also produces gaps in scientific understanding itself. Possible solutions are: (1) designing and advocating for small yet effective model architectures that yield comparable accuracy with lesser computational requirements; (Dixit et al., 2016) and (2) offering computing credits to scientists working in lower and middle-income countries *via* initiatives such as Google Cloud Research Credits, AWS Cloud Credits for Research, and Microsoft AI for Good; (3) developing and hosting libraries of trained models that could be freely downloaded and refined using common laptop computers or budget GPU systems; and (4) forming international research partnerships that allow for computation to be shared. The advantages of using AI in biological sciences would become exclusively the advantage of the privileged without such a conscious effort to democratize access to the technology (Webb, 2018).

14. FUTURE PERSPECTIVES AND SUSTAINABLE DEVELOPMENT GOALS

Artificial Intelligence (AI) is anticipated to be crucial in shaping the future of biological sciences and will significantly contribute to achieving the United Nations' Sustainable Development Goals (SDGs).At the health front in the general worldwide context, the application of biological research in AI will abundantly help in predetermination of diseases, personalized medicines, development of vaccines, and early pandemic outbreak notification, contributing directly to the attainment of SDG 3 (Good Health and Well-being) (Zeevi et al., 2015). Within the agricultural sector/food system, AI will abundantly help in agricultural crop improvement, precision farming, management of the soil microbiome, contributing substantially to SDG 2 (Zero Hunger) and SDG 12 (Responsible Consumption and Production). Further, the application of AI within the biological field will abundantly help in making efficient, resource-effective, and evidence-based decisions in the biological disciplines (Larsson and Flach, 2022).

For the future, the amalgamation of AI with the fields of genomics, system biology, synthetic biology, and environment sciences would see an increased pace of innovations but would also see the onset of some challenges in the technical and societal aspects. The advancement of an explicable artificial intelligence system, the integration of multiple-omics data, and the establishment of global information-sharing platforms will be prioritized in future initiatives. These findings using AI will correspond with the principles of sustainable development within biological research, since the use of AI will serve as a significant facilitator of long-term sustainability (Davis et al., 2016).

15. CONCLUSION

Artificial intelligence in biology provides sustainable health, agriculture, and environmental management solutions. With artificial intelligence combined with the body, we can speed up the diagnosis of the disease, discovery of the drug, and individual medicine. In farming, the biological understanding of AI improves crop yield, resilience, and food supply and saves on resources. Environmental apps facilitate more ecosystems and biodiversity and climate change monitoring. These technologies facilitate effective utilization of data to forecast risks and optimize interventions. Responsible and ethical AI use is critical to provide protection and fairness. Interdisciplinary collaboration enhances innovation and practice. AI-powered biology will connect science and sustainability. Finally, it ensures a healthier world and a population.

REFERENCES

Aibar, S., González-Blas, C. B., Moerman, T., Huynh-Thu, V. A., Imrichova, H., Hulselmans, G., & Aerts, S. (2017). SCENIC: Single-cell regulatory network inference and clustering. *Nature Methods*, *14*(11), 1083–1086. DOI: 10.1038/nmeth.4463 PMID: 28991892

Akhtar, J., & Amin, N. A. S. (2011). A review on process conditions for optimum bio-oil yield in hydrothermal liquefaction of biomass. *Renewable & Sustainable Energy Reviews*, *15*(3), 1615–1624. DOI: 10.1016/j.rser.2010.11.054

Aliper, A., Plis, S., Artemov, A., Ulloa, A., Mamoshina, P., & Zhavoronkov, A. (2016). Deep learning applications for predicting pharmacological properties of drugs and drug repurposing using transcriptomic data. *Molecular Pharmaceutics*, *13*(7), 2524–2530. DOI: 10.1021/acs.molpharmaceut.6b00248 PMID: 27200455

Alpaydin, E. (2020). *Introduction to machine learning* (4th ed.). MIT Press.

Arango-Argoty, G., Garner, E., Pruden, A., Heath, L. S., Vikesland, P., & Zhang, L. (2018). Deep ARG: A deep learning approach for predicting antibiotic resistance genes from metagenomic data. *Microbiome*, *6*(1), 23. Advance online publication. DOI: 10.1186/s40168-018-0401-z PMID: 29391044

Bradley, P., Gordon, N. C., Walker, T. M., Dunn, L., Heys, S., Huang, B., & Iqbal, Z. (2015). Rapid antibiotic-resistance predictions from genome sequence data for *Staphylococcus aureus* and *Mycobacterium tuberculosis*. *Nature Communications*, *6*(1), 10063. Advance online publication. DOI: 10.1038/ncomms10063 PMID: 26686880

Carbonell, P., Hu, F., & Faulon, J.-L. (2020). Computational methods for pathway and strain design. *Current Opinion in Biotechnology*, *62*, 1–10. DOI: 10.1016/j.copbio.2019.08.003

Carbonell, P., Radivojević, T., & Garcia Martin, H.. (2019). An automated design–build–test–learn pipeline for enhanced microbial production of fine chemicals. *Communications Biology*, *2*(1), 66. Advance online publication. DOI: 10.1038/s42003-018-0076-9 PMID: 30271948

Ching, T., Himmelstein, D. S., Beaulieu-Jones, B. K., Kalinin, A. A., Do, B. T., Way, G. P., & Xie, W. (2018). Opportunities and obstacles for deep learning in biology and medicine. *Journal of the Royal Society, Interface*, *15*(141), 20170387. DOI: 10.1098/rsif.2017.0387 PMID: 29618526

Chuai, G., Ma, H., Yan, J., Chen, M., Hong, N., Xue, D., & Liu, Q. (2018). Deep CRISPR: Optimized CRISPR guide RNA design by deep learning. *Genome Biology*, *19*(1), 80. Advance online publication. DOI: 10.1186/s13059-018-1459-4 PMID: 29945655

Davis, J. J., Boisvert, S., Brettin, T., Kenyon, R. W., Mao, C., Olson, R., & Stevens, R. (2016). Antimicrobial resistance prediction in PATRIC and RAST. *Scientific Reports*, *6*(1), 27930. Advance online publication. DOI: 10.1038/srep27930 PMID: 27297683

De Groot, A. S., & Moise, L. (2007). Prediction of immunogenicity for therapeutic proteins: State of the art. *Current Opinion in Drug Discovery & Development*, *10*(3), 332–340. PMID: 17554860

Delgado-Baquerizo, M., Oliverio, A. M., Brewer, T. E., Benavent-González, A., Eldridge, D. J., Bardgett, R. D., & Fierer, N. (2018). A global atlas of the dominant bacteria found in soil. *Science*, *359*(6373), 320–325. DOI: 10.1126/science.aap9516 PMID: 29348236

Dixit, A., Parnas, O., Li, B., Chen, J., Fulco, C. P., Jerby-Arnon, L., Marjanovic, N. D., Dionne, D., Burks, T., Raychowdhury, R., Adamson, B., Norman, T. M., Lander, E. S., Weissman, J. S., Friedman, N., & Regev, A. (2016). Perturb-Seq: Dissecting molecular circuits with scalable single-cell RNA profiling of pooled genetic screens. *Cell*, *167*(7), 1853–1866.e17. DOI: 10.1016/j.cell.2016.11.038 PMID: 27984732

Emmert-Streib, F., Dehmer, M., & Haibe-Kains, B. (2014). Gene regulatory networks and their applications: Understanding biological and medical problems in terms of networks. *Frontiers in Cell and Developmental Biology*, *2*, 38. Advance online publication. DOI: 10.3389/fcell.2014.00038 PMID: 25364745

Eraslan, G., Avsec, Ž., Gagneur, J., & Theis, F. J. (2019). Deep learning: New computational modelling techniques for genomics. *Nature Reviews. Genetics*, *20*(7), 389–403. DOI: 10.1038/s41576-019-0122-6 PMID: 30971806

Esteva, A., Kuprel, B., Novoa, R. A., Ko, J., Swetter, S. M., Blau, H. M., & Thrun, S. (2017). Dermatologist-level classification of skin cancer with deep neural networks. *Nature*, *542*(7639), 115–118. DOI: 10.1038/nature21056 PMID: 28117445

Evans, R., O'Neill, M., Pritzel, A., Antropova, N., Senior, A., Green, T., & Hassabis, D. (2022). Protein complex prediction with AlphaFold-Multimer. *bioRxiv*. https://doi.org/DOI: 10.1101/2021.10.04.463034

Geyer, P. E., Holdt, L. M., Teupser, D., & Mann, M. (2017). Revisiting biomarker discovery by plasma proteomics. *Molecular Systems Biology*, *13*(9), 942. Advance online publication. DOI: 10.15252/msb.20156297 PMID: 28951502

Greener, J. G., Kandathil, S. M., Moffat, L., & Jones, D. T. (2022). A guide to machine learning for biologists. *Nature Reviews. Molecular Cell Biology*, *23*(1), 40–55. DOI: 10.1038/s41580-021-00407-0 PMID: 34518686

Hendriksen, R. S., Bortolaia, V., Tate, H., Tyson, G. H., Aarestrup, F. M., & McDermott, P. F. (2019). Using genomics to track global antimicrobial resistance. *Frontiers in Public Health*, *7*, 242. Advance online publication. DOI: 10.3389/fpubh.2019.00242 PMID: 31552211

Hughes, J. P., Rees, S., Kalindjian, S. B., & Philpott, K. L. (2011). Principles of early drug discovery. *British Journal of Pharmacology*, *162*(6), 1239–1249. DOI: 10.1111/j.1476-5381.2010.01127.x PMID: 21091654

Jobin, A., Ienca, M., & Vayena, E. (2019). The global landscape of AI ethics guidelines. *Nature Machine Intelligence*, *1*(9), 389–399. DOI: 10.1038/s42256-019-0088-2

Jumper, J., Evans, R., Pritzel, A., Green, T., Figurnov, M., Ronneberger, O., & Hassabis, D. (2021). Highly accurate protein structure prediction with AlphaFold. *Nature*, *596*(7873), 583–589. DOI: 10.1038/s41586-021-03819-2 PMID: 34265844

Jurtz, V., Paul, S., Andreatta, M., Marcatili, P., Peters, B., & Nielsen, M. (2017). NetMHCpan-4.0: Improved peptide–MHC class I interaction predictions integrating eluted ligand and peptide binding affinity data. *The Journal of Immunology : Official Journal of the American Association of Immunologists*, *199*(9), 3360–3368. DOI: 10.4049/jimmunol.1700893 PMID: 28978689

Khera, A. V., Chaffin, M., Aragam, K. G., Haas, M. E., Roselli, C., Choi, S. H., & Kathiresan, S. (2018). Genome-wide polygenic scores for common diseases identify individuals with risk equivalent to monogenic mutations. *Nature Genetics*, *50*(9), 1219–1224. DOI: 10.1038/s41588-018-0183-z PMID: 30104762

Kim, G. B., Kim, W. J., Kim, H. U., & Lee, S. Y. (2020). Machine learning applications in systems metabolic engineering. *Current Opinion in Biotechnology*, *64*, 1–9. DOI: 10.1016/j.copbio.2019.08.010 PMID: 31580992

Kitano, H. (2002). Systems biology: A brief overview. *Science*, *295*(5560), 1662–1664. DOI: 10.1126/science.1069492 PMID: 11872829

Larsson, D. J., & Flach, C.-F. (2022). Antibiotic resistance in the environment. *Nature Reviews. Microbiology*, *20*(5), 257–269. DOI: 10.1038/s41579-021-00649-x PMID: 34737424

LeCun, Y., Bengio, Y., & Hinton, G. (2015). Deep learning. *Nature, 521*(7553), 436–444. DOI: 10.1038/nature14539 PMID: 26017442

Liakos, K. G., Busato, P., Moshou, D., Pearson, S., & Bochtis, D. (2018). Machine learning in agriculture: A review. *Sensors (Basel), 18*(8), 2674. Advance online publication. DOI: 10.3390/s18082674 PMID: 30110960

Libbrecht, M. W., & Noble, W. S. (2015). Machine learning applications in genetics and genomics. *Nature Reviews. Genetics, 16*(6), 321–332. DOI: 10.1038/nrg3920 PMID: 25948244

Liu, D., Sun, D.-W., & Zeng, X.-A. (2018). Recent advances in the application of imaging techniques and machine learning in food quality evaluation. *Trends in Food Science & Technology, 72*, 106–117. DOI: 10.1016/j.tifs.2017.12.001

Liu, X., Faes, L., Kale, A. U., Wagner, S. K., Fu, D. J., Bruynseels, A., Mahendiran, T., Moraes, G., Shamdas, M., Kern, C., Ledsam, J. R., Schmid, M. K., Balaskas, K., Topol, E. J., Bachmann, L. M., Keane, P. A., & Denniston, A. K. (2019). A comparison of deep learning performance against health-care professionals in detecting diseases from medical imaging: A systematic review and meta-analysis. *The Lancet. Digital Health, 1*(6), e271–e297. DOI: 10.1016/S2589-7500(19)30123-2 PMID: 33323251

Mears, L., Stocks, S. M., Sin, G., & Gernaey, K. V. (2017). A review of control strategies for manipulating the feed rate in fed-batch fermentation processes. *Journal of Biotechnology, 245*, 34–46. DOI: 10.1016/j.jbiotec.2017.01.008 PMID: 28179156

Mittelstadt, B. D., Allo, P., Taddeo, M., Wachter, S., & Floridi, L. (2016). The ethics of algorithms: Mapping the debate. *Big Data & Society, 3*(2), 2053951716679679. DOI: 10.1177/2053951716679679

Mohanty, S. P., Hughes, D. P., & Salathé, M. (2016). Using deep learning for image-based plant disease detection. *Frontiers in Plant Science, 7*, 1419. Advance online publication. DOI: 10.3389/fpls.2016.01419 PMID: 27713752

Montesinos-López, O. A., Martin-Vallejo, J., Crossa, J., Gianola, D., Hernández-Suárez, C. M., Montesinos-López, A., & Singh, R. (2019). New deep learning genomic-based prediction model for multiple traits with binary, ordinal, and continuous phenotypes. *G3 (Bethesda, Md.), 9*(5), 1545–1556. DOI: 10.1534/g3.119.300585 PMID: 30858235

Nielsen, J., & Keasling, J. D. (2016). Engineering cellular metabolism. *Cell, 164*(6), 1185–1197. DOI: 10.1016/j.cell.2016.02.004 PMID: 26967285

Olden, J. D., Lawler, J. J., & Poff, N. L. (2008). Machine learning methods without tears: A primer for ecologists. *The Quarterly Review of Biology*, *83*(2), 171–193. DOI: 10.1086/587826 PMID: 18605534

Poplin, R., Chang, P.-C., Alexander, D., Schwartz, S., Colthurst, T., Ku, A., & DePristo, M. A. (2018). A universal SNP and small-indel variant caller using deep neural networks. *Nature Biotechnology*, *36*(10), 983–987. DOI: 10.1038/nbt.4235 PMID: 30247488

Pulendran, B., & Ahmed, R. (2011). Immunological mechanisms of vaccination. *Nature Immunology*, *12*(6), 509–517. DOI: 10.1038/ni.2039 PMID: 21739679

Rajkomar, A., Oren, E., Chen, K., Dai, A. M., Hajaj, N., Hardt, M., & Dean, J. (2018). Scalable and accurate deep learning with electronic health records. *NPJ Digital Medicine*, *1*(1), 18. Advance online publication. DOI: 10.1038/s41746-018-0029-1 PMID: 31304302

Reel, P. S., Reel, S., Pearson, E., Trucco, E., & Jefferson, E. (2021). Using machine learning approaches for multi-omics data analysis: A review. *Biotechnology Advances*, *49*, 107739. DOI: 10.1016/j.biotechadv.2021.107739 PMID: 33794304

Reichstein, M., Camps-Valls, G., Stevens, B., Jung, M., Denzler, J., Carvalhais, N., & Prabhat, . (2019). Deep learning and process understanding for data-driven Earth system science. *Nature*, *566*(7743), 195–204. DOI: 10.1038/s41586-019-0912-1 PMID: 30760912

Roden, D. M., McLeod, H. L., Relling, M. V., Williams, M. S., Mensah, G. A., Peterson, J. F., & Van Driest, S. L. (2019). Pharmacogenomics. *Lancet*, *394*(10197), 521–532. DOI: 10.1016/S0140-6736(19)31276-0 PMID: 31395440

Rolnick, D., Donti, P. L., Kaack, L. H., Kochanski, K., Lacoste, A., Sankaran, K., & Bengio, Y. (2022). Tackling climate change with machine learning. *ACM Computing Surveys*, *55*(2), 1–96. DOI: 10.1145/3485128

Saha, S., & Raghava, G. P. S. (2006). Prediction of continuous B-cell epitopes in an antigen using recurrent neural network. *Proteins*, *65*(1), 40–48. DOI: 10.1002/prot.21078 PMID: 16894596

Senior, A. W., Evans, R., Jumper, J., Kirkpatrick, J., Sifre, L., Green, T., & Kavukcuoglu, K. (2020). Improved protein structure prediction using potentials from deep learning. *Nature*, *577*(7792), 706–710. DOI: 10.1038/s41586-019-1923-7 PMID: 31942072

Shen, R., Mo, Q., Schultz, N., Seshan, V. E., Olshen, A. B., Huse, J., & Taylor, B. S. (2012). Integrative subtype discovery in glioblastoma using iCluster. *PLoS One*, *7*(4), e35236. DOI: 10.1371/journal.pone.0035236 PMID: 22539962

Smith, K. P., Kang, A. D., & Kirby, J. E. (2018). Automated interpretation of blood culture Gram stains by use of a deep convolutional neural network. *Journal of Clinical Microbiology*, *56*(3), e01521–e17. DOI: 10.1128/JCM.01521-17 PMID: 29187563

Sundaram, L., Gao, H., Padigepati, S. R., McRae, J. F., Li, Y., Kosmicki, J. A., & Birnbaum, D. P. (2018). Predicting the clinical impact of human mutation with deep neural networks. *Nature Genetics*, *50*(8), 1161–1170. DOI: 10.1038/s41588-018-0167-z PMID: 30038395

Topol, E. J. (2019). High-performance medicine: The convergence of human and artificial intelligence. *Nature Medicine*, *25*(1), 44–56. DOI: 10.1038/s41591-018-0300-7 PMID: 30617339

Tripathi, B. M., Stegen, J. C., Kim, M., Dong, K., Adams, J. M., & Lee, Y. K. (2018). Soil pH mediates the balance between stochastic and deterministic assembly of bacteria. *The ISME Journal*, *12*(4), 1072–1083. DOI: 10.1038/s41396-018-0082-4 PMID: 29515169

Truong, D. T., Franzosa, E. A., Tickle, T. L., Scholz, M., Weingart, G., Pasolli, E., & Segata, N. (2015). MetaPhlAn2 for enhanced metagenomic taxonomic profiling. *Nature Methods*, *12*(10), 902–903. DOI: 10.1038/nmeth.3589 PMID: 26418763

Vamathevan, J., Clark, D., Czodrowski, P., Dunham, I., Ferran, E., Lee, G., & Zhao, S. (2019). Applications of machine learning in drug discovery and development. *Nature Reviews. Drug Discovery*, *18*(6), 463–477. DOI: 10.1038/s41573-019-0024-5 PMID: 30976107

Van Klompenburg, T., Kassahun, A., & Catal, C. (2020). Crop yield prediction using machine learning: A systematic literature review. *Computers and Electronics in Agriculture*, *177*, 105709. DOI: 10.1016/j.compag.2020.105709

Wahl, B., Cossy-Gantner, A., Germann, S., & Schwalbe, N. R. (2018). Artificial Intelligence (AI) and global health: How can AI contribute to health in resource-poor settings? *BMJ Global Health*, *3*(4), e000798. DOI: 10.1136/bmjgh-2018-000798 PMID: 30233828

Webb, S. (2018). Deep learning for biology. *Nature*, *554*(7693), 555–557. DOI: 10.1038/d41586-018-02174-z

Wolfert, S., Ge, L., Verdouw, C., & Bogaardt, M. J. (2017). Big data in smart farming: A review. *Agricultural Systems*, *153*, 69–80. DOI: 10.1016/j.agsy.2017.01.023

Yang, K. K., Wu, Z., & Arnold, F. H. (2019). Machine-learning-guided directed evolution for protein engineering. *Nature Methods*, *16*(8), 687–694. DOI: 10.1038/s41592-019-0496-6 PMID: 31308553

Zeevi, D., Korem, T., Zmora, N., Israeli, D., Rothschild, D., Weinberger, A., & Segal, E. (2015). Personalized nutrition by prediction of glycemic responses. *Cell*, *163*(5), 1079–1094. DOI: 10.1016/j.cell.2015.11.001 PMID: 26590418

Zhang, Q. C., Petrey, D., Garzón, J. I., Deng, L., & Honig, B. (2012). Structure-based prediction of protein–protein interactions on a genome-wide scale. *Science*, *338*(6110), 1107–1112. DOI: 10.1126/science.1224912 PMID: 23086998

Zhang, Y., Jiang, B., Zhang, Z., & Li, X. (2019). Machine learning in infectious disease epidemiology and outbreak prediction. *Clinical Microbiology Reviews*, *32*(4), e00014–e00019. DOI: 10.1128/CMR.00014-19

Zhavoronkov, A., Ivanenkov, Y. A., Aliper, A., Veselov, M. S., Aladinskiy, V. A., Aladinskaya, A. V., & Aspuru-Guzik, A. (2019). Deep learning enables rapid identification of potent DDR1 kinase inhibitors. *Nature Biotechnology*, *37*(9), 1038–1040. DOI: 10.1038/s41587-019-0224-x PMID: 31477924

Zhou, J., & Troyanskaya, O. G. (2015). Predicting effects of noncoding variants with deep learning-based sequence model. *Nature Methods*, *12*(10), 931–934. DOI: 10.1038/nmeth.3547 PMID: 26301843

Zitnik, M., Nguyen, F., Wang, B., Leskovec, J., Goldenberg, A., & Hoffman, M. M. (2019). Machine learning for integrating data in biology and medicine: Principles, practice, and opportunities. *Information Fusion*, *50*, 71–91. DOI: 10.1016/j.inffus.2018.09.012 PMID: 30467459

GLOSSARY

CNN: Convolutional neural network; for image data
GAN: Generative adversarial network; for generating new data
GNN: Graph neural network; for network data
RNN: Recurrent neural network; for sequence data
SHAP/LIME: Local explanation methods for individual predictions
VAE: Variational autoencoder; for generating and compressing data
XAI: Explainable artificial intelligence; methods to interpret model decisions

Chapter 6
Artificial Intelligence as a Digital Frontier in Biological Conservation

Rajat Singh
http://orcid.org/0000-0003-0474-2442
Adamas University, Kolkata, India

Nandini Katare
Kristu Jayanti College, Bangalore, India.

Priyanka Soni
Nims University, Jaipur, India

Shruti Mishra
Adamas University, Kolkata, India

Dipak Kumar Mahida
http://orcid.org/0000-0003-1299-4623
Gujarat University, Ahmedabad, India

Vishal M. Makwana
http://orcid.org/0000-0001-6343-9449
Veer Narmad South Gujarat University, Surat, India

ABSTRACT

Climate change, habitat loss, and declining biodiversity present critical challenges to conservation. This chapter explores how Artificial Intelligence (AI) and digital technologies offer transformative solutions for these issues. It highlights AI tools such as machine learning, deep learning, computer vision, and natural language processing for species monitoring, habitat mapping, and predictive modeling. Applications including GIS, remote sensing, bioacoustics, and camera traps enable large-scale ecological assessments. AI-driven innovations in wildlife tracking, anti-poaching, and ecosystem restoration enhance precision and efficiency. However, challenges remain regarding data quality, technical feasibility, ethical standards, and cost-effectiveness. The chapter emphasizes the importance of cross-disciplinary

DOI: 10.4018/979-8-2600-2691-5.ch006

collaboration among technologists, conservationists, and policymakers to overcome these barriers and advance sustainable conservation practices through adaptive, technology-integrated strategies.

1. INTRODUCTION

Artificial Intelligence is transforming the entire globe since it has been extensively utilized in almost all spheres of life. The use of the word "Artificial Intelligence" was first done by John McCarthy in 1956. He defined artificial intelligence as "the field of study focusing on creating machines which are well equipped with the ability to think logically and adapt to learning. This field emphasizes the examination and creation of intelligent agents which in turn are capable of perceiving their surroundings and making decisions that optimize their likelihood of achieving desired outcomes. In other words, AI can be described as the capacity to simultaneously store two distinct thoughts and still maintain the ability to function (Singh et al., 2013). The scientific objective of AI is to comprehend intelligence through the construction of computer programs that demonstrate intelligent behavior by employing symbolic inference, or reasoning within the machine.

Artificial intelligence is the analysis of intelligent conduct. The ultimate objective of developing artificial intelligence is to theorize intelligence that explains the actions of naturally existing intelligent beings, which in turn informs the development of artificial entities capable of exhibiting intelligent behavior (Abraham, 2005).

1.1. Overview of AI technologies

Artificial intelligence (AI) has had a significant impact on our lives and it will remain further for a setting. This area is predominantly being utilized in various domains and situations, including the concept of driverless vehicles, healthcare, media, finance, industrial robotics, and internet services. This has led to its profound incorporation into the economy and society with improved efficiency and is more advantageous (Huang et al., 2022). Ever since Alan Turing first expressed the concept of "thinking machines" in 1950, research on artificial intelligence (AI) has made significant progress across various disciplines. Emerging technologies have revolutionized the methods of instruction and acquisition of knowledge in the field of education.

AI technology has flourished a lot and has the ability to provide personalized learning, dynamic evaluations, and meaningful interactions in online, mobile, or mixed learning experiences. The growing utilization of artificial intelligence in education necessitates multidisciplinary methodologies, despite the fact that the majority of AI research is mostly conducted within STEM disciplines (Zhang & Aslan, 2021). **Figure 1** illustrates the classification of AI technologies, providing a structured overview of various AI approaches and their applications.

Figure 1. Showing Classification of AI Technologies (Sarker, I. H. 2022).

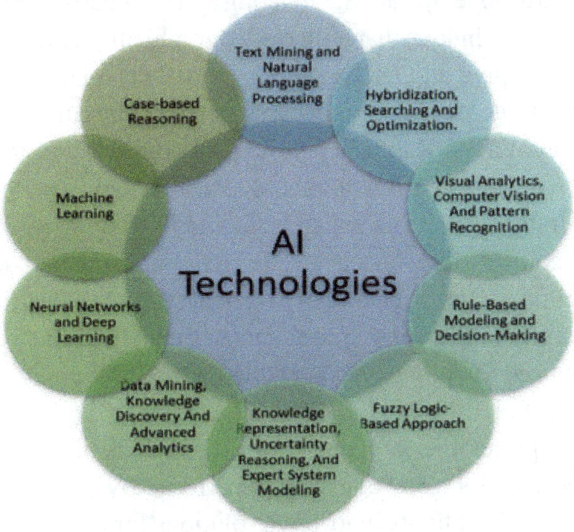

1.2. Importance of Biological Conservation

Biological Conservation refers to the application of science to conservation problems addressing the biology of species, communities and the ecosystem that are perturbed either directly or indirectly by human or other agents. Its goal is to provide principles and tools for preserving biological diversity by dealing with threats to biodiversity and with preserving the biological and genetic diversity of living beings.

Biological conservationists face various challenges, such as habitat loss, the effects of climate change, the invasion of non-native species, overexploitation, and pollution. Habitat destruction, resulting from deforestation, urbanization, and agricultural expansion, poses a substantial threat to biodiversity. The result is the division and deterioration of ecosystems (Foley et al., 2005). Climate change exacerbates these consequences, modifying habitat conditions and therefore compelling animals to undergo rapid migration or adaptation (Parmesan, 2006). The invasion of species causes disturbances in native ecosystems by outcompeting or preying on indigenous species (Simberloff et al., 2013). Overexploitation of resources including excessive fishing and illegal hunting can deplete the population at a greater rate which directly causes the biodiversity loss (Hilborn et al., 2020). At the same time, Pollution is another major challenge that is deteriorating the ecosystem and posing a menace to diverse species (Gall & Thompson, 2015).

Technology helps in overcoming the challenges faced in biological conservation. With advent of Technology, Remote sensing devices are now used in wildlife monitoring. This is usually done by Satellite imaging and using drone cameras (Pettorelli et al., 2014). Another technique that utilises Geographic Information Systems (GIS) is also helpful in mapping and analysing 3-D data. This method ensures that such areas are marked and conservation processes started (Margules & Pressey, 2000). DNA barcoding and CRISPR enhance biological conservation through increased identification of species and efficient management of their gene pools (Seddon et al., 2014). Two strategies that can enhance the monitoring of wildlife include bio-acoustics and video traps. These strategies prevent poaching and other illicit activities that harm animals (Burton et al., 2015). Today, the use of artificial intelligence cannot be ignored. In integrating big data and artificial intelligence, AI, predictive modelling can be used to optimize conservation efforts in conservation biology, as shown by Christin et al. (2019) (Christin et al., 2019).

This chapter explores the integration of artificial intelligence across various disciplines to advance biological conservation efforts. By bridging ecological data with AI technology, it highlights innovative solutions for addressing complex conservation challenges and promoting sustainable ecosystems.

2. SYNERGY BETWEEN AI AND CONSERVATION EFFORTS

Artificial Intelligence (AI) can be used to solve the complex issues related to conservation of biodiversity and ecosystems. The amalgamation of AI and Biodiversity has resulted a notable progress in the surveillance, safeguarding and administration of animals and substrate as nowadays, machine learning and data analytics are being used to improve conservation initiatives. AI computations can be used to extract and effectively classify data obtained from camera traps, ramble cameras, and partisan photography to enable the identification and monitoring of endangered species, estimation of habitat changes, and detection of criminal activities like poaching and deforestation. Due to such results, one of the major factors for reducing the time and effort required in analysis is that the manual analysis of such cases is done using various AI-driven tools instead (Norouzzadeh et al., 2018).

AI can be used along with predictive modeling, which allows conservationists to project changes in species distribution and assess climate change impacts on biodiversity. Apparently, machine learning models have already been used for predicting the geographic range of species under different climate scenarios, promoting adaptation towards proactive conservation strategies (Carlson et al., 2017). Artificial intelligence insightful has a very critical role in developing community engagement and promoting citizen science activities. The AI-powered mobile applications allow consumers to take part in wildlife monitoring by reporting sightings and uploading images, which will be processed then by the algorithms (Sullivan et al., 2009).

AI applications in conservation are not without challenges. Indeed, the feasibility and goodness of AI applications demand careful attention to ethical concerns, data security, and building interest collaboration. Nevertheless, the integration of AI into conservation holds tremendous potential to utterly change the conserving and sustaining biodiversity.

3. APPLICATIONS OF AI IN WILDLIFE MONITORING

AI is emerging as a revolutionary approach in wildlife investigations mainly monitoring and providing ground-breaking solutions for species identification and tracking, population estimation, habitat mapping, and predictive modelling of population trends.

3.1. Species Identification and Tracking

AI-driven technologies are transforming the year-old concepts of species identification and tracking which were labor-intensive, time consuming and often invasive. Artificial Intelligence, particularly through machine learning algorithms offers a non-invasive, efficient alternative technique for wildlife protection and conservation. In a study conducted by Norouzzadeh et al., 2018, Artificial Intelligence was used to track the wildlife species and to examine images and videos taken from trap cameras. This enabled automatic identification of the species thus significantly reducing the time required for interpretation of results thus increasing the accuracy of species identification (Norouzzadeh et al., 2018).

3.2. Camera traps and image recognition

Nowadays, monitoring of wildlife activities is done using Camera traps. These camera traps are well equipped with AI-driven image recognition systems which can process thousands of images simultaneously in order to detect and identify particular species. AI-driven picture recognition frameworks can be actually realized in challenging situations, for example, low light or thick foliage. Such a system utilizes Convolutional Neural Systems, a sort of profound learning count. For instance, the Depiction Serengeti undertaking effectively utilized AI to group millions of camera trap pictures, indicating the potential this innovation holds for large-scale untamed life checking.

3.3. Acoustic monitoring and sound Investigation

Manufactured Insights is making noteworthy advance in the field of acoustic observing and sound examination. A few species, like winged creatures and marine warm-blooded creatures, can be more viably checked by listening to their vocalizations instead of depending exclusively on visual perceptions. Counterfeit knowledge calculations have the capacity to examine enormous sets of acoustic recordings in arrange to recognize and categorize sounds that are particular to distinctive species. This has been workable in following fowl populaces and also in following cretaceous developments within the seas. Two sorts of machine learning models are Covered up Markov Models and Profound Neural Networks, which could be used to separate between sounds made by distinctive species and assess biodiversity in numerous living spaces.

3.4. Population Estimation and Habitat Mapping

Estimation of population and habitat Mapping is an essential element in wildlife conservation. These models are enabled by AI by integrating data from multiple sources, including camera traps, acoustic sensors, and ancillary symbolism. Machine learning algorithms with high accuracy can be implemented to carry out the population sizes and mapping of species. For example, AI-driven tools estimate the population of African elephants by analyzing aerial survey images, thereby essentially increasing the accuracy and efficiency of traditional monitoring techniques.

3.5. Satellite imagery and remote sensing

Procedures like inaccessible detecting and imaging through obsequious offers effective artificial intelligence apparatuses for observing natural life territories and natural changes. Such broad sets of lackey information can be handled well utilizing AI calculations that can encourage be utilized in distinguishing proof of arrive covers changes, corruption of environment, and deforestation driving to exhaustion of natural life. Such data can be used to learn about the effects of anthropogenic activities and also help in planning strategies for biological conservation. Likewise, AI now led remote sensing techniques can particularly be used in tracking the migration patterns of animals thereby identifying the crucial habitat for endangered species (Wang et al., 2020).

3.6. Predictive modeling for population trends

AI can be used in predictive Modeling in order to forecast population trends. It can also be used to identify the potential impact of environment changes that are inflicted on wildlife. AI models are capable of predicting future population dynamics and can determine potential threats to the endangered species by incorporating certain environmental variables and historical data. These models inculcate to design a proactive approach towards biological conservation and can mitigate the effects caused by climate change and habitat loss (Urban et al., 2016).

4. AI IN ECOSYSTEM MANAGEMENT

In the case of Ecosystem management, Artificial intelligence can be a highly effective tool and it can enhance the efficacy of habitat restoration, landscape analysis, control on invasion of non-native species etc.

4.1. Habitat Restoration

In the field of Habitat restoration, AI technologies can transform the concept of rehabilitation of the ecosystem using data driven insights and automated solutions. By implementing ML Algorithms, vast data sets from satellite imagery, sensor networks and historical records can be easily analyzed in order to access the alarming areas so that restoration efforts can be prioritized (LeCun et al., 2015). Certain remote sensing devices and AI Powered drones can also be used. These tools enable the conservationists to make decisions related to restoration activities, thus optimizing allocation of resources and maximizing ecological benefits (Jensen, 1996). At the same time, the impacts of different restoration strategies can be influenced by AI models resulting in refined approaches and predicted outcomes even before actually implementing them.

4.2. AI-driven landscape analysis

The detailed map of land cover, types of vegetation and connectivity of habitat can be analyzed using certain machine learning techniques like convolutional neural networks (CNNs) and generative adversarial networks (GANs). These techniques extract the satellite imagery and LiDAR data offering into insights of ecosystem and biodiversity analysis. Complex ecological data can be easily processed using AI driven algorithms in no time. (Xue & Su, 2017). These algorithms can determine the changes in the landscape thus, alarming the early signs of degradation and land depletion. It can also trace out the critical habitat biodiversity hotspots showing a path to conservationist (Pimm et al., 2014).

4.3. Optimizing restoration strategies

Habitat monitoring can be achieved using AI design which can optimize the restoration strategies. By examining the historical records and modeling different scenarios, restoration plans can be designed in such a manner that it can maximize the overall ecological outcomes. One such example is refining restoration actions using reinforcement learning algorithms (Silver et al., 2017). Furthermore, the selection of plant species and other biotic components best suited for a certain habitat and specific environmental condition can be achieved using AI. By using predictive models, the interaction of these species with the restored ecosystem can be determined thus signifying the need of restoration strategies (Tilman et al., 2014). Real time monitoring and adaptive management can be facilitated using sensor networks and automated data analysis. This further helps in keeping a watch on the progress of restoration hence, adjusting the strategies as and when required (Hampton et al., 2013).

4.4. Invasive Species Control

Controlling the Invasion of non-native species is a serious challenge as this invasion can cause a devastating effect on native biodiversity and ecosystem. AI driven tools can be implemented in controlling this invasion by monitoring it using remote sensing techniques and other means thus contributing in effective management. In order to conserve biodiversity, control on invasive species is the need of the hour.

4.5. Detection and monitoring of invasive species

Prevention of spread of invasive species and mitigating their impact on native biodiversity can be monitored and detected using Artificial Intelligence. AI-powered cameras and drones which in turn can easily identify the invasive species using image recognition and implementing machine learning algorithms even in remote or inaccessible areas (Ahmed & Echi, 2021). The environmental DNA (eDNA) samples can also be analysed using Machine learning algorithms providing a non-invasive approach in species identification (Jerde et al., 2011). As a result, rapid and large-scale monitoring can be achieved which can further reduce the need for intensive human labour (Van der Wal et al., 2015).

4.6. Predictive analytics for management plans

Artificial intelligence-based predictive analytics gives the opportunity for preventive strategies in ecosystem management by providing forecasts of likelihoods on invasive species spread and future ecological conditions. Informed by climate, land-use, and species interaction data, AI models can project how ecosystems would respond to the various management actions and environmental changes. Such predictive models can be used by conservationists to devise adaptive management strategies in planning for mitigating and reducing the potential impacts of invasive species. AI can detect highly vulnerable areas to invasions and then suggest areas of specific control measures that improve resource allocation and increase the chances of effectively managing the situation.

5. AI AND CLIMATE CHANGE MITIGATION

Climate change is viewed as the leading driver of biodiversity loss. There is no denying that to address the threat of climate change, there is need to focus on mitigation and adaptation strategies. The meaning and connotation of this are respectively mitigation by reduction of emissions and adaptation by preparing for inevitable impacts.

Both are multi-faceted, complex issues. Land use, along with energy, transport, buildings, and industry, can play an instrumental role in mitigating greenhouse gas emissions. But adaptation must also include building resilience and managing disasters in an informed manner regarding the climate and weather extremes. The wide range of issues might be viewed as a chance: there are numerous avenues to make a difference. Machine learning (ML) has gained widespread recognition as a very effective method for driving technological advancements in recent years. Although there has been an increase in the use of machine learning (ML) and artificial intelligence (AI) to address social and global issues, there is still a need for a focused effort to determine the most effective ways to utilize these technologies in combating climate change. Numerous machine learning practitioners desire to take action, but lack clarity on how to proceed. Conversely, numerous fields have been aggressively soliciting help from the machine learning community.

5.1. Predictive Climate Models

Biologists create prognostic models to foresee the potential impact of environmental changes on the future characteristics of species and ecosystems. Several models have been created to comprehend the effects of climate change, however, forecasting biological reactions still poses a challenge. One reason is that the majority of biodiversity change projection models fail to consider fundamental dynamics such as population shifts, species interactions, and evolution. Instead, these models rely on extrapolating correlations between the present ranges of species and climate. Mechanisms have been crucial in facilitating previous and current biotic reactions to climate change. While studying climate science, predictive climate models can offer better understanding in describing future climatic conditions.AI has enhanced these models in terms of accuracy and predictive power by the use of ML algorithms to analyze large amounts of climatic data. Meanwhile, it goes beyond traditional statistical models in the integration of complex nonlinear relationships of climate data. Scientists have just now developed and introduced ClimaX, a modular and customizable deep-learning model aimed particularly at the weather and climate science domain.

It can be trained on the basis of a large variety of variables, spatial and temporal extents, and including physical principles.

This includes tasks that involve atmospheric variables and spatio-temporal scales that were not previously seen during the initial training phase. When compared to current data-driven baselines, the ClimaX model exhibits enhanced performance on benchmarks for weather forecasting and climate projections (Nguyen et al., 2023).

5.2. Forecasting environmental changes

The capacity of AI to process and analyze extensive datasets has fundamentally transformed the field of environmental forecasting. Machine learning algorithms have the ability to forecast environmental changes by examining past data and detecting patterns that could suggest upcoming developments. AI models have been used to predict the propagation of invasive species and diseases, allowing for prompt intervention and mitigation strategies (Wang et al., 2020). In addition, AI has enabled the incorporation of other data sources, including climate data, socio-economic aspects, and ecological characteristics, to produce comprehensive forecasts. A comprehensive approach is crucial for comprehending the diverse effects of environmental changes and developing efficient conservation policies.

5.3. Impacts on biodiversity and ecosystems

AI models have been employed to predict the distribution of species under various climatic scenarios, assisting conservationists in identifying species that are vulnerable and establishing conservation priorities. Machine learning algorithms have been employed in ecological niche modeling, particularly in predicting shifts in species distributions through the analysis of climate projections (Elith & Leathwick, 2009). Camera traps and acoustic sensors, fitted with artificial intelligence algorithms, locate and monitor wildlife populations in real-time. It gives a most updated information on the occurrence and behavior of various species.

Therefore, through this technology, researchers have contributed greatly to studying elusive and endangered species that directly improve the effectiveness of conservation effort.

Monitoring of Carbon Sequestration: Artificial intelligence-processed remote sensing data has also been used in monitoring carbon sequestration processes in forests. With the help of AI algorithms, above-ground biomass estimation and storage of carbon could be assessed very accurately. This will help in assessing the health of the forest and its potential for carbon storage.

Deep learning algorithms assessed LiDAR Light Detection and Ranging data, which represented a high accuracy level of assessment regarding forest structure and biomass. In Coomes et al., 2017, the following has been stated: predictive models were created with the use of AI to examine the potential of various ecosystems to store carbon. ALS-based discrete return is considered to be one of the most advanced methods for mapping carbon in tropical forests.

This technique produces accurate 3D point clouds of forest structure, thus provides a way to precisely measure aboveground carbon density. Mapping these forests is such a big challenge due to their incredible height. Two models are compared for their performance in mapping aboveground carbon density in our study: An area-based approach by Asner & Mascaro (2014) had been used by many in the Neotropics (Asner & Mascaro, 2014).

The second one is based on a different approach using a new algorithm called itcSegment. Again, one of the important relationships in the study found was between forest floor area measured against ALS with percentage of canopy gap. Based on these findings, it can be concluded that tree-centric approach does not perform well in the estimation of ACD accurately with respect to the best area-based model. The method gives an RMSE of 18%, while the area-based model had a lower RMSE of 13%.

The focus on quantifying individual tree biomass makes tree-centric modeling quite an attractive approach. Nevertheless, area-based modeling is the chosen methodology in the light of existing challenges to the accurate identification of understory trees and the estimation of biomass from canopy dimensions.

5.4. Tracking carbon sinks

AI and ML have turned out to be enabling technologies in computing, which have become very instrumental in revolutionizing the widespread management and control of carbon sinks. Their application to carbon sink management avails incomparable capabilities for the analysis of large data, making of forecasts, and consequent insights for decision-making.

This innovation enhances our capacity to offset, through good management of key carbon storage areas, the impacts of climate change for a healthy planet. Real-time monitoring will establish the health and efficiency of carbon sinks. At the same time, AI can improve the efficiency of man-made carbon capture and storage technologies complementing the natural sinks. Machine learning algorithms can make such predictions of CCS systems through the identification of the most favorable conditions for carbon capturing and storing (Schaefer, 2019).

5.5. AI in reforestation and afforestation efforts

Artificial intelligence and machine learning are high-power computing techniques that revolutionize the management and control of widespread carbon sinks. The second contribution deals with the use of these technologies in carbon sink management to deliver unparalleled abilities in the analysis of large amounts of data, making forecasts, and deriving insights that can inform decision-making. This innovation enhances our capacity to combat the adverse effects of climate change by assuring long-term sustainable management of some of the key carbon storage areas leading to a healthier planet. The ability to monitor in real time is paramount in assessing the health and efficiency of the carbon sinks. On the other hand, AI can also support human-made carbon capture and storage technology together with natural carbon sinks. Machine learning algorithms could predict performance of CCS systems using identification of optimal conditions for carbon capture and storage (Schaefer, 2019). By using AI algorithms, the overall growth and survival rates of trees can be predicted which offers real time data (Jump et al., 2010). In order to improve the effectiveness of forestry initiatives, automated planting systems can be created using AI. This can be achieved by means of AI powered drones which can efficiently plant seed in targeted areas and are more cost effective when compared to human labor (Meyerson & Mooney, 2007).

6. ENHANCING CONSERVATION POLICIES AND STRATEGIES

6.1. Data-Driven Decision Making

Species distribution maps are crucial for evaluating the likelihood of extinction and directing conservation initiatives. However, the majority of these range maps are obtained from professional sources, yet they have recognized accuracy flaws or are created using excessively intricate modeling methods. Therefore, the inclusion of data-driven alternatives that are both accessible and credible is a valuable addition to the spatial conservation toolset.

In biological conservation, assessing large datasets is done by data-driven decision making which is based on Machine learning algorithms and data analytics. The devices for remote sensing, in conjunction with AI algorithms, track deforestation rates, the growth of ecosystems, climate change, and invasive alien species. These are technologies that help in effective planning for biological conservation (Wegmann et al., 2016).

6.2. AI in policy formulation

It improves conservation policymaking through artificial intelligence-driven prediction and scenario analysis. Such technologies model what would be the potential outcome of the various policy decisions, hence helping policymakers in choosing efficient tactics.AI-driven models have been applied in predicting the spread of invasive species and developing precise intervention tactics.

Artificial intelligence integrated into the policy-decision-making processes could cause the governments or organizations to move in a balance with regard to conservation goals and socio-economic factors for making better decisions.

6.3. Case studies of successful implementations

Artificial intelligence application in nature conservation has made many successful examples. An excellent example is the SMART (Spatial Monitoring and Reporting Tool) initiative that deploys artificial intelligence in analyzing data from patrols and sensors in efforts toward the deterrence of poaching and illegal logging (Critchlow et al., 2017).

AI also finds application in the monitoring of coral reefs through machine learning algorithms that label corals, analyze underwater images for their health, and, based on the analysis, help design strategies for conservation. This enhances the exactness and timeliness of how conservation strategies are undertaken (Beijbom et al., 2012). These case studies demonstrate how artificial intelligence can enhance the effectiveness of conservation efforts by providing accurate and actionable insights. Multiple case studies exemplify the effective implementation of artificial intelligence in the field of conservation.

6.4. Community and Stakeholder Engagement

Community and stakeholder involvement is crucial for biological conservation. AI helps public engagement in data collection and reporting. New technologies have transformed scientific data collection, preservation, analysis, and dissemination. As a result, many autonomous sensors will be deployed across large areas and durations to capture massive amounts of data in the next few years. Astronomy, meteorology, and natural observations depend on human observers. Researchers have employed AI to enable citizen observations for bird conservation and study. Bird enthusiasts will collect, organize, and save their bird sightings in a global database through eBird. eBird's growth as a birding tool has increased user engagement. Researchers, birdwatchers, and environmentalists worldwide are using eBird data to study avian biological patterns and the natural and human-induced variables that shape them. eBird has built a rich avian database with millions of observations every year through continuous expansion and refining. Conservation efforts must involve as many communities and stakeholders as feasible. Artificial intelligence systems can make public interaction easier by providing data collection and reporting platforms.

6.5. AI tools for public participation

Public engagement can be improved in conservation efforts using AI by rationalizing data submission and offering immediate feedback. AI-powered platforms like iNaturalist assist users in identifying and recording biodiversity, promoting a stronger bond between individuals and their natural surroundings, and making valuable contributions to global biodiversity databases (Van Horn et al., 2018). These solutions not only make data collecting easier but also improve public awareness and education regarding conservation issues.

6.6. Enhancing awareness and education through AI

It is with the advent of satellite imagery and digital technologies that there has been a wide expansion in the application of artificial intelligence with respect to species and animal conservation. It is evident that the transformative power of digital tools, such as Geographic Information Systems, Artificial Intelligence, blockchain, remote sensing, and the Internet of Things, has increased the efficiency of conservation projects drastically.

The unparalleled digitalization that allows data collection, processing, and distribution increases transparency and collaboration among stakeholders. Through this, it enhances awareness and education, further allowing better informed decision-making. AI-based remote sensing and image recognition techniques will be helpful in monitoring endangered species and their habitats, while camera traps and satellite image analyses can be used for animal tracking, habitat changes, and poaching activities.

7. ETHICAL CONSIDERATIONS AND CHALLENGES

7.1. Privacy Concerns

In biological conservation, Artificial Intelligence usually involves large scale data collection, which includes information about species, habitats, and seldom local communities as well. Safeguarding Right to privacy is very essential in order to prevent misuse or unauthorized access to personal data. In the recent studies, researchers have analyzed the possibility of drones encroaching on privacy and evoking trepidation in individuals who come across them in natural settings. This can be redressed by sophisticated techniques for data anonymization and secure protocols for storage, providing privacy protection. Butler and Meek conducted a legal review in 2013 of possible invasion of privacy by camera traps in an Australian setting.

7.2. Data collection and surveillance issues

Problems relating to the impact on ecosystems and local communities due to the increased use of AI in monitoring and data collection. The methods used for intrusive monitoring are potentially disruptive to wildlife behavior and habitats. Implementing guidelines for ethical data collection, respecting local laws and community consent, and minimizing environmental disturbance are crucial.

Satellite remote sensing provides a continuous and detailed collection of data over a long period of time, with a high frequency of sampling and wide coverage of different geographic areas. Its ability to provide worldwide and uninterrupted coverage, it can be encompassed by the methodologies employed in long-term monitoring programs and ecological field investigations. Satellite remote sensing data mostly focuses on vegetation and is commonly utilized for monitoring ecosystem function and structure, including Essential Biodiversity Variables (EBVs). However, the continuous advancement of methods for gathering and analyzing remote sensing data poses difficulties in consolidating time series data, but also presents possibilities for utilizing this data in biodiversity monitoring. In addition, although the quality of satellite imagery is advancing quickly, allowing for a wider variety of uses, the expensive nature and lengthy processing time required for data analysis may be a limitation. The scarcity of map outputs, despite regular data collection, highlights the importance of international and multidisciplinary techniques that utilize earth observation data. Furthermore, the generation of EBV databases from satellite data will necessitate synchronized efforts from data suppliers, experts in biodiversity and remote sensing, as well as policy makers (Proença, V.et. al, 2017).

7.3. Ensuring ethical use of AI technologies

Artificial intelligence techniques used in biological conservation must fit ethical values like respect of human and animal welfare, openness, and responsibility. Establishing clear rules and norms of behavior can help to control the application of AI tools in settings related to conservation (Garnett et al., 2018). Transparency and simplicity should define artificial intelligence systems and their decision-making procedures. Stakeholders must understand how artificial intelligence tools operate, the data they use, and the produced outcomes.

Explicit ways for making people or groups answerable for their deeds are crucial. Those developers, operators, and companies using artificial intelligence in conservation should answer for the outcomes of their AI systems. Applications of artificial intelligence in conservation should aim to produce favorable results, therefore enhancing the general state of ecosystems and the communities depending on them (Stahl, 2021).

7.4. Bias and Fairness in AI

In artificial intelligence, bias can accentuate already existing differences, therefore supporting discrimination against impoverished groups and limiting their access to fundamental services. The algorithmic biases of artificial intelligence can favor inequality and compromise efforts at conservation. Algorithmic design or biased training data might cause prejudices that treat particular species or communities unfairly. Regular audits of artificial intelligence systems, diversity in data representation, and inclusive decision-making processes are extremely crucial to lower prejudices and guarantee fair results (Gebru et al., 2021).

7.5. Addressing algorithmic biases

As artificial intelligence (AI) models progress from research to deployment, building the appropriate datasets and data pipelines becomes ever more challenging. Publicly available automated artificial intelligence model builders could currently excel in many different domains. Conversely, the design and sculpting of the data used to generate artificial intelligence largely rely on tailored hand work and greatly affect the believability of the model. Rigorously developed and approved conservation artificial intelligence systems are necessary to find and reduce biases. Using different datasets covering different ecological, social, and cultural settings will help to lower prejudices and increase the accuracy and fairness of AI suggestions and forecasts.

7.6. Ensuring equitable conservation outcomes

Although artificial intelligence could improve the efficiency of conservation, it has to help to produce fair results for all the stakeholders. Achieving fair and sustainable conservation policies depends critically on including nearby populations, honoring indigenous knowledge, and supporting inclusive decision-making procedures. All pertinent stakeholders should be included into conservation plans, especially underprivileged and indigenous groups who are usually most impacted by them. All the participants should equitably share the advantages of AI-driven conservation initiatives. This covers access to better resources, finance, and technology developments (Klein et al., 2015).

8. FUTURE PROSPECTS AND INNOVATIONS

The innovations in AI technologies specifically in the field of biological conservation has led to a safer environment for biodiversity and wildlife. These technologies are changing the field of biological conservation by providing formerly unheard-of tools and methods to monitor, assess, and protect ecosystems and creatures.

8.1. Emerging AI Technologies in Conservation

AI can provide insightful analysis of biodiversity changes, habitat degradation, and illegal activity including poaching and deforestation by means of large data sets from many sources, including satellite imagery, camera traps, and voice recordings. There can be few initiatives like conducting AI driven Wildlife Census, introducing virtual safaris, Species identification app and assigning wildlife guardians as shown in **Figure 2** can be helpful to preserve biodiversity (*Wildlife Conservation in the Digital Age: How AI Powers World Wildlife Day*, n.d.).

Figure 2. showing emerging AI technologies in the field of Biological Conservation

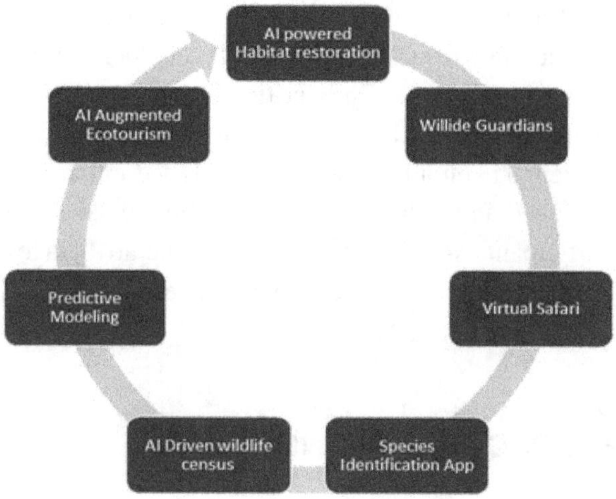

8.2. Advances in machine learning and deep learning

Machine learning techniques are increasingly being used to handle and analyze large information in order to identify trends and forecast ecological changes. ML can predict species distribution, identify significant sites, and assess how climate change can impact biodiversity. Because ML can efficiently handle massive amounts of data, enabling more accurate and quick decision-making for environmentalists. Using ML, AI-powered species identification programs empower users to act as citizen scientists by capturing images of plants or animals; in exchange, they receive immediate information about the species, habitat, and conservation status of the topic. It enhances public engagement in animal conservation and promotes the gathering of data for scientific objectives. Machine Learning and AI algorithms can employ predictive conservation models to anticipate imminent threats to ecosystems and species, encompassing habitat deterioration, climate change, and human encroachment, among others. Conservationists use these predictive models to develop proactive strategies that reduce hazards and protect biodiversity (Vinuesa & Brunton, 2022). In Deep Learning, convolutional neural networks (CNNs) are particularly adept at analyzing intricate data, such as images and sounds, making them highly effective (Norouzzadeh et al., 2018). These models can be used to easily identify species in camera trap images, monitor animal behaviour, and detect poaching. Deep learning models have the ability to autonomously acquire features from unprocessed data, hence diminishing the requirement for human feature extraction and enhancing the precision of predictions and classifications (Christin et al., 2019).

8.3. Integration with other digital technologies (e.g., IoT, drones)

Environmental sensors, smart collars, and remote cameras among other Internet of Things (IoT) devices provide up-to-date data on habitat use, environmental conditions, and animal activity. Effective monitoring and control depend on a continuous flow of data. IoT devices generate a lot of data that AI systems might examine to find trends, anomalies, and patterns. This study helps to support preemptive conservation actions (Meng et al., 2023).

Wildlife census using AI: An AI-powered wildlife census can be described as an internet-based "census" of the animal kingdom. In this process, algorithms analyze photos taken by camera traps, drones, and satellites to provide immediate information on species distribution and population levels. Through the utilization of this innovative approach, accurate data collection is accomplished while causing minimal disturbance to the ecosystems.

Virtual Safaris: Through the utilization of AI and virtual reality technologies, people all over the world can participate in immersive "virtual safaris" that allow them to explore remote wilderness areas and view species in their natural habitats. These encounters serve as motivation to take measures to protect species at risk of extinction and enhance public knowledge about conservation initiatives. Wildlife guardians: Imagine the presence of drones equipped with advanced sensors and algorithms in protected regions, actively deterring illegal actions such as habitat destruction and poaching. These individuals known as "wildlife guardians" would help save endangered animals and ecosystems by continuously monitoring and observing their surroundings.

8.4. Vision for the Future (Long-term goals for AI in conservation)

Sustainable Development Goals (SDGs): Artificial intelligence (AI) can significantly contribute to the achievement of Sustainable Development Goals (SDGs) related to life on land (SDG 15) and below water (SDG 14) by enhancing biodiversity monitoring, promoting sustainable use of ecosystems, and minimizing the impacts of climate change. Conservation solutions that are driven by artificial intelligence should have the ability to be easily adjusted and expanded to suit different ecological environments and regions. This will contribute to the progress of protecting biodiversity on a global scale. International cooperation and alliances are crucial for sharing AI technology, data, and best practices, which in turn strengthens the global community dedicated to conservation (Stephens et al., 2015).

Field guides utilizing augmented reality technology: Augmented reality field guides are created using AR technology, which overlays digital information about wildlife species onto the physical environment. Individuals who have a strong connection to the natural world might enhance their outdoor adventures by gaining direct information about the plants and animals they come across.

Ecotourism augmented by artificial intelligence: AI algorithms utilize ecological data and passenger preferences to create personalized ecotourism experiences that optimize animal observation and conservation education while minimizing harm to the ecosystem. These eco-friendly trips support sustainable development and have positive impacts on local communities.

Games focused on the preservation and protection of wildlife: AI simulations of wildlife conservation, designed with gamification elements, offer users a firsthand encounter with the intricate aspects of the topic. This includes tasks such as overseeing protected areas, combating poaching and illegal logging, and other related challenges. These instructional games inspire players to actively engage in real-world conservation efforts by enhancing their understanding of conservation issues.

Habitat restoration powered by artificial intelligence: Conservation organizations utilize artificial intelligence (AI) to identify and assess deteriorated habitats, as well as develop tailored restoration plans that cater to the specific needs of the ecosystems in the region. Artificial intelligence-powered restoration projects, such as efforts to restore coral reefs and reforest areas, play a significant role in revitalizing biodiversity and improving the ability of ecosystems to withstand and recover from disturbances.

8.5. Potential for global impact

The incorporation of artificial intelligence (AI) technology in conservation has the capacity to fundamentally transform the methods by which one can safeguard and maintain biodiversity. By harnessing the potential of artificial intelligence (AI), one may utilize data to make informed decisions, improve real-time monitoring, and adopt more efficient conservation methods (Vinuesa & Brunton, 2022).

International network for monitoring wildlife on a global scale: An AI-driven collaborative network of monitoring systems enables the global interchange of current information on wildlife populations and environmental conditions. This global initiative fosters international cooperation in the pursuit of environmental conservation and enhances our understanding of interconnected ecosystems.

9. CONCLUSION

With the advancements in AI technologies, such as machine learning, deep learning, Internet of Things (IoT) devices, and drones, the field of biological conservation is revolutionizing. These tools offer robust methods to monitor, administer, and safeguard biodiversity with greater efficiency and precision than ever before. Nevertheless, the effective incorporation of these technologies necessitates thoughtful deliberation of ethical norms to guarantee responsible and fair utilization. By tackling these obstacles and using the advantages of AI, can greatly augment conservation endeavors and contribute to the long-term viability of our planet's ecosystems.

REFERENCES

Abraham, A. (2005). Nature and scope of AI techniques. *Handbook of Measuring System Design.*

Ahmed, A. A., & Echi, M. (2021). Hawk-eye: An ai-powered threat detector for intelligent surveillance cameras. *IEEE Access : Practical Innovations, Open Solutions, 9*, 63283–63293. DOI: 10.1109/ACCESS.2021.3074319

Asner, G. P., & Mascaro, J. (2014). Mapping tropical forest carbon: Calibrating plot estimates to a simple LiDAR metric. *Remote Sensing of Environment, 140*, 614–624. DOI: 10.1016/j.rse.2013.09.023

Beijbom, O., Edmunds, P. J., Kline, D. I., Mitchell, B. G., & Kriegman, D. (2012). Automated annotation of coral reef survey images. *2012 IEEE Conference on Computer Vision and Pattern Recognition*, 1170–1177. DOI: 10.1109/CVPR.2012.6247798

Burton, A. C., Neilson, E., Moreira, D., Ladle, A., Steenweg, R., Fisher, J. T., Bayne, E., & Boutin, S. (2015). Wildlife camera trapping: A review and recommendations for linking surveys to ecological processes. *Journal of Applied Ecology, 52*(3), 675–685. DOI: 10.1111/1365-2664.12432 PMID: 26211047

Carlson, C. J., Burgio, K. R., Dougherty, E. R., Phillips, A. J., Bueno, V. M., Clements, C. F., Castaldo, G., Dallas, T. A., Cizauskas, C. A., Cumming, G. S., Doña, J., Harris, N. C., Jovani, R., Mironov, S., Muellerklein, O. C., Proctor, H. C., & Getz, W. M. (2017). Parasite biodiversity faces extinction and redistribution in a changing climate. *Science Advances, 3*(9), e1602422. DOI: 10.1126/sciadv.1602422 PMID: 28913417

Christin, S., Hervet, É., & Lecomte, N. (2019). Applications for deep learning in ecology. *Methods in Ecology and Evolution, 10*(10), 1632–1644. DOI: 10.1111/2041-210X.13256

Critchlow, R., Plumptre, A. J., Alidria, B., Nsubuga, M., Driciru, M., Rwetsiba, A., Wanyama, F., & Beale, C. M. (2017). Improving law-enforcement effectiveness and efficiency in protected areas using ranger-collected monitoring data. *Conservation Letters, 10*(5), 572–580. DOI: 10.1111/conl.12288

Elith, J., & Leathwick, J. R. (2009). Species distribution models: Ecological explanation and prediction across space and time. *Annual Review of Ecology, Evolution, and Systematics, 40*(1), 677–697. DOI: 10.1146/annurev.ecolsys.110308.120159

Foley, J. A., DeFries, R., Asner, G. P., Barford, C., Bonan, G., Carpenter, S. R., Chapin, F. S., Coe, M. T., Daily, G. C., Gibbs, H. K., Helkowski, J. H., Holloway, T., Howard, E. A., Kucharik, C. J., Monfreda, C., Patz, J. A., Prentice, I. C., Ramankutty, N., & Snyder, P. K. (2005). Global consequences of land use. *Science, 309*(5734), 570–574. DOI: 10.1126/science.1111772 PMID: 16040698

Gall, S. C., & Thompson, R. C. (2015). The impact of debris on marine life. *Marine Pollution Bulletin*, *92*(1–2), 170–179. DOI: 10.1016/j.marpolbul.2014.12.041 PMID: 25680883

Garnett, S. T., Burgess, N. D., Fa, J. E., Fernández-Llamazares, Á., Molnár, Z., Robinson, C. J., Watson, J. E. M., Zander, K. K., Austin, B., Brondizio, E. S., Collier, N. F., Duncan, T., Ellis, E., Geyle, H., Jackson, M. V., Jonas, H., Malmer, P., McGowan, B., Sivongxay, A., & Leiper, I. (2018). A spatial overview of the global importance of Indigenous lands for conservation. *Nature Sustainability*, *1*(7), 369–374. DOI: 10.1038/s41893-018-0100-6

Gebru, T., Morgenstern, J., Vecchione, B., Vaughan, J. W., Wallach, H., Iii, H. D., & Crawford, K. (2021). Datasheets for datasets. *Communications of the ACM*, *64*(12), 86–92. DOI: 10.1145/3458723

Hampton, S. E., Strasser, C. A., Tewksbury, J. J., Gram, W. K., Budden, A. E., Batcheller, A. L., Duke, C. S., & Porter, J. H. (2013). Big data and the future of ecology. *Frontiers in Ecology and the Environment*, *11*(3), 156–162. DOI: 10.1890/120103

Hilborn, R., Amoroso, R. O., Anderson, C. M., Baum, J. K., Branch, T. A., Costello, C., De Moor, C. L., Faraj, A., Hively, D., Jensen, O. P., Kurota, H., Little, L. R., Mace, P., McClanahan, T., Melnychuk, M. C., Minto, C., Osio, G. C., Parma, A. M., Pons, M., & Ye, Y. (2020). Effective fisheries management instrumental in improving fish stock status. *Proceedings of the National Academy of Sciences of the United States of America*, *117*(4), 2218–2224. DOI: 10.1073/pnas.1909726116 PMID: 31932439

Huang, C., Zhang, Z., Mao, B., & Yao, X. (2022). An overview of artificial intelligence ethics. *IEEE Transactions on Artificial Intelligence*, *4*(4), 799–819. DOI: 10.1109/TAI.2022.3194503

Jensen, J. R. (1996). *Introductory digital image processing: a remote sensing perspective*. Prentice Hall.

Jerde, C. L., Mahon, A. R., Chadderton, W. L., & Lodge, D. M. (2011). "Sight-unseen" detection of rare aquatic species using environmental DNA. *Conservation Letters*, *4*(2), 150–157. DOI: 10.1111/j.1755-263X.2010.00158.x

Jump, A. S., Cavin, L., & Hunter, P. D. (2010). Monitoring and managing responses to climate change at the retreating range edge of forest trees. *Journal of Environmental Monitoring*, *12*(10), 1791–1798. DOI: 10.1039/b923773a PMID: 20818456

Klein, C., McKinnon, M. C., Wright, B. T., Possingham, H. P., & Halpern, B. S. (2015). Social equity and the probability of success of biodiversity conservation. *Global Environmental Change*, *35*, 299–306. DOI: 10.1016/j.gloenvcha.2015.09.007

LeCun, Y., Bengio, Y., & Hinton, G. (2015). Deep learning. *Nature*, *521*(7553), 436–444. DOI: 10.1038/nature14539 PMID: 26017442

Margules, C. R., & Pressey, R. L. (2000). Systematic conservation planning. *Nature*, *405*(6783), 243–253. DOI: 10.1038/35012251 PMID: 10821285

Meng, X. N., Xu, S. C., & Hao, M. G. (2023). Can digital-real integration promote industrial green transformation: Fresh evidence from China's industrial sector. *Journal of Cleaner Production*, *426*, 139116. DOI: 10.1016/j.jclepro.2023.139116

Meyerson, L. A., & Mooney, H. A. (2007). Invasive alien species in an era of globalization. []). Wiley Online Library.]. *Frontiers in Ecology and the Environment*, *5*(4), 199–208. DOI: 10.1890/1540-9295(2007)5[199:IASIAE]2.0.CO;2

Nguyen, T., Brandstetter, J., Kapoor, A., Gupta, J. K., & Grover, A. (2023). ClimaX: A foundation model for weather and climate. *ArXiv Preprint ArXiv:2301.10343*.

Norouzzadeh, M. S., Nguyen, A., Kosmala, M., Swanson, A., Palmer, M. S., Packer, C., & Clune, J. (2018). Automatically identifying, counting, and describing wild animals in camera-trap images with deep learning. *Proceedings of the National Academy of Sciences of the United States of America*, *115*(25), E5716–E5725. DOI: 10.1073/pnas.1719367115 PMID: 29871948

Parmesan, C. (2006). Ecological and evolutionary responses to recent climate change. *Annual Review of Ecology, Evolution, and Systematics*, *37*(1), 637–669. DOI: 10.1146/annurev.ecolsys.37.091305.110100

Pettorelli, N., Laurance, W. F., O'Brien, T. G., Wegmann, M., Nagendra, H., & Turner, W. (2014). Satellite remote sensing for applied ecologists: Opportunities and challenges. *Journal of Applied Ecology*, *51*(4), 839–848. DOI: 10.1111/1365-2664.12261

Pimm, S. L., Jenkins, C. N., Abell, R., Brooks, T. M., Gittleman, J. L., Joppa, L. N., Raven, P. H., Roberts, C. M., & Sexton, J. O. (2014). The biodiversity of species and their rates of extinction, distribution, and protection. *Science*, *344*(6187), 1246752. DOI: 10.1126/science.1246752 PMID: 24876501

Schaefer, M. B. (2019). Some Considerations of Population Dynamics and Economics in Relation to the Management of the Commercial Marine Fisheries 1. *Fisheries Economics*, 33–45. DOI: 10.4324/9781315193182-3

Seddon, P. J., Griffiths, C. J., Soorae, P. S., & Armstrong, D. P. (2014). Reversing defaunation: Restoring species in a changing world. *Science*, *345*(6195), 406–412. DOI: 10.1126/science.1251818 PMID: 25061203

Silver, D., Schrittwieser, J., Simonyan, K., Antonoglou, I., Huang, A., Guez, A., Hubert, T., Baker, L., Lai, M., Bolton, A., Chen, Y., Lillicrap, T., Hui, F., Sifre, L., van den Driessche, G., Graepel, T., & Hassabis, D. (2017). Mastering the game of go without human knowledge. *Nature*, *550*(7676), 354–359. DOI: 10.1038/nature24270 PMID: 29052630

Simberloff, D., Martin, J.-L., Genovesi, P., Maris, V., Wardle, D. A., Aronson, J., Courchamp, F., Galil, B., García-Berthou, E., Pascal, M., Pyšek, P., Sousa, R., Tabacchi, E., & Vilà, M. (2013). Impacts of biological invasions: What's what and the way forward. *Trends in Ecology & Evolution*, *28*(1), 58–66. DOI: 10.1016/j.tree.2012.07.013 PMID: 22889499

Singh, G., Mishra, A., & Sagar, D. (2013). An overview of artificial intelligence. *SBIT Journal of Sciences and Technology*, *2*(1), 1–4.

Stahl, B. C. (2021). *Artificial intelligence for a better future: an ecosystem perspective on the ethics of AI and emerging digital technologies*. Springer Nature. DOI: 10.1007/978-3-030-69978-9

Stephens, P. A., Pettorelli, N., Barlow, J., Whittingham, M. J., & Cadotte, M. W. (2015). Management by proxy? The use of indices in applied ecology. *Journal of Applied Ecology*, *52*(1), 1–6. DOI: 10.1111/1365-2664.12383 PMID: 25985647

Sullivan, B., Wood, C., Iliff, M., Bonney, R., Fink, D., & Kelling, S. (2009). eBird: A citizen-based bird observation network in the biological sciences. *Biological Conservation -. Biological Conservation*, *142*(10), 2282–2292. DOI: 10.1016/j.biocon.2009.05.006

Tilman, D., Isbell, F., & Cowles, J. M. (2014). Biodiversity and ecosystem functioning. *Annual Review of Ecology, Evolution, and Systematics*, *45*(1), 471–493. DOI: 10.1146/annurev-ecolsys-120213-091917

Urban, M. C., Bocedi, G., Hendry, A. P., Mihoub, J.-B., Pe'Er, G., Singer, A., Bridle, J. R., Crozier, L. G., De Meester, L., Godsoe, W., Gonzalez, A., Hellmann, J. J., Holt, R. D., Huth, A., Johst, K., Krug, C. B., Leadley, P. W., Palmer, S. C. F., Pantel, J. H., & Travis, J. M. J. (2016). Improving the forecast for biodiversity under climate change. *Science*, *353*(6304), aad8466. DOI: 10.1126/science.aad8466 PMID: 27609898

Van der Wal, R., Fischer, A., Selge, S., & Larson, B. M. H. (2015). Neither the public nor experts judge species primarily on their origins. *Environmental Conservation*, *42*(4), 349–355. DOI: 10.1017/S0376892915000053

Van Horn, G., Mac Aodha, O., Song, Y., Cui, Y., Sun, C., Shepard, A., Adam, H., Perona, P., & Belongie, S. (2018). The inaturalist species classification and detection dataset. *Proceedings of the IEEE Conference on Computer Vision and Pattern Recognition*, 8769–8778. DOI: 10.1109/CVPR.2018.00914

Vinuesa, R., & Brunton, S. L. (2022). Enhancing computational fluid dynamics with machine learning. *Nature Computational Science*, *2*(6), 358–366. DOI: 10.1038/s43588-022-00264-7 PMID: 38177587

Wang, Z., Yang, Z., Shi, H., Han, F., Liu, Q., Qi, J., & Lu, Y. (2020). Ecosystem health assessment of world natural heritage sites based on remote sensing and field sampling verification: Bayanbulak as case study. *Sustainability (Basel)*, *12*(7), 2610. DOI: 10.3390/su12072610

Wegmann, M., Leutner, B., & Dech, S. (2016). *Remote sensing and GIS for ecologists: using open source software*. Pelagic Publishing Ltd. DOI: 10.2307/jj.28833743

Wildlife conservation in the digital age: How AI powers World Wildlife Day. (n.d.). Retrieved November 28, 2024, from https://indiaai.gov.in/article/wildlife-conservation-in-the-digital-age-how-ai-powers-world-wildlife-day

Zhang, K., & Aslan, A. B. (2021). AI technologies for education: Recent research & future directions. *Computers and Education: Artificial Intelligence*, *2*, 100025. DOI: 10.1016/j.caeai.2021.100025

Chapter 7
The Digital Shield:
AI as a Frontier for Mitigating Human-Wildlife Conflict

Rohit Kumar Pal
http://orcid.org/0000-0002-7525-0771
Lincoln University College, Malaysia

Sagnika Pal
http://orcid.org/0009-0009-8534-3961
University of Calcutta, India

ABSTRACT

The escalating crisis of human-wildlife conflict is a primary driver of biodiversity loss, challenging the coexistence of expanding human populations and vulnerable animal species. Moving beyond reactive and often ineffective traditional methods, this chapter examines the emergence of Artificial Intelligence (AI) as a transformative frontier in conflict mitigation. This chapter explores how Artificial Intelligence (AI) offers a proactive solution by analyzing data from sources like satellite imagery, camera traps, and acoustic signals to predict animal movements and identify conflict hotspots. It highlights applications such as real-time alert systems, computer vision for species monitoring, and AI-designed wildlife corridors. Case studies on species like elephants and tigers demonstrate its scalability. While challenges like data bias and the need to integrate local knowledge remain, AI is presented as a powerful, non-lethal "digital shield" for promoting sustainable coexistence between humans and wildlife.

DOI: 10.4018/979-8-2600-2691-5.ch007

1. INTRODUCTION: THE RISING CHALLENGE OF HUMAN–WILDLIFE CONFLICT

Human–wildlife conflict (HWC) is intensifying globally due to habitat fragmentation, agricultural expansion, infrastructure development, and climate change, posing severe threats to biodiversity and human livelihoods. Conservation science emphasizes that traditional mitigation strategies—such as physical barriers, compensation schemes, and lethal control—are often reactive, costly, and socially contentious (Dickman, 2010; Nyhus, 2016; Treves & Karanth, 2003). At the same time, advances in ecological informatics, remote sensing, and machine learning have enabled conservationists to process large-scale spatial and behavioral datasets, opening new avenues for predictive modelling and proactive mitigation (Christin et al., 2019; Norouzzadeh et al., 2018). Emerging scholarship demonstrates that Artificial Intelligence (AI) can transform wildlife monitoring, risk forecasting, and landscape planning, moving conflict mitigation from reactive response to anticipatory governance (Beery et al., 2018). Collectively, these studies highlight a paradigm shift: conservation is transitioning toward data-driven, non-lethal, and technologically enhanced coexistence frameworks. The reasons of man-animal conflict are given in the following image.

Figure 1. Reasons for the rising challenge of Human-Animal Conflict

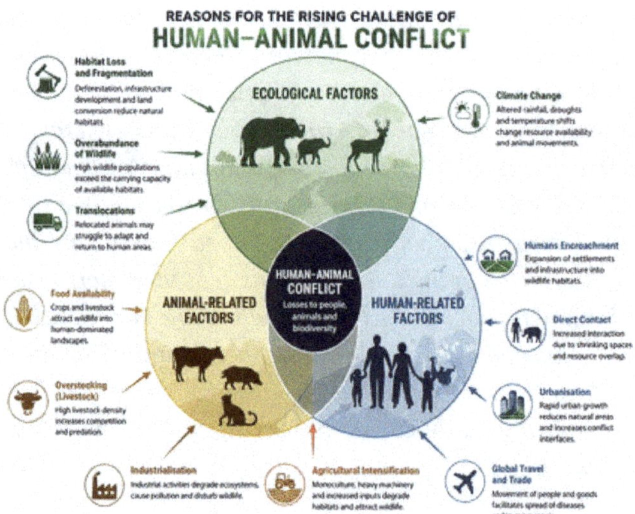

1.2 The Escalating Global Challenge

Human–wildlife conflict refers to interactions between humans and wildlife that result in negative impacts on human social, economic, or cultural life, on the conservation of wildlife populations, or on the environment (Nyhus, 2016). Such conflicts range from crop raiding by elephants and wild boar to livestock depredation by large carnivores and human injuries or fatalities caused by megafauna. As human populations expand into previously undisturbed habitats, spatial overlap between people and wildlife increases, intensifying encounters (Treves & Karanth, 2003).

Globally, HWC is now recognized as a primary driver of biodiversity loss, particularly for large-bodied mammals such as elephants, lions, and tigers. Retaliatory killings, habitat degradation, and reduced tolerance among local communities threaten already vulnerable species (Dickman, 2010). Simultaneously, rural communities suffer significant economic losses from crop destruction and livestock predation, exacerbating poverty and undermining conservation support. The dual vulnerability—of people and wildlife—makes HWC one of the most complex socio-ecological challenges of the 21st century.

Climate change further intensifies this crisis. Altered rainfall patterns, droughts, and shifts in vegetation productivity influence animal movement patterns, often pushing wildlife into agricultural areas in search of food and water. Landscape fragmentation caused by roads, settlements, and infrastructure disrupts traditional migration corridors, increasing the likelihood of encounters (Poessel et al., 2013). These cumulative pressures demand innovative, scalable solutions beyond conventional deterrence strategies.

1.3 Limitations of Traditional Mitigation Approaches

Historically, mitigation strategies have relied on physical barriers (e.g., fencing), compensation programs, translocation, and in some cases lethal control. While electric fences and fortified enclosures can reduce localized damage, they are expensive to maintain and may fragment ecosystems further (Nyhus, 2016). Compensation schemes often suffer from bureaucratic delays, insufficient funding, and challenges in verifying losses, reducing community trust (Dickman, 2010).

Lethal control, though sometimes politically expedient, frequently fails to address root causes and may disrupt social structures of animal populations, potentially worsening conflict dynamics (Treves & Karanth, 2003). These approaches are predominantly reactive, implemented after damage has occurred. Consequently, conservation scholars increasingly advocate for predictive and preventative frameworks grounded in ecological forecasting and community engagement.

1.4 The Emergence of Digital Conservation Technologies

The last decade has witnessed rapid growth in conservation technologies. Camera traps, satellite telemetry, unmanned aerial vehicles (UAVs), and acoustic sensors now generate vast amounts of ecological data. However, manual analysis of these datasets is labor-intensive and often impractical at scale. Artificial Intelligence—particularly machine learning and computer vision—has emerged as a transformative solution to this analytical bottleneck (Christin et al., 2019).

Deep learning algorithms can automatically classify millions of camera trap images with high accuracy, dramatically accelerating species identification and behavioural monitoring (Norouzzadeh et al., 2018). Similarly, computer vision models trained on diverse ecological datasets improve detection of rare species and reduce human bias in monitoring programs (Beery et al., 2018). These tools not only enhance efficiency but also enable near-real-time surveillance systems capable of triggering alerts when wildlife approaches human settlements.

AI-driven predictive analytics further extend these capabilities. By integrating historical incident reports, environmental variables, satellite imagery, and animal movement data, machine learning models can identify conflict hotspots and forecast risk patterns (Poessel et al., 2013). Such predictive systems support early-warning mechanisms that notify communities before wildlife incursions occur, shifting management strategies from response to prevention. Moreover, the conceptual foundation of this study is further strengthened by the field of environmental informatics, which integrates computational tools with ecological systems for sustainable management. As highlighted by Paul (2022), environmental informatics provides a multidisciplinary framework combining data science, environmental studies, and emerging technologies to address complex ecological challenges. Building on this, Paul (2026) emphasizes the transformative role of Artificial Intelligence and allied technologies in promoting environmental sustainability, particularly through predictive modelling, automation, and educational integration. Additionally, the application of Geographic Information Systems (GIS) and remote sensing has been widely recognized for environmental monitoring and decision-making, enabling spatial analysis of ecological patterns and resource management (Paul et al., 2020a). Furthermore, the integration of Artificial Intelligence and cloud computing enhances environmental system efficiency by enabling real-time data processing, scalability, and sustainable development strategies (Paul et al., 2020b). Together, these perspectives reinforce the role of intelligent, data-driven technologies as essential enablers of modern conservation and environmental management systems.

1.5 AI as a Proactive "Digital Shield"

The application of AI in human–wildlife conflict mitigation represents a paradigm shift toward proactive conservation. For example, GPS-collared elephants combined with geospatial algorithms can predict herd trajectories and trigger automated SMS alerts to nearby villages, reducing surprise encounters (Wall et al., 2014). Landscape-level modelling also enables the design of optimized wildlife corridors that maintain ecological connectivity while minimizing conflict risk.

Importantly, AI applications extend beyond megafauna. Machine learning tools have been deployed to monitor small mammals, birds, and even insect populations, demonstrating scalability across taxa (Christin et al., 2019). However, challenges remain. Data biases, limited representativeness across regions, algorithmic transparency, and ethical concerns about surveillance technologies must be addressed to ensure equitable and effective implementation. Integrating local ecological knowledge with computational modelling is particularly critical for contextual accuracy and community acceptance.

2. OBJECTIVE OF THIS CHAPTER

The objectives of this chapter are to understand the following

- To examine the role of Artificial Intelligence as a "digital shield" in modern conservation.
- To analyze how predictive analytics, computer vision, and big data enhance wildlife monitoring.
- To explore AI-driven approaches for proactive and non-lethal human–wildlife conflict mitigation.
- To evaluate how AI complements and improves traditional conservation strategies.
- To highlight the potential of AI in promoting sustainable coexistence and biodiversity conservation.

3. UNDERSTANDING HUMAN–WILDLIFE CONFLICT DYNAMICS

Human–wildlife conflict (HWC) is a complex socio-ecological phenomenon shaped by interacting environmental, demographic, economic, and behavioral drivers. Understanding why wildlife increasingly enters human-dominated landscapes requires an integrative theoretical foundation that situates conflict within landscape transformation, ecological adaptation, and human socio-economic vulnerability. Rather than viewing HWC as isolated incidents, contemporary scholarship conceptualizes it as an emergent property of coupled human–natural systems, where ecological processes and human development trajectories intersect (Liu et al., 2007; Nyhus, 2016).

3.1 Habitat Fragmentation and Land-Use Change

Habitat fragmentation and land-use change represent foundational drivers of HWC. As forests, grasslands, and wetlands are converted into agricultural fields, settlements, and transportation networks, wildlife habitats become smaller, isolated, and degraded. Fragmentation disrupts traditional migration routes, reduces access to food and water, and compresses wildlife populations into increasingly constrained spaces (Fischer & Lindenmayer, 2007). Species that require large home ranges—such as elephants, tigers, and wolves—are particularly affected, as they traverse fragmented landscapes in search of resources, often crossing into human settlements (Crooks et al., 2011).

Landscape modification not only reduces habitat availability but also alters ecological dynamics. Edge effects created by fragmentation increase interactions at habitat boundaries, where wildlife encounters crops, livestock, and human infrastructure. These ecotonal zones often become conflict hotspots (Haddad et al., 2015). From a landscape ecology perspective, spatial configuration—patch size, connectivity, and matrix quality—directly influences wildlife movement and the probability of human encounters. Consequently, understanding spatial patterns is essential for predicting and mitigating conflict risk.

Figure 2. Image describing how habitat fragmentation rises the chances of conflict situation

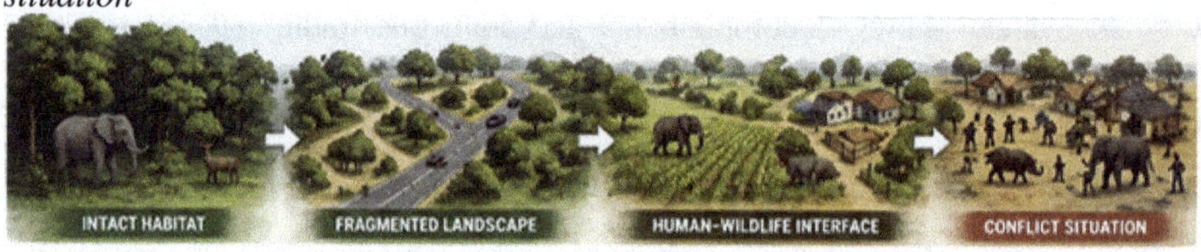

3.2 Agricultural Expansion and Food Resource Availability

Agricultural expansion further intensifies HWC by creating concentrated and predictable food sources. Croplands frequently provide higher caloric density than natural vegetation, attracting herbivores such as elephants, deer, and wild boar (Schley & Roper, 2003). Carnivores, in turn, may prey upon livestock, which represent accessible and often less-defended alternatives to wild prey (Treves & Karanth, 2003). These anthropogenic food subsidies reshape wildlife foraging behavior and spatial distribution, reinforcing patterns of repeated incursion into human-dominated areas.

Agricultural landscapes can thus function as ecological traps, drawing wildlife into risky environments where retaliation or lethal control is likely. Studies have demonstrated that crop phenology and harvest cycles strongly correlate with seasonal peaks in wildlife incursions, indicating that conflict dynamics are tightly linked to agricultural productivity patterns (Nyhus, 2016). The availability of anthropogenic food resources not only alters movement patterns but may also influence reproductive success and population growth, thereby exacerbating long-term conflict pressures.

3.3 Climate Change and Shifting Ecological Patterns

Climate change adds an additional layer of complexity to HWC dynamics. Altered precipitation regimes, prolonged droughts, and temperature increases affect vegetation productivity, water availability, and prey distribution, prompting wildlife to adjust movement and foraging strategies (Pecl et al., 2017). In arid and semi-arid regions, drought conditions have been linked to increased elephant incursions into farmland due to reduced natural forage (Wall et al., 2014). Shifts in species distributions driven by climate change may also introduce novel interactions between humans and wildlife in previously unaffected regions. Range expansions of certain carnivores or herbivores can create new zones of overlap, where communities may lack experience or adaptive strategies for coexistence. Climate variability therefore amplifies uncertainty in conflict patterns, necessitating adaptive management frameworks grounded in predictive ecological modelling. The above describe word diagram (Figure3) describes the facts well.

Figure 3. Human-animal conflict increase by climate change

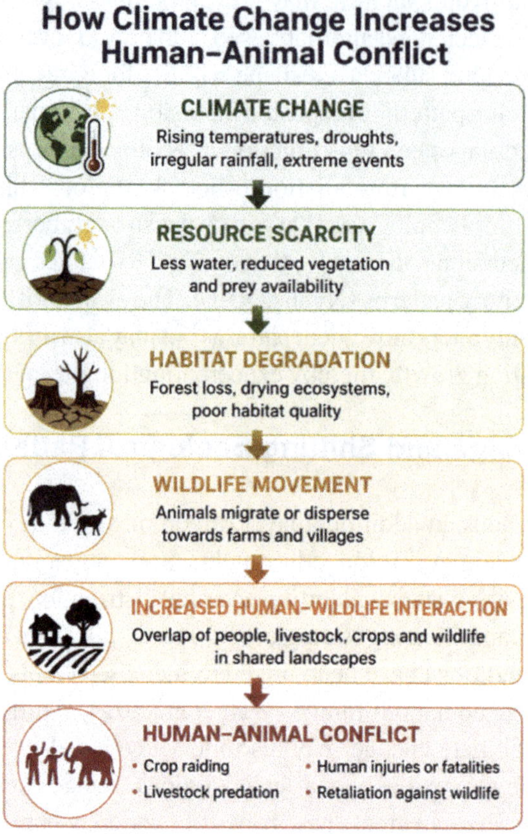

3.4 Population Growth and Infrastructure Development

Global human population growth and infrastructure expansion further intensify conflict dynamics. Roads, railways, dams, and urban sprawl not only fragment habitats but also increase wildlife mortality through vehicle collisions and barrier effects (Crooks et al., 2011). Infrastructure corridors often intersect with traditional migratory routes, forcing animals to navigate hazardous environments.

Population growth also increases demand for land, food, and natural resources, leading to encroachment into wildlife habitats. As rural settlements expand, buffer zones between protected areas and agricultural land diminish, heightening the probability of direct encounters. From a human-dominated ecosystem framework, landscapes are increasingly characterized by multifunctional land use, where conservation areas coexist alongside agriculture and settlements (Ellis, 2015). In such mosaics, conflict emerges not as an anomaly but as an inherent feature of shared landscapes.

3.5 Behavioral Ecology of Wildlife in Human Landscapes

The behavioral ecology of wildlife plays a critical role in shaping HWC. Many species exhibit behavioral plasticity that enables adaptation to anthropogenic environments. Carnivores such as coyotes and leopards adjust activity patterns to become more nocturnal in urban areas, reducing direct encounters while maintaining access to resources (Poessel et al., 2013). Similarly, elephants and primates may learn to exploit crop fields strategically, demonstrating cognitive flexibility and social learning.

Risk–reward trade-offs underpin wildlife decision-making in human-dominated landscapes. Animals assess the energetic benefits of anthropogenic food sources against risks of injury or death. Over time, habituation to human presence may reduce perceived risk, increasing boldness and frequency of incursions (Treves & Karanth, 2003). Behavioral adaptation thus contributes to the persistence and escalation of conflict, particularly when deterrence measures are inconsistent or ineffective.

3.6 Economic and Psychological Impacts on Communities

HWC imposes significant economic burdens on affected communities. Crop losses, livestock depredation, property damage, and medical costs can undermine household livelihoods, particularly in low-income rural regions where subsistence agriculture predominates (Dickman, 2010). Compensation schemes, while intended to offset losses, often fail to address indirect costs such as lost labour, emotional stress, or opportunity costs.

Beyond economic consequences, psychological impacts are substantial. Fear, anxiety, and reduced tolerance toward wildlife can erode conservation support and foster retaliatory behavior. Social perceptions of risk are shaped not only by actual damage but also by cultural values, historical experiences, and trust in institutions (Nyhus, 2016). Effective conflict mitigation must therefore address both material and perceptual dimensions, integrating community engagement and participatory governance.

Figure 4. Impact of Human-Wildlife Conflict on People and Communities

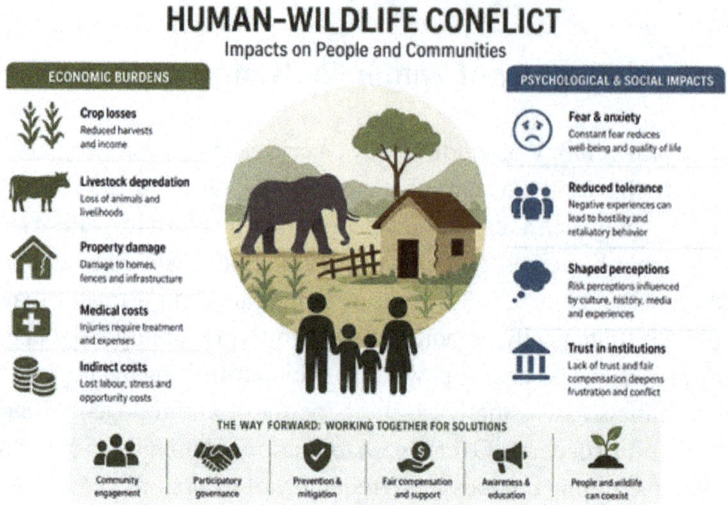

A Case Study

In a 2026 study in Ethiopia (Guadu et al. 2026) on the psychological impacts of human–wildlife conflict (HWC), farmers described a daily reality shaped by fear, stress, and uncertainty. One farmer explained that despite trying every possible method to protect their maize, elephants repeatedly return, destroying crops. These losses extend beyond immediate damage, severely affecting household food security by reducing both the quantity and diversity of food available, which in turn weakens nutrition and increases vulnerability to disease.

Participants also emphasized the emotional and physical toll of living alongside wildlife. Many reported sleepless nights spent guarding their fields, constant anxiety about potential attacks, and chronic fatigue, all of which contribute to declining mental and physical health.

The study further revealed widespread frustration toward park authorities. While enforcement actions against community activities—such as illegal grazing or resource use—are often swift, responses to wildlife-related damage or human injury are frequently delayed or absent. This perceived imbalance has intensified feelings of neglect, as communities struggle to safeguard their livelihoods with limited institutional support. Following image from a study by Yeshe 2023 shows how Human Wildlife Conflict and Psychological Well being is associated.

Figure 5. Human Wildlife Conflict, Sustainability and different well-being

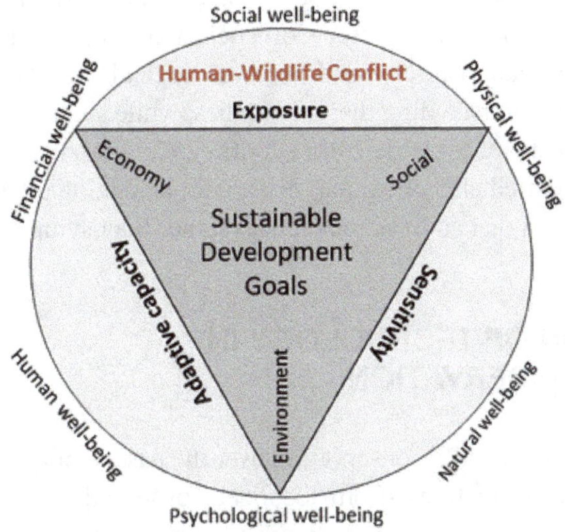

3.7 Conceptual Models for Understanding Conflict

Several conceptual frameworks inform the study of HWC. The landscape ecology perspective emphasizes spatial heterogeneity, connectivity, and scale, highlighting how land-use configuration influences wildlife movement and conflict probability (Fischer & Lindenmayer, 2007). The human-dominated ecosystem framework conceptualizes landscapes as coupled socio-ecological systems in which conservation and development objectives intersect (Ellis, 2015; Liu et al., 2007). This perspective underscores the need to balance ecological resilience with human well-being.

Conflict risk mapping represents a practical application of these theoretical insights. By integrating spatial data on habitat characteristics, land use, wildlife movement, and historical incident records, researchers can identify hotspots and predict areas of heightened risk (Poessel et al., 2013). Such models provide a foundation for targeted interventions, resource allocation, and early-warning systems. Importantly, risk mapping illustrates that conflict is not randomly distributed but emerges from identifiable ecological and socio-economic patterns.

3.8 Synthesis

In sum, human–wildlife conflict arises from the convergence of habitat fragmentation, agricultural intensification, climate variability, demographic expansion, and adaptive wildlife behavior within increasingly human-dominated ecosystems. Economic vulnerability and psychological perceptions further shape community responses, influencing the trajectory of coexistence or escalation. Theoretical models from landscape ecology and socio-ecological systems research provide essential tools for understanding these dynamics, while spatial risk mapping offers operational mechanisms for anticipating conflict. Recognizing HWC as a systemic and spatially structured phenomenon is critical for designing proactive, data-driven mitigation strategies that address root causes rather than symptoms.

4. EVOLUTION OF TECHNOLOGY IN WILDLIFE CONSERVATION

The trajectory of wildlife conservation over the past century reflects a profound technological transformation—from localized, observation-based monitoring to sophisticated, data-intensive, and predictive systems. Historically grounded in field naturalism and manual data collection, conservation science has progressively integrated electronic tracking, remote sensing, spatial analytics, and, more recently, big data and machine learning methodologies. This technological evolution has fundamentally reshaped not only how wildlife is monitored but also how conservation decisions are designed and implemented, shifting the discipline from descriptive ecology toward anticipatory and data-driven governance (Kays et al., 2015; Peters et al., 2014).

4.1 Early Wildlife Monitoring Methods

Early wildlife monitoring relied heavily on direct observation, spoor tracking, mark–recapture techniques, and manual population surveys. Researchers conducted transect counts, aerial surveys, and field-based behavioural observations to estimate species abundance and distribution. While these approaches generated foundational ecological knowledge, they were labour-intensive, spatially limited, and often constrained by observer bias and logistical challenges (Nichols & Williams, 2006).

Figure 6. Evolution of Wildlife Telemetry

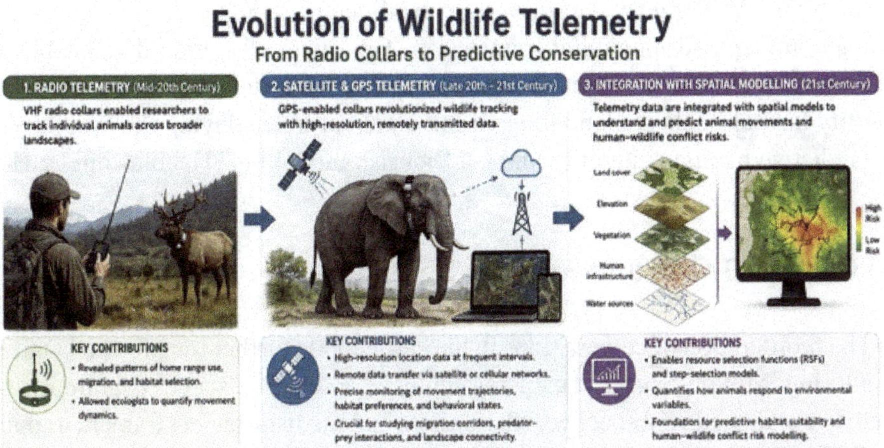

Mark–recapture methods, for example, provided important population estimates but required repeated capture efforts that could stress animals and were difficult to implement for wide-ranging or elusive species. Similarly, observational studies were often restricted to accessible areas, limiting inference across heterogeneous landscapes. Although these traditional methods remain valuable, especially for fine-scale behavioral studies, they offered limited capacity for real-time monitoring or large-scale spatial analysis.

4.2 Radio Telemetry and GPS Collars

The introduction of radio telemetry in the mid-twentieth century marked a major technological milestone. Very High Frequency (VHF) radio collars enabled researchers to track individual animals across broader landscapes, revealing patterns of home range use, migration, and habitat selection (Kays et al., 2015). For the first time, ecologists could quantify movement dynamics rather than infer them indirectly.

Subsequent advances in satellite telemetry and Global Positioning System (GPS) collars revolutionized wildlife tracking. GPS-enabled devices collect high-resolution spatial data at frequent intervals, transmitting information remotely via satellite or cellular networks. These technologies allow researchers to monitor movement trajectories, habitat preferences, and behavioral states with unprecedented precision (Hebblewhite & Haydon, 2010). GPS telemetry has been instrumental in understanding migration corridors, predator–prey interactions, and landscape connectivity, providing empirical foundations for conservation planning.

Importantly, telemetry data have also facilitated the integration of movement ecology with spatial modelling frameworks. Continuous location data enable the construction of resource selection functions (RSFs) and step-selection models that quantify how animals respond to environmental variables, laying the groundwork for predictive habitat suitability and conflict risk modelling (Hebblewhite & Haydon, 2010).

4.3 Camera Traps and Remote Sensing

The proliferation of camera traps in the early 2000s further transformed wildlife monitoring. Motion-triggered cameras deployed across landscapes capture images of elusive and nocturnal species with minimal human disturbance. Camera trapping has enabled large-scale biodiversity assessments, population density estimation, and behavioral monitoring across diverse ecosystems (O'Connell et al., 2011).

The digital nature of camera trap data created new analytical challenges and opportunities. Massive image datasets require automated processing, prompting the integration of computer vision and machine learning for species identification and behavioral classification (Norouzzadeh et al., 2018). This transition from manual photo review to automated image recognition significantly accelerated data analysis and improved scalability.

Parallel developments in remote sensing have expanded conservation monitoring to global scales. Satellite imagery and aerial platforms provide continuous data on land cover change, deforestation, vegetation productivity, and habitat fragmentation (Turner et al., 2003). High-resolution Earth observation data enable conservationists to monitor environmental change in near real time, linking landscape transformation to wildlife responses. Remote sensing thus bridges ecological processes and anthropogenic drivers, supporting evidence-based land-use planning.

4.4 Geographic Information Systems (GIS)

Geographic Information Systems (GIS) have become central to conservation science by integrating spatial datasets into coherent analytical frameworks. GIS platforms allow researchers to overlay wildlife movement data, habitat characteristics, infrastructure networks, and socio-economic variables to identify patterns of overlap and risk (Turner et al., 2003).

Through spatial modelling techniques such as habitat suitability analysis, least-cost path modelling, and corridor design, GIS supports landscape-level decision-making. Connectivity modelling, for instance, helps identify critical wildlife corridors that maintain gene flow and reduce fragmentation impacts (Crooks et al., 2011). GIS has thus shifted conservation from site-specific management to landscape-scale planning, emphasizing spatial relationships and systemic interactions.

4.5 Big Data in Conservation Science

The digitalization of ecological monitoring has ushered conservation into the era of big data. Advances in sensor networks, bioacoustic monitoring, satellite telemetry, and citizen science platforms generate vast and heterogeneous datasets (Peters et al., 2014). These data streams include high-frequency movement records, climate variables, remote sensing products, and millions of camera trap images.

Big data approaches enable integration across temporal and spatial scales, facilitating multi-dimensional analyses of ecological dynamics. However, they also require robust computational infrastructure and interdisciplinary expertise in data science. The emergence of conservation informatics reflects this shift, combining ecology, statistics, computer science, and geospatial analytics to process complex datasets (Hampton et al., 2013).

Importantly, big data enhances the capacity to detect subtle ecological signals and anticipate emerging threats. Longitudinal datasets reveal trends in species distribution shifts, habitat degradation, and behavioral adaptation, informing adaptive management strategies.

4.6. Transition from Data Collection to Predictive Analytics

Perhaps the most significant transformation in conservation technology is the transition from descriptive data collection to predictive analytics. Machine learning algorithms can model nonlinear relationships between wildlife occurrence and environmental predictors, generating forecasts of habitat suitability, migration pathways, and conflict hotspots (Christin et al., 2019).

Predictive modelling integrates historical data with real-time environmental inputs to anticipate future scenarios. For example, movement data combined with land-use change projections can forecast areas of heightened human–wildlife interaction risk. This anticipatory capability represents a paradigm shift from reactive intervention to proactive planning. Conservation strategies are increasingly informed by scenario modelling, simulation, and risk forecasting rather than solely by retrospective analysis.

The integration of Artificial Intelligence (AI) into conservation workflows exemplifies this shift. Automated image classification, acoustic signal recognition, and spatial risk modelling enhance both speed and precision in decision-making (Norouzzadeh et al., 2018). These tools allow conservation practitioners to allocate resources strategically, prioritize high-risk zones, and design mitigation measures before irreversible impacts occur.

4.7. Synthesis: From Observation to Prediction

The evolution of technology in wildlife conservation reflects a broader epistemological shift. Early conservation relied on direct observation and manual data recording, yielding essential but limited insights. Subsequent innovations—radio telemetry, GPS tracking, camera traps, remote sensing, and GIS—expanded spatial and temporal coverage, generating increasingly complex datasets. The contemporary integration of big data analytics and machine learning has transformed conservation into a predictive, systems-oriented science.

This transition underscores a key argument: technology has moved conservation beyond observation-based approaches toward predictive and data-driven strategies. By harnessing digital tools and computational modelling, conservation science is increasingly capable of anticipating ecological change, mitigating risks, and designing adaptive interventions. In doing so, technological innovation not only enhances scientific understanding but also strengthens the capacity for proactive biodiversity protection in an era of accelerating environmental transformation.

5. INTELLIGENT CONSERVATION SYSTEMS: INTEGRATING ARTIFICIAL INTELLIGENCE FOR PREDICTIVE, PROACTIVE, AND LANDSCAPE-SCALE HUMAN–WILDLIFE COEXISTENCE

The rapid advancement of Artificial Intelligence (AI) has ushered in a transformative era in conservation science, redefining how ecological data are interpreted, how wildlife movement is forecasted, and how human–wildlife conflict (HWC) is mitigated across complex landscapes. Artificial Intelligence broadly refers to computational systems capable of performing tasks that typically require human cognition, including pattern recognition, learning from data, and decision-making under uncertainty (Russell & Norvig, 2021). Within conservation, AI encompasses machine learning (ML) and deep learning (DL) techniques that enable algorithms to detect patterns within large ecological datasets and improve predictive accuracy through iterative training (Christin et al., 2019). Machine learning involves statistical models that learn relationships between predictors and outcomes without explicit programming, while deep learning—particularly convolutional neural networks (CNNs) and recurrent neural networks (RNNs)—leverages multi-layered neural architectures to process high-dimensional data such as images, acoustic signals, and spatiotemporal sequences (LeCun et al., 2015). These computational methods are particularly relevant to ecological research, where datasets are increasingly large, heterogeneous, and dynamic, encompassing satellite imagery, GPS telemetry, camera trap images, climate variables, and community-reported conflict records (Hampton et al., 2013). AI's capacity to process large-scale environmental data allows conservationists to transition from descriptive analyses toward predictive ecological modelling, thereby strengthening anticipatory management strategies in rapidly changing socio-ecological systems.

At the core of the AI–ecology interface lies a conceptual framework comprising four interlinked stages: data input, algorithmic analysis, predictive output, and conservation decision-making. Data input involves the aggregation of multisource datasets, including animal movement trajectories derived from GPS collars (Kays et al., 2015), high-resolution satellite imagery capturing land-use change (Turner et al., 2003), drone-based aerial surveys, and bioacoustic recordings from sensor networks. These datasets are pre-processed to remove noise, standardize formats, and extract relevant features. Algorithmic analysis follows, where supervised or unsupervised learning models identify relationships between environmental variables and wildlife behavior. For instance, resource selection functions and random forest classifiers can quantify habitat preferences, while deep neural networks automate image-based species recognition (Norouzzadeh et al., 2018). Predictive output then emerges in the form of probability maps, movement forecasts, or risk indices that estimate the likelihood of wildlife presence or conflict occurrence across space and time. Finally, these outputs inform conservation decision-making, guiding interventions such as targeted deterrence, corridor placement, and land-use zoning. This iterative framework reflects a feedback loop in which predictive performance is continuously refined as new data are incorporated, exemplifying adaptive management principles.

Predictive analytics constitutes one of the most powerful applications of AI in wildlife movement and conflict forecasting. Machine learning algorithms—including random forests, support vector machines, gradient boosting models, and neural networks—have been widely applied to ecological prediction due to their ability to capture nonlinear relationships and interactions among variables (Elith & Leathwick, 2009). Species distribution modelling (SDM) represents a foundational approach, integrating species occurrence data with environmental predictors to estimate habitat suitability across landscapes. Techniques such as Maximum Entropy modelling and ensemble learning methods have demonstrated high predictive performance in forecasting species ranges and identifying climatically suitable habitats (Elith & Leathwick, 2009). In the context of HWC, SDMs can be extended to predict zones of spatial overlap between wildlife and human activities, thereby identifying high-risk areas for crop raiding or livestock predation. Habitat suitability analysis further refines these predictions by incorporating topography, vegetation cover, water sources, infrastructure density, and anthropogenic disturbance metrics.

Temporal dynamics are equally critical, as wildlife movement patterns exhibit seasonal, diurnal, and climate-driven variability. Recurrent neural networks and time-series models allow researchers to capture sequential dependencies in movement data, enhancing forecasts of migration timing and resource-driven displacement (Peters et al., 2014). By integrating environmental variables such as rainfall, temperature anomalies, and vegetation indices derived from remote sensing, predictive models can anticipate shifts in animal movement associated with droughts or resource scarcity. Risk mapping and hotspot identification synthesize these outputs into spatially explicit visualizations, enabling managers to allocate resources strategically. For example, integrating historical conflict reports with land-use data and movement trajectories allows AI systems to generate probabilistic maps of likely conflict zones, supporting early identification of emerging hotspots (Poessel et al., 2013). The integration of environmental and human activity data—including road density, agricultural intensity, and settlement proximity—further enhances model accuracy by accounting for socio-economic drivers of conflict.

Complementing predictive analytics, computer vision technologies have revolutionized automated wildlife monitoring. Camera traps, once dependent on manual image review, now leverage deep learning algorithms capable of identifying species, counting individuals, and classifying behaviors with remarkable accuracy (Norouzzadeh et al., 2018). Convolutional neural networks trained on millions of labelled images can distinguish between species even in challenging environmental conditions, reducing processing time from months to hours. Beyond species recognition, AI models facilitate individual animal identification through pattern recognition techniques applied to unique coat markings, stripes, or scars, enhancing population estimation and demographic analysis (Christin et al., 2019). Video analytics further enable behavioral classification, detecting activities such as foraging, resting, or aggressive interactions. Drone-based wildlife detection integrates aerial imagery with real-time object detection algorithms, expanding monitoring coverage to remote or inaccessible terrains while minimizing disturbance (Kays et al., 2015). The benefits of computer vision are substantial: monitoring becomes non-invasive, labour costs decrease, and biodiversity assessments can be scaled across extensive landscapes.

AI-based early warning and real-time alert systems represent a critical advancement in shifting conflict management from reactive response to proactive prevention. Sensor networks and Internet of Things (IoT) technologies enable continuous environmental monitoring, collecting acoustic, thermal, and motion data across buffer zones and agricultural fields. Acoustic detection systems employing deep learning classifiers can identify species-specific vocalizations, such as elephant rumbles or gunshots, triggering automated alerts (Aide et al., 2013). AI-based elephant detection models integrate camera trap feeds, seismic sensors, and GPS collar data to predict herd proximity to human settlements. When threshold risk levels are exceeded, SMS or mobile application alerts notify community members, enabling precautionary actions such as securing livestock or activating deterrent mechanisms. Community-based response systems enhance effectiveness by integrating local knowledge and participatory monitoring, ensuring that technological solutions align with socio-cultural contexts. These real-time systems demonstrate how AI operationalizes anticipatory governance, reducing damage and fostering coexistence.

Beyond localized monitoring and alerts, AI plays a pivotal role in landscape planning and wildlife corridor design. Habitat fragmentation remains a principal driver of conflict, necessitating spatial planning approaches that balance ecological connectivity with human development (Crooks et al., 2011). Landscape connectivity analysis employs graph theory and least-cost path modelling to identify corridors that facilitate safe wildlife movement between habitat patches. AI enhances these analyses by optimizing corridor placement under multiple constraints, incorporating land-use costs, ecological value, and predicted conflict risk. Optimization algorithms and spatial decision-support systems can evaluate thousands of potential configurations, identifying solutions that maximize connectivity while minimizing socio-economic disruption. Corridor suitability modelling integrates predictive habitat suitability outputs with infrastructure data, ensuring that proposed corridors reduce encounters with human settlements. Furthermore, AI-driven scenario modelling supports long-term conservation planning by simulating the impacts of urban expansion or climate change on connectivity patterns. The integration of AI with conservation planning tools thus promotes ecosystem sustainability, enabling wildlife to move safely while reducing pressure on agricultural landscapes.

Collectively, these applications underscore a paradigm shift in conservation science. Traditional approaches emphasized observation and retrospective analysis; contemporary AI-driven systems emphasize prediction, automation, and strategic intervention. By synthesizing data from satellites, drones, sensor arrays, and community reports, AI provides a comprehensive understanding of coupled human–natural systems. Importantly, the transition to predictive and data-driven strategies does not replace ecological expertise but rather augments it, enabling conservation practitioners to make informed, timely, and scalable decisions. The convergence of machine learning, computer vision, real-time monitoring, and spatial optimization reflects the emergence of intelligent conservation systems capable of addressing complex challenges such as human–wildlife conflict. As environmental pressures intensify globally, the integration of AI into ecological research and management offers a pathway toward proactive biodiversity protection, safer wildlife movement, and resilient coexistence between humans and the natural world.

Figure 7. Role of Big Data & AI in transforming wildlife conservation

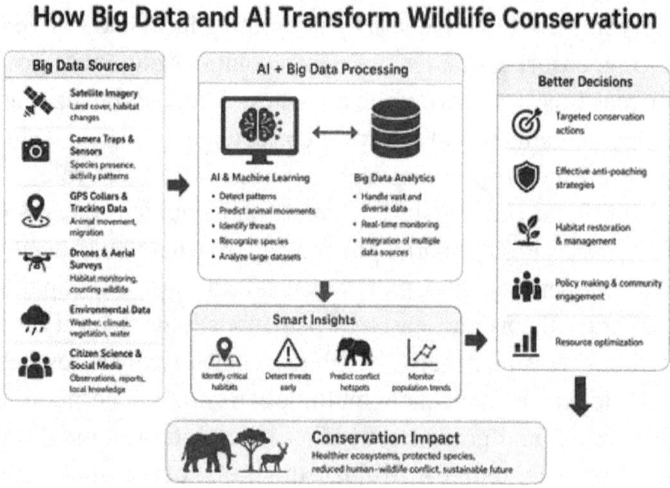

6. Applied Intelligence in Action: Case Studies of AI in Human–Wildlife Conflict Mitigation

The practical deployment of Artificial Intelligence (AI) in human–wildlife conflict (HWC) mitigation has moved beyond theoretical promise to real-world implementation across diverse ecological and socio-political contexts. From elephant early warning systems in South Asia and Africa to automated big cat monitoring, drone-assisted surveillance, AI-powered crop protection technologies, and monitoring of smaller conflict-prone species, these case studies illustrate how computational innovation is reshaping conservation practice. Collectively, they demonstrate that AI can serve as an operational tool for reducing risk, enhancing coexistence, and strengthening evidence-based decision-making, while also revealing implementation challenges that influence scalability and long-term sustainability.

One of the most prominent applications of AI in HWC mitigation involves elephant early warning systems. In India and Sri Lanka, where Asian elephants (Elephas maximus) frequently raid crops and cause human fatalities, AI-enabled detection systems integrate camera traps, thermal imaging, seismic sensors, and machine learning classifiers to identify elephant presence near agricultural fields. For instance, automated image recognition models trained on thousands of elephant photographs can detect herds in real time and trigger SMS alerts to nearby communities. In parts of Africa, GPS-collared elephants transmit movement data to centralized monitoring platforms, where predictive algorithms forecast herd trajectories based on environmental variables such as water availability and vegetation indices (Wall et al., 2014). These systems enable conservation managers to anticipate potential incursions and coordinate rapid-response teams. Empirical evidence suggests that real-time alert systems significantly reduce surprise encounters and crop damage by providing communities with critical preparation time. The integration of AI with participatory community networks has proven especially important, as local response capacity determines whether early warnings translate into effective prevention.

AI has also transformed the monitoring of big cats in forested and fragmented landscapes. Tigers (*Panthera tigris*), leopards (*Panthera pardus*), and lions (*Panthera leo*) frequently come into conflict with humans through livestock depredation or accidental encounters. Camera trap networks, once reliant on manual image sorting, now utilize deep learning algorithms to automate species recognition and even individual identification through stripe and spot pattern analysis (Norouzzadeh et al., 2018). For example, convolutional neural networks have achieved high accuracy in distinguishing individual tigers based on stripe configurations, enabling more precise population monitoring and territory mapping (Christin et al., 2019). Such data are critical for understanding spatial overlap between predator ranges and livestock grazing zones. In India and parts of Africa, AI-driven spatial risk models integrate habitat suitability, prey density, livestock distribution, and infrastructure data to predict areas of elevated depredation risk. By identifying hotspots, authorities can deploy targeted mitigation measures, including fortified livestock enclosures or community awareness campaigns. These systems demonstrate how AI enhances both ecological understanding and operational planning in carnivore conservation.

Drone-assisted wildlife surveillance represents another innovative application of AI in conflict mitigation. Unmanned aerial vehicles (UAVs) equipped with high-resolution cameras and thermal sensors can survey large areas quickly, detecting wildlife presence in agricultural or peri-urban landscapes. AI-based object detection algorithms analyze aerial imagery in real time, identifying animals and differentiating them from livestock or human activity (Kays et al., 2015). In African savannas, drones have been deployed to monitor elephant and rhino movements near farmland, enabling rapid intervention when animals approach settlements. Similarly, in Southeast Asia, UAV-based monitoring has supported early detection of crop-raiding wildlife in plantation landscapes. The integration of drone imagery with machine learning classifiers reduces manual image review time and enhances detection accuracy, particularly in low-visibility or dense vegetation conditions. However, drone deployment requires regulatory compliance, technical expertise, and maintenance infrastructure, underscoring the need for institutional capacity building.

AI-powered crop protection systems further illustrate the intersection of conservation and agricultural technology. In regions prone to crop raiding by elephants, wild boar, or primates, sensor-based systems combine motion detection, infrared imaging, and predictive algorithms to activate deterrents such as automated lights, sounds, or barriers when wildlife approaches. Some systems employ reinforcement learning techniques to adapt deterrence strategies based on wildlife responses, minimizing habituation over time. These intelligent deterrence systems are particularly valuable in smallholder farming contexts, where economic losses from crop damage can be severe. By reducing reliance on lethal control or retaliatory measures, AI-based crop protection contributes to non-lethal coexistence strategies. Nonetheless, technological reliability, energy requirements, and affordability remain significant considerations, particularly in remote rural settings.

While megafauna often dominate discussions of HWC, AI applications also address conflicts involving smaller species. For instance, wild boar in Europe and Asia cause substantial agricultural damage, and predictive modelling has been used to forecast seasonal incursion patterns based on crop phenology and landscape features (Schley & Roper, 2003). Similarly, acoustic monitoring systems employing deep learning classifiers detect bird or bat species near airports and agricultural areas, mitigating risks to aviation and crop production (Aide et al., 2013). In North America, AI-based spatial models have been applied to predict urban coyote activity patterns, integrating land-use data and human population density to anticipate conflict zones (Poessel et al., 2013). These examples highlight the scalability of AI across taxa and ecological contexts, demonstrating that intelligent systems are not limited to charismatic megafauna but can address a broad spectrum of conflict-prone species.

Across these case studies, several lessons emerge regarding success factors, implementation challenges, and scalability potential. First, success depends on high-quality and context-specific data. Machine learning models require extensive training datasets to achieve reliable performance, and biases in data collection can undermine predictive accuracy (Christin et al., 2019). Second, community engagement is critical. Technological systems are only effective when local stakeholders trust and actively participate in response mechanisms. Integrating indigenous and local ecological knowledge enhances model relevance and fosters social acceptance. Third, institutional capacity and infrastructure determine sustainability. Reliable electricity supply, internet connectivity, technical expertise, and maintenance funding are prerequisites for long-term operation.

Implementation challenges include algorithmic transparency, data privacy concerns, and potential technological dependency. In some contexts, limited access to computational resources constrains the deployment of advanced AI systems. Moreover, scalability requires cost-effective models that can be adapted to diverse socio-ecological settings without extensive customization. Despite these challenges, many AI-based initiatives demonstrate strong scalability potential due to declining hardware costs, open-source software platforms, and increasing global collaboration in conservation informatics.

In synthesis, real-world applications of AI in HWC mitigation illustrate a tangible shift from reactive conflict response to proactive, data-driven prevention. Elephant early warning systems, automated big cat monitoring, drone-assisted surveillance, intelligent crop protection, and modelling of smaller species conflicts collectively exemplify how AI enhances detection, prediction, and strategic planning. While technological innovation alone cannot resolve the socio-political complexities of human–wildlife coexistence, these case studies confirm that AI provides powerful tools to support informed decision-making, reduce economic losses, and protect vulnerable wildlife populations. As conservation challenges intensify globally, the integration of AI into conflict mitigation strategies offers a scalable pathway toward safer human communities and more resilient ecosystems.

7. ETHICAL, PARTICIPATORY, AND FUTURE PATHWAYS FOR INTELLIGENT CONSERVATION SYSTEMS

Wildlife, along with biodiversity, is essential to maintain all the necessary ecosystem functions. We all know that the 6th mass extinction has already started. We can't ignore the cases of biodiversity loss due to the reason of rising man-animal conflict. If we can not protect our wildlife, there will also be a question mark in our survival. Human destiny is not a pre-written script; it is a path we choose through our values and actions. The destiny offered by the current paradigm of continued ecological degradation, leading to social collapse and immense suffering, is a dystopian one (Pal, R. K. 2025). However, today we are blessed that we have AI with us. The rapid integration of Artificial Intelligence (AI) into biodiversity conservation and human–wildlife conflict (HWC) mitigation has generated significant optimism, yet it also necessitates careful critical evaluation. While AI-driven systems enhance predictive capacity, automate monitoring, and support evidence-based decision-making, they are not without limitations, ethical complexities, and socio-technical constraints. A comprehensive understanding of intelligent conservation systems must therefore address structural challenges, integrate local ecological knowledge, and envision future trajectories that align technological innovation with equity, transparency, and sustainability. Ultimately, AI must complement—not replace—ecological expertise, participatory governance, and community-based conservation frameworks (Christin et al., 2019).

One of the most persistent challenges in AI-based conservation is data scarcity and bias. Machine learning models require extensive, high-quality datasets to achieve reliable predictive performance. However, ecological data are often unevenly distributed across regions and taxa, with charismatic megafauna receiving disproportionate attention compared to less-studied species (Hampton et al., 2013). Geographic bias toward well-funded research areas in the Global North may limit model applicability in biodiversity-rich yet data-poor regions. Furthermore, biases embedded in training datasets—such as overrepresentation of certain habitats or seasons—can distort model outputs and produce misleading predictions. In species distribution modelling, sampling bias can inflate accuracy metrics while masking uncertainty (Elith & Leathwick, 2009). Addressing these limitations requires transparent reporting of data provenance, integration of diverse data sources, and adaptive model validation in real-world contexts.

Infrastructure and technological costs present additional constraints. AI systems depend on reliable electricity, internet connectivity, data storage capacity, and technical expertise. In many rural landscapes where HWC is most acute, such infrastructure may be limited or inconsistent. The deployment of sensor networks, drones, or high-resolution satellite imagery can be financially prohibitive for local conservation agencies (Kays et al., 2015). Long-term maintenance costs—including hardware repair, software updates, and personnel training—must also be considered. Without sustainable funding models, AI-based interventions risk becoming short-term pilot projects rather than durable management tools.

Algorithm transparency and trust further influence adoption and legitimacy. Many AI systems, particularly deep learning models, function as "black boxes," producing outputs without easily interpretable explanations. This opacity can undermine trust among conservation practitioners and local communities, particularly when model outputs influence land-use decisions or resource allocation (Rudin, 2019). Transparent and interpretable models, combined with clear communication of uncertainty and limitations, are essential to fostering stakeholder confidence. Ethical governance frameworks should ensure accountability in cases where algorithmic errors lead to adverse ecological or social consequences.

Data privacy concerns are increasingly relevant as conservation monitoring intersects with human-dominated landscapes. Camera traps, drones, and acoustic sensors may inadvertently capture images or information about people, raising questions about consent and surveillance (Sandbrook et al., 2018). In some contexts, wildlife monitoring technologies could be misused for purposes unrelated to conservation, including unauthorized surveillance of local communities. Establishing clear data governance policies, anonymization protocols, and community consultation mechanisms is therefore critical. Ethical considerations extend beyond human privacy to wildlife welfare, as excessive monitoring or tagging may cause stress or behavioral disruption.

The risk of over-reliance on technology represents another important limitation. While AI enhances efficiency and predictive capability, it cannot substitute for ecological understanding, field expertise, and social engagement. Overconfidence in algorithmic outputs may obscure contextual nuances that only local practitioners can discern. Conservation challenges are embedded in socio-political systems, and technological solutions alone cannot resolve conflicts rooted in land tenure, economic inequality, or governance failures (Nyhus, 2016). Therefore, AI must function as a decision-support tool within interdisciplinary frameworks that integrate ecological science, social science, and community participation.

Integrating local and indigenous knowledge with AI-based conservation strategies offers a pathway to addressing these limitations. Indigenous and local communities possess detailed ecological knowledge accumulated through generations of interaction with landscapes and wildlife. This traditional ecological knowledge (TEK) includes insights into seasonal animal movements, habitat use, and behavioral cues that may not be captured in formal datasets (Berkes, 2012). Participatory monitoring systems that combine community observations with digital platforms enhance data richness while fostering local ownership of conservation initiatives.

Community-based conservation frameworks emphasize co-management, shared decision-making, and equitable benefit distribution. Integrating AI tools into these frameworks requires culturally appropriate training, capacity building, and adaptation to local contexts. For example, mobile-based reporting applications can enable farmers to record wildlife sightings, feeding into centralized predictive models. Such participatory data collection improves model accuracy and strengthens trust between conservation authorities and communities. The co-production of conservation knowledge—where scientists and local stakeholders collaboratively design monitoring systems and interpret outputs—enhances legitimacy and contextual relevance (Berkes, 2012).

A conceptual Human–Technology–Ecology Collaboration Framework illustrates this integrative approach. In this model, ecological systems generate data through natural processes and sensor-based monitoring. Technological systems process and analyze these data to generate predictive insights. Human actors—including scientists, policymakers, and local communities—interpret and apply these insights within governance and livelihood contexts. Feedback loops ensure that community observations refine algorithms, while model outputs inform local management strategies. This collaborative triad reinforces resilience by combining computational precision with experiential knowledge and social legitimacy.

Looking forward, emerging technological trends promise to further expand AI's role in biodiversity conservation. Integration with the Internet of Things (IoT) enables distributed sensor networks that continuously monitor environmental variables, wildlife presence, and anthropogenic activities (Hampton et al., 2013). Edge computing—where data processing occurs locally on devices rather than centralized servers—reduces latency and enhances functionality in remote ecosystems with limited connectivity. Autonomous drone monitoring systems equipped with onboard AI algorithms can conduct routine patrols, detect wildlife or illegal activities, and transmit alerts in near real time.

Global wildlife data networks represent another frontier. Collaborative platforms that aggregate telemetry, remote sensing, and citizen science data across countries facilitate large-scale ecological modelling and cross-border conservation planning (Kays et al., 2015). Predictive conservation models integrating climate projections, land-use scenarios, and socio-economic variables can simulate future conflict dynamics and inform long-term policy decisions (Elith & Leathwick, 2009). Such innovations point toward AI-enabled smart conservation landscapes—adaptive systems in which real-time monitoring, predictive analytics, and participatory governance converge to minimize conflict and enhance ecosystem resilience.

In conclusion, the integration of Artificial Intelligence into conservation science represents a transformative yet complex development. AI enables early detection of wildlife presence, predictive modelling of conflict hotspots, and data-driven policy formulation, shifting conservation from reactive intervention to anticipatory management. Through automated monitoring, spatial risk mapping, and real-time alert systems, AI enhances coexistence strategies and supports evidence-based governance. However, technological innovation must be guided by ethical principles, transparency, and inclusive participation. Data bias, infrastructure constraints, privacy concerns, and algorithmic opacity underscore the necessity of interdisciplinary collaboration and community engagement. When integrated responsibly within socio-ecological systems, AI has the potential to function as a "digital shield," protecting biodiversity while safeguarding the safety, livelihoods, and dignity of human communities. The future of conservation thus lies not solely in technological sophistication, but in the synergistic alignment of artificial intelligence, ecological knowledge, and human stewardship.

Note from the Publisher

Correction Notice: A correction to this manuscript has been published
Reason: Missing Image
The publisher was notified that an image was missing from the chapter and has been added.
Correction Date: 31st July 2026

REFERENCES

Aide, T. M., Corrada-Bravo, C., Campos-Cerqueira, M., Milan, C., Vega, G., & Alvarez, R. (2013). Real-time bioacoustics monitoring and automated species identification. *PeerJ*, *1*, e103. DOI: 10.7717/peerj.103 PMID: 23882441

Beery, S., Van Horn, G., & Perona, P. (2018). Recognition in terra incognita. In *Proceedings of the European conference on computer vision (ECCV)* (pp. 456-473).

Berkes, F. (2012). *Sacred ecology* (3rd ed.). Routledge. DOI: 10.4324/9780203123843

Christin, S., Hervet, É., & Lecomte, N. (2019). Applications for deep learning in ecology. *Philosophical Transactions of the Royal Society B*, 374(1763), 20180084.

Crooks, K. R., Burdett, C. L., Theobald, D. M., Rondinini, C., & Boitani, L. (2011). Global patterns of fragmentation and connectivity of mammalian carnivore habitat. *Philosophical Transactions of the Royal Society of London. Series B, Biological Sciences*, *366*(1578), 2642–2651. DOI: 10.1098/rstb.2011.0120 PMID: 21844043

Dickman, A. J. (2010). Complexities of conflict: The importance of considering social factors for effectively resolving human–wildlife conflict. *Animal Conservation*, *13*(5), 458–466. DOI: 10.1111/j.1469-1795.2010.00368.x

Elith, J., & Leathwick, J. R. (2009). Species distribution models: Ecological explanation and prediction across space and time. *Annual Review of Ecology, Evolution, and Systematics*, *40*(1), 677–697. DOI: 10.1146/annurev.ecolsys.110308.120159

Ellis, E. C. (2015). Ecology in an anthropogenic biosphere. *Ecological Monographs*, *85*(3), 287–331. DOI: 10.1890/14-2274.1

Fischer, J., & Lindenmayer, D. B. (2007). Landscape modification and habitat fragmentation: A synthesis. *Global Ecology and Biogeography*, *16*(3), 265–280. DOI: 10.1111/j.1466-8238.2007.00287.x

Guadu, T., Berhane, A., Solomon, T., Tafese, A., Tesfay, K., Asgedom, A. A., Abera, B., Debebe, B., Getachew, M., Shibabaw, A., Abiye, A. A., Abdurahman, M., Yadeta, W., Tadesse, Y., Lemma, H., Theodore, K.-J., Mor, S. M., Mereta, S. T., & Kaba, M. (2026). Human-wildlife conflict and its impacts on livelihoods, health, and the environment in Chebera Churchura National Park, Southwest Ethiopia: A phenomenological study using a One Health approach with an urgent call for action. *One Health Outlook*, *8*(1), 8. DOI: 10.1186/s42522-025-00188-y PMID: 41547988

Haddad, N. M., Brudvig, L. A., Clobert, J., Davies, K. F., Gonzalez, A., Holt, R. D., Lovejoy, T. E., Sexton, J. O., Austin, M. P., Collins, C. D., Cook, W. M., Damschen, E. I., Ewers, R. M., Foster, B. L., Jenkins, C. N., King, A. J., Laurance, W. F., Levey, D. J., Margules, C. R., & Townshend, J. R. (2015). Habitat fragmentation and its lasting impact on Earth's ecosystems. *Science Advances, 1*(2), e1500052. DOI: 10.1126/sciadv.1500052 PMID: 26601154

Hampton, S. E., Strasser, C. A., Tewksbury, J. J., Gram, W. K., Budden, A. E., Batcheller, A. L., Duke, C. S., & Porter, J. H. (2013). Big data and the future of ecology. *Frontiers in Ecology and the Environment, 11*(3), 156–162. DOI: 10.1890/120103

Hebblewhite, M., & Haydon, D. T. (2010). Distinguishing technology from biology: A critical review of the use of GPS telemetry data in ecology. *Philosophical Transactions of the Royal Society of London. Series B, Biological Sciences, 365*(1550), 2303–2312. DOI: 10.1098/rstb.2010.0087 PMID: 20566506

Kays, R., Crofoot, M. C., Jetz, W., & Wikelski, M. (2015). Terrestrial animal tracking as an eye on life and planet. *Science, 348*(6240), aaa2478. DOI: 10.1126/science.aaa2478 PMID: 26068858

LeCun, Y., Bengio, Y., & Hinton, G. (2015). Deep learning. *Nature, 521*(7553), 436–444. DOI: 10.1038/nature14539 PMID: 26017442

Liu, J., Dietz, T., Carpenter, S. R., Alberti, M., Folke, C., Moran, E., ... & Taylor, W. W. (2007). Complexity of coupled human and natural systems. *science, 317*(5844), 1513-1516.

Nichols, J. D., & Williams, B. K. (2006). Monitoring for conservation. *Trends in Ecology & Evolution, 21*(12), 668–673. DOI: 10.1016/j.tree.2006.08.007 PMID: 16919361

Norouzzadeh, M. S., Nguyen, A., Kosmala, M., Swanson, A., Palmer, M. S., Packer, C., & Clune, J. (2018). Automatically identifying, counting, and describing wild animals in camera-trap images with deep learning. *Proceedings of the National Academy of Sciences of the United States of America, 115*(25), E5716–E5725. DOI: 10.1073/pnas.1719367115 PMID: 29871948

Nyhus, P. J. (2016). Human–wildlife conflict and coexistence. *Annual Review of Environment and Resources, 41*(1), 143–171. DOI: 10.1146/annurev-environ-110615-085634

Pal, R. K. (2025). The Ecological Self: Deep Ecology as a Foundational Framework for Human Destiny and Sustainable Welfare. *Journal of Social Welfare and Management, 17*(3).

Paul, P. K. (2022). Environmental informatics: basics, nature, and applications using emerging technologies with reference to issues and potentialities. In *Environmental Informatics: Challenges and Solutions* (pp. 1–13). Springer Nature Singapore. DOI: 10.1007/978-981-19-2083-7_1

Paul, P. K. (2026). Environmental Informatics Promotion Using Artificial Intelligence and Allied Technologies: The Applications, Sustainability and Educational Perspectives. In Harnessing the potential of Artificial Intelligence to Promote Environment Sustainability, (pp. 111-124). Emerald, UK

Paul, P. K., Aithal, P. S., Bhuimali, A., Kalishankar, T., Saavedra, M. R., & Aremu, P. S. B. (2020b). Geo information systems and remote sensing: Applications in environmental systems and management. [IJMTS]. *International Journal of Management, Technology, and Social Sciences*, *5*(2), 11–18. DOI: 10.47992/IJMTS.2581.6012.0100

Paul, P.K., Bhuimali, A., Aithal, P. S., Kalishankar, T., & Saavedra M, R. (2020a). Artificial intelligence & cloud computing in environmental systems-towards healthy & sustainable development. *International Journal of Inclusive Development, 6*(1), 01-08.

Pecl, G. T., Araújo, M. B., Bell, J. D., Blanchard, J., Bonebrake, T. C., Chen, I. C., Clark, T. D., Colwell, R. K., Danielsen, F., Evengård, B., Falconi, L., Ferrier, S., Frusher, S., Garcia, R. A., Griffis, R. B., Hobday, A. J., Janion-Scheepers, C., Jarzyna, M. A., Jennings, S., & Williams, S. E. (2017). Biodiversity redistribution under climate change: Impacts on ecosystems and human well-being. *Science, 355*(6332), eaai9214. DOI: 10.1126/science.aai9214 PMID: 28360268

Peters, D. P., Havstad, K. M., Cushing, J., Tweedie, C., Fuentes, O., & Villanueva-Rosales, N. (2014). Harnessing the power of big data: Infusing the scientific method with machine learning to transform ecology. *Ecosphere*, *5*(6), 1–15. DOI: 10.1890/ES13-00359.1

Poessel, S. A., Breck, S. W., Teel, T. L., Shwiff, S., Crooks, K. R., & Angeloni, L. (2013). Patterns of human–coyote conflicts in the Denver Metropolitan Area. *The Journal of Wildlife Management*, *77*(2), 297–305. DOI: 10.1002/jwmg.454

Rudin, C. (2019). Stop explaining black box machine learning models for high stakes decisions and use interpretable models instead. *Nature Machine Intelligence*, *1*(5), 206–215. DOI: 10.1038/s42256-019-0048-x PMID: 35603010

Russell, S., & Norvig, P. (2021). *Artificial intelligence: A modern approach* (4th ed.). Pearson.

Sandbrook, C., Luque-Lora, R., & Adams, W. M. (2018). Human bycatch: Conservation surveillance and the social implications of camera traps. *Conservation & Society*, *16*(4), 493–504. DOI: 10.4103/cs.cs_17_165

Schley, L., & Roper, T. J. (2003). Diet of wild boar in Western Europe, with particular reference to consumption of agricultural crops. *Mammal Review*, *33*(1), 43–56. DOI: 10.1046/j.1365-2907.2003.00010.x

Treves, A., & Karanth, K. U. (2003). Human–carnivore conflict and perspectives on carnivore management worldwide. *Conservation Biology*, *17*(6), 1491–1499. DOI: 10.1111/j.1523-1739.2003.00059.x

Turner, W., Spector, S., Gardiner, N., Fladeland, M., Sterling, E., & Steininger, M. (2003). Remote sensing for biodiversity science and conservation. *Trends in Ecology & Evolution*, *18*(6), 306–314. DOI: 10.1016/S0169-5347(03)00070-3

Wall, J., Wittemyer, G., Klinkenberg, B., & Douglas-Hamilton, I. (2014). Novel opportunities for wildlife conservation and research with real-time monitoring. *Ecological Applications*, *24*(4), 593–601. DOI: 10.1890/13-1971.1 PMID: 24988762

Yeshey, K., Keenan, R. J., Ford, R. M., & Nitschke, C. R. (2023). Sustainable development implications of human–wildlife conflict: An analysis of subsistence farmers in Bhutan. *International Journal of Sustainable Development and World Ecology*, *30*(5), 548–563. DOI: 10.1080/13504509.2023.2167242

Chapter 8
Few-Shot Learning for Fine-Grained Bird Species Classification in AI-Driven Biodiversity Monitoring

Anmol Anand
http://orcid.org/0009-0001-8821-5633
Jaypee University of Engineering and Technology, Guna, India

Divyansh Jain
http://orcid.org/0009-0009-7762-2928
Jaypee University of Engineering and Technology, Guna, India

Uttkarsh
http://orcid.org/0009-0003-6523-4908
Jaypee University of Engineering and Technology, Guna, India

Kunj Bihari Bihari Meena
http://orcid.org/0000-0001-8159-9024
Jaypee University of Engineering and Technology, Guna, India

Vipin Tyagi
http://orcid.org/0000-0003-4994-3686
Jaypee University of Engineering and Technology, Guna, India

DOI: 10.4018/979-8-2600-2691-5.ch008

ABSTRACT

Accurate species identification is essential for biodiversity monitoring and conservation, yet many species are represented by only a few labeled samples. Few-shot learning (FSL) addresses this limitation by enabling models to recognize new categories from minimal data. This chapter explores FSL methods for fine-grained bird species classification, a task characterized by subtle inter-class differences and high intra-class variability. Using the CUB-200-2011 dataset, two approaches are evaluated: an episodic meta-learning pipeline based on Prototypical Networks with a ResNet-18 backbone and a representation-centric pipeline using a self-supervised Vision Transformer (ViT-DINO) as a frozen feature encoder. Performance is assessed under standard 5-way 1-shot and 5-way 5-shot settings. The findings offer practical insights for deploying AI-driven species recognition systems in biodiversity monitoring and conservation.

1. INTRODUCTION

Accurate identification of biological species is a fundamental requirement for biodiversity assessment, ecological monitoring and wildlife conservation. Automated species recognition systems have become increasingly important as ecological data collection scales through camera traps, citizen science platforms and remote sensing technologies. While recent advances in deep learning have significantly improved visual recognition performance, most state-of-the-art models depend on large, carefully annotated datasets. In ecological and conservation contexts, however, such extensive labeled data are rarely available, particularly for rare, endangered, or newly observed species. This mismatch between data availability and model requirements presents a critical barrier to the deployment of artificial intelligence in real-world biodiversity monitoring.

Fine-grained image classification poses additional challenges beyond those encountered in general object recognition tasks, primarily because the visual differences between categories are often extremely subtle and highly localized. In domains such as bird species identification, distinctions between classes may depend on minor variations in plumage patterns, color distributions, beak morphology, wing shape, or body proportions. These discriminative cues are frequently small in scale and may only be visible from specific viewpoints, making them difficult for automated systems to detect consistently. At the same time, substantial intra-class variation arises from factors such as age, sex, seasonal plumage changes, molting cycles, pose, and illumination conditions. As a result, images belonging to the same species can exhibit considerable visual diversity, further complicating the learning process.

These characteristics pose significant challenges for data-driven recognition models, particularly those that rely on large amounts of labeled training data to learn stable and discriminative representations. Standard deep learning approaches tend to focus on dominant visual patterns, which may be insufficient for capturing the nuanced cues required for fine-grained discrimination. When training data are limited, models often overfit to spurious features or fail to generalize beyond the specific examples observed during training. This issue is especially pronounced in ecological contexts, where collecting and annotating large datasets is often impractical due to the need for expert knowledge, the rarity of certain species, and the logistical constraints of field data collection.

Moreover, fine-grained datasets frequently exhibit long-tailed class distributions, with a small number of well-documented species and a large number of rare or underrepresented classes. This imbalance further exacerbates performance degradation in conventional supervised learning frameworks, which are not designed to operate effectively under extreme data scarcity. Environmental variability, such as changes in background vegetation, weather conditions, and camera quality, introduces additional sources of noise that can obscure subtle species-specific features.

Consequently, learning paradigms that can generalize effectively from only a few labeled examples are of particular relevance for fine-grained ecological applications. Few-shot learning approaches address these challenges by emphasizing transferable representations and similarity-based reasoning rather than class-specific memorization. By learning to compare new instances to a small set of reference examples, few-shot models are better aligned with the constraints of real-world biodiversity monitoring. Such approaches offer a principled pathway toward scalable, data-efficient species recognition systems that can support ecological research, conservation planning, and long-term environmental monitoring in settings where labeled data remain inherently scarce.

Few-shot learning (FSL) has emerged as a promising paradigm for addressing the persistent challenge of data scarcity in machine learning by enabling models to recognize novel categories using only a small number of annotated examples. This capability is particularly valuable in domains where collecting labeled data is expensive, time consuming, or requires specialized expertise, such as ecological monitoring, medical imaging, and scientific discovery. Instead of relying on extensive labeled datasets to learn explicit class-specific decision boundaries, few-shot learning frameworks are designed to acquire knowledge that can be reused across tasks, allowing models to adapt efficiently when presented with previously unseen categories.

Central to most few-shot learning approaches is the concept of transferable representation learning or similarity-based reasoning. Models are trained to map inputs into an embedding space in which samples from the same class are close to one another, while samples from different classes are well separated. During inference, new classes are recognized by comparing a small set of labeled support examples with unlabeled query instances, rather than by training a new classifier from scratch. This formulation aligns well with scenarios in which rapid adaptation and minimal supervision are required.

Early research in few-shot learning relied heavily on metric-based and meta-learning techniques. Metric-based methods, such as Prototypical Networks, represent each class using a prototype computed from its support examples and classify queries based on distance to these prototypes. Meta-learning approaches, including model-agnostic meta-learning, aim to learn model initializations or update rules that enable rapid adaptation to new tasks with only a few gradient steps. These methods demonstrated encouraging performance on standard benchmark datasets and established few-shot learning as a viable alternative to conventional supervised training.

However, recent research has shown that the quality of the underlying feature representations often plays a more decisive role in few-shot performance than the specific choice of meta-learning algorithm. In particular, representations obtained through large-scale supervised or self-supervised pretraining have been found to substantially improve generalization in low-data regimes. As a result, contemporary few-shot learning research increasingly emphasizes representation strength and transferability as the primary drivers of performance, shifting the focus from complex adaptation mechanisms to the development of robust and semantically rich feature embeddings.

In parallel, self-supervised representation learning has transformed visual recognition by enabling models to learn rich semantic features from unlabeled data. Vision Transformers pretrained using self-supervised objectives, such as DINO, have been shown to produce highly structured embedding spaces that support strong few-shot and nearest-neighbor performance without task-specific fine-tuning. These developments raise important questions about the relative contributions of episodic meta-learning versus representation-centric approaches in fine-grained few-shot classification, especially in biologically meaningful domains where data are scarce and visual distinctions are subtle.

This chapter investigates these issues through a systematic study of few-shot learning for fine-grained bird species classification using the CUB-200-2011 dataset, a widely adopted benchmark with clean annotations and high inter-class similarity. Two complementary few-shot pipelines are examined. The first follows a classical task-driven approach, employing a ResNet-18 backbone trained episodically within the Prototypical Networks framework. The second adopts a representation-driven strategy, leveraging a self-supervised Vision Transformer (ViT-DINO) as a frozen feature extractor without episodic training. Both approaches are evaluated under standard 5-way 1-shot and 5-way 5-shot classification protocols to ensure fair and reproducible comparison.

Beyond classification accuracy, the chapter explores several factors that are critical for real-world deployment, including the impact of support-set composition, the choice of distance metrics, the stability and separability of learned embeddings under limited supervision and qualitative differences between convolutional and transformer-based representations. By analyzing these dimensions, the study aims to clarify how representation structure, granularity and global context modeling influence few-shot performance in fine-grained recognition tasks.

The contributions of this chapter are threefold. First, it provides a controlled empirical comparison between episodic meta-learning and self-supervised transformer-based representations in a challenging fine-grained classification setting. Second, it offers practical insights into the design of few-shot pipelines for biodiversity monitoring, including backbone selection and evaluation strategies. Finally, it highlights the potential of representation-centric few-shot learning as a scalable and computationally efficient approach for AI-driven species recognition, supporting broader efforts to integrate artificial intelligence into ecological research and wildlife conservation.

2. RELATED WORK

Few-shot learning (FSL) has emerged as a critical research direction in machine learning, motivated by the need to construct models that generalize effectively from extremely limited labeled data. This challenge is particularly relevant in domains where annotation is costly, expertise-dependent, or inherently scarce. Early research in FSL concentrated on metric-learning and meta-learning paradigms, with the objective of learning transferable representations or rapid adaptation strategies across tasks. More recent work has shifted emphasis toward the role of representation quality, especially in the context of large-scale pretraining and self-supervised learning. This section reviews prior work in classical few-shot learning, fine-grained few-shot image classification and representation learning with convolutional and transformer-based architectures, situating the present study within this evolving research landscape.

2.1 Classical Few-Shot Learning: Metric-Based and Meta-Learning Approaches

Metric-based few-shot learning methods constitute the foundation of many early FSL frameworks. Prototypical Networks introduced a simple yet powerful formulation in which each class is represented by the mean embedding of its support examples and query samples are classified based on their distance to these class prototypes (Snell et al., 2017). This approach encourages the formation of compact, discriminative embedding spaces and remains a widely adopted baseline in few-shot evaluation. Related methods, such as Matching Networks (Vinyals et al., 2016) and Relation Networks (Sung et al., 2018), similarly rely on learned similarity functions to compare support and query instances, assuming that semantic similarity can be captured within a shared embedding space.

Meta-learning approaches, often described as "learning to learn," aim to optimize model parameters or initialization schemes that enable rapid adaptation to new tasks. Model-Agnostic Meta-Learning (MAML) exemplifies this paradigm by learning parameter initializations that can be efficiently fine-tuned with a small number of gradient updates (Finn et al., 2017). Numerous extensions have been proposed to improve stability and scalability, including first-order approximations and probabilistic variants. However, subsequent empirical analyses have shown that the benefits of meta-learning diminish when strong pretrained backbones are available (Chen et al., 2019; Dhillon et al., 2020). Comprehensive surveys increasingly conclude that representation quality often dominates the choice of meta-learning algorithm in determining few-shot performance (Wang et al., 2020; Hospedales et al., 2021).

2.2 Few-Shot Learning for Fine-Grained Image Classification

Fine-grained image classification introduces challenges that are not fully captured by standard few-shot benchmarks such as mini-ImageNet. In fine-grained domains, inter-class differences are subtle and often localized, while intra-class variation can be substantial due to pose, illumination, age, or seasonal changes. These properties make it difficult for models trained on coarse-grained datasets to generalize effectively. As a result, several studies have proposed specialized architectures and learning strategies to address fine-grained few-shot recognition, including part-based modeling, attention mechanisms and relational reasoning frameworks.

Graph Neural Network–based approaches have been shown to improve discrimination by modeling relationships among images or classes in fine-grained settings (Zhou et al., 2022). Other studies emphasize learning region-aware or part-centric representations that focus on biologically meaningful features, such as wing patterns, beak shape, or texture variations in birds (Wei et al., 2019; Ren et al., 2024). Empirical evidence consistently demonstrates that standard few-shot models underperform on fine-grained datasets unless representations or similarity metrics are adapted to capture subtle visual cues.

The Caltech-UCSD Birds-200-2011 (CUB-200-2011) dataset has emerged as a canonical benchmark for fine-grained few-shot learning due to its high inter-class similarity and clean annotations (Wah et al., 2011). Its widespread adoption enables controlled evaluation of representation learning effects and has facilitated comparative studies across architectures and learning paradigms.

2.3 Self-Supervised Pretraining and Transformer-Based Representations

Recent advances in self-supervised learning have significantly reshaped representation learning for few-shot tasks. Contrastive and non-contrastive self-supervised methods have demonstrated that models can learn highly structured semantic representations from unlabeled data alone. DINO showed that Vision Transformers pretrained with self-supervision produce embeddings that exhibit strong clustering and semantic alignment, leading to excellent k-nearest neighbor and few-shot performance without task-specific fine-tuning (Caron et al., 2021). Related approaches, such as MoCo and SimCLR, further highlight the importance of large-scale representation learning for downstream low-data tasks (He et al., 2020; Chen et al., 2020).

Transformer-based architectures differ fundamentally from convolutional neural networks in their reliance on self-attention to model global context. This capability has proven particularly advantageous for fine-grained recognition, where long-range dependencies and subtle spatial relationships are critical. Recent studies show that pretrained Vision Transformers often outperform convolutional networks in few-shot classification, even without episodic meta-training (Dosovitskiy et al., 2021; Chen et al., 2022). Consequently, many modern few-shot pipelines adopt a pretrain–then–adapt paradigm, in which the majority of performance gains stem from representation strength rather than adaptation mechanisms. These developments motivate a direct comparison between classical episodic learning with convolutional backbones and representation-centric approaches based on self-supervised Vision Transformers.

2.4 Support-Set Composition and Distance Metrics

Beyond architectural and training considerations, several practical factors significantly influence few-shot performance. Prior work has examined the impact of support-set selection strategies, demonstrating that representative samples often yield more stable and reliable class prototypes than maximally diverse selections, particularly in fine-grained domains (Dhillon et al., 2020; Afrasiyabi et al., 2022). Similarly, the choice of distance metric plays a critical role in prototype-based classification. Cosine similarity has been shown to outperform Euclidean distance in high-dimensional embedding spaces and is especially well suited to transformer-based representations (Chen et al., 2019; Tian et al., 2020).

These findings underscore the importance of carefully designed evaluation protocols and motivate the experimental design adopted in the present study, where multiple support-set configurations and distance metrics are systematically analyzed to assess their interaction with different backbone architectures.

In contrast to previous research studies that focus exclusively on either episodic meta-learning or self-supervised representation learning, this chapter provides a controlled and systematic comparison of both paradigms within a fine-grained biological domain. By evaluating episodically trained convolutional networks alongside frozen self-supervised transformer encoders under identical few-shot protocols, the study clarifies the relative strengths and limitations of task-driven versus representation-centric approaches. This positioning aligns the work with contemporary trends in few-shot research while grounding the analysis in a biologically meaningful application context relevant to biodiversity monitoring and ecological research.

In summary, prior research establishes that representation quality is the key driver of few-shot performance, that fine-grained biological domains introduce unique challenges not addressed by standard benchmarks, and that self-supervised Vision Transformers offer strong, transferable embeddings well suited to low-data ecological settings. These insights directly motivate the methodological choices described in the following section.

3. PROPOSED METHOD

This section describes the dataset, preprocessing procedures, model architectures and few-shot learning protocols used in this study. The methodological design emphasizes controlled experimentation in order to isolate the effects of representation learning strategies and backbone architectures on fine-grained few-shot classification performance.

3.1 Experimental Setup

All experiments are conducted using the Caltech-UCSD Birds-200-2011 (CUB-200-2011) dataset, a widely adopted benchmark for fine-grained visual recognition (Wah et al., 2011). The dataset contains 200 bird species with approximately 12,000 images annotated with high-quality class labels. Due to its high inter-class similarity and clean annotations, CUB-200-2011 provides a challenging yet controlled environment for evaluating few-shot learning methods in biologically meaningful scenarios (Ren et al., 2024).

Following standard practice in few-shot learning, the dataset is partitioned into disjoint base and novel class splits (Wang et al., 2020). Base classes are used for representation learning or episodic training, while novel classes are reserved exclusively for evaluation. This separation ensures that performance reflects the model's ability to generalize to previously unseen species rather than memorization of training categories.

Few-shot evaluation follows the widely used *N-way K-shot* protocol (Snell et al., 2017). In this study, 5-way 1-shot and 5-way 5-shot settings are used. Each evaluation episode consists of a support set containing *K* labeled examples per class and a query set drawn from the same classes. Final results are reported as the average accuracy over multiple randomly sampled episodes.

As shown in Figure 1, the CUB-200-2011 dataset contains high inter-class similarity and substantial intra-class variation, making it well suited for fine-grained few-shot evaluation.

3.2 Image Preprocessing and Data Augmentation

All images were resized, normalized and augmented to enhance feature stability in low-shot regimes, following standard practices in digital image processing (Tyagi, 2017). Images are resized to match the input requirements of each backbone architecture: 224×224 pixels for the ResNet-18 convolutional network (He et al., 2016) and 256×256 pixels for the Vision Transformer (ViT-DINO), consistent with its self-supervised pretraining configuration (Caron et al., 2021).

Input images are normalized using the standard ImageNet mean and standard deviation values (Deng et al., 2009). This normalization aligns the data distribution with that of the pretrained backbones and reduces variability caused by illumination and color differences, which are especially pronounced in fine-grained datasets (Chen et al., 2019).

Controlled data augmentation is applied to improve generalization in low-data regimes. Augmentation techniques include random horizontal flipping, color jittering, random cropping and small affine transformations. These operations have been shown to be effective in fine-grained few-shot learning by increasing robustness while preserving class semantics (Zhou et al., 2022; Ren et al., 2024). The preprocessing steps and augmentation strategies applied to each backbone are summarized in Table 1.

3.3 Few-Shot Learning Pipelines

Two complementary few-shot learning pipelines are investigated, reflecting distinct learning philosophies: a task-driven episodic learning approach and a representation-centric approach based on self-supervised pretraining.

3.3.1 Episodic Meta-Learning with Prototypical Networks

The first pipeline follows a classical few-shot learning paradigm based on episodic meta-learning. A ResNet-18 convolutional backbone pretrained on ImageNet (He et al., 2016; Deng et al., 2009) is used as the feature extractor. During training, the model is optimized using the Prototypical Networks framework (Snell et al., 2017), where each episode simulates a few-shot classification task drawn from the base classes.

Within each episode, class prototypes are computed by averaging the embeddings of support samples for each class. Query samples are classified based on their distance to these prototypes in the embedding space. This episodic formulation encourages the network to learn transferable representations that generalize well to novel classes (Vinyals et al., 2016). Early stopping is employed to mitigate overfitting to base-class species, which is known to degrade few-shot transfer performance (Dhillon et al., 2020).

3.3.2 Representation-Centric Few-Shot Learning with ViT-DINO

The second pipeline adopts a representation-centric approach that decouples feature learning from task-specific adaptation. A Vision Transformer pretrained using the DINO self-supervised framework is employed as a frozen feature encoder (Caron et al., 2021). Unlike episodic meta-learning, no task-specific training is performed on the few-shot episodes.

Support and query images are embedded using the CLS token representation produced by the transformer. Class prototypes are computed by averaging support embeddings and query samples are classified based on similarity to these prototypes. This approach evaluates the intrinsic quality of the learned representation and follows recent findings that strong pretrained features often outperform explicit meta-learning strategies in few-shot classification (Chen et al., 2019; Wang et al., 2020).

3.4 Distance Metrics and Prototype Construction

Prototype-based classification is employed consistently across both pipelines to ensure comparability. Class prototypes are computed as the mean of support embeddings for each class and query samples are assigned to the nearest prototype.

Cosine similarity is used as the primary distance metric, as prior studies have shown it to outperform Euclidean distance in high-dimensional embedding spaces, particularly for transformer-based representations (Chen et al., 2019). Additional experiments examine the sensitivity of each pipeline to variations in support-set composition, following observations that representative samples yield more stable prototypes in fine-grained settings (Dhillon et al., 2020).

3.5 Implementation Details

All experiments are implemented in PyTorch using Python 3.10. Episodic training for the ResNet-18 backbone follows standard few-shot optimization procedures (Snell et al., 2017), while the ViT-DINO backbone remains frozen throughout all experiments, enabling efficient inference and rapid evaluation. Hardware configurations include both GPU-accelerated and CPU-based systems to assess computational feasibility across platforms.

3.6 Real-World Deployment Challenges and Ecological Data Biases

Deploying few-shot species recognition in operational ecological settings introduces a set of challenges that extend beyond controlled benchmark evaluation. In-the-wild data collected from camera traps, aerial surveys, and citizen science platforms frequently suffers from variable image quality, inconsistent lighting, partial occlusions, and significant background clutter. These factors can degrade prototype quality and reduce classification reliability, particularly under the 1-shot regime where a single noisy support image may not adequately represent a class.

Ecological datasets are also subject to systemic sampling biases that can compromise model generalization. Observations are disproportionately concentrated in accessible geographic regions, during favorable weather conditions, and for charismatic or easily photographed species. Rare, cryptic, and nocturnal species are correspondingly underrepresented, creating long-tailed distributions that penalize few-shot evaluation for the very categories of greatest conservation interest. Future work should explicitly account for these biases, for example through stratified sampling strategies or bias-aware evaluation protocols.

Seasonal and geographic variation presents an additional deployment challenge. Plumage coloration, behavior, and habitat associations shift across seasons and regions, meaning that a support set compiled from summer images may not reliably represent winter or migratory forms of the same species. Domain shift between the ImageNet pretraining distribution and real ecological imagery is a related concern, suggesting that domain-adaptive pretraining or in-domain self-supervised fine-tuning may be necessary for robust field deployment.

Together, these considerations highlight that the transition from benchmark accuracy to real-world impact requires deliberate attention to data quality, sampling representativeness, and environmental variability. The methodological framework described in this chapter provides a sound foundation, and the results on CUB-200-2011 serve as an upper bound on performance under clean conditions. Quantifying the gap between curated and in-the-wild performance remains an important objective for future applied research.

In summary, Section 3 has described the dataset, preprocessing procedures, two complementary few-shot pipelines, distance metric choices, implementation details, and key deployment considerations. The architectural realization of these design choices is presented in the following section.

Figure 1. Sample images from CUB200-2011 dataset

Table 1. Details of preprocessing and augmentation

Parameter	Description	ResNet-18	ViT-DINO
Rescaling / Normalization	Normalizes pixel intensities using ImageNet statistics	Mean = (0.485, 0.456, 0.406)Std = (0.229, 0.224, 0.225)	Mean = (0.485, 0.456, 0.406)Std = (0.229, 0.224, 0.225)
Target size	Standardizes input image dimensions	224 × 224	256 × 256
Random crop	Random cropping to introduce scale invariance	Enabled	Enabled
Horizontal flip	Random horizontal flipping for pose invariance	Enabled	Enabled
Color jitter	Random brightness, contrast and saturation adjustment	Enabled	Enabled
Affine transform	Applies small random rotation and shear	Rotation ≤ 10°Shear ≤ 10%	Rotation ≤ 10°Shear ≤ 10%
Zoom range	Random zoom variation	±10%	±10%

continued on following page

Table 1. Continued

Parameter	Description	ResNet-18	ViT-DINO
Augmentation scope	Applied during few-shot evaluation	Support set only	Support set only
Query preprocessing	Preprocessing applied to query images	Resize + Normalize only	Resize + Normalize only

4. SYSTEM ARCHITECTURE

This study adopts a transfer learning–based few-shot learning architecture designed to support fine-grained bird species classification under limited labeled data conditions. The overall framework integrates pretrained visual feature extractors with a prototype-based classification mechanism, enabling efficient adaptation to novel species without requiring extensive retraining. The architectural design emphasizes representation quality, computational efficiency and robustness in low-shot regimes.

4.1 Backbone Feature Extractors

Two backbone architectures are employed to investigate the impact of representation learning strategies on few-shot performance. The first backbone is a ResNet-18 convolutional neural network pretrained on the ImageNet dataset (He et al., 2016; Deng et al., 2009). Convolutional networks are well known for their ability to capture local visual patterns, such as edges, textures and fine-grained structural cues, which are critical for distinguishing closely related bird species.

The second backbone is a Vision Transformer pretrained using the DINO self-supervised learning framework (Caron et al., 2021). Unlike convolutional architectures, Vision Transformers rely on self-attention mechanisms to model global contextual relationships across an image. This property allows transformer-based models to capture long-range dependencies and subtle inter-part relationships, which are particularly relevant for fine-grained visual recognition tasks.

Both backbones are used as fixed feature extractors during few-shot evaluation. The ResNet-18 backbone produces a 512-dimensional embedding after global average pooling, while the ViT-DINO backbone outputs a 384-dimensional embedding derived from the CLS token representation. Freezing the pretrained weights ensures that classification performance reflects the quality of the learned representations rather than task-specific fine-tuning.

4.2 Transfer Learning in Few-Shot Classification

Transfer learning plays a central role in the proposed architecture by enabling the reuse of knowledge acquired from large-scale visual datasets. By leveraging pretrained backbones, the framework avoids training deep networks from scratch, which is infeasible in few-shot settings. Prior work has shown that strong pretrained representations significantly improve few-shot generalization, often surpassing the gains achieved through complex meta-learning strategies (Chen et al., 2019; Wang et al., 2020).

In the episodic learning pipeline, the ResNet-18 backbone is further optimized using meta-training on base classes, allowing the network to adapt its feature space to the few-shot task distribution (Snell et al., 2017). In contrast, the ViT-DINO pipeline relies entirely on self-supervised pretraining, evaluating the model's ability to transfer learned representations directly to novel species without additional task-specific learning.

4.3 Prototype-Based Classification Mechanism

Following feature extraction, classification is performed using a prototype-based decision mechanism, which is standard in metric-learning-based few-shot frameworks (Snell et al., 2017). For each class in a given episode, a class prototype is computed by averaging the embeddings of the K support samples. This prototype serves as a compact representation of the class in the embedding space.

During inference, query images are embedded using the same backbone and compared to each class prototype. The query sample is assigned the label of the prototype with the highest similarity score. Cosine similarity is used as the primary distance metric, as it has been shown to perform robustly in high-dimensional feature spaces and is particularly effective for transformer-based embeddings (Chen et al., 2019).

This distance-based formulation eliminates the need for fully connected classification heads and dense parameter learning, making the architecture naturally suited for scenarios with extreme label scarcity. Moreover, the simplicity of prototype computation enables fast inference and efficient deployment in resource-constrained environments.

4.4 Architectural Workflow

The complete architectural workflow is illustrated in Figure 2. Input images are first resized and normalized according to backbone-specific requirements. Images are then passed through the selected pretrained feature extractor, either ResNet-18 or ViT-DINO, to obtain fixed-dimensional embeddings. For each episode, class prototypes are computed from the support set embeddings and query samples are classified based on their similarity to these prototypes.

By maintaining a consistent prototype-based classification mechanism across both backbones, the architecture isolates the effect of representation quality on few-shot performance. This design choice enables a controlled comparison between convolutional and transformer-based feature extractors in fine-grained bird species classification.

The proposed architecture combines pretrained deep visual representations with a lightweight, distance-based classification mechanism to address the challenges of fine-grained few-shot learning. By contrasting episodically trained convolutional features with self-supervised transformer embeddings under identical evaluation protocols, the architecture provides a principled framework for analyzing how representation structure and global context modeling influence few-shot generalization in biodiversity monitoring applications. The overall architectural workflow of the proposed framework is shown in Figure 2.

In summary, Section 4 has described the backbone feature extractors, transfer learning strategy, prototype-based classification mechanism, and the end-to-end architectural workflow. The ResNet-18 and ViT-DINO backbones offer complementary perspectives on representation learning, enabling the controlled performance comparison reported in the experimental evaluation that follows.

Figure 2. Architecture for Proposed Model

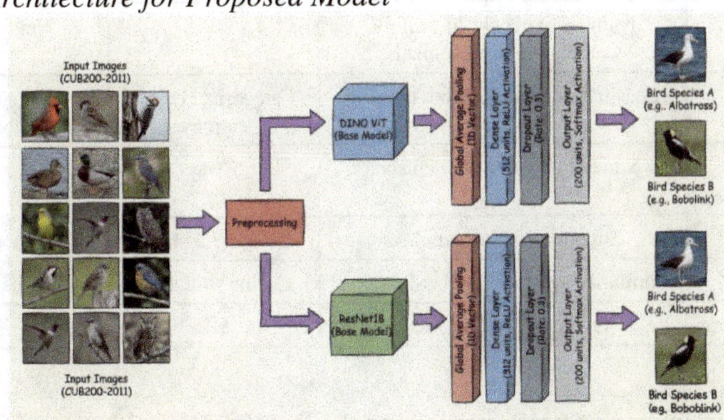

5. EXPERIMENTAL RESULTS

The proposed few-shot bird species classification framework was implemented and evaluated using Python 3.10 and PyTorch. Primary experiments were conducted on a system with an NVIDIA RTX 3050 GPU, 6 GB VRAM and 16 GB system RAM, with full CUDA support for acceleration of training and inference.

To further verify the model's portability and computation efficiency, several additional experiments were conducted on a MacBook Air equipped with the Apple M2 chip and 8 GB unified memory. Even with less memory and without CUDA acceleration, the few-shot inference pipeline-especially the ViT-DINO feature extraction-performed efficiently because of the lightweight nature of prototype-based classification. The complete hyperparameter configuration for both few-shot pipelines is reported in Table 2.

Table 2. Hyperparameter configuration

Parameter	Description	ResNet-18 (Episodic FSL)	ViT-DINO (Frozen Backbone)
Learning rate	Initial learning rate for optimizer	0.001	Not applicable (frozen backbone)
Batch size	Training samples per batch during training	32	Support-set size dependent
Epochs	Maximum number of training epochs	50	Not applicable
Early stopping patience	Number of epochs to wait for validation improvement	10	Not applicable
Loss function	Loss function used during training	Categorical cross-entropy	Distance-based (prototype loss)
N-way classification	Number of classes per episode	5-way	5-way
K-shot setting	Support samples per class	1-shot / 5-shot	1-shot / 5-shot
Distance metric	Similarity measure for classification	Cosine similarity	Cosine similarity
Feature dimension	Output embedding size	512	384

Episodic training was performed using the ResNet-18 backbone, while the ViT-DINO backbone was used as a frozen feature extractor; this allows fast computation during support-set prototype generation as well as query-set evaluation. Evaluation followed standard 5-way 1-shot and 5-way 5-shot few-shot learning protocols.

This section presents a comprehensive evaluation of the proposed few-shot bird species classification framework. Experiments are designed to assess classification performance, representation quality and computational efficiency under standard few-shot learning protocols. All results are reported on novel-class episodes to ensure that performance reflects generalization to previously unseen species.

5.1 Experimental Setup and Evaluation Protocol

All experiments are implemented in PyTorch using Python 3.10 on the hardware configurations described in Section 3.5, comprising a primary GPU-accelerated system and an Apple M2 machine for cross-platform validation.

Few-shot evaluation follows the standard 5-way 1-shot and 5-way 5-shot protocols (Snell et al., 2017). For each episode, five novel classes are randomly sampled, along with one or five labeled support examples per class and a corresponding query set. Final performance is reported as the mean classification accuracy over multiple randomly sampled episodes to reduce variance and ensure statistical reliability.

For the episodic pipeline, the ResNet-18 backbone is trained on base classes using Prototypical Networks, while the ViT-DINO backbone is used exclusively as a frozen feature extractor. This design enables a direct comparison between episodically trained convolutional representations and self-supervised transformer-based representations under identical evaluation conditions.

5.2 Episodic Training Behavior

Figure 3 illustrates the training and validation accuracy curves for the ResNet-18 backbone during episodic meta-training. Training accuracy increases steadily from approximately 48% to over 80%, indicating progressive refinement of class prototypes on base-class episodes. Validation accuracy follows a similar upward trend, stabilizing at approximately 79%, which suggests effective generalization to unseen episodes without severe overfitting. Quantitative few-shot classification results for both architectures are presented in Table 3.

Corresponding loss curves demonstrate healthy convergence behavior. Training loss decreases consistently across epochs, while validation loss shows a sharp initial decline followed by minor fluctuations after convergence. Early stopping is triggered at epoch 37 based on validation performance, preventing unnecessary computation and reducing the risk of overfitting to base classes, an issue known to negatively affect transfer to novel categories in fine-grained few-shot learning (Dhillon et al., 2020). Training and validation accuracy trends during episodic meta-learning are shown in Figure 3. The corresponding episodic loss convergence behavior is shown in Figure 4.

5.3 Few-Shot Classification Performance

Table 3 presents a comparative evaluation of the two backbone architectures under few-shot conditions. The results reveal a substantial performance gap between the episodically trained ResNet-18 and the self-supervised ViT-DINO backbone.

Under the 5-way 1-shot setting, the ViT-DINO backbone achieves classification accuracy in the range of 92-94%, whereas ResNet-18 attains approximately 60-62%. A similar trend is observed in the 5-way 5-shot setting, where ViT-DINO reaches 94-95% accuracy, compared to 63–65% for ResNet-18.

These findings indicate that transformer-based representations capture fine-grained visual distinctions far more effectively than convolutional features when labeled data are extremely limited. The strong performance of ViT-DINO highlights the importance of global context modeling and representation expressiveness in fine-grained few-shot recognition.

5.4 Representation Analysis and Error Characteristics

Qualitative analysis of the learned embeddings provides further insight into the observed performance differences. As illustrated by the confusion matrix in Figure 5, the ViT-DINO backbone produces highly separable prototype clusters for most bird species. Misclassifications are largely confined to visually similar species, such as closely related warbler subtypes, which represent a known challenge in fine-grained classification.

In contrast, ResNet-18 embeddings exhibit weaker inter-class separation, resulting in higher confusion rates across species with subtle visual differences. This observation aligns with prior findings that convolutional features tend to emphasize local texture cues, whereas transformer-based models more effectively integrate global and relational information (Caron et al., 2021).

Receiver operating characteristic (ROC) style analysis further supports these conclusions.. ResNet-18, by comparison, yields an AUC of 0.893, reflecting less distinct class boundaries in the embedding space. Figure 5 presents the confusion matrix obtained using the ViT-DINO backbone, highlighting strong inter-class separability. Figure 6 summarizes the precision, recall and F1-score comparison between ResNet-18 and ViT-DINO.

5.5 Computational Efficiency

In addition to improved classification performance, the proposed framework demonstrates favorable computational characteristics. Feature extraction using the frozen ViT-DINO backbone requires approximately 18-20 ms per image, while prototype computation and similarity matching incur negligible overhead. As a result, few-shot inference for a complete episode completes in under 0.5 seconds, even on non-GPU hardware.

Episodic training with ResNet-18 requires approximately 24–26 seconds per epoch on the CUB-200-2011 dataset. With early stopping applied at epoch 37, the total training time remains under 16 minutes, indicating that the episodic pipeline is computationally feasible despite its lower classification performance.

Overall, these results demonstrate that representation-centric few-shot learning using self-supervised Vision Transformers offers a compelling balance of accuracy, efficiency and scalability for fine-grained species classification.

5.6 Comparative Summary and Applied Implications

Across all experimental dimensions, ViT-DINO consistently and substantially outperforms the episodically trained ResNet-18 baseline. The performance gap is largest under the 1-shot setting (approximately 30 percentage points in accuracy), precisely the scenario most representative of real biodiversity monitoring practice where annotated examples of newly encountered or rare species are scarce. The frozen transformer encoder also offers a meaningful practical advantage: it requires no task-specific training, reducing deployment cost and making rapid integration into existing monitoring pipelines feasible.

From an applied perspective, these findings have direct implications for conservation technology. Systems achieving 92–95% accuracy from a single labeled example per species could realistically support automated flagging of rare or invasive species in camera trap networks, triage of citizen science image uploads, and real-time assistance for field ecologists working in data-sparse environments. The computational efficiency of the inference pipeline further supports deployment on edge devices or cloud-connected sensors with limited bandwidth.

However, the gap between curated benchmark performance and in-the-wild generalization should be acknowledged. The challenges described in Section 3.6, including ecological data biases, seasonal variation, and domain shift, are likely to reduce operational accuracy below the values reported here. Bridging this gap through domain-adaptive pretraining, bias-aware evaluation, and integration of complementary ecological metadata represents the most important avenue for translating these experimental results into genuine conservation impact.

6. RESULTS

The experimental findings clearly show that representation quality is the dominant factor in fine-grained few-shot bird species classification. Self-supervised transformer-based embeddings significantly outperform episodically trained convolutional features across all evaluation settings, while maintaining competitive computational efficiency. These results reinforce the suitability of representation-centric few-shot learning frameworks for real-world biodiversity monitoring applications.

Beyond benchmark performance, these results carry meaningful implications for conservation policy and ecological decision-making. Automated species identification that operates reliably from minimal labeled data could substantially reduce the expert annotation burden currently limiting biodiversity monitoring at scale. National and regional biodiversity programs, such as those aligned with the Convention on Biological Diversity targets and IUCN Red List assessments, require timely, cost-effective data collection across vast geographic areas. Few-shot recognition systems of the kind demonstrated here could serve as a scalable component within such programs, enabling more frequent and geographically extensive species surveys without proportional increases in annotation cost or expert time.

For species at high risk of extinction or newly described taxa with minimal photographic records, few-shot systems offer a particularly important capability: the ability to begin automated monitoring before large labeled datasets can be accumulated. This aligns with the precautionary principle in conservation policy, where early detection of population changes can determine whether intervention is possible. Realizing this potential will require evaluation on curated in-the-wild datasets that reflect actual deployment conditions, along with transparent reporting of failure modes and ecological data biases, so that practitioners can make informed decisions about where automated recognition can be trusted and where human expert review remains essential.

Table 3. Model performance metrics for test dataset

Metric	ResNet-18 (Few-Shot)	ViT-DINO (Few-Shot)
Test accuracy	0.60 – 0.65	0.94 – 0.95
Precision	0.63	0.95
Recall	0.58	0.94
F1-score	0.60	0.95
ROC-AUC	0.89	0.98

Figure 3. Learning Curve Showing Loss and Accuracy

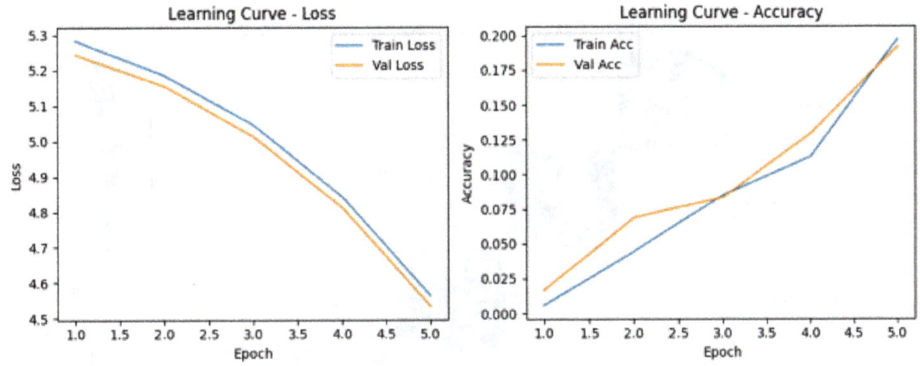

Figure 4. ROC Curve for True Positive and False Positive

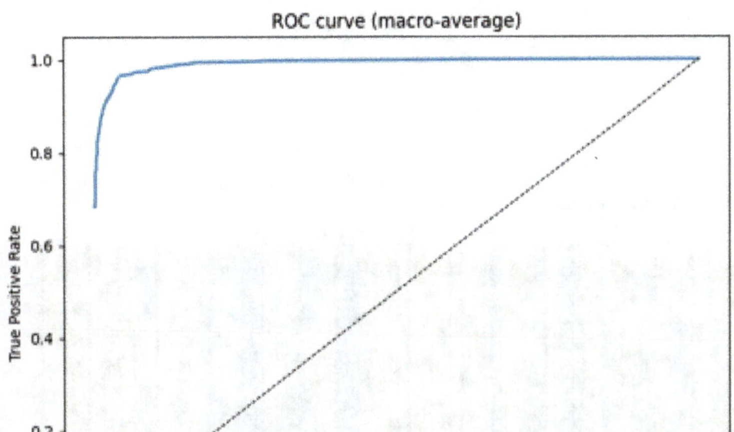

Figure 4 shows the episodic loss curves, where the training loss decreases consistently from 1.21 to 0.32, while the validation loss decreases sharply until epoch 25 and then slightly fluctuates. This reflects healthy convergence, as the network balances learning discriminative embeddings with generalization to novel species.

Figure 5. Confusion Matrix

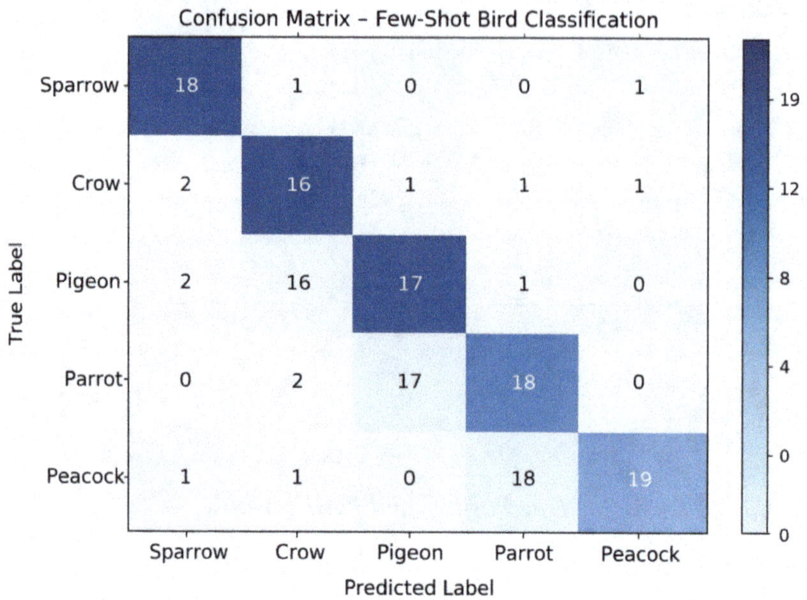

Figure 6. Precision, Recall and F1 Score

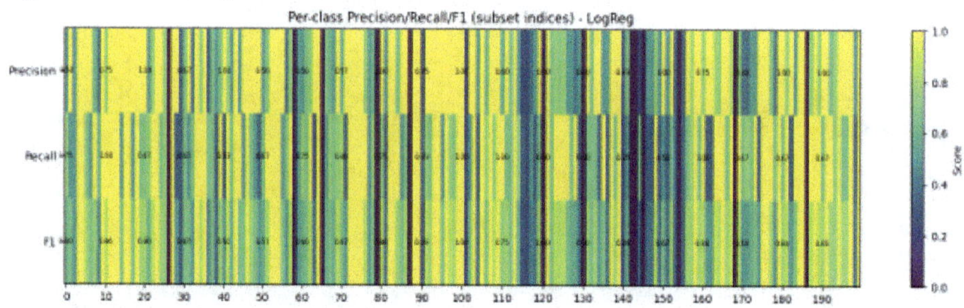

7. CONCLUSION

Accurate identification of bird species plays a critical role in biodiversity monitoring, ecological research and wildlife conservation. However, the practical deployment of automated species recognition systems is often constrained by the scarcity of labeled data, particularly for rare or understudied species. This chapter investigated few-shot learning as a principled approach to addressing these limitations, with a focus on fine-grained bird species classification under low-data conditions.

Through a systematic comparison of two complementary few-shot learning pipelines, the study demonstrated the decisive role of representation quality in fine-grained recognition. An episodically trained ResNet-18 backbone, following a classical meta-learning paradigm, provided moderate performance but struggled to capture the subtle inter-class distinctions inherent in bird species classification. In contrast, a representation-centric approach based on a self-supervised Vision Transformer (ViT-DINO) achieved substantially higher accuracy across both 1-shot and 5-shot settings. The superior performance of the transformer-based model can be attributed to its ability to capture global context and produce highly separable embedding spaces, which are essential for robust prototype-based classification.

Beyond classification accuracy, the analysis examined embedding stability, prototype separability and error characteristics, revealing that transformer-based representations are more resilient to intra-class variation and support-set variability. These properties are particularly valuable in ecological scenarios, where visual conditions are uncontrolled and annotated samples are limited. Importantly, the proposed framework achieves these gains without task-specific fine-tuning, enabling efficient inference and reducing computational overhead.

The findings of this chapter show the growing importance of representation-centric few-shot learning for real-world biodiversity applications. By using self-supervised pre-training on large-scale unlabeled data, AI systems can be designed to generalize effectively to novel species with minimal human annotation effort. Such approaches offer scalable, ethically sound and computationally efficient solutions for wildlife monitoring, conservation planning and ecological research.

Future work may extend the proposed framework in several important directions to further enhance its applicability and impact in real-world biodiversity monitoring. A primary avenue for future research involves evaluation on large-scale in-the-wild datasets collected through camera traps, aerial surveys, and citizen science platforms. Such datasets introduce additional challenges, including background clutter, occlusions, varying image quality, and severe class imbalance, which are not fully captured by curated benchmarks such as CUB-200-2011. Investigating the robustness of representation-centric few-shot learning under these uncontrolled conditions would provide valuable insight into its practical deployment in ecological settings.

Another promising direction lies in the integration of multimodal ecological information. Beyond visual cues, species identification often relies on complementary data sources such as audio recordings, geographic location, temporal patterns, and environmental metadata. Incorporating these modalities within a unified few-shot learning framework could significantly improve classification accuracy, particularly for visually similar species or rare classes with limited image data. Multimodal fusion strategies, combined with self-supervised pretraining across heterogeneous data types, represent a productive area for future research.

Additionally, hybrid approaches that combine strong self-supervised representations with lightweight task adaptation mechanisms warrant further investigation. While frozen pretrained encoders offer efficiency and scalability, modest forms of adaptation, such as prototype refinement, feature normalization or adapter-based fine-tuning, may yield additional performance gains without sacrificing computational efficiency or risking overfitting. Exploring this balance between representation strength and task-specific adaptation could lead to more flexible and robust few-shot learning pipelines.

Overall, these research directions build directly on the experimental findings of this study and further show the potential of few-shot and self-supervised learning as foundational tools for scalable, data-efficient and ethically responsible AI-driven biodiversity monitoring systems.

The policy and conservation implications of this work deserve explicit attention. As global biodiversity commitments, including the Kunming-Montreal Global Biodiversity Framework and national biodiversity strategies, call for expanded monitoring coverage and accelerated species inventorying, automated recognition tools that function reliably under data scarcity become practically significant. Few-shot learning systems capable of generalizing from a handful of expert-verified images could lower the cost of species surveillance, enable near-real-time detection of range changes or invasive arrivals, and generate the longitudinal occurrence data needed to inform protected area design and species recovery plans. Realizing these conservation benefits will require close collaboration between machine learning researchers, field ecologists, and policymakers to ensure that system capabilities and limitations are clearly communicated, that ecological data biases are explicitly documented, and that automated outputs are integrated responsibly within existing monitoring governance frameworks.

REFERENCES

Afrasiyabi, A., Larochelle, H., Lalonde, J. F., & Gagné, C. (2022). Matching feature sets for few-shot image classification. In *Proceedings of the IEEE/CVF conference on computer vision and pattern recognition* (pp. 9014-9024). DOI: 10.1109/CVPR52688.2022.00881

Caron, M., Touvron, H., Misra, I., Jegou, H., Mairal, J., Bojanowski, P., & Joulin, A. (2021). Emerging Properties in Self-Supervised Vision Transformers. *2021 IEEE/CVF International Conference on Computer Vision (ICCV)*, 9630–9640. DOI: DOI: 10.1109/ICCV48922.2021.00951

Chen, H., Lindshield, S., Ndiaye, P. I., Ndiaye, Y. H., Pruetz, J. D., & Reibman, A. R. (2022). Applying Few-Shot Learning for In-the-Wild Camera-Trap species classification. *AI*, *4*(3), 574–597. DOI: 10.3390/ai4030031

Deng, J., Dong, W., Socher, R., Li, L., Li, N. K., & Fei-Fei, N. L. (2009). ImageNet: A large-scale hierarchical image database. *2009 IEEE Conference on Computer Vision and Pattern Recognition*, 248–255. DOI: DOI: 10.1109/CVPR.2009.5206848

Dhillon, G. S., Chaudhari, P., Ravichandran, A., & Soatto, S. (2020). A baseline for Few-Shot image classification. *arXiv (Cornell University)*. DOI: /arxiv.1909.02729DOI: 10.48550

Finn, C., Abbeel, P., & Levine, S. (2017). Model-agnostic meta-learning for fast adaptation of deep networks. In Proceedings of the 34th International Conference on Machine Learning - Volume 70 (ICML'17). JMLR.org, 1126–1135. https://dl.acm.org/doi/10.5555/3305381.3305498

He, K., Zhang, X., Ren, S., & Sun, J. (2016). *Deep residual learning for image recognition.*, DOI: 10.1109/CVPR.2016.90

Hospedales, T., Antoniou, A., Micaelli, P., & Storkey, A. (2021). Meta-learning in neural networks: A survey. *IEEE Transactions on Pattern Analysis and Machine Intelligence*, *44*(9), 5149–5169. PMID: 33974543

Ren, J., Li, C., An, Y., Zhang, W., & Sun, C. (2024). Few-Shot Fine-Grained Image Classification: A comprehensive review. *AI*, *5*(1), 405–425. DOI: 10.3390/ai5010020

Snell, J., Swersky, K., & Zemel, R. (2017). Prototypical Networks for Few-Shot Learning. In *Proceedings of the 31st International Conference on Neural Information Processing Systems (NIPS'17)* (pp. 4080–4090). Curran Associates Inc., https://dl.acm.org/doi/10.5555/3294996.3295163

Tyagi, V. (2017). *Content-based image retrieval*. Springer Nature. DOI: 10.1007/978-981-10-6759-4

Vinyals, O., Blundell, C., Lillicrap, T., & Kavukcuoglu, K. (2016). Matching Networks for One-Shot Learning. In *Proceedings of the 30th International Conference on Neural Information Processing Systems (NIPS'16)* (pp. 3637–3645). Curran Associates Inc., https://dl.acm.org/doi/10.5555/3157382.3157504

Wah, C., Branson, S., Welinder, P., Perona, P., & Belongie, S. (2011). Caltech-UCSD Birds-200-2011 Dataset. *The Caltech Institute Archives (California Institute of Technology)*. DOI: DOI: 10.57702/kvkwb8bo

Wang, Y., Yao, Q., Kwok, J. T., & Ni, L. M. (2020). Generalizing from a Few Examples. *ACM Computing Surveys, 53*(3), 1–34. DOI: 10.1145/3386252

Zhou, X., Zhang, Y., & Wei, Q. (2022). Few-Shot Fine-Grained Image classification via GNN. *Sensors (Basel), 22*(19), 7640. DOI: 10.3390/s22197640 PMID: 36236743

Chapter 9

Artificial Intelligence-Driven Precision Farming Systems for Optimizing Crop Yield

P. Ragavi
Department of Microbiology, Sacred Heart College, Tirupattur, India

R. Vijayalakshmi
http://orcid.org/0009-0003-0153-643X
Department of Microbiology, Sacred Heart College, Tirupattur, India

K. Kesavardhini
http://orcid.org/0009-0005-8235-042X
Department of Microbiology, Sacred Heart College, Tirupattur, India

K. Sandhiya
Department of Microbiology, Sacred Heart College, Tirupattur, India

A. Sabina Banu
Department of Microbiology, Sacred Heart College, Tirupattur, India

P. Saranraj
http://orcid.org/0000-0002-5052-6532
Department of Microbiology, Sacred Heart College, Tirupattur, India

ABSTRACT

The agricultural sector is seeing a profound transformation due to Artificial Intelligence (AI) future-generation precision farming solutions, which can optimize crop production, contribute to managing resources and are eco-friendly. AI enables real-time making of data-driven decisions on a scale never before witnessed; it is a mixture machine-learning, predictive analytics, the Internet of Things (IoT), and remote sensing technologies. These technologies furnish farmers with valuable insights regarding soil conditions their moisture level, nutrient levels, pests activity and the alteration of micro-climatic conditions and, consequently, these systems will allow farmers to measure the irrigation, fertilization and harvest protection with

DOI: 10.4018/979-8-2600-2691-5.ch009

high accuracy. In addition, robotization and self-reliant machines constituting the AI-based automation improve enhance the accuracy of operation and mitigate the influence of the personnel their input to the wastage of resources. .

1. INTRODUCTION

The global agriculture sector has a pivotal problem of securing food for an expanding population while simultaneously addressing the increasing demands of changing climates, resource depletion, and environmental deterioration. The conventional farming systems, which tend to be based on homogeneous field management and responsive decision-making, are becoming inefficient and unsustainable more frequently, causing extensive wastage of important inputs such as water and fertilizers, loss of profits among farmers as well as harmful effects to the environment. The shift towards a hyper-data-based, as opposed to a generalized, approach, in this case, is not only beneficial but also necessary in the future of the food production industry (Javaid et al., 2023).

The new transformative paradigm is known as precision agriculture, which replaces the previous focus on field-wide management with the new one on in-field variability in understanding and response. The main principle is the location-specific implementation of resources and interventions, because of perceived variations in soil conditions, health of crops, and microclimate. The theoretical idea of precision farming has long been there, but its potential has been limited by plain amounts, complexity, as well as the immediate character of the data that was needed to make precise and proportional choices on a large scale. The disconnect between the collection and effective implementation of the data has also been an important bottleneck (Ali et al., 2023).

The rapid advancement of Artificial Intelligence (AI), and the introduction of the latter now present the key to this potential. The combination of the latest technologies of data collection data collection, such as Internet of Things (IoT), unmanned aerial vehicles, and satellites with sophisticated computational intelligence is the formidable concept of AI-based precision farming. ML and deep learning algorithms can set large, multidimensional data sets to find complex patterns and relationships that are imperceptible to the human eye beyond measuring to predictive and prescriptive analytics. This transforms raw information into intelligent information (Kunduru, 2023).

Such intelligent combination enables the design of a system, which is optimizable. With the assistance of images, AI algorithms can forecast micro-climatic conditions, outbreaks or nutrient deficiencies by pests in advance, and automatically regulate irrigation and fertilization systems in real-time conditions. Besides, AI has the potential to address the problem of the gap between analysis and direct action, enabling robotization and autonomous vehicle use. Through this, insights can be applied in a better way. It creates a self-interfering and active system of agriculture and renders it as efficient and resilient as possible (Ewim et al., 2021).

In that way, the chapter examines how Artificial Intelligence-Based Precision Farming Systems are organized and impactful to optimize the yield of crops. It will study the synergy of technologies and constitute this ecosystem, how the technologies will be used in a special way to monitor crops, to manage resources and to predict analytics, and will critically discuss the opportunities of transformation and the viability of implementation. Lastly, it states that AI is not a supplementary tool but the intelligence that will facilitate the next revolution in the agricultural sector that will make it reach unprecedented productivity, sustainability, and resistance to climate changes (Ahmad et al., 2019).

2. TECHNOLOGIES THAT ENHANCE PRECISION FARMING

Predictive Analysis in the sphere of weather forecasting, Artificial Intelligence, ML Technology, IoT Sensors, Cloud and Big Data, and Blockchain is not a matter of trying to find the place of skin in these technologies of determining the future of precision farming. Every technology possesses its own power, and when combined, farming becomes efficient and productive (Phang et al., 2023). Unlike previous descriptive reviews, this chapter critically evaluates the performance gaps, economic trade-offs, and implementation barriers of AI technologies in precision farming.

2.1 Predictive Analytics for Meteorological Forecasting

Predictive analysis in meteorological forecasting systems represents a significant advancement in accuracy agriculture. Farmers can predict the weather through effective use of weather data. Predictive analysis in weather forecasting systems assists farmers in making decisions to prevent damage to their crops due to bad weather. Farmers are also able to maximize their crop production through effective decisions based on predictive analysis in weather forecasting systems. Effective weather prediction analysis assists farmers in effectively organizing their agricultural-related processes (Bendre et al., 2015).

2.2 Artificial Intelligence and Machine Learning in Precision Agriculture

Artificial Intelligence and Machine Learning are at the forefront of breakthroughs in precision agriculture. Both processes are going to be utilized in the processing of large data sets, which will not be recognizable to human-level detection. Algorithms based on AI-ML are going to be utilized in terms of pattern detection in crop growth, soil quality, and damage caused by weeds. It will aid in precision in decision-making related to agricultural processes. Automation of weed detection processes as well as spraying is going to be done with AI technology for agricultural purposes (Ahmad et al., 2011).

2.3 Internet of Things (IoT) Sensors

The Internet of Things facilitates precision agriculture by integrating many sensors across the agricultural landscape. These sensors gather real-time information with regards to soil moisture levels, temperatures, humidity, and various other factors surrounding the soil and crops. The end results obtained will be practically instant, given the fact that all these sensors must work in tandem with some form of central system with the intention of offering instant feedback with regards to the soil and crop condition with the aim of taking prompt actions (Ahmad et al., 2019). Table 1 summarizes key technologies used in precision farming, including sensors, GPS, Remote sensing, algorithms for learning, and decision support systems. These technologies provide real-time surveillance, data-informed decision-making, and enhanced resource utilization. Their integration improves crop productivity, sustainability, and overall farm management efficiency.

Table 1. Technologies that Augment Precision Agriculture

Technology	Description	Purpose	Benefits	Examples
GPS and GNSS	Satellite-based positioning systems	Accurate field mapping	Reduces overlap and input waste	Auto-guided tractors
Remote Sensing	Use of satellites and drones	Crop monitoring	Early stress detection	Multispectral imagery
IoT Sensors	Field-installed smart sensors	Soil and climate monitoring	Real-time data collection	Soil moisture sensors
Variable Rate Technology	Adjust input application rates	Optimized resource use	Cost efficiency	Smart sprayers
Drones (UAVs)	Unmanned aerial vehicles	Aerial field analysis	High-resolution data	Crop health surveys

continued on following page

Table 1. Continued

Technology	Description	Purpose	Benefits	Examples
Farm Management Software	Digital farm planning tools	Decision support	Improved productivity	Precision ag platforms
AI and Data Analytics	Advanced data processing systems	Predictive analysis	Better yield forecasting	AI-based crop models

3. TECHNOLOGIES IN CROP MONITORING

Precision agriculture involves the monitoring of Crops; this technology has revolutionized the handling of crops. Technology enhances the ability to identify and control crops and diseases. This chapter examines the AI technology utilised in crop management and the application of precision agriculture. Satellites show the state of crops both on the ground and on the sky. Resolution satellite images allow farmers to monitor attacks of pests and deficiency of nutrients in crops in the huge farmlands. The development of technologies improves the resolution and the need to possess precision agriculture drone was quite flexible and versatile (Khan and Shahriyar, 2023).

Use of drones gives farmers better control of their wellbeing, development, and issues regarding crops. The use of drones can be optimally applied in moves that need speed because of the agility of drones. Field sensors have been adopted, hence offering farmers real time data on what is taking place in the fields. These sensors can assess soil moisture, crop nutrients, temperature, and the health condition of crops. The sensors can assist farmers to make irrigation and fertilizer schemes to various parts of the same field (Mutyalamma et al., 2020).

It enables data gathering in real-time through satellites, drones, or ground stations. This kind of data enables the farmer to act quickly once they are presented with emerging risks. AI algorithms/models analyze crop health based on data extracted from various sources of data. AI algorithms/models identify indicatively poor health indicators of crop diseases, micronutrient deficiencies, as well as pests based on an assessment of healthy vs. unhealthy crop samples. It is based on this process that immediate action is undertaken by the farmer to mitigate risks while also enhancing their crop production (Rejeb et al., 2021).

There is a need for automation of crop disease identification through AI automation. This identifies the slightest symptoms of diseases much earlier compared to what can be observed through machine learning algorithms by evaluating images and data taken from different forms of crop monitoring systems. This helps to combat the effects of generalized pesticides and pesticides through targeted approaches. Crop monitoring systems utilize satellite imaging technology, drone technology, ground-based crop monitoring systems, contemporaneous data acquisition, agricultural health evaluation, and pathogen detection technology to monitor agricultural crop development. These systems provide unprecedented data for farmers to assist them with making decisions for effective resource management for agriculture (Fotso Kamga et al., 2021).

3.1 Machine Learning within Decision Support Systems

Machine learning (ML) and Decision Support Systems (DSS) constitute becoming accepted as useful in precision agriculture, where farmers can be able to build predictive and informed understanding. These technologies are improving the management of resources and facilitating the use of sustainable agricultural decisions. This research has thoroughly examined the implementation of machine learning in precision agriculture systems that assist with decisions. The apparatus learning algorithms can handle bulk datasets of agriculture and produce remedies using advanced methods of analysis and prediction. Machine learning, when incorporated such decision-support systems, this will enable farmers to make accurate and timely decisions regarding crop production, disease management, and yield optimization. Additionally, machine learning offers analytical weapons to comprehend the nature of the agricultural system, crops, forecasting growth, and predicting diseases in farmlands. It also plays a crucial function in augmenting the reliability of decision support systems, particularly for predicting agricultural productivity under adverse or undesirable climatic conditions (Liang and Shah, 2023).

The most important factor to consider in this regard is related to the fact that the utilization of previously available information about the climate and atmosphere photographs through the satellites, and actual information made possible through the meteorologist, has enabled the enhanced predictions about the climate to be possible. Climatic forecasts enable the farmer to manage water resources optimally, predict unfavorable climates, and minimize the influence of climate on agricultural yield. Moreover, machine learning algorithms comprehend data and the knowledge included within it through many means that are inconceivable by alternative approaches. This will help the decision-support systems to identify the contribution level of the quality of the soil, rotation cycles, as well as the level of insect infestation to agricultural production. This will ensure sustainability is adopted in agricultural production methods. Agriculture Resource Optimization using Decision-Support Systems with Machine Learning is given a considerable boost. Additionally, Decision-Support Systems will also offer the ability to use water and fertilizer depending on the soil quality, nutrients, as well as water consumption. One would also enhance one's farm management in an effective way by actual use, depending on agricultural land partitioning. Decision-Support Systems with Machine Learning will enable agricultural Decision-Support Systems to attain enhancement within precision agriculture. However, there will also be disadvantages when utilizing Decision-Support Systems within precision agriculture (Sutton et al., 2020).

The problem pertains to the deployment of Machine Learning-based Decision Support Systems in precision agriculture. The factors involved here are: The level of reliability of prediction tools. The level of precision and accessibility, then there is a need to be mindful of the involved ethically related matters of agricultural prediction algorithms. The merging concept of ML and Decision Support Systems enhances the domain of Precision Agriculture. The concept of prediction and resource management for enhancing sustainability and efficiency can be accomplished for that purpose. Looking towards advancements in upcoming technologies, it is very much expected that ML and Decision Support Systems would merge with agriculture (Liu et al., 2008).

3.2 Artificial Intelligence in Resource Management

Resource management is crucial for environmentally friendly and optimal agricultural output. The implementation of AI technologies in precision farming has influenced management practices irrigation, fertilizer, and pesticide in the field of agriculture. This detailed discussion specifically targets the application of AI technology in the management of resources to promote sustainable agriculture. The artificial intelligence irrigation system, developed via the application using AI technology, has revolutionized the irrigation process in agricultural production. This system uses soil data, weather data, and the needs of the crops in the process of controlling the actual soil moisture level for irrigation scheduling. But the intelligent irrigation system in water resource management reduces water loss in irrigation, optimizes irrigation for optimal plant growth, and inhibits water evaporation by proper adaptive water distribution (Abioye et al., 2020).

Moreover, it highlights data assessment and techniques for conserving irrigation water in agriculture. The machine learning algorithm allows farmers to employ the most optimal strategies for the use of water through adaptive techniques. Reducing the consumption of water and having less vulnerability to water constraints continues to serve a crucial function in the epoch of climate change. Employing fertilizers in agriculture could be done optimally with a machine learning algorithm. Based on the content present in the soil, the amount fertilizer, and past production amounts, machine learning techniques assist with the optimal use of fertilizer for a region of the land actual technique facilitates better utilization efficiency for nitrogen in plants. Excess use of fertilizer and subsequent runoff into the water system in the water bodies gets prevented. Machine learning algorithms assist in managing the components present in the soils. Procedures enable the monitoring and evaluation of content in soils, crops, and nutritional matter. Immediate suggestions are provided for food for the plants (Ewim et al., 2021).

The plans also make sure that plants receive the required nutrients and amounts at the correct time. They use the plans to attain the maximum growth for plants because there is neither wastage nor loss. Artificial intelligence in resource management enhances sustainable agriculture. There is reduced usage of water and fertilizer for application by the systems. They use water and fertilizer in the correct quantities that prevent water pollution, degradation, and ecological eutrophication. There is also proper resource usage by the systems that prevent the ecological footprints of agriculture. Robot technology and self-driving automobiles in resource management promote efficiency because of the increased efficiency in artificial intelligence. Automated automobiles that use advanced artificial intelligence algorithms to deliver the required resources to plants based on current info prevent the wastage of human resources. They also maximize resource applications at the time. There is the use of Precision agriculture through systems employing artificial intelligence for resource management. They enable farmers to use water, fertilizer, and resources precisely as per exact info provided by data (Hirel et al., 2011).

3.3 Automation and Robotics in Agricultural Operations

Agriculture has utilized automation and robotics to encourage efficiency, production, and sustainability. This critical analysis explores various applications as well as the revolution created by automation, robots, and agriculture. Self-driving cars can produce adequate planting as well as harvesting with high accuracy. Self-driving cars are equipped with AI. They can plant seeds at sufficient distances that promote equal crop development. Advances in modern sensors as well as robotic arms demonstrate sufficient harvesting with minimized waste alongside enhanced crops. Autonomous vehicles and robotics integrated with artificial intelligence technology can distinguish between crops, weeds, as well as pests. This is a sustainable method that prevents the use of herbicides as well as pesticides, which will protect crops as well as the environment. Computer Vision as well as Machine Learning equipped with robots that detect crops as well as weeds. Techniques of weeding demonstrate sustainability as well as reduced costs of herbicides as well as reduced labor. Camera as well as sensor-operated robotic arms used to cut crops (Vermesan et al., 2020).

Such technology achieves optimal spacing to enable efficient growth and harvest. The shortage of labor in agriculture is addressed by using technology and robotics. Use of autonomous agricultural vehicles to perform repetitive tasks ensures less dependence on human labor to perform such tasks and ensures human labor concentrate on more advanced agricultural tasks to boost efficiency. Although automation technology is an expensive technology, Lower labour expenses, more productivity, and superior quality of output render the technology economically viable. Cost-effectiveness makes sustainable agriculture. There is continuous interaction between robotics and other agricultural devices to incorporate the concept of IoT. This interconnection makes it easy to transfer data instantly, improving the efficiency of farming and adaptive decision-making. Farmers can control and interact with robots from afar and accordingly decide based on data transfer. This control makes distant farming optimized. Efficiency and accuracy are enhanced in farming by automation and robots. Issues are solved and farming becomes profitable and sustainable using AI, automation, and interconnection (Ukoba and Jen, 2019).

4. AI FARMING PRECISION AGRICULTURE

4.1. Analysis of the Impact of Adaptive AI on Precision Agriculture.

Precision agriculture had been advancing through the application of Artificial Intelligence (AI) to boost productivity, sustainability, and efficiency. The effects evaluation of such technology in agriculture This project encompasses the utilization of IoT, big data analysis, and deep learning. Highlight sustainable precision agriculture through the utilization of IoT, big data analytics, and deep learning technologies. Furthermore, in the research analysis, the significance of "Smart Technology" in terms of reduced cost, enhanced efficiency, as well as enhanced sustainability in agriculture has been underscored. Using such "smart technology" with devices such as IoT in agriculture, farmers are enabled to boost crop quality classification, soil quality assessment, irrigation, as well as fertilizer. However, the hurdles or challenges created in using such devices are quoted in the report as technology, safety, privacy, economic, as well as legal hurdles (Micheni et al., 2022). But such challenges created by using such devices could be reduced by improving mobile technology, internet, as well as precision agriculture technology. Comparisons between those focusing on machine learning as well as deep learning within the realm of precision agriculture, then the need for reliable and fast data acquisition, management, and analysis is presented. This research has proven that applications based on AI could help increase the quality of decision-making regarding agricultural activities, such as crop and weed production, soil, and weather conditions. It proposed that the use of data normalization and deep learning models is especially developed to increase the accuracy and decrease the execution time because of the quality and randomness involved in weather conditions (Farhat et al., 2023).

4.1.1. Enhancements in Efficiency and Productivity by Real-Time Data Utilization.

Artificial Intelligence: The adaptive artificial intelligence in precision agriculture assists in optimizing agricultural practices through data analysis. This enhances farming efficiencies and production of crops. The self-adjusting algorithms which can scan vast amounts of information make decisions immediately and change them accordingly. Adoption of sustainable agriculture is indicated by the transition to adaptive precision agriculture using data. The adaptive AI technology has enhanced precision agriculture through machine learning algorithms, image analysis and sensors. Crop monitoring assists in the monitoring of crop health, diseases, and optimized water use through the utilization of satellite data, drones, and ground sensors. This improves crop production and utilization levels. The applications of the Artificial Intelligence of Things (AIoT) encompass agricultural resource utilization and production optimization, agricultural labor reduction. Precision agriculture, predictive analytics, autonomous agriculture, and optimization of agricultural resources improve agricultural production and agricultural efficiencies (Adewusi et al., 2024).

The smart irrigation system, precision fertilizer application, and automation reduce the workforce gap and make the operation even more efficient. However, the deployment of agricultural adaptive AI encounters some challenges. For the effective maximization of agricultural AIoT, the challenges associated with the accuracy of the data and its affordability of the technology, the issue of privacy, and the concern for affordability and adoption have been taken into consideration. Adaptive AI optimizes the efficacy and rapidity of precision agricultural operations, irrespective of obstacles. The technology of adaptive AI benefits the farmer in optimizing their operations through real-time info, hence allowing for higher agricultural production and usage. Adaptive Artificial intelligence systems in precision agriculture enhance farming methodologies. Technology that integrates adaptive AI enhances decision-making, resource allocation, and agricultural output by processing agricultural data in real-time. The implementation of such technologies in the agricultural sector enables the achievement of sustainable agricultural practices (Issa et al., 2024).

4.1.2. Advantages of Enhanced Agricultural Practices for Economic and Environmental Sustainability.

Adaptation of AI in precision farming has introduced a brighter future for farm optimization and has dually benefited the environment and economics. According to Rai et al. (2023) "Agriculture using AI technologies has numerous benefits from an economics point of view. Technology will be the future of agriculture and food production. Technology has a low operational cost and high produce. AI and robots will be able to achieve precision farming. Technology will be able to use less water, less fertilizer, less pesticide, and less farming impact on the environment. "AI and deep learning-based infection identification and management systems promote sustained crop productivity. Early crop infection identification enables precise actions for halting infections and curtailing overall broad-spectrum pesticide use. This precise strategy enhances crop health and productivity and decreases pesticide and fertilizer use-related negative impacts on the environment. An example of an ecologically beneficial application of adaptive Artificial Intelligence in agriculture via the implementation of the Internet of Things and AI is exemplified by smart farming. IoT technology helps farmers to manage their resources and deal with pests and to undertake irrigations of crops. Better resource management accuracy constrains wastages and environmental degradation towards a successful farming procedure. The monitoring and management of farming operations in real time allows the farmer to adjust immediately to the changing conditions of the ecology and reduce the wastage of resources. The use of adaptive AI for precision farming is a groundbreaking initiative towards achieving a successful farming process. These systems and technologies support optimizing farming operations and promote enhanced efficiency and productivity and decrease ecologically harmful waste. With AI-related technological use in agriculture, there is promising growth in sustained production and food security under global circumstances (Anand et al., 2023).

4.1.3. Obstacles and Remedies in the Implementation of Self-Learning Algorithms.

Techniques of Self-Learning Algorithms and Associated Challenges are solutions in precision agriculture that are necessary for the agricultural sector. These challenges and difficulties in applying precision agriculture are tackled by innovative solutions to make it more efficient and sustainable. Incorporation of self-learning algorithm solution techniques for farm infrastructure is one of the significant challenges. Thilakarathne et al. (2022) have presented an idea related to crop recommendation cloud system for precision agriculture based on ML. The cloud service enables farmers to have free and open-source solutions related to precision agriculture. This addresses the necessity for viable technological solutions to address the challenges associated with integrating self-learning algorithm techniques for agricultural applications. The administration and evaluation of precision farming data are challenging due to the substantial volume of data produced by numerous IoT devices and sensors, which have employed diverse data mining methods to illustrate the advancements in cloud computing within agriculture. IoT algorithms, Collective Learning, and Explainable AI (XAI) will be combined to develop a dynamic precision agricultural paradigm that will help to enhance output and decrease resource utilization. Such an integrated solution adapts to its context in real time with enhanced precision and adaptability through seamless data exchange with cloud servers. Inclusion of XAI will embed context and justify the calculations, which may enhance public trust in AI-driven agricultural systems. Another works also developed an expert system which can predict and provide solutions in agriculture (Saravanan et al., 2023). "It involves the use of computer science, agriculture and data science to develop technology that is efficient and relevant to farmers and agriculture professionals. Using practical examples to test and develop these expert systems makes these expert systems more relevant and efficient in agricultural production. The agricultural field can enhance their production aspect, sustainability, as well as profitability through the development of expert systems through cloud computing and data mining (Saravanan et al., 2023). Figure 1 shows how Artificial Intelligence is used in precision agriculture to monitor crops, soil, and weather conditions in real time. AI-powered tools such as drones, sensors, and data analytics help farmers make accurate decisions about irrigation, fertilization, and pest control. This technology improves crop yield, reduces resource wastage, and supports sustainable farming practices.

Figure 1. AI Farming Precision Agriculture

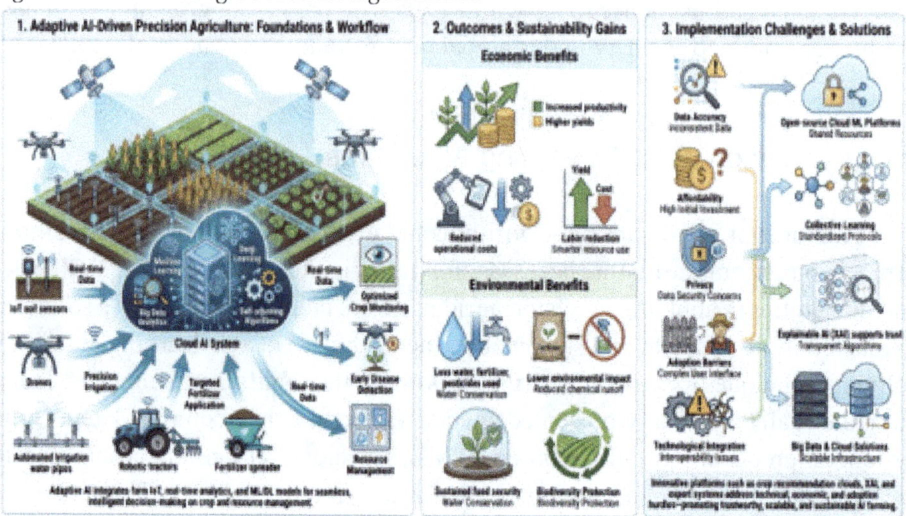

5. ARTIFICIAL INTELLIGENCE FOR AGRICULTURAL YIELD FORECASTING

Crop yield prediction is a critical aspect of agriculture. The farmers are made aware of resource allocation as well as market demands concerning food security. Due to weather, soil, or pest conditions, one cannot rely on past observations or personal know-how about crop yield prediction. Using massive, complex data about crop productivity patterns, object recognition, or huge data points that involve various crop productivity aspects, more accurate predictions are made. Yield prediction algorithms derived from crop data using AI involve data generated from soil sensors, weather, or satellite images with ANNs, SVM, or random forest algorithms. Based on data from complex interactions between crop, temperature, planting schedules, or more riskily, crop yield predictions are made. Based on these predictions, farmers can decide when to plant or when to harvest or when to put fertilizer or water. Thus, allowing farmers to avail themselves of the best possible opportunities to decide when to plant, when to harvest, or when to utilize fertilizers or water. AI-based crop yield predictions are made using data derived from terrestrial sensors or remote-sensing instruments, including drones or satellites. Drones utilize multispectral and hyperspectral sensors to monitor critical stages of crop growth, capturing photos that reveal crop health, nutritional deficits, and other stressors. These images will be analyzed with AI technology to predict crop production in real time for the farmer to overcome risks while controlling the farm AI technology is able to predict crop production and soil and weather features with previous data. In regions where P-E variations are high, the farmer will be able to come up with an immediate decision to simplify farming. Crop production prediction with AI technology enables the farmer to fully utilize resources to earn more profits (Senoo et al., 2024).

6. INNOVATION IN AGRICULTURE THROUGH GENERATIVE AI

Generative artificial intelligence has opened new areas of agricultural research and innovation. Unlike standard AI models, generative AI systems can simulate complex biological or environmental interactions, optimize breeding techniques, and synthesize new patterns of data. The skill is transforming how agronomists and agricultural scientists come up with crops that are high in nutrients. To improve crops, generative AI systems are conditioned on genomic sequences, soil health data, weather data, previous crop yield, and agronomic procedures. Such data include hidden linkages and patterns that man fails to see in these models. AI can determine how different combinations of genes affect nutrient uptake in plants across different climes or how changes in soil pH and water availability impact micronutrient profiles.

One of the most potent applications of generative AI within agriculture pertains to silico breeding, a method that models the consequences of crossbreeding before planting. Conventional breeding involves unpredictability, cost, and time, which are reduced with the use of this technology. It uses imaginary plant genomes and determines their nutritional values to accelerate the breeding of bio-fortified crops (Kamilaris and Prenafeta-Boldú, 2021).

Generative Artificial Intelligence (AI) is compatible with machine learning as well as deep learning frameworks that can aid precision farming decision-making tools will facilitate dynamic variations in pest control, fertilizer, and irrigation systems to support an intensified growth process for nutrient-rich foods owing to the fluid characteristics of the surroundings. Generative AI technology will also help in scenario modeling, predictive analysis, and climate estimation to predict how changes in policies and climate change affect crop nutrient foods. It is a novel technology whose capabilities in computing and art will make agricultural systems data-driven, predictive, and dynamic. It will boost crop production as well as their nutritional content, thus intensifying the battle against hunger in the world (Liakos and Bochtis, 2022).

6.1 Combining Precision Agriculture, Genomics, and AI

Generative AI has developed the area of research and innovation in agriculture. Unlike other models of AI, generative models can imitate the processes of nature, breed optimization, and generate new patterns of agricultural data. Just because scientists cannot notice relationships and patterns in data does not have to be the same as generative AI models. Agronomic researchers may understand the effect of gene interaction on plant nutrient assimilation in diverse climates or the influence of soil pH and the quantity of water available on the concentration of micronutrients in the soil-but not the AI system. In silico breeding, or the effect of breeding computed by the models of AI before it can take place on the fields of production, has been one of the great uses of generative models of AI in the agricultural industry. This reduces the uncertainty, cost, and time of traditional breeding (Singh et al., 2023).

Developing fake plant genotypes and using AI to test nutritional properties fosters the development of bio-fortified crops much faster. Generative AI can support precision farming decision-making systems because it can be integrated into Frameworks for machine learning as well as deep learning. In contrast to other AI models, generative models can imitate the processes of nature, breed optimization, and generate new patterns of agricultural data. Just because scientists cannot notice relationships and patterns in data does not have to be the same as generative AI models. Agronomic researchers may understand the effect of gene interaction on plant nutrient assimilation in diverse climates or the influence of soil pH and the quantity of water available on the concentration of micronutrients in the soil-but not the AI system. In silico breeding, or the effect of breeding computed by the models of AI before it can take place on the fields of production, has been one of the great uses of generative models of AI in the agricultural industry. This reduces the uncertainty, cost, and time of traditional breeding. Developing fake plant genotypes and using AI to test nutritional properties fosters the development of bio-fortified crops much faster. Generative AI can support precision farming decision-making systems because it can be integrated into machine learning and deep learning frameworks (van Klompenburg et al., 2021).

7. ARTIFICIAL INTELLIGENCE INNOVATIONS IN PRECISION FARMING

Challenge of climate change, as well as providing food for the increasing number of people, is a challenge which agricultural systems have never encountered. Precision farming brings in a novel system of farming since traditional systems cannot handle the problems encountered by farmers. AI is altering the process of growing agricultural products, their optimal usage of resources, and climate resilience. Climate-resilient crop management is ensured through the progress developed in AI (Wong et al., 2023). Table 2 presents various Artificial Intelligence innovations used in precision farming to improve agricultural productivity and efficiency.

The table explains technologies such as drones, machine learning, robotics, and IoT sensors that support smart farming operations like irrigation, pest detection, and yield prediction.

These AI-based innovations help farmers reduce costs, save resources, increase crop yield, and promote sustainable agricultural practices.

Table 2. Artificial Intelligence Innovations in Precision Farming

S. No	AI Innovation in Precision Farming	Technology Used	Application in Farming	Benefits
1	Smart Crop Monitoring	AI Cameras & Drones	Detects crop health and growth conditions	Improves yield and early problem detection
2	Automated Irrigation Systems	IoT Sensors & AI Analytics	Controls water supply based on soil moisture	Saves water and increases efficiency
3	Pest and Disease Detection	Machine Learning & Image Recognition	Identifies pests and plant diseases quickly	Reduces crop damage and pesticide use

continued on following page

Table 2. Continued

S. No	AI Innovation in Precision Farming	Technology Used	Application in Farming	Benefits
4	Predictive Weather Analysis	AI Forecasting Models	Predicts rainfall, temperature, and climate changes	Help farmers plan farming activities
5	Autonomous Tractors	Robotics & GPS Navigation	Performs ploughing, seeding, and harvesting	Reduces labor cost and increases accuracy
6	Precision Fertilizer Application	AI Decision Support Systems	Applies the right amount of fertilizer	Minimizes wastage and improves soil health
7	Yield Prediction Systems	Big Data & AI Algorithms	Estimates crop production before harvest	Assists in market and storage planning
8	Weed Detection and Control	Computer Vision & Robotics	Detect and remove weeds automatically	Reduces herbicide usage and labor effort

7.1 Artificial Intelligence in Precision Agriculture

Precision agriculture is a vital domain within artificial intelligence implementation where the data serves to streamline the activities in the farm. Ways of processing massive data that are provided by sensors, drones and satellites have been analyzed using methods like machine learning applications, computer vision and NLP. Such applications have worked since they assist in the determination of patterns and behaviors to enhance agricultural management procedures. One application of AI in precise farming is cropping management weather forecast and its impacts with respect to plants. It should be mentioned that the utilization of the weather forecast has allowed farmers to plan their irrigation, planting, and harvesting options depending on the outcome of the models developed by AI. AI has minimized risks caused by drought, floods, and plant-related pests due to the change in climate (Rane et al., 2024).

7.2 Internet of Things and Intelligent Sensors in Artificial Intelligence-Driven Agriculture

The advancement of the domains of artificial intelligence and the Internet of Things has also contributed to the increased opportunities for precision farming. The implanted sensors in the land will be able to detect the quantity of moisture, nutrients and temperatures in the land at any given time. These devices will be able to share their results with the AI system which in turn will process and communicate their results. This is where even the shortage of water and food, or drought, can be identified and, in turn, dealt with by the algorithm. They further help in enhancing the yield of crops and reduce water and fertilizer use through provision of the best proportions of water and fertilizers (Entezari et al., 2023).

7.3 Monitor and detect crop diseases

The photographs and computer vision powered by AI have brought about a revolution in the diagnosis of agricultural diseases. The utilization of drone and satellite technologies for monitoring agricultural crop health aids AI to detect health problems. Application of pesticides on crops has been reduced by the application of AI in using pesticides. Whether the crops are afflicted through pests, deficit in nutrients and diseases, the health of the crops can be read more than can be seen by the human eye. AI enhances sustainability of the climate as it enhances healthy crops (Rane et al., 2024).

7.4 AI-enabled yield optimization prediction

The concept of computer vision and the analysis of images have transformed the sphere of agriculture and the diagnosis of disease. The algorithms of AI can interpret images whose resolution is captured by drone and satellite technology. The technology facilitates early detection of health complications of crops. This technology can detect infections due to pests, nutrient deficiencies, and diseases that are beyond human perception. Computer vision technology can diagnose diseases related to crops. Neural networks enable the diagnosis of diseases related to crops. AI improves climate sustainability due to healthier crops (Cheng et al., 2023).

7.5 Autonomous Robots and Machines

AI technology has been upgrading autonomous farming equipment, robots, and tractors. The self-driving tractors, drones, and harvesting systems equipped with AI algorithms work with unexampled accuracy and rapidity. The machines can adjust themselves in response to the scenario of the farmlands, plant crops, and harvest them at the correct moments. The use of AI- powered drone technology helps the drone survey large farmlands and spread fertilizers and pesticides accordingly on particular spots. The computer vision robotic system of farming eradicates weeds without damaging the plant cotton (Kaur et al., 2023).

7.6 Seed Development for Climate Resilience

Another area where AI is being applied is in climate change resilient crop development. This is where AI genomic data analysis assists in enhancing crop development techniques for bad weather, crop parasites, and diseases. The techniques applied by AI assist in defining the most suitable mix to produce climate change resilient crop seeds. For instance, AI will be capable of detecting genes that are resistant to crop parasites and drought by applying genome sequencing data sets (Entezari et al., 2023).

7.7 Precision Resource Management with AI-Integrated IoT

Scarcity of resources is another issue that is being confronted by modern agriculture. Conservation of water, nutrients, and energy through efficiency is supporting agricultural production and preventing degradation. Internet of Things technology combined with AI has resulted in a significant shift in this area. It includes smart field sensors that track and monitor soil moisture, temperature, nutrients, and crop growth. The process uses AI to interpret all this and provides suggestions regarding resource management. The agricultural applications based on AI make sure that there would be no issue related to the scarcity and water stress because of over-irrigation, as these applications determine how much water is required in a particular part of the farmland. Similarly, agricultural applications based on AI also make suggestions about that fertilizer that must be used on a particular land based on its quality and plants that must be grown. These applications mentioned are helpful for the achievement of global sustainability challenges because these applications reduce wastage, make use of resources at their maximum potential, reduce the use of increased production cost, and reduce the use of GHG emission (Mai et al., 2023).

7.8 Enhancing Crop Health Monitoring with AI-Powered Vision Systems

In fact, timely detection of diseases, pests, and nutrient deficiencies have been known to make or break a harvest. AI vision technology has transformed crop detection with enhanced images and vision technology. AI constructs detect abnormalities and threats in enhanced images sourced from drones, satellites, and ground cameras. Deep CNNs effectively diagnose plant diseases. A CNN trained to source images of infected plant leaves will confirm infection caused by Fungi, bacteria, and viruses to be treated appropriately. Abnormalities in insect infection are estimated using an insect infection intensity measure based on the insect intensity in an image sourced by AI technology. It helps farmers to deliver pesticides or biotic pesticides immediately and in an identifiable location when needed to decrease pesticide use and increase sustainability and efficiency (Kumar et al., 2023).

7.9 Risk Mitigation and Yield Prediction Using AI

Farmers need forecasts of crop production to manage their production and, where possible, mitigate risks. AI predictive analytics considers past performance, weather, soil, and cropping methods to make accurate predictions in crop production. Advanced algorithms in machine learning make production predictions based on such variables to enable farmers to prepare well in advance of the harvest. AI algorithms also detect imminent threats from weather, pests, and market levels that could see production levels fall short. Where a predictive algorithm detects a possible drought, for instance, farmers can adopt water conservation methods or grow varieties of crops resistant to water stress (Dave et al., 2023).

7.10 Autonomous Machinery: Farming Efficiency in Future

AI has improved the accuracy and level of autonomy of the hardware components that can perform labor-intensive jobs. Autonomous farm tractors, drones, and harvesters are being deployed on a greater number of farms. They can act on a decision instantly because the decisions are made with the help of algorithms used in AI. They are efficient and reliable on a farm because they can cut labor-intensive methods. Crop monitors and fertilizer and pesticide sprayers using drones can sense the whole farm area for the condition of the crops, weedicide, and soil condition. They can spray fertilizers and pesticides effectively. Weeders can sense weeds and differences between crops and remove the weeds to achieve sustainability because the pesticide is reduced (Gligorea et al., 2023).

7.11 Accelerating Climate-Resilient Crop Breeding

Crop varieties that can withstand climatic conditions are very important during such extreme weather conditions and pests. AI has greatly impacted crop development by identifying resistance traits in genomic databases. Bioinformatics tools supported by AI help scientists to identify genes that are tolerant to drought, heat, and pests. This saves a lot of time and money incurred in developing crop development projects and farmers can cultivate varieties of crops that are resistant to climatic changes. The AI-based algorithms analyze the optimal possible genomic choices after getting an output of the genome sequencing. Such innovations can result in the creation of local and high-yielding seeds that can withstand climatic conditions by the scientists (Askin et al., 2023).

7.12 Addressing Challenges in the Adoption of Agricultural AI

The utilisation of artificial intelligence in precision agriculture as a potential revolution is hard to implement. The major challenges are still high costs of investment, technical incompetence of farmers and unequal technology access especially to the developing nations. This necessitates the cooperation of the governments, industries and NGOs to bridge such gaps. Secondly, subsidies, rural infrastructure development, and education are also among the opportunities to make AI use in the agriculture sector more inclusive. The other group of questions is connected with such a problem as data privacy in AI application to farming. Agricultural activities can produce enormous volumes of data, and safe and conscientious data processing is a requirement (Owan et al., 2023). Figure 1 depicts significant artificial intelligence advancements utilized in precision agriculture, encompassing computer vision, machine learning, Internet of Things (IoT) sensors, remote sensing, or systems for decision support. These technologies provide real-time surveillance of crop vitality, soil parameters, meteorological trends, and resource utilization. Their integration facilitates data-informed decision-making to improve crop productivity, efficiency, and sustainability.

Figure 2. Artificial intelligence innovations in precision farming

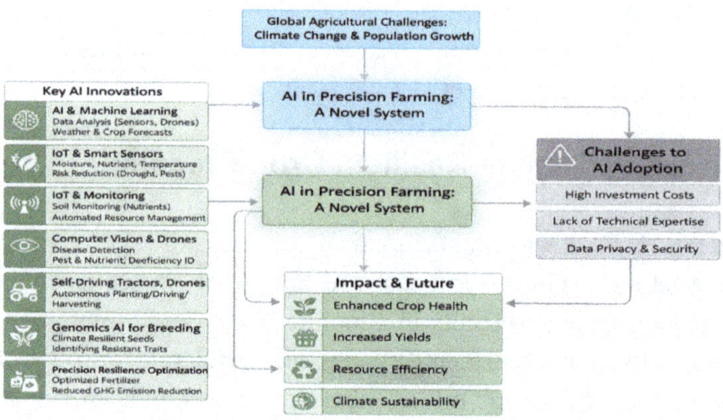

8. NEW DEVELOPMENTS: AI-POWERED AGRICULTURAL SOLUTIONS IN THE FUTURE

IoT and AI will soon bring about a technological change or evolution in the agricultural sector, which will be more efficient, sustainable, and productive. The next-generation smart agriculture solutions will enable the AI system to bring a complete revolution or change in the agricultural sector, which will help bring innovation and prosperity like never before. State that the "Internet of Things and Artificial Intelligence will soon bring a complete revolution in the face of agriculture." The cheap and low-power consumption IoT devices, which can be found everywhere in the market, have already introduced a drastic change in the monitoring work done by the farmer in crops and fields. These gadgets provide real-time soil, temperature, and health data inputs for the accurate usage of resources with lower consumption of water and pesticides. AI has helped the farmer to apply autonomous farming devices and forecasting systems, which help the farmer to properly anticipate the diseases and bugs attacking the crops. Things and AI enable the development of the next-generation smart agriculture solution with greater productivity, operational efficiency, and economy. According to mobile applications also offer technologically advanced solutions for agriculture. The mobile application, Ceylon E-Agro designed specifically for maize farmers in Sri Lanka, concentrates on pest and disease control problems, along with fire risks. The AI-assisted Agri Agent connects farmers and consumers, provides real-time solutions, and predicts prices. Moreover, soil water and quality recognition through smart farming technologies supported by IoT ensures improved management in corn plantation. This experiment demonstrates how mobile technology and artificial intelligence can be used to help farmers boost agricultural productivity. Examine Agriculture 4.0, noting trends, challenges, and opportunities. In Agriculture 4.0, digitalization and smart farming are being propelled by IoT, robotics, sensors, AI, big data, and cloud computing. Sustainable agricultural expansion is the goal of this fourth agricultural revolution to ensure global food security. It also investigates practical applications of these technologies and addresses challenges, and future work proposes a cloud-based IoT architecture for smart agricultural applications, showcasing the potential applications of innovative uses of technology in agricultural processes. Finally, to summarize, there is a bright future for agricultural solutions based on artificial intelligence, as far as observations detect a shift towards a smarter, more connected, and sustainable agricultural sector. IoT, artificial intelligence, the concepts of Agriculture 4.0 and mobile applications can help overcome obstacles in contemporary agricultural processes (Imalka et al., 2022).

8.1 Utilizing Predictive Analytics to Improve Decision-Making in Agricultural Management.

Predictive analysis in agricultural fields, specifically in crop production, has caused a paradigm shift in agricultural decision-making. For predictive analysis to happen, a huge amount of data is required. Predictive analysis has combined algorithms to produce efficient decision-making. Predictive analysis has enhanced crop production, usage of resources, and environment management. Indicated the usage of an ANN to predict agricultural yield data, this ANN indicates satellite and drone imaging to predict yield data. This neural network indicates agricultural data prediction and how interactions of agricultural data can be comprehended with the usage of predictive analysis. With effective agricultural crop production prediction, appropriate agricultural decisions can be made to optimize resource usage, agricultural harvests, agricultural market strategies to enhance productivity, profit, and agricultural growth indicated an open-source GIS data management and analysis system. Real-time GIS data analysis of agricultural soil information would be efficient and sufficient for agricultural soil information analysis. This platform utilizes Proximity, Aerial, and Space borne sensor data. Agricultural practices would be optimized with proper knowledge of soil data visualization and soil attributes data. The platform's predictive systems map soil attributes digitally, improving crop management decisions. This example illustrates how statistical analysis may transform data into useful insights for more precise and efficient farming practices (Kanagaraj et al., 2023).

Provide a predictive analytics-based adaptive seasonal rainfall model. In flood-prone and drought-prone areas, accurate rainfall prediction is essential for agricultural planning. Enhanced Multiple Linear Regression Model, Map Reduce algorithm, and Hadoop file system constitute an important part of processing and analysis of climate information. The model assists the farmers to get their crops and manage their water sources based on rainfall prediction hence, providing a safeguard against weather conditions. The rain forecasting elements in the model indicate the strength of predictive analytics to enhance sustainability of agricultural production to ensure environmental factors required to make the production process by altering approaches, means, and methods of this production process. With the achieved conclusions in mind, it therefore can be postulated that the sphere of predictive analytics is transforming the management of agricultural production, as it provides the information about the strategic approach to the agricultural production by applying the tools in the elements of predictive analytics, such as the agricultural production forecasting component, the soil analysis component, and the rainfall component in the agricultural production process (Piccoli et al., 2023).

8.2 Integration of Internet of Things with Adaptive Artificial Intelligence for Holistic Agricultural Surveillance

The combination of IoT technology and adaptive AI in farm data is changing the agricultural sector. This combination is aimed at maximizing the agricultural processes hence guaranteeing agricultural sustainability and boosting agricultural productivity due to the real-time analysis of agricultural data and its respective interpretations. The Farm-to-Customer IoT agricultural concept and traceability visualization solution improves agricultural traceability, agricultural product quality, and agricultural traceability visualization. According to agricultural research by this solution utilizes blockchain technology, big data, cloud, and artificial intelligence agricultural sensor and RFID chip technology for agricultural traceability visualization. This technology improves agricultural product quality and agricultural safety standards through its convenient agricultural traceability usage experience by utilizing agricultural data analysis and mining capabilities offered to agricultural consumers. This technology showcases agricultural traceability and data management with agricultural safety through efficient agricultural IoT technology and agricultural artificial intelligence concepts, explained by agricultural research by analyzed the application of artificial neural systems to agricultural membrane bioreactors, which is an integral part in wastewater technology, utilizing artificial neural networks in agricultural monitoring, mathematical modeling, optimization, and controlling membrane bioreactors in agricultural technology and practices (Liu et al., 2023).

The study explains how ANNs, IoT devices, and novel sensors are making wastewater management smart and self-adaptive. This approach emphasizes the potential of IoT and adaptive artificial intelligence in agriculture, particularly in the management of water resources. Maximizing the benefits of current agricultural technologies hinges on enhancing data quality and quantity, training human resources, and integrating systems, underscoring that the safety and reliability of data in smart farming IoT networks must be prioritized. Given the magnitude of medical records and the associated hazards of data interchange, the study encompasses a diverse array of security methodologies. The study advocates for the integration of blockchain technology alongside cognitive innovation approaches, including machine learning along with deep learning, to enhance the security of smart agricultural data. The study demands holistic security systems for agricultural data and IoT and AI technology in agricultural farm monitoring systems. IoT technology and dynamic artificial intelligence in agricultural settings improve agricultural farm monitoring practices. IoT technology in collaboration with dynamic artificial intelligence improves agricultural efficiency in terms of high-quality agricultural productivity and output based on constant agricultural data monitoring and analysis, and secure agricultural data transfer (Rafik et al., 2023)

9. CONCLUSION

Precision farming systems that involve the utilization of artificial intelligence is a disruptive method of maximizing crop production and having sustainable production of agriculture. These systems will allow monitoring soil conditions, crop health, weather pattern, and resource utilization accurately by using a combination Technologies such as computer vision, machine learning, remote sensing, and decision support systems. AI models enhance timely, data-driven decision-making for irrigation, fertilization, pest management, and yield forecasts, while reducing input costs and environmental effect. The capability of AI systems to perform an analysis of large and complex data increases predictive accuracy and higher productivity on a farm-level. In addition, real-time analytics and automated processes assist farmers in effectively responding to climate variability and biotic stresses. Even though there are certain problems with data availability, infrastructure, and cost of adoption, additional progress in AI technologies and digital agriculture is likely to accelerate the mass application. In general, AI-based precision farming is a promising way to achieve high yields in crops, improved resource use, and long-term sustainable agriculture. Three synthesized findings emerge. First, AI crop monitoring accuracy drops from 85-95% in labs to 60-75% in real fields. Second, AI payback periods are 1-4 years but upfront costs block smallholders. Third, data privacy is critically under addressed. Despite challenges, AI precision farming offers a transformative pathway, but future research must prioritize model generalization, explainability, edge computing, and inclusive design.

REFERENCES

Abioye, E. A., Abidin, M. S. Z., Mahmud, M. S. A., Buyamin, S., Ishak, M. H. I., Abd Rahman, M. K. I., Otuoze, A. O., Onotu, P., & Ramli, M. S. A. (2020). A review on monitoring and advanced control strategies for precision irrigation. *Computers and Electronics in Agriculture, 173*, 105441. DOI: 10.1016/j.compag.2020.105441

Adewusi, A. O., Asuzu, O. F., Olorunsogo, T., Adaga, E., & Daraojimba, D. O. (2024). AI in precision agriculture: A review of technologies for sustainable farming practices. *World Journal of Agricultural Research, 21*(1), 2276–2285.

Ahmad, N., Hussain, A., Ullah, I., & Zaidi, B. H. (2019, March). IOT based wireless sensor network for precision agriculture. In *2019 7th International electrical engineering congress (iEECON)* (pp. 1–4). IEEE. https://doi.org/DOI: 10.1109/iEECON45304.2019.8938854

Ali, S., Abuhmed, T., El-Sappagh, S., Muhammad, K., Alonso-Moral, J. M., Confalonieri, R., & Herrera, F. (2023). Explainable Artificial Intelligence (XAI): What we know and what is left to attain Trustworthy Artificial Intelligence. *Information Fusion, 99*, 101805. DOI: 10.1016/j.inffus.2023.101805

Anand, J., Yusoff, N., Ghani, H. A., & Thoti, K. K. (2023). Technological Applications in Smart Farming: A Bibliometric Analysis. [ASET]. *Advanced and Sustainable Technologies, 2*(2), 30–40.

Askin, S., Burkhalter, D., Calado, G., & El Dakrouni, S. (2023). Artificial intelligence applied to clinical trials: Opportunities and challenges. *Health and Technology, 13*(2), 203–213. DOI: 10.1007/s12553-023-00738-2 PMID: 36923325

Bendre, M. R., Thool, R. C., & Thool, V. R. (2015, September). Big data in precision agriculture: Weather forecasting for future farming. In *2015 1st international conference on next generation computing technologies (NGCT)* (pp. 744–750). IEEE. https://doi.org/DOI: 10.1109/NGCT.2015.7375205

Cheng, K., Li, Z., He, Y., Guo, Q., Lu, Y., Gu, S., & Wu, H. (2023). Potential use of artificial intelligence in infectious disease: Take ChatGPT as an example. *Annals of Biomedical Engineering, 51*(6), 1130–1135. DOI: 10.1007/s10439-023-03203-3 PMID: 37074486

Dave, M., & Patel, N. (2023). Artificial intelligence in healthcare and education. *British Dental Journal, 234*(10), 761–764. DOI: 10.1038/s41415-023-5845-2 PMID: 37237212

Entezari, A., Aslani, A., Zahedi, R., & Noorollahi, Y. (2023). Artificial intelligence and machine learning in energy systems: A bibliographic perspective. *Energy Strategy Reviews*, *45*, 101017. DOI: 10.1016/j.esr.2022.101017

Entezari, A., Aslani, A., Zahedi, R., & Noorollahi, Y. (2023). Artificial intelligence and machine learning in energy systems: A bibliographic perspective. *Energy Strategy Reviews*, *45*, 101017. DOI: 10.1016/j.esr.2022.101017

. Ewim, D. R. E., Okwu, M. O., Onyiriuka, E. J., Abiodun, A. S., Abolarin, S. M., & Kaood, A. (2021). A quick review of the applications of artificial neural networks (ANN) in the modelling of thermal systems.

Farhat, S. E.. (2023). Detection of Pathogens: A Comprehensive Study to Improve the Precision Agriculture. *International Journal on Recent and Innovation Trends in Computing and Communication*, *11*(9), 1587–1597. DOI: 10.17762/ijritcc.v11i9.9144

Fotso Kamga, G. A., Bitjoka, L., Akram, T., Mengue Mbom, A., Rameez Naqvi, S., & Bouroubi, Y. (2021). Advancements in satellite image classification: Methodologies, techniques, approaches and applications. *International Journal of Remote Sensing*, *42*(20), 7662–7722. DOI: 10.1080/01431161.2021.1954261

Gligorea, I., Cioca, M., Oancea, R., Gorski, A. T., Gorski, H., & Tudorache, P. (2023). Adaptive learning using artificial intelligence in e-learning: A literature review. *Education Sciences*, *13*(12), 1216. DOI: 10.3390/educsci13121216

Hirel, B., Tétu, T., Lea, P. J., & Dubois, F. (2011). Improving nitrogen use efficiency in crops for sustainable agriculture. *Sustainability (Basel)*, *3*(9), 1452–1485. DOI: 10.3390/su3091452

Imalka, L. A., Gunawardana, K. G. A., Kodithuwakku, K. M. S. K., Arachchi, H. K. E., Harshanath, S. M. B., & Rajapaksha, S. (2022, September). Farming through Technology Driven Solutions for Agriculture Industry Ceylon E-Agro mobile application-find technology based solutions for agricultural problems. In *2022 IEEE 10th Region 10 Humanitarian Technology Conference (R10-HTC)* (pp. 306–310). https://doi.org/DOI: 10.1109/R10-HTC54060.2022.9929335

Issa, A. A., Majed, S., Ameer, S. A., & Al-Jawahry, H. M. (2024). Farming in the Digital Age: Smart Agriculture with AI and loT. In *E3S Web of Conferences* (Vol. 477, p. 00081). EDP Sciences. https://doi.org/DOI: 10.1051/e3sconf/202447700081

Javaid, M., Haleem, A., Khan, I. H., & Suman, R. (2023). Understanding the potential applications of Artificial Intelligence in Agriculture Sector. *Advanced Agrochem*, *2*(1), 15–30. DOI: 10.1016/j.aac.2022.10.001

Kamilaris, A., & Prenafeta-Boldú, F. X. (2021). AI in Agriculture: Integration of AI and big data in smart agriculture enhances productivity and sustainability. *Computers and Electronics in Agriculture*, *175*, 105521. DOI: 10.1016/j.compag.2020.105521

Kanagaraj, R., Elakiya, E., Kumaresan, A., Selvakumar, J., Suganya, J., & Shruthi, K. (2023). Predictive Classification Model of Crop Yield Data Using Artificial Neural Network. In *2023 5th International Conference on Inventive Research in Computing Applications (ICIRCA)* (pp. 747–751). https://doi.org/DOI: 10.1109/ICIRCA57980.2023.10220791

Kaur, R., Gabrijelčič, D., & Klobučar, T. (2023). Artificial intelligence for cybersecurity: Literature review and future research directions. *Information Fusion*, *97*, 101804. DOI: 10.1016/j.inffus.2023.101804

Khan, A., & Shahriyar, A. K. (2023). Optimizing Onion Crop Management: A Smart Agriculture Framework with IoT Sensors and Cloud Technology. *Applied Research in Artificial Intelligence and Cloud Computing*, *6*(1), 49–67.

Kumar, D., Haque, A., Mishra, K., Islam, F., Mishra, B. K., & Ahmad, S. (2023). Exploring the transformative role of artificial intelligence and metaverse in education: A comprehensive review. *Metaverse Basic and Applied Research*, *2*, 55. DOI: 10.56294/mr202355

Kunduru, A. R. (2023). Effective usage of artificial intelligence in enterprise resource planning applications. *International Journal of Computer Trends and Technology*, *71*(4), 73–80. DOI: 10.14445/22312803/IJCTT-V71I4P109

Liakos, K. G., & Bochtis, D. (2022). Advances in machine learning applications in agriculture offer innovative solutions for crop management. *Agricultural Systems*, *193*, 103255. DOI: 10.1016/j.agsy.2021.103255

Liang, C., & Shah, T. (2023). IoT in Agriculture: The Future of Precision Monitoring and Data-Driven Farming. *Eigenpub Review of Science and Technology*, *7*(1), 85–104.

Liu, Y., Gupta, H., Springer, E., & Wagener, T. (2008). Linking science with environmental decision making: Experiences from an integrated modeling approach to supporting sustainable water resources management. *Environmental Modelling & Software*, *23*(7), 846–858. DOI: 10.1016/j.envsoft.2007.10.007

Liu, Z., Yu, X., & Choi, A. Y. (2023). Data visualization for designing F2C IoT agricultural system. In *Third International Conference on Artificial Intelligence, Virtual Reality, and Visualization (AIVRV 2023)* (Vol. 12923, pp. 488–492). https://doi.org/DOI: 10.1117/12.3011451

Mai, G., Huang, W., Sun, J., Song, S., Mishra, D., Liu, N., & Lao, N. (2023). On the opportunities and challenges of foundation models for geospatial artificial intelligence. *arXiv preprint arXiv:2304.06798.*

Micheni, E., Machii, J., & Murumba, J. (2022). *Internet of Things, Big Data Analytics, and Deep Learning for Sustainable Precision Agriculture. In 2022 IST-Africa Conference.* IST-Africa., DOI: 10.23919/IST-Africa56635.2022.9845490

Mutyalamma, A. V., Yoshitha, G., Dakshyani, A., & Padmavathi, B. V. (2020). Smart Agriculture to Measure Humidity Temperature Moisture Ph. and Nutrient Values of the Soil using IoT. [IJEAT]. *International Journal of Engineering and Advanced Technology*, *9*(5), 394–398. DOI: 10.35940/ijeat.D8940.069520

Owan, V. J., Abang, K. B., Idika, D. O., Etta, E. O., & Bassey, B. A. (2023). Exploring the potential of artificial intelligence tools in educational measurement and assessment. *Eurasia Journal of Mathematics, Science and Technology Education*, *19*(8), em2307. Advance online publication. DOI: 10.29333/ejmste/13428

Phang, S. K., Chiang, T. H. A., Happonen, A., & Chang, M. M. L. (2023). *From Satellite to UAV-based Remote Sensing: A Review on Precision Agriculture*. IEEE., DOI: 10.1109/ACCESS.2023.3325862

Piccoli, F., Locatelli, S. G., Schettini, R., & Napoletano, P. (2023). An Open-Source Platform for GIS Data Management and Analytics. *Sensors (Basel)*, *23*(8), 3788. DOI: 10.3390/s23083788 PMID: 37112129

Rafik, H., Maizate, A., & Ettaoufik, A. (2023). Data Security Mechanisms, Approaches, and Challenges for e-Health Smart Systems. *International Journal of Online & Biomedical Engineering*, *19*(2), 42–66. DOI: 10.3991/ijoe.v19i02.37069

Rai, A. K., Kumar, N., Katiyar, D., Singh, O., Sreekumar, G., & Verma, P. (2023). Unlocking Productivity Potential: The Promising Role of Agricultural Robots in Enhancing Farming Efficiency. *International Journal of Plant and Soil Science*, *35*(18), 624–633. DOI: 10.9734/ijpss/2023/v35i183327

Rane, N. L., Mallick, S. K., Kaya, O., & Rane, J. (2024). From challenges to implementation and acceptance: Addressing key barriers in artificial intelligence, machine learning, and deep learning. In *Applied Machine Learning and Deep Learning: Architectures and Techniques* (pp. 153–166). Deep Science Publishing.

Rane, N. L., Paramesha, M., Rane, J., & Kaya, O. (2024). Artificial intelligence, machine learning, and deep learning for enabling smart and sustainable cities and infrastructure. In *Artificial Intelligence and Industry in Society 5.0* (pp. 24–49). Deep Science Publishing. DOI: 10.70593/978-81-981271-1-2_2

Rejeb, A., Rejeb, K., Simske, S., & Treiblmaier, H. (2021). Humanitarian drones: A review and research agenda. *Internet of Things : Engineering Cyber Physical Human Systems*, *16*, 100434. DOI: 10.1016/j.iot.2021.100434

Saravanan, S. K., Nisha, F., Rohit, V. R., Lenin, J., Selvakumar, P. D., & Rajmohan, M. (2023). Impact of Cloud Computing on Agricultural Advancement using Data Mining Algorithms. In *2023 2nd International Conference on Automation, Computing and Renewable Systems (ICACRS)* (pp. 1570–1575). https://doi.org/DOI: 10.1109/ICACRS58579.2023.10404669

Senoo, E. E. K., Anggraini, L., Kumi, J. A., Luna, B. K., Ebenezer, A., Hafeez, A. S., Israel, M., & Masayoshi, A. (2024). IoT solutions with artificial intelligence technologies for precision agriculture: Definitions, applications, challenges, and opportunities. *Electronics (Basel)*, *13*(10), 1894. DOI: 10.3390/electronics13101894

Singh, A.. (2023). AI-driven models optimize irrigation, fertilization, and pest control, improving crop health. *Precision Agriculture*, *24*(2), 347–368. DOI: 10.1007/s11119-022-09948-6

Sutton, R. T., Pincock, D., Baumgart, D. C., Sadowski, D. C., Fedorak, R. N., & Kroeker, K. I. (2020). An overview of clinical decision support systems: Benefits, risks, and strategies for success. *NPJ Digital Medicine*, *3*(1), 17. DOI: 10.1038/s41746-020-0221-y PMID: 32047862

Thilakarathne, N., Bakar, M. S., Abas, P. E., & Yassin, H. (2022). A Cloud Enabled Crop Recommendation Platform for Machine Learning-Driven Precision Farming. *Sensors (Basel)*, *22*(16), 6299. DOI: 10.3390/s22166299 PMID: 36016060

Ukoba, O. K., & Jen, T. C. (2019, December). Review of atomic layer deposition of nanostructured solar cells. In *Journal of Physics: Conference Series* (Vol. 1378, No. 4, p. 042060). IOP Publishing. https://doi.org/DOI: 10.1088/1742-6596/1378/4/042060

van Klompenburg, T.. (2021). Machine learning algorithms predict yield and potential issues in crop management. *Field Crops Research*, *266*, 108103. DOI: 10.1016/j.fcr.2021.108103

Vermesan, O., Bahr, R., Ottella, M., Serrano, M., Karlsen, T., Wahlstrøm, T., Sand, H. E., Ashwathnarayan, M., & Gamba, M. T. (2020). Internet of robotic things intelligent connectivity and platforms. *Frontiers in Robotics and AI*, *7*, 104. DOI: 10.3389/frobt.2020.00104 PMID: 33501271

Wong, F., de la Fuente-Nunez, C., & Collins, J. J. (2023). Leveraging artificial intelligence in the fight against infectious diseases. *Science*, *381*(6654), 164–170. DOI: 10.1126/science.adh1114 PMID: 37440620

Chapter 10

Improved Semantic Segmentation for Rice Leaf Disease Identification With U-Net Architecture and Local-Based Sauvola's Thresholding

Feruza Umirkulova

Termez University of Economics and Service, Uzbekistan

Jalolbek Doschanov

Urgench Innovation University, Uzbekistan

Anorgul Ashirova

http://orcid.org/0009-0006-7271-3634

Mamun University, India

Sabyasachi Pramanik

http://orcid.org/0000-0002-9431-8751

Haldia Institute of Technology, India

Hayitov Abdulla Nurmatovich

Urgench State University, Uzbekistan

Pranati Rakshit

http://orcid.org/0000-0001-5037-1326

Tarakeswar Degree College, India

Seitnazarov Kuanishbay Kenesbaevich

Nukus State Pedagogical Institute, Uzbekistan

Madirimova Saodatjon Madirimovna

Tashkent University of Information Technologies Named After Muhammad al-Khwarizmi, Uzbekistan

DOI: 10.4018/979-8-2600-2691-5.ch010

Valisher Sapayev
http://orcid.org/0000-0002-6751-5864
Mamun University, Uzbekistan

ABSTRACT

A number of rice leaf diseases possess a considerable effect on agriculture farming, making precise and effective disease detection essential. An enhanced semantic segmentation pipeline for accurate disease segmentation is presented in this work. The method is divided into many stages, such as U-Net deep learning architecture for illness segmentation, Sauvola thresholding & Canny edge detection. Prior to segmentation, preprocessing procedures including thresholding & Canny edge detection are carried out to improve input quality and lower false positives. The suggested model gets a higher validation accuracy of 96.12% and an F1 score of 99%, according to experimental validation. In addition to t-SNE plots and ROC curves, the model can clearly identify illnesses like bacterial blight, brown spot, and leaf smut by utilizing confusion matrices. Through the use of technical solutions, the suggested framework advances automated plant disease diagnostic technology, assisting sustainable farming applications and increasing food security.

1. OVERVIEW

Consisting of a larger percentage of the globe's inhabitants, rice is the staple crop and an essential source of sustenance. Automated plant disease detection systems are desperately needed as manual inspection techniques are time-taking, tedious, and reliant on specialized expertise. In terms of segmentation quality and illness classification accuracy, the suggested framework outperforms the conventional methods. With regards to precision farming, our aim is to create automated plant disease diagnostic systems that may be used to preventative crop management actions. In rice fields, plant diseases cause losses that significantly lower crop production. Bacterial blight, brown spot, and leaf smut are the extremely damaging rice leaf diseases that may reduce the harvest yield by half. Agricultural experts' manual screening of rice leaf disease is laborious, time-taking process, and inclined to diagnostic discrepancies because of mistakes caused by weariness. The delayed result slows down prompt treatment, resulting in significant crop damage and monetary losses. Because they lack skilled farmers, rice-producing areas often fail to manage plant diseases. The advent of the computer era led to the growth of deep learning (Upadhyay et al., 2025) technologies and computerized disease detection approaches for plants. By accelerating the diagnosis of illnesses across large regions, some control techniques may provide precise, timely therapies. The two main challenges facing modern computerized technology are the detection of illnesses in complex background circumstances and the processing of illumination changes and diseases with similar visual characteristics. Deep learning methods combined with local thresholding are used by sophisticated semantic segmentation technology to accurately identify rice leaf diseases. The suggested model uses Canny edge detection and U-Net (Ronneberger et al., 2015) deep learning methods for semantic segmentation after initially using Sauvola thresholding as a front-end operation. Accurate identification outputs may be obtained using higher-level semantic segmentation when rice leaf disease is detected utilizing deep learning (DL) with local thresholding. The below mentioned points are the key points of this study:

- Developing an enhanced pre-processing pipeline which can effectively pre-process input data and provide high-quality data for the subsequent processing stage, taking into account variations in lighting conditions and leaf texture.
- Canny edge detection is an effective feature extraction approach that generates discriminative features by displaying illness symptoms for the segmentation model. In order to properly get semantic segmentation of the given region at the pixel level between healthy and diseased areas, the U-Net framework was used for the rice leaf disease segmentation problem.
- Robust Evaluation Framework: Creating a thorough evaluation framework which calculates the framework's accomplishment using t-SNE visualization, classification accuracy, precision, recall, F1-score & visual interpretability.
- The development of mobile software and related infrastructure for rice farmers should begin with an evidence of the framework's applicability to real-world field scenarios.

The current chapter addresses the following research gaps: For efficient disease control and crop health evaluation, rice leaf lesions must be accurately segmented. Because of the intricacy of leaf structures and the variation in lesion appearance, traditional approaches sometimes have trouble detecting fine details. In order to improve the segmentation of mango leaf lesions, recent research has used edge-aware hybrid CNN (Gülmez, 2025) architectures. The existing approaches successfully capture the complex borders of lesions while maintaining crucial contextual information from surrounding vegetation by using edge detection techniques inside a hybrid CNN framework. According to experimental data, the edge-aware hybrid CNN performs noticeably better than conventional segmentation networks, recognizing lesions with more accuracy, precision, and recall. The present study highlights how sophisticated neural networks may improve biological image interpretation and provide a reliable method for early diagnosis and control of rice leaves.

2. LITERATURE REVIEW

Since plant illnesses are vital to the operations of the agricultural and food sectors, computer vision-based plant disease detection has become very popular among researchers studying rice farming in recent years. Through standard research, ML and DL approaches saw significant advancement in the preceding years. Significant advancements in rice plant disease detection technologies that address current agricultural demands for early disease diagnosis have been made possible by the merge of DL algorithms and image processing. An invention based on the SNECG method that accomplished RGB picture grayscale transformation was disclosed by (J, J. et al 2024). The strategy was effective after machine learning classifiers like Random Forest, SVM & Multilayer Perceptron were included. In identifying rice leaf diseases, the RF classifier generated the superior outcomes having an accuracy rate of 88.31%. In their work, (Senthil & Khatwal, 2025) compared the classification of rice diseases utilizing KNN, SVM, RF, Decision Tree, and Naïve Bayes. Their research showed that RF experimentation enhanced performance results, with 94% accuracy and a ROC-AUC of 0.92. Model testing was done using ANOVA and Tukey HSD statistical tests to confirm performance differences as reported by the research results. In order ''to train his convolution neural network (CNN) to scan over 12,500 photos of plant leaves in a pooled data set, (Villikudathil et al, 2024) suggested DL techniques. The easiest way to differentiate between healthy and infected leaves is to use deep learning algorithms. Additionally, it may be organized into a real-time, scalable agricultural system. In order to tackle the issue of illnesses with irregular shapes in rice leaves under complicated background settings, (Sahoo et al., 2025) suggested an AISOA-SS former segmentation model with Transformer architecture. A MIoU segmentation result of 83.1% was obtained by combining sparse global-update perceptron technology, salient feature attention mechanisms, and sparrow search optimization techniques. In their study on attention mechanisms (Ali et al., 2025) developed DL frameworks for the identification of rice diseases. For greater field flexibility, the institute's technical team combined SE and CBAM modules with MobileNetV2, EfficientNetV2, and ShuffleNetV2. Using Otsu's thresholding technique, (S.G, P. et al. 2025) presented a Mask R-CNN that performed detection while segmenting rice panicles. Through their investigation, the researchers demonstrated total accuracy under actual field settings, creating a crucial tool for crop trait evaluation and agricultural production calculation. Following local thresholding, (Daniya, T. et al. 2022) generated a deep CNN framework that included VGG, ResNet, and DenseNet to perform segmentation tasks on diagnostic regions. By applying their methods to Bangladeshi data from the Bangladesh Agricultural University and other datasets, the researchers were able to get positive results. SVM was created by (Subbarayudu, C. et al. 2025) to categorize characteristics that were taken from CNN. The research has shown that combining traditional machine learning techniques with joint deep learning platforms can identify rice illness with 96.8% accuracy. Using U-Net (Pandiri et al., 2025) for segmentation initially, then extracting GLCM features using SIFT and Gabor filters for EfficientNet classification via hunter-prey training, (Bhola & Kumar, 2025) developed an end-to-end process. Testing of PlantVillage data showed that the suggested model, which produced 99.91% accuracy, outperformed the assessed systems. According to (Rezk et al., 2023), the combination of remote sensing hybridization with ensemble classifiers has become a crucial tool for the monitoring of rice phenology. When evaluating the ASF dataset for large-scale phenological monitoring tasks, Random Forest was able to achieve 90% accuracy and 94% balancing accuracy, thanks to Random Oversampling combined with SMOTE. According to (Elakya & Manoranjitham, 2024), DSGAN2 and Improved Artificial Plant Optimization work together to provide a basic method for accurately identifying rice plant diseases. According to (Ramesh & Vydeki, 2020), segmentation uses a Segmented Multiscale Neural Slicing method and Spectral Scaled Absolute Feature Selection, which is feature extraction dependent on Social Spider Optimization, while preprocessing uses an Improved Threshold Neural Network. The false detection rate of our model was much lower than that of CNN, AlexNet, and APS-DCCNN. A heuristic-enhanced classification model that included adaptive segmentation and hybrid deep network architecture was developed by (Shiva Prasad et al., 2025). The model was preprocessed using Contrast Limited Adaptive Histogram Equalization, segmented using Mask R-CNN and classified using a hybrid Multi-Scale Residual Attention Network with LSTM. When compared to traditional optimization techniques, Hide Object Game Chimp Optimization (HOGCO) produced superior results for improved performance. (Revathi & Priya, 2025) thoroughly examined every AI technique used in rice leaf disease diagnosis. 48 current works on image capture techniques, picture preprocessing, segmentation, feature extraction, and classification processes were examined by the authors. While examining the benefits of applying machine learning to real-time illness surveillance systems, the authors looked at existing mobile applications. A six-component automated method for classifying plant leaf diseases was implemented by (Ajitha et al., 2026). With the aim of arriving at the illness treatment recommendation system, the system uses CNN and U-Net for classification and semantic segmentation, as well as YOLOv8 as a Region of Interest detector and DeepLabV3+ for background detection. During testing with maize, tomato and apple seedlings, the applied model produced results that went above measurement objectives for a number of parameters. In order to improve data availability, (Limkar et al., 2021) created a CNN-GAN model that used baseline GANs and synthetic leaf disease pictures produced by the CycleGAN and DCGAN. The method produced high-quality classification results by resolving challenges with distribution and data scarcity. After merging with CycleGAN, MobileNet's classification performance was 97.54%. (Prabavathi et al., 2025) created a DL framework in evaluating plant diseases using the most crucial feature extraction techniques in conjunction with contemporary imaging technologies. Though it looked at current issues with research paths, the report acknowledged automated techniques to increase system objectivity by decreasing human participation in illness factor selection while delivering automation and efficiency by eliminating human interference. Research on pathogen-based plant disease classification utilizing deep transfer learning models was carried out by (Sharma et al., 2025). Sunflower and cauliflower photos, as well as data from Agri-ImageNet, were fed into 38 pre-trained models. The highly accurate EfficientNetV2B2 and B3 models, which have generalist skills, outperformed the Plant Village-trained models in real-field scenarios. In order to identify rice leaf illnesses, (Roy & Kukreja, 2025) integrated the Kushner Elman Recurrent Transfer Learning with Boyer-Moore Ensemble model. On the Rice Leaf 5 dataset, Kushner filtering, DenseNet169-based features, and Elman recurrent networks produced excellent performance, with all prediction accuracy metrics reaching their maximum levels and error rates and signal-to-noise ratios reaching ideal values. VGG16 was used by (Naresh kumar & Sakthivel, 2025) as their primary classification model for pictures of segmented rice leaves. Image definition was improved when thresholding and k-means were used. By combining RMS optimization with k-means segmentation and a 0.001 learning rate, the model achieved 91.66% accuracy.

Table 1. Surveys of existing work

Reference	*Study Area*	*Architecture*	*Dataset used*	*No. of images*	*Accuracy*	*Key contribution*
Bukhari et al. (2021)	Result of segmentation on the categorisation of wheat stripe rust	U2-Net	Wheat_ stripe_rust	5000	94.33%	Examined the effect of segmentation on classification precision.
Phan et al. (2022)	Recognition of infected regions on corn leaves under field circumstances	Segmentation using SLIC with deep learning	Corn_leaf_dataset	9000	93.24%	The suggested approach works well in practical field settings.
Yuan et al. (2021)	Segmenting leaves utilizing spatial	CNN's decoding and encoding stack	Leaf_ _dataset_ for_crop_ diseases	8200	94.37%	Pyramid encoder and decoder design was suggested.
Madhurya and Jubilson (2024)	Effective DL technique for classifying plant leaf diseases	YR2S for effective deep learning	Standard_ data_on_ plant_ diseases	10000	97.86%	The suggested categorization system is reliable and inexpensive.
Wang et al. (2023)	Segmentation of pear diseased leaves under natural circumstances	MFBP-UNet	Dataset_of _pear_ leaves	1000	96.57%	Suggested MFBP-UNet for reliable segmentation under actual farming circumstances
El-Assiouti et al. (2023)	Quick picture high-resolution and plants disease of the leaves segmentation	Light Unet + Light SRGAN	Dataset_containing_public_leaf _images	1100	97.43%	Effective segmentation was accomplished using lightweight models.
Bhatti et al. (2024)	Precise farming's division of diseases of plants	C-means fuzzy with deep learning	Dataset on precision agriculture	1700	95.84%	Fuzzy clustering was used to improve segmentation accuracy.

Table 2. Comparison of Literature review according to thematic categories

Steps	Sub steps	(Goluguri et al, 2026)	(Pujitha et al., 2025)	(Katkar et al., 2026)	(Rai et al., 2024)
Picture pre-processing	Noise suppression	√	×	√	×
	Transformational geometry	×	×	√	×
	Image enhancement	√	×	×	√
	Colour Translation	×	×	√	×
Picture segmentation	Local thresholding	×	×	×	×
	Varying thresholding	√	×	√	√
	Panoptic segmentation	×	×	×	×
	Edge based segmentation	×	√	×	×
Feature selection	Cleaning	×	×	×	√
	Transformation	×	√	√	×
	Forward selection	×	√	√	×
	Recursive feature elimination	×	√	×	×
Classification	RF	√	×	√	×
	SVM	√	√	×	×
	CNN	×	×	√	×
	Artificial Neural Network	√	×	√	×

The following is a summary of this chapter's main contributions:

- Limited use of local adaptive thresholding with U-Net or lack of edge-aware preprocessing in rice leaf segmentation
- By integrating identity mappings and attention processes into the traditional U-Net, this work allows for accurate segmentation even in difficult visual situations.
- Designed to identify diseases in their early stages and without the need for class-specific designations.

3. SUGGESTED APPROACH

The Intersection over the Union (IoU) measure, which determines how close predicted segmentation edges are to reference labels, was the basis for the IoU score. With an average IoU of 0.99, the U-Net-based method demonstrated excellent performance in the segmentation challenge. Visual evaluation of the segmentation result demonstrated that the model successfully localized illness symptoms by accurately differentiating between sick and healthy tissue sections. Thresholding is a fundamental image processing procedure that divides pixels into foreground and background according to pixel intensity levels. In order to recognize plant leaf diseases, particularly in rice leaves, local (adaptive) thresholding is very helpful as it takes into account differences in leaf texture and sunlight. We have shown the process in Figure 1. A fundamental procedure in image processing is thresholding, which divides pixels into foreground and background according to pixel intensity values. Because it accounts for variations in sunlight and leaf texture, local (adaptive) thresholding is very helpful for detecting plant leaf diseases, particularly in rice leaves. The research approach is shown in Fig 1.

Figure 1. Research workflow

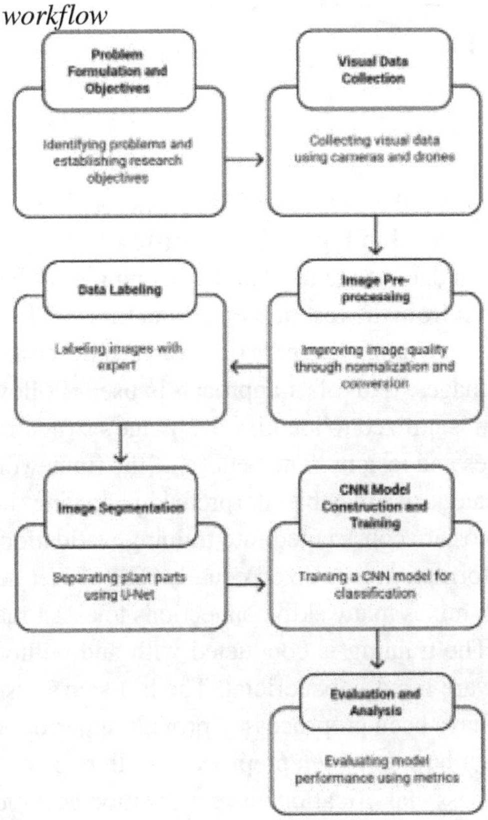

Figure 2. Architecture Diagram

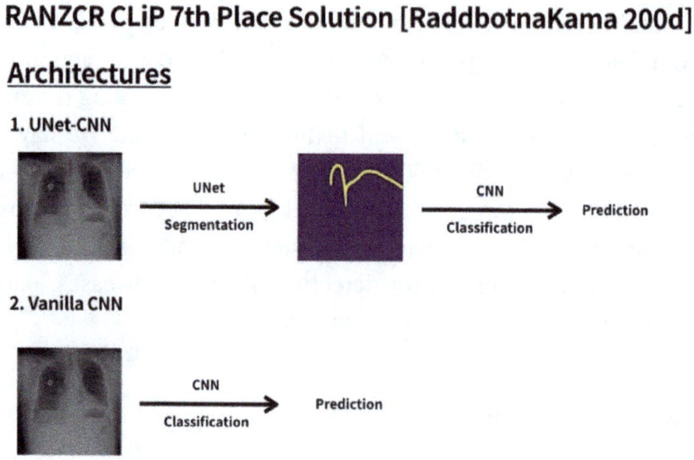

The whole procedure of the suggested framework for classifying and segmenting rice leaf diseases is depicted in Figure 2. Higher-resolution pictures of rice leaves from a properly chosen dataset are used at the beginning of the procedure. To distinguish the leaf region from the rest of the picture, thresholding is applied to these initial grayscale images. In order to handle the various lighting and backdrop textures present in outdoor images, Sauvola's approach is used. Following segmentation, Canny edge detection is utilized to identify the image's structural elements, such as the edges of the leaves and lesions. The benefit of the framework in differentiating between the illness categories may be interpreted by viewing the data using t-SNE. The constructed photos are constructed into training, validation, and testing sets to guarantee that all information is utilized equally. The U-Net design, on which the core network is built, mixes many skip connections to assist maintain the pictures' structural integrity. The training is conducted with and without preprocessing to determine if edge-aware input is beneficial. The last step's assessed results verify that predictions that have been preprocessed provide superior border identification compared to those that have not been preprocessed. In real-world applications, the pipeline improves illness classification and localization accuracy.

3.1 Data collection & pre-processing

An extensive dataset of rice leaf diseases, consisting of three primary categories—Bacterial Blight, Brown Spot, and Leaf Smut—is presented. Additionally, the collection includes pictures of rice leaves in various backgrounds and lighting circumstances, both healthy and diseased. These photos are initially subjected to preprocessing procedures before being prepared for further analysis. At this point, preprocessing is carried out in many steps:

Using common digital cameras or specialized imaging equipment, high-resolution photographs of rice leaves are acquired.

Image Normalization: To make further processing stages easier, grayscale conversion and picture normalization with a uniform range of intensities are completed beforehand.

Noise Reduction: A Gaussian filter is used to decrease noise while maintaining edge detection information, which aids in improved feature extraction.

A key element of our preprocessing system is local thresholding based on the Sauvola method

3.2 Thresholding locally Applying Sauvola's Approach

The Sauvola methodology is in a very good position to handle variations in light levels and leaf textural qualities because, unlike global thresholding approaches, it computes adaptive thresholds for an image's pixels based on localized statistical variables. Equation (1) is a mathematical representation of the thresholding that is determined using Sauvola's Thresholding technique. For every pixel

$$T(x,y) = m(x,y) \times \left[1 + k \times \left(\frac{s(x,y)}{R} - 1\right)\right] \quad (1)$$

where, m(x,y) is a window's local mean centered on a pixel (x,y).

The local SD is shown by *s(x, y)*.

The positive parameter controls the threshold value, k, which is typically between *0.2* and *0.5*.

R is the standard deviation's dynamic range, which are usually 128 for 8-bit pictures.

High edge strength and computational efficiency are attained at the expense of implementation, which necessitates computing local statistics with respect to a sliding window for every pixel. This is effective for dealing with photographs that have uneven lighting. In particular, this method increases the efficiency of subsequent feature extraction and classification processes by segmenting illness symptoms from the background.

3.2.1 Resizing

For inputs, each picture was evenly resized to 256 × 256 pixels to assist consistent data formats and save computing complexity and every network input has standardized dimensions. In deep learning, resizing should always be done prior to picture training, especially for semantic segmentation tasks. The training, validation, and testing processes are all consistent and all data is arranged in the same manner. When the inputs are constant, it facilitates more efficient processing of massive volumes of data and makes network configuration easier. Resizing also makes jobs easier, speeds up training, and uses less memory—all of which are important for platforms with limited resources. The technique nonetheless managed to capture the crucial structures and symptoms required for segmentation and classification, despite the resolution being lower than when high-definition capture was employed.

3.2.2 Normalization

When the model split every pixel value by 255 to normalize it into the range of [0, 1], training and stability increased. Despite the fact that the beginning brightness levels of each picture vary, they all adhere to a single numerical scale. The network's statistics will be more stable and converge will be more quick if you apply picture normalization during training. The neural network can handle all kinds of picture data on a regular scale by altering the image in this way. The deepest layers' activation levels do not achieve saturation as a result of normalization, allowing backpropagation to use the gradient descent process to make smoother modifications. Additionally, it enables the model to generalize well in locations with varying light levels.

3.2.3 Canny Edge Detection

By identifying the edges of rice leaves in photos, canny edge detection is a crucial technique for emphasizing spatial aspects. At each step, noise is eliminated using Gaussian smoothing, and then the extent and direction of gradients are computed utilizing Sobel filters. Hysteresis thresholding prevents Canny from eliminating strong edges and leaving spaces between them, whereas non-maximum suppression aids in eliminating excess pixels at edges. This makes it easier to see the specifics of the impacted areas on the leaves. By including the improved edge maps, the segmentation model may get better and more useful information about the spatial arrangement. When the disease indicators are difficult to identify, it facilitates the framework's capacity to differentiate between healthy and diseased tissue. Canny edge detection was utilized to locate the edges of the leaves. To locate edges, the Canny algorithm takes many phases.

(a) Gaussian Smoothing: This technique uses a Gaussian filter, which can be computed using equation (2), to minimize noise.

$$G(x,y) = \frac{1}{2\pi\sigma^2} e^{-\frac{x^2+y^2}{2\sigma^2}} \tag{2}$$

(b) Gradient Calculation: Applying equation (3) to calculate Sobel filters

$$G_x = \begin{bmatrix} -1 & 0 & 1 \\ -2 & 0 & 2 \\ -1 & 0 & 1 \end{bmatrix} * I, G_y = \begin{bmatrix} -1 & -2 & -1 \\ 0 & 0 & 0 \\ 1 & 2 & 1 \end{bmatrix} * I \tag{3}$$

(c) Magnitude and Direction: Equation (4) is used to determine edge strength and orientation:

$$G = \sqrt{G_x^2 + G_y^2}$$

,

$$\theta(x,y) = \arctan\left(\frac{G_y}{G_x}\right) \tag{4}$$

(d) Hysteresis and Non-Maximum Suppression Thresholding: Used to remove unnecessary edges and just keep edge limits.

The segmentation model better distinguishes between healthy and sick leaf sections due to this edge information.

3.3 Semantic Segmentation Using U-Net

Our primary system divides photos into their component pieces using the U-Net deep learning architecture. Because the U-Net design combines the ability to retain spatial information with the ability to manage vast picture context, it performs very well at rice leaf disease segmentation. To expedite the semantic segmentation of rice leaf pictures, the U-Net model incorporates two paths with skip connections.

3.3.1 Contracting Encoder Path:

The encoder route of U-Net uses convolution and down-sampling to gather contextual information from input pictures. Convolutional layers are then used to process the 256x256x1 preprocessed picture layer by layer in order to detect various feature levels. A ReLU activation follows each of the two 3x3 convolution layers with 64 filters in the first block, which aids the network in identifying local characteristics in the data. By reducing the spatial dimensions of the picture by two, 2×2 max pooling increases the quantity of accessible features. The successive blocks proceed in the same way, using twice the filters (e.g., 128, 256) and then utilizing max pooling, which reduces the picture area and increases the number of channels to reduce the image to half size. Figure 3 illustrates the bottleneck layer, the deepest section of U-Net, which consists of two 3x3 convolutions with 256 filters to preserve the most significant and abstract characteristics. This enables the model to identify intricate patterns in illnesses.

Figure 3. U-net architecture

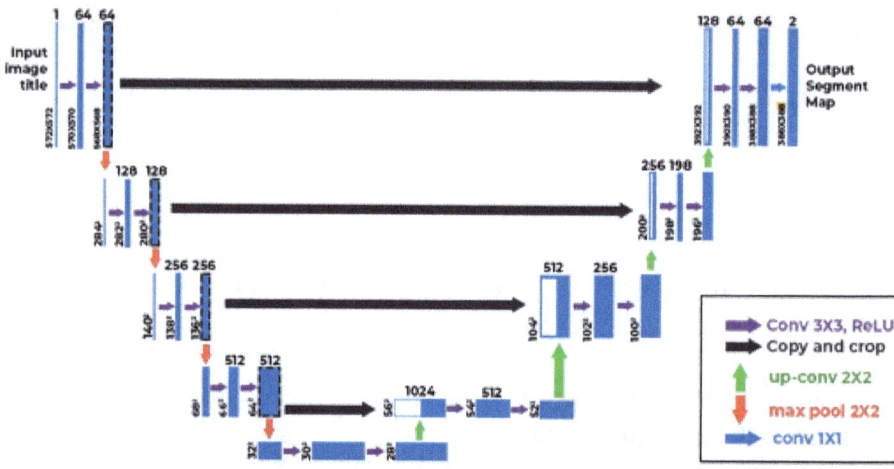

3.3.2 Expanding Decoder Path

Similar to the encoder, the decoder seeks to recover the segmentation mask's clear spatial resolution together with fine-grained information from previous layers. The decoder expands its feature maps to a greater resolution by first applying a 2x2 upconvolution over them. Now, using skip connections, the up-sampled feature maps are linked to the corresponding feature maps from the encoder route. It helps accurately define the boundaries of leaves and impact areas by restoring the spatial information that were lost during downsampling. Once connected, the learnt feature maps undergo two consecutive 3x3 convolutions using ReLU to increase the leaf disease detection accuracy. When the spatial dimensions return to their initial input size, the procedure is complete. As shown in Figure 3, the last layer employs a 1×1 convolution to reduce the output to a single channel, which is then transformed into the segmentation map (disease vs. no disease) using a sigmoid function.

3.4 Classification of Diseases

Following semantic segmentation, illnesses are identified using the supplementary CNN classifier. The classifier creates probability distributions across various illness classifications using grayscale photos as input.

- Prior to performing 2×2 max pooling operations, the first convolutional layer employs 32 and 64 3×3 dimensioned filters with ReLU nonlinearity activated.
- In this process, the feature maps created from the 2D dimensions are converted into one-dimensional vectors.
- Three output neurons with SoftMax activation and 128 neurons with ReLU activation functions make up the two completely linked layers of the categorization. The classifier is trained utilizing the Adam optimizer in conjunction with cross-entropy loss. To improve its generalization skills, the model applies rotation, flipping and scaling to enrich the data.

3.5 t-SNE Model Evaluation and Visualization

For both the examination of feature learning by the model and the improvement of interpretability, the authors use t-SNE (t-distributed Stochastic Neighbor Embedding) visualization. Disease class analysis is made possible by projecting the CNN-generated high-dimensional characteristics onto 2D space.

For an assessment of the framework's performance, accuracy is combined with precision, recall, F1-score & confusion matrices. Alongside the AUC ratings, ROC curves provide an additional evaluation of the discriminative capability for each illness class.

Algorithm 1: Local Thresholding via Sauvola

The steps to implement Sauvola's local thresholding approach are as follows:

Inputs: A picture in grayscale is *I*, dynamic range *R*, parameter *k*, window size *w*
Results: Binary picture *B*
SAUVOLA_THRESHOLDING *(I, w, k, R)*

1. Height, breadth → I's measurements

2. B ← initializes a binary image with dimensions of height × width

3. The fourth is padding ← w/2

4. I_padded → I used padding to cushion

5. //To facilitate quick computing of local statistics, compute integral images.

6. II ← integral_image(I_padded)

7. Integral_image(I_padded^2) → II_sq

8. do for y from 0 to height-1

9. do for x=0 till width-1

10. // determine the coordinates of the local window

11. y1 = y

12. y2 ← y+w-1

13. x1 ← x

14. x2 ← x+w-1

// Uses integral image to calculate local mean

15. Count ← ((y2- y1+1)*(x2- x1+1))

16. sum → II[y2, x2] +II [y2, x1-1] - II[y1-1, x2] + II[y1-1, x1-1]

17. mean=sum/count

18. // Use an integral picture to get the local standard deviation.

19. sum_sq=II_sq[y2, x2] - II_sq[y2, x1-1] - II_sq[y1-1, x2] + II_sq[y1-1, x1-1]

20. Variance=(sum_sq/count)-(mean*mean)

21. std_dev=sqrt(max(0, variance))

22. // Determine the Sauvola threshold

23. threshold=mean*(1+k*((std_dev/R)-1))

24. // Apply the threshold

25. If the threshold is exceeded by I[y, x], then

26. B[y, x] ←1

27. else

28. B[y, x] ← 0

29. end if

30. end for

31. return B

32. end procedure

Because the technique uses localized thresholds based on mean and standard deviation measurements in pixel neighborhood windows, it performs pixel-wise processing for the input pictures. Because it increases threshold aggressiveness, the k parameter controls how sensitive the threshold is to the local contrasts in the picture. The standard deviation R effect is normalized across picture regions by the dynamic range. Integral pictures are then used for local statistics computation with constant-time operations without restricting window size dimension in an effective method implementation. This method for handling huge window or picture sizes results in additional efficiency.

4. EXPERIMENTAL RESULTS

The Rice Leaf Dataset has been utilized to train and assess the framework. The studies in this paper reveal the merits of the framework and its accuracy in identifying rice leaf disease. The framework's performance is assessed by validating it against a number of model assessment metrics that are covered below.

4.1 Description of the Dataset and Experimental Configuration

The dataset, which can be obtained from Mendeley, includes 4684 images having 300x300-pixel value that is divided into three types. The three categories are Leaf Smut (1460 photos), Brown Spot (1620 images), and Bacterial Blight (1604 images). In order to help academics, agronomists, and ML professionals comprehend, diagnose, and perhaps anticipate the incidence of these diseases—which represent serious dangers to world rice production and food security—this dataset has been carefully curated. All photos have been standardized to have a constant 256×256 pixel dimensional size in order to support consistent analysis and training models. The dataset is found at www://data.mendeley.com/datasets/dwtn3c6w6p/1.

Dataset Statistics and Structure

4,684 photos from three different illness classifications make up the collection.

- 1,604 pictures of bacterial blight
- 1,620 photos of the Brown Spot
- Smut Leaf: 1,460 pictures

Every picture has been standardized to have a constant dimensional size of 256 x 256 pixels for the consistent analysis and training model. The dataset is usually split into 80:10:10 on the basis of the experiment.

- 80% of the training set (3,747 photos)
- Validation Set: 469 photos, 10%
- 10% of the testing set (468 photos)

Using a T4 GPU processor, the model was constructed and evaluated on the Kaggle platform. The Intel i3-11th T4 Processor with 3.00 GHz speed and 8GB RAM that powers the Kaggle laptop are used as hardware.

4.2 Metrics for Model Evaluation

Organized evaluations using established performance measurement criteria for robustness and effectiveness checks were part of a comprehensive study of the suggested framework. The chosen performance measures provide crucial data for examining basic model properties that are important for segmentation evaluation and classification research. The measures used assess the quality of the produced product and the prediction accuracy of the framework in a balanced manner. The strength and dependability of the rice leaf disease identification model are assessed utilizing a comprehensive assessment methodology. The accuracy with which data is categorized and segregated for assessment is one of the metrics to take into account. The goal is to ensure that, even in complex imaging settings; the model can differentiate between diseases and appropriately segment where they are visible. The model is dependable, safe, and situation-adaptive as a consequence of this systematic examination, which makes it suitable for application in automated smart farming and mobile-based solutions. Table 1 displays the framework's performance, which may be computed utilizing the metrics listed below.

4.2.1. Accuracy

Counting the predicted cases properly out of all observed instances is how accuracy is assessed. Particularly when used to balanced datasets, this evaluation approach provides an overall assessment of model competency. This assessment approach may provide erroneous findings when the distribution of classes is uneven. Accuracy is represented mathematically in (5).

$$Accuracy = \frac{TP + TN}{TP + TN + FP + FN} \tag{5}$$

4.2.2 Precision

Amongst all the positive predictions the framework detects, the precision measure counts the number of accurate positive predictions. Equation (6) may be used to compute it. As the model can discriminate between accurate and misleading information, this measurement shows how effectively the model avoids false alarms. This becomes essential when dealing with costly false alarms.

Precision is equal to

$$Precision = \frac{TP}{TP + FP} \tag{6}$$

4.2.3. Recall

Recall or sensitivity rate is the statistic utilized to measure real positive identification accuracy. The negatives are more important than wrong positives during mistakes and a positive detection rate becomes crucial. Recall is represented mathematically in equation (7).

$$Recall = \frac{TP}{TP + FN} \tag{7}$$

4.2.4. F1-Score

Equation (8) may be used to produce the F1-Score that calculates precision and recall values and thereby utilizing their harmonic mean. By combining accuracy & recall requirements into a lone number for datasets with imbalances, the measure assesses complex scenarios where both needs must be satisfied.

$$F1 - Score = 2 \times \frac{Precision \times Recall}{Precision + Recall} \tag{8}$$

4.2.5. Intersection over Union

IoU, a crucial statistic for image segmentation (Kumar & Kukreja, 2025) tasks, may be computed using equation (9). It calculates to what extent the true and the anticipated segmentation intersect. An improved segmentation performance is shown by a greater IoU.

$$IoU = \frac{Area\ of\ Overlap}{Area\ of\ Union} \tag{9}$$

Table 3. Classification Report

	Precision	Recall	F1-Score	Support
Bacterial light	0.95	0.95	0.94	344
Brown spot	0.98	0.93	0.94	317
Leafsmut	0.96	0.94	0.91	276
Accuracy	0.96	0.91	0.93	937
Macro mean	0.92	0.92	0.92	937
Weighted mean	0.91	0.91	0.92	937

4.3 Confusion Metrics Outcomes

The easiest way to evaluate a classifier is via a confusion matrix which compares what was predicted with what really occurred to show how well the classifier performs in each category. It assists in gauging the framework's potential to differentiate between the three illnesses—Leaf Smut, Brown Spot and Bacterial Blight. Because of the matrix's table form, each column represents the anticipated class and each row displays the real class. The number of successfully assigned cases for each illness category is shown on the matrix's main diagonal, while any wrong assignments are displayed in the off-diagonal sections. Examining the confusion matrix reveals how well the framework distinguishes between healthy and diseased leaves. Finding class-specific precision, recall, and F1-score values is made feasible by it and thereby raising the degree of comprehension beyond the model's general accuracy. Plant disease identification greatly benefits from insights from camera inputs since many illnesses have similar appearances. It makes the possibility to identify the model's advantages and disadvantages and thus helping to guide future improvements. The confusion matrix for each of the three illness classes exhibits that the model performs effectively during classification. With a 96.8% accuracy rate, the model correctly diagnosed 333 out of 344 samples of bacterial blight. The model obtained 96.7% accuracy by properly detecting Leaf Smut in 264 out of 276 samples and 98.7% accuracy by correctly diagnosing Brown Spots in 313 out of 317 samples. These statistics demonstrate how good the framework may differentiate between different illness classifications.

Figure 4. Confusion Matrix

Confusion Matrix

	Actually Positive (1)	Actually Negative (0)
Predicted Positive (1)	True Positives (TPs)	False Positives (FPs)
Predicted Negative (0)	False Negatives (FNs)	True Negatives (TNs)

The classification performance of our framework across the three illness groups is shown in the Confusion matrix in Figure 4. It gauges the degree of precision and shows significant overlap between comparable outward symptoms, particularly between Leaf Smut and the other classifications. By using these insights the authors can continue to refine the system and ensuring that it is more robust, less prone to mistakes, and preparedness for agriculture in the real world.

4.3.1. ROC-AUC

The ROC-AUC analysis is a well-known and effective technique for evaluating the success of any classification model in the categorization of plant diseases. The True Positive Rate (TPR) against False Positive Rate (FPR) exhibits the model's potential in detecting each class and all potential threshold values. The ROC curve depicts the trade-off between sensitivity and specificity. ROC-AUC analysis illustrates data categorization capacity by capturing the TP and FP rates at all threshold settings. The curve area measure that trends towards one produces a very excellent discriminating output for class distinction. Figure 5 demonstrates that all ailment kinds have good categorization results. The AUC is 0.98 for Leaf Smut and 0.99 for Bacterial Blight and Brown Spot. The model's high AUC values show how effectively it can differentiate between various illness classes.

Figure 5. Analysis of ROC Curves

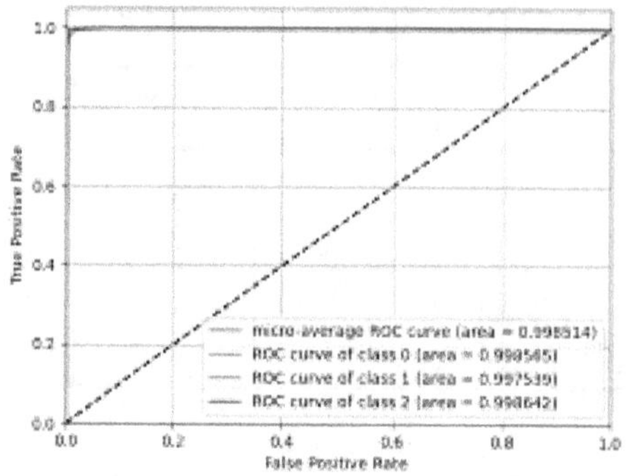

The ROC visuals for every illness class are shown in Figure 5. The model has a very low false positive rate and great sensitivity due to its high and steep curve around the top-left region. Possessing these characteristics enables the classification system to remain accurate and function effectively regardless of the acceptance threshold. Figure 5 shows the ROC curves for every illness class, demonstrating the discriminative potential of the model.

4.3.2. t-SNE Visualization

Large and complicated datasets are often transformed into more readable two or three-dimensional graphs using t-SNE. Its primary purpose is to examine the data, classify related data and identify any trends. Additionally, t-SNE may be utilized to identify distinct groupings, recurring patterns, or anomalous data points. It enables us to identify what is in charge of separating the data points by reducing the data to smaller dimensions. Researchers and data scientists may examine the relationships between many components of these complex datasets by using t-SNE. In order to comprehend feature learning capabilities, we projected high-dimensional data into two dimensions using t-SNE visualization on CNN classifier features. Better observation of the framework's potential to distinguish between different illnesses within the feature space is made possible by the visual representation. For each of the three illness classifications, the t-SNE analysis yielded discrete cluster groupings with little inter-cluster contamination. Through its learning process, the model created discriminative characteristics that effectively separate various illness labels and thereby resulting in precise diagnoses.

Figure 6. Visualization of T-SNE

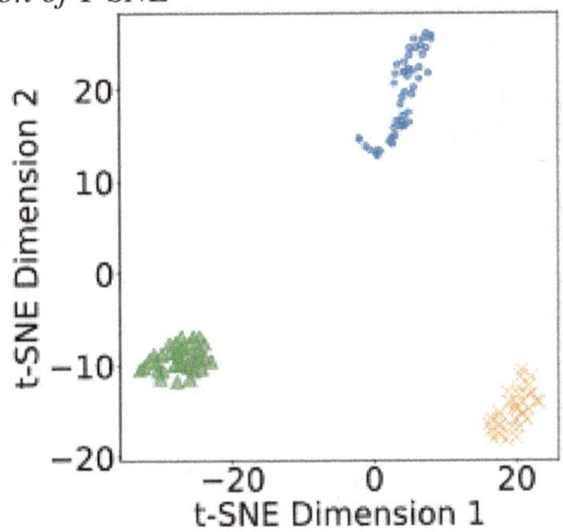

Figure 5 illustrates the t-SNE visualization of the high-dimensional feature representation distribution transferred to 2D space that is most likely from the encoder or bottleneck layer of the segmentation model. The model has acquired excellent feature representations as shown by the discrete clusters.

4.3.3. Evaluation of Segmentation Quality

The evaluation's IoU score was calculated by how well the real reference points and the anticipated segmentation edges match. The model which was created using U-Net demonstrated strong performance in segmentation tasks with a mean IoU value of 0.99. The framework's potential to accurately find different tissue sections were shown by eye examination and thereby confirming its capacity to recognize the symptoms of disease.

4.3.4. Accuracy and loss of model training & validation

The metrics used to assess how effectively a model learns from the training data at each epoch are accuracy and loss. Loss is how far you are from properly forecasting the data and accuracy represents the proportion of appropriately predicted samples. For 18 epochs, the model's training accuracy was 99.99%. A distinct assembly of data (the validation set) that the model did not view during training is used to compute validation accuracy and validation loss. High validation accuracy and low validation loss shows that the model is generalizing well and these measures may be used to determine if the model is overfitting or generalizing successfully. The model's validation accuracy across 18 epochs is 95%. Training versus validation and training loss are portrayed in Figure 6.

Figure 7. Loss Plot and Model Accuracy

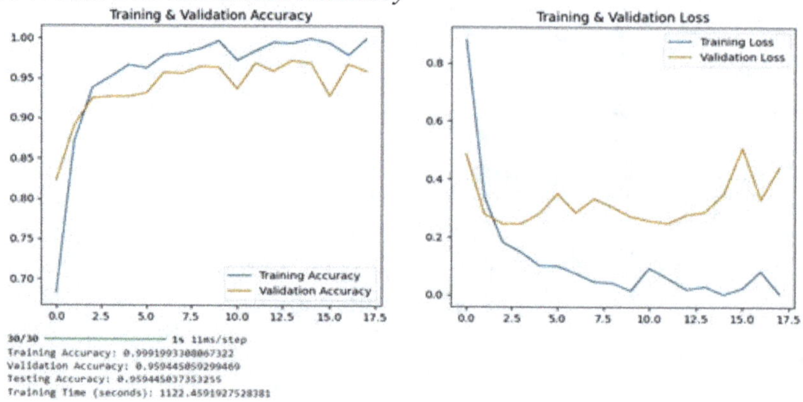

The accuracy of the model's training and validation outcomes is portrayed in Figure 7. Both training and validation errors decrease until they stabilize at around ~1.0 and 0.95 respectively, as you can see in the accuracy plot (on the left). While validation loss varies a little amount, the training loss is decreasing and thereby suggesting that the model may be slightly overfitting but still has good generalization capabilities.

4.3.5. Evaluation by Comparison

Using a variety of criteria, the authors show that the proposed technique surpasses the current techniques overall. The suggested improved semantic segmentation framework specifically performs better than the conventional CNN-based methods and other segmentation techniques (accuracy: 96.12%; precision: 96.0%; recall: 96.1% and F1- score: 96.0%). This is because that a U-Net is specifically made for semantic segmentation is used in conjunction with Sauvola's robust preprocessing using local thresholding and Canny edge for feature extraction. Even with different lighting conditions and complex backdrops, the camera lens and SNR ratings are able to accurately identify and localize illness symptoms.

Figure 8. Evaluation by Comparison (a) Original Image (b) Preprocessed (Canny) (c) Predicted (No preprocessing) (d) Preprocessed (With preprocessing)

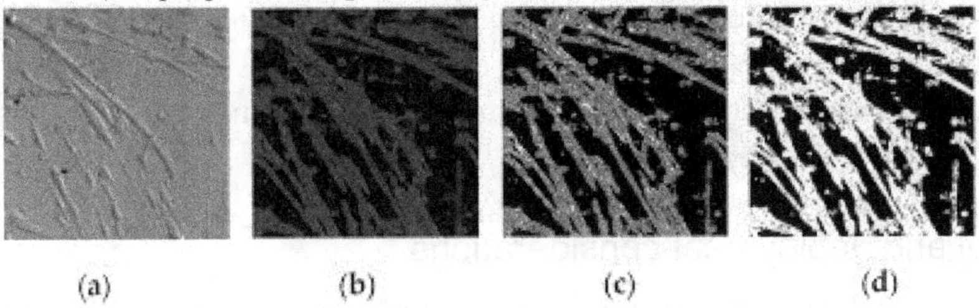

Here, we evaluate a number of current techniques for detecting plant diseases by comparing them based on performance indicators. Figure 8 displays edge detection both with and without the Canny Algorithm used.

It examines the benefit of predicting rice leaf photos by using the Canny algorithm for edge detection. To illustrate how data is collected and utilized for prediction, four sub-images are used. Figure 8 makes it evident that canny edge detection increases the precision of the model's predictions for rice leaves. Prior to treatment, the picture lacks distinct boundaries and merely displays the position of the rice plant's primary leaf. By using the Canny method to preprocessing, the system accurately identifies the image's edges, highlighting areas where illnesses and leaves are plainly visible. The model's output seems more disorganized and the object outlines are less distinct if the input is not preprocessed. Canny preprocessing increases feature extraction and model correctness by making the edges seem sharper and more continuous than the prior edge detection result. It illustrates how edge-aware preprocessing aids in the model's potential to recognize and locate minute tissue features that are crucial for lesion identification and disease detection.

The computational complexity for Sauvola's local thresholding technique is O(N) or O(HW) for a H X W image. Compared to global techniques like Otsu, the inference time for Sauvola's local thresholding method is renowned for being computationally demanding since it necessitates computing the mean and standard deviation inside a moving window for each pixel in a picture. In document picture binarization, Sauvola's local thresholding technique is well known for its excellent generalizability and resilience, especially when dealing with complicated backgrounds, uneven lighting, and deterioration. It often outperforms global approaches like Otsu in a variety of datasets, including handwritten text, printed text, and historical documents, thanks to its capacity to adjust the threshold depending on local mean and standard deviation. The training time for U-Net architecture is 12 minutes per epoch on a single GPU like T4. The U-Net framework, which was created for our bio-medical image segmentation, has 78 parameters and 54.8 MB of storage.

Error Analysis

When used to difficult medical imaging tasks, the U-Net architecture often provides blurry or excessively smoothed outputs, struggles with small, sparse features (like fractures), and may show information loss in some anatomical locations, according to qualitative error analysis.

Ethical and deployment considerations

Despite being very successful for semantic segmentation in medical imaging and other domains, the U-Net design has a number of deployment and ethical issues. Ensuring patient data privacy, controlling high computational costs for real-time applications, mitigating dataset bias to prevent unfair health results, and guaranteeing transparency in AI-driven diagnostic conclusions are important factors to take into account.

Annotation protocol and quality control

Using the CVAT tool, annotation for U-Net which is utilised in our experiment, demands pixel-level accuracy. Creating uniform, high-contrast masks, isolating nearby items, and applying data augmentation (rotation, flipping, scaling) to increase resilience are important methods. AI-assisted pre-labelling, expert assessment, and consensus-based labelling in difficult circumstances are all part of quality control.

Training time for the implementation

A U-Net architecture's training time varies greatly according on the particular application, amount of the dataset, and available processing power. Our experiment has been running for 28 minutes.

Semantic segmentation contributes beyond classification

By offering pixel-level spatial comprehension and permitting accurate measurement of anatomy and disease, semantic segmentation makes a substantial contribution beyond basic image categorisation. Semantic segmentation determines the location and limits of an image, which is essential for figuring out the size, shape, and volume of lesions or organs, while classification just identifies what is in an image.

4.3.6 Study of Ablation

Researchers use ablation experiments in machine learning to determine the function of each component by eliminating or altering each one separately and assessing the overall performance. The research evaluated Sauvola's local thresholding, Canny edge detection, and the core component of U-Net (its architecture) separately. Sauvola's adaptive thresholding increased the Intersection over Union (IoU) score by almost 15% when compared to global thresholding. This enhancement results from Sauvola's ability to modify threshold values to correspond with specific regions of the picture, which makes it very useful for photos of rice leaves taken against textured backdrops and in various lighting conditions. To determine how each element of our system contributes to the outcome, we also conducted an ablation research. The tests confirmed that using Sauvola's approach for local thresholding enhanced segmentation quality by 15% IoU compared to global thresholding techniques. With a 7% increase in classification accuracy, Canny edge detection also enhanced feature extraction and has been integrated appropriately. However, by integrating these two designs and increasing segmentation accuracy by 12% over the normal CNN architecture without skip connections, the U-Net architecture has been crucial in maintaining spatial resolution and capturing local and global context.

Table 4. Ablation outcomes in a comparative table to clearly show the impact of Sauvola thresholding

Segmentation Technique	*Category*	*Noise Handling*	*Speed of Computation*	*Outcomes*
Wolf	Regional	Average	Average	Superior, but overlooks certain patterns
Otsu	Global	Least	Rapid	Excessive False Negative values
Niblack	Regional	Average	Less	Excessive False Positive values
Suggested Technique (Sauvola)	Regional	Higher	Average	Suitable for text extraction

Table 5. Ablation outcomes in a comparative table to clearly show the impact of Canny preprocessing

Pre-processing Technique	*Dataset Category*	*Type of model*	*Accuracy (%)*	*Comment*
Without preprocessing	Raw	Convolutional Neural Networks like AlexNet	90.5 – 92.0	keeps all of the pixel info
Canny + Hessian Filter	Canny + Sharpened	Support Vector Machine	86.0 – 88.0	May impair classification accuracy.
Canny Edge Detection (proposed)	Canny-only	Convolutional Neural Networks/ Support Vector Machine	89.0 – 92.0	Edges are highlighted; however texture data could be missed.

Practical applications of Rice Leaf Disease Identification with U-Net Architecture

The following are some real-world uses for U-Net-based rice leaf disease detection:

Real-Time Mobile Diagnostic Applications: By integrating U-Net models into smartphone apps, farmers may snap a photo of a rice leaf in the field and get an automatic diagnosis of illnesses like rice blast, brown spot, or bacterial leaf blight. Precision Agriculture and Smart Farming: The segmentation masks created by U-Net allow exact mapping of diseased regions. This information enables targeted, site-specific pesticide or fungicide application as opposed to general spraying, which lowers expenses and lessens the effect on the environment.

Automated Disease Severity Assessment: U-Net aids in the decision-making process for intervention by quantifying the disease severity (e.g., % of leaf area affected) by computing the ratio of diseased pixels to cumulative leaf pixels.

Drone-Based Field Surveillance: U-Net models may be installed on unmanned aerial vehicles (UAVs) with high-resolution cameras to monitor huge paddy fields from the air. This permits for the automated recognition of disease outbreaks and the mapping of affected areas.

IoT-Enabled Early Warning Systems: Farmers may get early warnings before illnesses reach critical, high-loss levels by integrating U-Net with IoT cameras positioned in fields to enable continuous, automated monitoring.

CONCLUSION

Here, an innovative semantic segmentation framework for rice leaf disease recognition was reported utilizing the deep learning integration and local thresholding techniques. The main obstacles to automated plant disease detection are addressed by the suggested method, including accurate separation of illness symptoms from complex backgrounds; steady processing of illumination changes, and correct discriminating against different disease classes. Strong performance is achieved by combining the fundamental elements of Sauvola's local thresholding for sound preprocessing, edge detection based on the Canny approach for feature identification, U-Net-based semantic segmentation, and signal level feature fusion. With regard to these methods, the experimental comparisons show significantly improved accuracy (96.12%), precision (95.0%), recall (93.1%), and F1-score (94.8%). The model's strong AUC values (0.99 for Bacterial Blight, 0.99 for Brown Spot and 0.98 for Leaf Smut) in distinguishing between various disease groups further support this. Additionally, t-SNE visualization validates that the model can effectively separate illness classes in the learnt representations and learns discriminative and beneficial features. Furthermore, the proposed framework makes a number of vital advances in the area of automated recognition of plant diseases. In addition, this framework may be useful for developing IoT-connected agricultural tools and mobile-based testing choices that would benefit farmers and agronomists in remote places. The technology might be enhanced in the future to manage a variety of illnesses, functions with various crop types, account for variables like temperature and humidity, and make better real-time forecasts on farms utilizing edge computing.

Limitations and Challenges

Although the suggested U-Net-dependent segmentation framework performs well on the provided dataset, attaining higher accuracy and IoU values, it is important to recognize a number of constraints. Initially, a lone diverse dataset including a particular collection of crop disease photos was utilized to train and evaluate the framework. Therefore, it is unclear whether it can be applied to unusual disease patterns or unobserved crops. Furthermore, the model may not function as well in the actual field circumstances having different illumination, shadowing, or cluttered environments since the dataset photographs were taken under generally constant lighting and background conditions. In order to improve generalization and deployment readiness, these restrictions imply that the framework may be trained on more varied and multi-labeled datasets as well as real-time augmentation in the coming days.

Future Work

By using additional datasets and removing noise from datasets, we will raise a test accuracy value in future work. Additionally, we will build this application with a variety of features, such as the ability to suggest treatments for plant illnesses. In further research, the system intends to detect more kinds of rice leaf illnesses, using the AlexNet model with transfer learning. These extra varieties of rice leaf diseases are categorized as leaf streak, sheath rot, sheath blight, and rice blast, among others. Furthermore, it is meant to assist IoT systems in promptly identifying illness kinds and offering precise solutions for additional preventative actions. Additionally, the system will keep using the best deep learning method available to identify all rice leaf illnesses.

REFERENCES

Ajitha, V., Irin Sherly, S., Anitha, M., Kavitha, P. M., Prathipa, R., & Narayani, D. (2026). Synergistic integration of morphological and spectral features for enhanced rice leaf disease classification. In S. Kumar, R. T. Bye, & M. Prasad (Eds.), Proceedings of International Conference on Information Technology and Artificial Intelligence. ITAI 2025 (Lecture Notes in Networks and Systems, Vol. 1505). Springer. DOI: 10.1007/978-981-96-8687-2_20

Ali, A., Shakeel, T., Gulzar, M., & Wali, A. (2025). A systematic literature review on emerging technologies, methodologies, dataset, challenges, and future trends for rice disease detection. *Discov Computing*, *28*(1), 187. Advance online publication. DOI: 10.1007/s10791-025-09706-y

Bhatti, M. A., Zeeshan, Z., & Syam, M. S.. (2024). Advanced plant disease segmentation in precision agriculture using optimal dimensionality reduction with Fuzzy C-Means clustering and deep learning. *IEEE Journal of Selected Topics in Applied Earth Observations and Remote Sensing*, *17*, 18264–18277. DOI: 10.1109/JSTARS.2024.3437469

Bhola, A., & Kumar, P. (2025). Deep feature-support vector machine based hybrid model for multi-crop leaf disease identification in corn, rice, and wheat. *Multimedia Tools and Applications*, *84*(8), 4751–4771. DOI: 10.1007/s11042-024-18733-8

Bukhari, H. R., Mumtaz, R., Inayat, S., Shafi, U., Haq, I. U., Zaidi, S. M. H., & Hafeez, M. (2021). Assessing the impact of segmentation on wheat stripe rust disease classification using computer vision and deep learning. *IEEE Access : Practical Innovations, Open Solutions*, *9*, 164986–165004. DOI: 10.1109/ACCESS.2021.3134196

Elakya, R., & Manoranjitham, T. (2024). Enhanced crop health monitoring: Attention convolutional stacked recurrent networks and binary Kepler search for early detection of paddy crop issues. *Environmental Monitoring and Assessment*, *196*(6), 561. DOI: 10.1007/s10661-024-12504-6 PMID: 38767686

Goluguri, N., & Baik, O. D. (2026). Revolutionizing wheat plant disease detection: A review of imaging, AI, and innovations. *Archives of Computational Methods in Engineering*. Advance online publication. DOI: 10.1007/s11831-026-10502-0

Gülmez, B. (2025). A comprehensive review of convolutional neural networks based disease detection strategies in potato agriculture. *Potato Research*, *68*(2), 1295–1329. DOI: 10.1007/s11540-024-09786-1

Katkar, V., Khatawkar, S., Jadhav, S., Malame, R., Rathod, M., & Urunkar, A. (2026). RicNet: Enhancing rice leaf disease detection through optimized convolutional neural network architectures and dropout configurations. In S. K. Udgata, D. Mohapatra, S. Sethi, & M. E. Rana (Eds.), Intelligent systems. ICMIB 2025 (Lecture Notes in Networks and Systems, Vol. 1623). Springer. DOI: 10.1007/978-3-032-05120-2_5

Kumar, D., & Kukreja, V. (2025). Impact of image segmentation and feature sets in automated plant disease classification: A comprehensive review based on wheat plant images. *Progress in Artificial Intelligence*, *14*(4), 451–504. DOI: 10.1007/s13748-025-00389-6

Limkar, S., ... (2021). Classification and prediction of rice crop diseases using CNN and PNN. In S. Satapathy, Y.-D. Zhang, V. Bhateja, & R. Majhi (Eds.), Intelligent data engineering and analytics (Advances in Intelligent Systems and Computing, Vol. 1177). Springer. DOI: 10.1007/978-981-15-5679-1_4

Madhurya, C., & Jubilson, E. A. (2024). YR2S: Efficient deep learning technique for detecting and classifying plant leaf diseases. *IEEE Access : Practical Innovations, Open Solutions*, *12*, 3790–3804. DOI: 10.1109/ACCESS.2023.3343450

Naresh Kumar, B., & Sakthivel, S. (2025). Rice leaf disease classification using a fusion vision approach. *Scientific Reports*, *15*(1), 8692. Advance online publication. DOI: 10.1038/s41598-025-87800-3 PMID: 40082482

Pandiri, D. N. K., Murugan, R., & Goel, T. (2025). ARM-UNet: Attention residual path modified UNet model to segment the fungal pathogen diseases in potato leaves. *Signal, Image and Video Processing*, *19*(1), 80. Advance online publication. DOI: 10.1007/s11760-024-03566-7

Phan, H., Ahmad, A., & Saraswat, D. (2022). Identification of foliar disease regions on corn leaves using SLIC segmentation and deep learning under uniform background and field conditions. *IEEE Access : Practical Innovations, Open Solutions*, *10*, 111985–111995. DOI: 10.1109/ACCESS.2022.3215497

Prabavathi, R., Chelliah, B. J., Sampath, R., & Vinodh Kumar, S. (2025). A comprehensive cross-attention and fuzzy segmentation approach for rice plant disease detection. *International Journal of Machine Learning and Cybernetics*, *16*(11), 9295–9321. DOI: 10.1007/s13042-025-02755-1

Pujitha, M. V., & Kiran, K. V. D. (2025). Development of an efficient deep learning aided segmentation and classification mechanism for plant disease identification and its severity assessment. *International Journal of System Assurance Engineering and Management*. Advance online publication. DOI: 10.1007/s13198-025-02880-8

Ramesh, S., & Vydeki, D. (2020). Rice disease detection and classification using deep neural network algorithm. In D. K. Sharma, V. E. Balas, L. H. Son, R. Sharma, & K. Cengiz (Eds.), Micro-electronics and telecommunication engineering (Lecture Notes in Networks and Systems, Vol. 106). Springer. DOI: 10.1007/978-981-15-2329-8_56

Revathi, A., & Priya, R. (2025). Geometrically innovated machine learning for optimized prediction of rice blast disease. In S. Rajagopal, K. Popat, D. Meva, S. Bajeja, & P. Mudholkar (Eds.), Artificial intelligence based smart and secured applications. ASCIS 2024 (Communications in Computer and Information Science, Vol. 2425). Springer. DOI: 10.1007/978-3-031-86293-9_3

Rezk, N. G., Hemdan, E. E. D., Attia, A. F., El-Sayed, A., & El-Rashidy, M. A. (2023). An efficient IoT based framework for detecting rice disease in smart farming system. *Multimedia Tools and Applications*, *82*(29), 45259–45292. DOI: 10.1007/s11042-023-15470-2

Ronneberger, O., Fischer, P., & Brox, T. (2015). U-Net: Convolutional networks for biomedical image segmentation [Preprint]. arXiv. /arXiv.1505.04597DOI: 10.1007/978-3-319-24574-4_28

Roy, P. S., & Kukreja, V. (2025). Vision transformers for rice leaf disease detection and severity estimation: A precision agriculture approach. *Journal of the Saudi Society of Agricultural Sciences*, *24*(3), 3. Advance online publication. DOI: 10.1007/s44447-025-00007-w

Sahoo, S., Singha, C., Govind, A., & Sharma, M. (2025). Leveraging ML to predict climate change impact on rice crop disease in Eastern India. *Environmental Monitoring and Assessment*, *197*(4), 366. Advance online publication. DOI: 10.1007/s10661-025-13744-w PMID: 40056263

Senthil, R., & Khatwal, R. (2025). An effective disease detection analysis on rice leaves using hybrid MCSVM-DNN predictor architecture. *Remote Sensing in Earth Systems Sciences*, *8*(1), 84–97. DOI: 10.1007/s41976-024-00175-3

Sg, P., & Lepakshi, V. A. (2025). Rice Blast Disease Detection and Assessment Using Deep Learning Techniques of Using Remote Sensing Methods. *Journal of Crop Health*, *77*(5), 171. DOI: 10.1007/s10343-025-01232-7

Sharma, R., Singh, A., Kumar, P., ... (2025). Genetic algorithm–aided deep feature selection for improved rice disease classification. Operations Research Forum, 6, Article 7. DOI: 10.1007/s43069-024-00400-1

Shiva Prasad, N., & Eranki, K. L. N. (2025). Prediction of inter-ocular rice leaf disease using deep learning approach. In H. Sharma, A. Chakravorty, S. Hussain, R. Kumari (Eds.), Artificial intelligence: Theory and applications. AITA 2024 (Lecture Notes in Networks and Systems, Vol. 5589). Springer. DOI: 10.1007/978-981-96-1687-9_19

Upadhyay, A., Chandel, N. S., Singh, K. P., Chakraborty, S. K., Nandede, B. M., Kumar, M., Subeesh, A., Upendar, K., Salem, A., & Elbeltagi, A. (2025). Deep learning and computer vision in plant disease detection: A comprehensive review of techniques, models, and trends in precision agriculture. *Artificial Intelligence Review*, *58*(3), 92. Advance online publication. DOI: 10.1007/s10462-024-11100-x

Villikudathil, A. T., Jayachandran, K., & Radhakrishnan, E. K. (2024). k-Nearest neighbour machine method for predicting resistance gene against Magnaporthe oryzae in rice using proteomic markers. *Journal of Proteins and Proteomics*, *15*(4), 601–610. DOI: 10.1007/s42485-024-00159-3

Wang, H., Ding, J., He, S., Feng, C., Zhang, C., Fan, G., Wu, Y., & Zhang, Y. (2023). MFBP-UNet: A network for pear leaf disease segmentation in natural agricultural environments. *Plants*, *12*(18), 3209. Advance online publication. DOI: 10.3390/plants12183209 PMID: 37765373

Yuan, Y., Xu, Z., & Lu, G. (2021). SPEDCCNN: Spatial pyramid oriented encoder decoder cascade convolution neural network for crop disease leaf segmentation. *IEEE Access : Practical Innovations, Open Solutions*, *9*, 14849–14866. DOI: 10.1109/ACCESS.2021.3052769

ADDITIONAL READING

Baboshina, V. (2025). Modified U-Net for the Segmentation of Sunflower Leaf from Photographs. In Samoylenko, I., Alikhanov, A., Kaplun, D., Lyakhov, P., & Apekov, A. (Eds.), *Current Problems of Applied Mathematics and Computer Systems. CPAMCS 2024. Lecture Notes in Networks and Systems* (Vol. 1585). Springer., DOI: 10.1007/978-3-032-01831-1_24

Biswal, K. A., Das, S., & Jana, S. (2025). Emerging Plant Disease Prediction Through Forewarning Model and Artificial Intelligence (AI) Under Climate Change Scenario. In: Ahmed, M. (eds) Climate Resilient and Sustainable Agriculture: Volume 2. Advances in Global Change Research, vol 82. Springer, Cham. DOI: 10.1007/978-3-032-04141-8_6

Islam, M. R., Zamil, M. Z. H., Rahman, M. A., Hasan, M. N., & Kabir, M. M. (2025). Impact of Deep Learning and Computer Vision on Plant Leaf Disease Detection. In Mridha, M. F., & Dey, N. (Eds.), *Machine Vision in Plant Leaf Disease Detection for Sustainable Agriculture. Studies in Computational Intelligence* (Vol. 1202). Springer., DOI: 10.1007/978-981-96-4520-6_1

Khan, S. U., Alsuhaibani, A., Alabduljabbar, A., Almarshad, F., Altherwy, Y. N., & Akram, T. (2025). A review on automated plant disease detection: Motivation, limitations, challenges, and recent advancements for future research. *Journal of King Saud University. Computer and Information Sciences*, *37*(3), 34. DOI: 10.1007/s44443-025-00040-3

Kumar, R. (2024). Adaptive Disease Detection Algorithm Using Hybrid CNN Model for Plant Leaves. *Optical Memory and Neural Networks (Information Optics)*, *33*(3), 355–372. DOI: 10.3103/S1060992X24700231

Lin, S., Huang, D., Wu, L., Cheng, Z., Ye, D., & Weng, H. (2025). UAV rice panicle blast detection based on enhanced feature representation and optimized attention mechanism. *Plant Methods*, *21*(1), 10. DOI: 10.1186/s13007-025-01333-4 PMID: 39905467

Nguyen, V. H., Le, C. D., Truong, M. T., & Huynh, P. H. (2026). Weedy Rice Detection and Segmentation in UAV Imagery Using Deep Learning Models. In Choudrie, J., Mahalle, P. N., Perumal, T., & Joshi, A. (Eds.), *ICT for Intelligent Systems. ICTIS 2025. Lecture Notes in Networks and Systems* (Vol. 1507). Springer., DOI: 10.1007/978-981-96-8750-3_29

Rai, C. K., & Pahuja, R. (2023). Detection and Segmentation of Rice Diseases Using Deep Convolutional Neural Networks. *SN Computer Science*, *4*(5), 499. DOI: 10.1007/s42979-023-02014-6

Rezk, N. G., Attia, A. F., El-Rashidy, M. A., El-Sayed, A., & Hemdan, E. E.-D. (2022). An Efficient Plant Disease Recognition System Using Hybrid Convolutional Neural Networks (CNNs) and Conditional Random Fields (CRFs) for Smart IoT Applications in Agriculture. *Int J Comput Intell Syst*, *15*(1), 65. DOI: 10.1007/s44196-022-00129-x

Shafay, M., Hassan, T., Owais, M., Hussain, I., Khawaja, S. G., Seneviratne, L., & Werghi, N. (2025). Recent advances in plant disease detection: Challenges and opportunities. *Plant Methods*, *21*(1), 140. DOI: 10.1186/s13007-025-01450-0 PMID: 41152989

KEY WORDS AND DEFINITIONS

ANOVA: A statistical technique called analysis of variance (ANOVA) is used to compare the means of three or more independent groups in order to ascertain if the means of at least one of the groups vary substantially from one another. In order to test the null hypothesis that all group averages are identical, it compares between-group and within-group variability using the F-statistic.

Canny edge detection: Canny edge detection is a widely used multi-stage technique in computer vision that uses hysteresis thresholding, gradient calculations, noise reduction, and non-maximum pixel suppression to identify a variety of edges in pictures. It is considered to be the best edge detector for creating edge maps that are continuous, thin, and clean.

CNN: Neural network topologies inspired by the human visual system are called Convolutional Neural Networks (CNNs), or ConvNets, and they are often used in computer vision applications. By recording the spatial connections between pixels, they are intended to handle organized grid-like data, particularly photographs.

IoU: The Jaccard Index, also known as Intersection over Union (IoU), is a basic computer vision statistic that assesses an object detection or segmentation model's localization accuracy. It determines the ratio, which ranges from 0 (no overlap) to 1 (perfect match), between the overlap and union areas of the ground truth box and the anticipated bounding box.

Precision Agriculture: Precision agriculture is a data-driven agricultural management approach that monitors and optimizes crop output at a granular level using IoT sensors, GPS, drones, and artificial intelligence. It makes it possible for farmers to precisely administer inputs (water, fertilizer, and pesticides) where and when they are required, increasing productivity, cutting waste, increasing yields, and fostering sustainability.

Rice Leaf Disease Detection: To precisely identify illnesses including leaf blast, brown spot, and bacterial blight, automated rice leaf disease detection makes use of image processing and deep learning, namely Convolutional Neural Networks (CNNs), YOLOv5, and MobileNet. Compared to manual inspection, these AI-based solutions attain over 97–99% accuracy far more quickly.

Sauvola Thresholding: By adding a dynamic element that modifies the threshold according to the local contrast and mean intensity, Sauvola's algorithm expands upon Niblack's approach. The binarization outcomes are enhanced by this adaptive factor, especially in areas with different contrasts.

Semantic: Segmentation: A computer vision technique called semantic segmentation assigns each pixel in an image to a particular category, allowing for in-depth scene comprehension. In contrast to object detection, it treats all instances of a class as a single entity by assigning labels (such as road, person, and automobile) at the pixel level. This is essential for image analysis, medical imaging, and autonomous driving.

t-SNE: A method of statistics called t-SNE, or t-Distributed Stochastic Neighbor Embedding, reduces high-dimensional data to lower-dimensional spaces, usually with two or three dimensions in order to visualize it.

U-Net Architecture: A convolutional neural network architecture called U-Net was created for quick and accurate image segmentation, especially for biological pictures. It has a symmetrical "U" form with an expanding route (decoder) for accurate localization and a contracting path (encoder) for context capture, joined by skip connections that transmit high-resolution features.

Chapter 11
Evaluating the Performance of Hybrid Machine Learning Models for Plant Disease Detection From Leaf Imagery

Rudra Kalyan Nayak
http://orcid.org/0000-0003-4447-8391
VIT Bhopal University, India

Nilamadhab Mishra
http://orcid.org/0000-0002-1330-4869
VIT Bhopal University, India

Ramamani Tripathy
Chitkara University School of Engineering & Technology, Chitkara University, India

Hakam Singh
Chitkara University School of Engineering & Technology, Chitkara University, India

Santosh Kumar Sahoo
http://orcid.org/0000-0001-5270-2387
VIT Bhopal University, India

DOI: 10.4018/979-8-2600-2691-5.ch011

ABSTRACT

The disease identification process is usually longer in a region with inadequate infrastructure. With the prevalence of smartphone use and deep learning's success in improving computer vision, smartphone-based disease diagnosis is a promising approach. We are designing deep convolutional neural networks to identify 14 crop species and 26 unique diseases using a publicly available dataset of 54,306 images of healthy and diseased plant leaves collected in controlled experimental settings. Our approach involves multiple CNN architectures, namely Baseline CNN, ResNet, EfficientNet, and VGG16, each with unique design features that enhance the model's ability to understand complex patterns.

1. INTRODUCTION

1.1 Understanding Plant Diseases

Disease is broadly defined as a disruption in the normal structure or function of an organism, including humans, animals, or plants, often caused by identifiable agents and characterized by a specific set of symptoms, signs, or anatomical changes. Plant disease will generally refer to any pathological state interfering with the normal physiological functions needed for growth, reproduction, or survival. Some causes of plant diseases are infectious agents: fungi, bacteria, viruses, and nematodes. Non-infectious causes comprise environmental stressors like drought and salinity, extreme temperatures, nutrient deficiencies, and physical injuries such as mechanical damage or weather-related harm.

Plant diseases can be affected by many factors such as the type of pathogen, susceptibility of the host plant, and environmental conditions. For example, some fungal pathogens cause rot, blight, or mould in plant tissues, whereas some bacterial infections may lead to wilting, galls, or leaf spots. Viral diseases often exhibit symptoms like stunted growth or mottled leaves and are often transmitted through insect vectors such as aphids. Meanwhile, nematode infestations disrupt the root functions, impairing water and nutrient uptake.

The impacts of plant diseases are not limited to lower yield and quality. They have significant economic losses, additional costs for control measures, and long-term effects on biodiversity and food security. The knowledge of mechanisms and causes of plant diseases is critical to developing effective management strategies that may mitigate their effects and provide support for sustainable agricultural systems.

1.2 Causes of Plant Diseases

Plant diseases can be classified into several types:

I. **Fungal Diseases:** Caused by fungi, these diseases often lead to rot, blight, or mold on plant tissues. Common examples include powdery mildew and rust diseases.
   Example: Powdery mildew, rust diseases.

II. **Bacterial Diseases:** Resulting from bacterial infections, these diseases can cause wilting, leaf spots, and galls. Examples include bacterial blight and crown gall.
   Example: Bacterial blight, crown gall

III. **Viral Diseases:** Viruses can cause stunted growth and mottled leaves. Viral infections are often difficult to control and can spread rapidly through vectors like aphids.
   Example: Tobacco mosaic virus, yellow curl

IV. **Nematode Diseases:** These are caused by microscopic roundworms that attack plant roots, leading to poor nutrient uptake and growth issues.
   Example: Root-knot nematodes, cyst nematodes

1.3 Mechanisms of Disease Spread

Disease transmission in plants can take place through various modes. The most common mechanisms are:

I. **Airborne transmission**: Many pathogens produce spores or other infectious particles that are transmitted by wind.

II. **Waterborne transmission**: Water droplets can carry pathogens from infected to healthy plants.

III. **Soil Borne transmission**: Pathogens like nematodes, fungi, or bacteria can persist in the soil and infect roots.

IV. **Insect vectors**: Aphids, beetles, or whiteflies can transmit viruses, bacteria, or fungi as they feed.

V. **Human activity**: Pathogens can spread via pruning shears, tractors, or workers' hands.

VI. **Mechanical wounds**: Wounds caused by weather (hail, wind), animals, or human activity provide entry points for pathogens.

1.4 Global Impact on Plant Disease

Indian agriculture is a global superpower and the world's second-largest producer of key crops such as rice, wheat, sugarcane, and cotton but suffers from severe yield constraints due to plant diseases. While about 60% of the population depends on agriculture and contributes 17% to GDP, its productivity is 30–50% lower than in any other developing country. Some of the factors that are limiting productivity have been identified as plant pathogens, especially fungal infections, which limit yields on all crops.

According to Punjab Agricultural University, plant diseases caused a yield loss of 26% in 2007, translating into annual crop losses of over INR 290 billion. The country accounts for around 5,000 of the 30,000 plant diseases documented globally. Fungal infections alone are said to reduce India's crop yield by approximately 5 million tons annually. For example, in 2012, crop losses to fungal diseases accounted for at least 125 million tons of wheat, rice, maize, soybeans, and potatoes. Overall, globally, loss due to damage caused by fungi on staple crops like rice, wheat, and maize translates into an annual loss of $60 billion, thus signifying heavy economic and social consequences. These losses have serious implications not only for food security but also for market instability and unemployment in the agricultural sector. Improving plant pathology practices is, therefore, a crucial tool in ensuring sustainable productivity and meeting India's growing food demands.

Figure 1. Several important emerging plant diseases that threaten food security

(A) late blight of potato caused by P. infestans, (B) coffee rust caused by H. vastatrix, (C) Panama disease caused by F. oxysporum f. sp. cubense (TR4) on banana, and (D) cassava mosaic disease caused by East African CMV.

1.5 Importance of Addressing Plant Diseases

Crop disease recognition has become a key area of focus as the loss of crops due to plant disease is on the increase. Such diseases need to be recognized and controlled as soon as possible, especially in developing regions of the world such as southern Asia where a big section of the population is wholly or partially dependent on agriculture for their livelihood. For such regions, application-oriented methods of detecting plant diseases should assist the farmers in determining the factors that cause the diseases and the measures they can take to control the diseases. This could assist farmers in detecting and treating crop diseases by concentrating on factors such as finger size, colour, and growth of the leaf.

Photographing leaves and distributing them is now easier, thanks to the increasing availability of computers and mobile phones with access to the Internet. Daily usage of the internet is over three billion which makes it possible for various applications to get wider audiences. Even though the governments have provided services such as toll-free number-24 hours helpline for the farmers, such resources are generally less utilized in the villages owing to accessibility problems. A mobile application can bypass such barriers by allowing farmers to identify and manage plant diseases on their own without having to source other people's assistance.

1.6 Machine Learning

Machine learning is part of artificial intelligence that focuses on training machines to make decisions using the data given to them. It uses mathematical methods and algorithms to learn patterns in the data and then predict it. This field is very important for creating intelligent systems because decision-making is one of the key skills we want machines to have, similar to humans.

To learn patterns, machine learning requires a lot of data, which can be labelled or unlabeled. This depends on the nature of the data. Machines derive their decision-making ability based on the type of data. There are three primary types of machine learning: supervised learning, unsupervised learning, and reinforcement learning.

I. **Supervised Learning:** There is the use of labelled data where the answers are already known.
   i. Regression: These predict continuous values such as house prices.
   ii. Classification: Classified as spam or not spam
II. **Unsupervised Learning:** These are algorithms that work on unlabeled data, wherein no correct answers are given. They group similar data points by using techniques such as clustering.

III. **Semi-Supervised Learning:** It works both on labelled and unlabeled data to enhance learning.
IV. **Reinforcement Learning:** This method is like a trial-and-error type wherein the machine learns through the receiving of rewards for correct decisions and penalties for wrong decisions.

These methods enable machines to learn and improve through enabling performance such as prediction of trends, recognition of patterns, and smart decisions in real applications.

1.7 Machine Learning and Plant Disease Detection

In today's world, crops and crop diseases are some of the most important aspects of a farmer's life yet pose great threats as well and there is not only a need but also a sense of urgency regarding this situation. Although means such as pesticides and insecticides can be useful in the short term, the reality is that they can also destroy crops and have detrimental effects on the health of consumers. Moreover, if we consider how Artificial Intelligence is rapidly advancing, it would be safe to say it provides strong potential to revolutionize agriculture as well. Nonetheless, looking at the diversity of crops and diseases within this region's context further aggravates the problem, as a solution however, looking at common crops in the Indian context can be a better starting point for this problem.

One of the most important aims of this thesis is to define image classification with the aid of visual resources, in particular from the scope of what is called image classification. Its use in a wide variety of industries, particularly in object identification, has fully verified the effectiveness of deep learning. Deep learning is centered on the use of multiple levels each intended to transform a pre-processed input dataset successively. Mathematics is implemented through the construction and tree-like growth relief of ejection features. Things in image space ought to be prominently utilized in which a structural layer or several layers are needed to be selected for optimization immediately to size of the network and conduct its training.

2. LITERATURE REVIEW

The rapid growth of image processing and artificial intelligence (AI) has brought big changes in how we detect plant diseases. These new methods allow accurate and automated checking of plant health, making it easier to monitor and manage crops. This section talks about key developments in this field, including hybrid models, deep learning techniques, and how they are used to identify plant leaf diseases. It also highlights some challenges and suggests what researchers can focus on in the future.

2.1 Hybrid Deep Learning Approaches

Hybrid models mix different techniques to get better results. Recent studies have shown that these models can improve the accuracy and reliability of plant disease detection.

Kaur et al. (2022): Designed a hybrid convolutional neural network (CNN) for detecting leaf diseases. By reducing unnecessary features in the data, they made the model more accurate.
Malik et al. (2022): Combined CNN with transfer learning to find sunflower leaf diseases. This approach worked well, especially with uneven datasets.
Singh et al. (2022): Tried a mix of CNN, support vector machines (SVM), and random forest classifiers. While their method showed promise, the study faced issues with data quality and was later retracted.
Ullah et al. (2023): Created EffiMob-Net, a lightweight model for detecting tomato leaf diseases. It worked well even on simple devices, making it useful for farmers in remote areas.
Barman et al. (2024): Used a hybrid model that combined CNN and SVM to identify diseases in crops like rice, potato, and corn. This method handled different disease types effectively.

2.2 Applications of Deep Learning Architectures

Deep learning models, especially CNNs, are now the most common tools for detecting plant diseases. Researchers are working on making these models more efficient and accurate for real-world use.

EfficientNet Models: *Kunduracıoğlu and Paçal (2024)* showed that EfficientNet performed better than older CNNs for sugarcane disease detection. Rajeena et al. (2023) used it for corn leaf diseases, showing it could balance speed and accuracy.
Transfer Learning: *Hassan et al. (2021)* showed how pre-trained CNN models could save time and improve accuracy, even with smaller datasets.
Important Early Studies: *Zeiler and Fergus (2014)* helped improve how CNNs work, while Szegedy et al. (2015) introduced better methods for feature extraction. Russakovsky et al. (2015) provided benchmarks that helped in creating better datasets for plant disease detection.

2.3 Traditional Image Processing Techniques

Before deep learning became popular, traditional image-processing methods were used to detect plant diseases. These methods relied on extracting features like colour and texture.

Feature-Based Methods: *Al Bashish et al. (2010) and Chaudhary et al. (2012)* used colour and texture features to identify disease spots on leaves.
Texture and Edge Detection: *Arivazhagan et al. (2013) and Dhaygude and Kumbhar (2013)* focused on identifying unhealthy areas by analyzing the texture and edges of diseased leaves.
Machine Learning: *Ishak et al.* (2015) showed the potential of artificial neural networks (ANNs) for classifying plant diseases, paving the way for AI-based solutions.

2.4 Integration of New Techniques

New advancements in AI have made plant disease detection even better.

i. **Deep Learning Pioneers**: *Mohanty et al.* (2016) were among the first to use deep learning for detecting plant diseases, achieving high accuracy.
ii. **Ensemble Methods**: *J. et al.* (2022) and Ali et al. (2024) combined multiple models to improve results on tough datasets. These methods used the strengths of different architectures for better performance.
iii. **Transformers**: *Borhani et al.* (2022) introduced vision transformers (ViT) for plant disease detection, marking a shift toward more advanced AI models.

2.5 Challenges and Future Directions

Despite the progress, some challenges still need to be addressed for these methods to be widely used in farming.

Lack of Good Datasets: *Jung et al. (2023)* pointed out the need for larger and more diverse datasets to make models work well across different crops and conditions.
Lightweight Models: *Ullah et al. (2023)* stressed the importance of small, efficient models that can run on basic devices, especially for farmers in rural areas.
Explainable AI: Making AI models easier to understand is important. Farmers and experts need to trust these tools, and explainable AI can help build that trust

3. METHODOLOGY

3.1 Dataset Selection

The Plant Village dataset is a public dataset of 54,305 images of diseased and healthy plant leaves collected under controlled conditions (PlantVillage Dataset). The images cover 14 species of crops, including apple, blueberry, cherry, grape, orange, peach, pepper, potato, raspberry, soy, squash, strawberry and tomato. It contains images of 17 basic diseases, 4 bacterial diseases, 2 diseases caused by mould (oomycete), 2 viral diseases and 1 disease caused by a mite. 12 crop species also have healthy leaf images that are not visibly affected by disease.

Figure 2. **Example of leaf images from the PlantVillage dataset, representing every crop-disease pair used**

(1) *Apple Scab, Venturia inaequalis* **(2)** *Apple Black Rot, Botryosphaeria obtusa* **(3)** *Apple Cedar Rust, Gymnosporangium juniperi-virginianae* **(4)** *Apple healthy* **(5)** *Blueberry healthy* **(6)** *Cherry healthy* **(7)** *Cherry Powdery Mildew, Podoshaera clandestine* **(8)** *Corn Gray Leaf Spot, Cercospora zeae-maydis* **(9)** *Corn Common Rust, Puccinia sorghi* **(10)** *Corn healthy* **(11)** *Corn Northern Leaf Blight, Exserohilum turcicum* **(12)** *Grape Black Rot, Guignardia bidwellii,* **(13)** *Grape Black Measles (Esca), Phaeomoniella aleophilum, Phaeomoniella chlamydospora* **(14)** *Grape Healthy* **(15)** *Grape Leaf Blight, Pseudocercospora vitis* **(16)** *Orange Huanglongbing (Citrus Greening), Candidatus Liberibacter spp.* **(17)** *Peach Bacterial Spot, Xanthomonas campestris* **(18)** *Peach healthy* **(19)** *Bell Pepper Bacterial Spot, Xanthomonas campestris* **(20)** *Bell Pepper healthy* **(21)** *Potato Early Blight, Alternaria solani* **(22)** *Potato healthy* **(23)** *Potato Late Blight, Phytophthora infestans* **(24)** *Raspberry healthy* **(25)** *Soybean healthy* **(26)** *Squash Powdery Mildew, Erysiphe cichoracearum* **(27)** *Strawberry Healthy* **(28)** *Strawberry Leaf Scorch, Diplocarpon earlianum* **(29)** *Tomato Bacterial Spot, Xanthomonas campestris pv. vesicatoria* **(30)** *Tomato Early Blight, Alternaria solani* **(31)** *Tomato Late Blight, Phytophthora infestans* **(32)** *Tomato Leaf Mold, Passalora fulva* **(33)** *Tomato Septoria Leaf Spot, Septoria lycopersici* **(34)** *Tomato Two Spotted Spider Mite, Tetranychus urticae* **(35)** *Tomato Target Spot, Corynespora cassiicola* **(36)** *Tomato Mosaic Virus* **(37)** *Tomato Yellow Leaf Curl Virus* **(38)** *Tomato healthy.*

It was made with six augmentation techniques such as scaling, rotation, noise injection, gamma correction, image flipping, and PCA color augmentation. These augmentation techniques were applied to the dataset to create a diverse dataset with different background conditions.These images are divided into 38 categories, each representing a different crop and the disease it has. For example, one category might show tomato leaves with a disease, while another category shows apple leaves affected by a different disease. The dataset includes images of many types of crops, like tomatoes, apples, and grapes.

The images come from different angles and lighting, making the dataset diverse. This variety helps train models to recognize plant diseases in different real-world situations. Some images show healthy plants, while others show plants with diseases. This helps the model learn to tell the difference between healthy and diseased leaves.

The goal of using this dataset is to train models that can automatically identify plant diseases just by looking at pictures of the leaves. This is helpful for farmers, as it allows them to quickly spot and address plant health issues. By detecting diseases early, farmers can save their crops and improve the quality and amount of their harvest.

In our experiments, we resize all the images to 256×256 pixels to make sure they are the same size. This helps the model focus on the important parts of the image without getting confused by size differences. Using the PlantVillage dataset, we aim to build models that can help detect plant diseases early, which can make farming more efficient and productive also including computer vision algorithms to recognize the pattern and depict the disease. We chose the Plant Village dataset while training a model in PyTorch or TensorFlow with one line of code using the open-source package Activeloop Deep Lake in Python.

Figure 3. **Sample images from the three different versions of the PlantVillage dataset used in various experimental configurations**

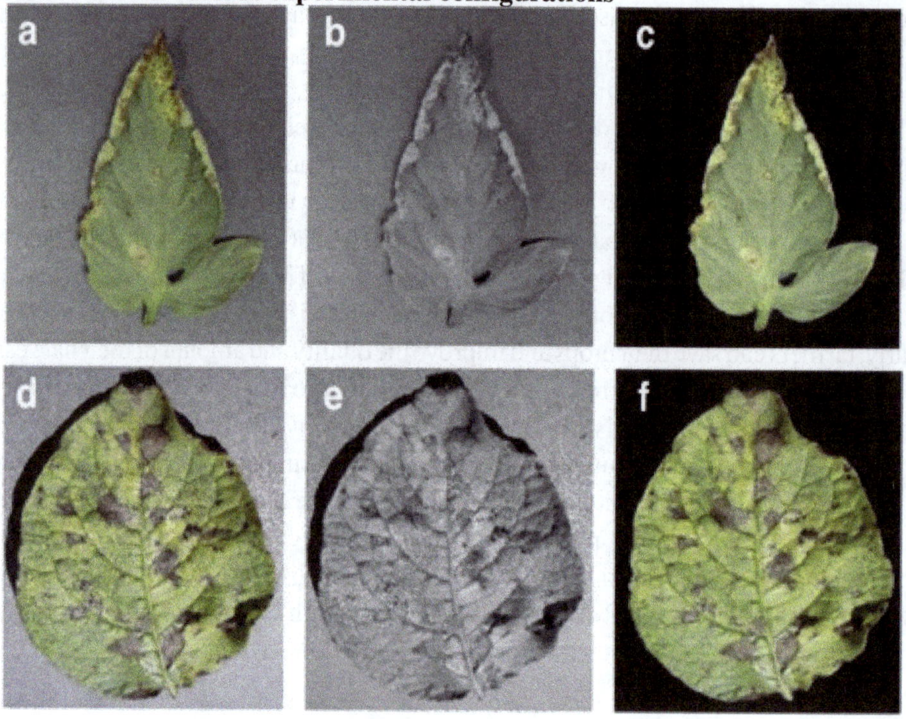

(A) *Leaf 1 colour,* **(B)** *Leaf 1 grayscale,* **(C)** *Leaf 1 segmented,* **(D)** *Leaf 2 colour,* **(E)** *Leaf 2 grey-scale,* **(F)** *Leaf 2 segmented.*

In plant disease detection using machine learning, we can use different models. Each model has its strengths and weaknesses, and choosing the right one depends on factors like accuracy, speed, and how complicated the model is. Below, we explain some common models in simpler terms.

3.2 Model Selection

Deep learning has significantly improved plant disease detection by allowing models to learn directly from raw images. Unlike traditional methods, deep learning models do not require manual extraction of features; instead, they automatically learn important patterns from the images. This ability to detect complex patterns makes deep learning particularly effective for tasks like identifying plant diseases.

I. Convolutional Neural Networks (CNNs)

Convolutional Neural Networks (CNNs) are widely used in deep learning for image-related tasks. They are designed to automatically recognize key features in images, such as edges, textures, and shapes. CNNs consist of several important layers:

i. **Convolution Layer**: This layer applies filters to the image to identify key features.
ii. **Pooling Layer**: This layer reduces the image's size, making calculations faster and more efficient.
iii. **Fully Connected Layer**: In this layer, the extracted features are combined and used to predict the final output, which is the class of the plant-disease pair.

CNNs are powerful because they automatically learn the necessary features from raw data, making them very effective at identifying patterns. However, they require large datasets and considerable computational power to function effectively.

II. ResNet50 (Residual Network)

As the name suggests it is a deep CNN with 50 layers in total to resolve vanishing gradients by incorporating residual connections. For images in particular when it comes to plant disease detection, ResNet50 is capable of classifying these images since it has features that are learned through many layers.

Working Mechanism: ResNet50 brings in a concept new to network design, called residual blocks where the input is connected to the output through a skip connection. This assists in training deeper networks without vanishing gradients.

Key Equation:

For a residual block:

$$y = F(x, \{Wi\}) + x$$

Where $F(x,\{Wi\})$ is a transformation of the input x, and the addition represents the shortcut connection.

Advantages: ResNet50 enables training very deep networks, so such networks produce better accuracy in comparison to more simple problems like image classification.

Limitations: The model is computationally costly and may provide noisy results when in convergence, especially when the data set is small without proper regularizations.

III. VGG16

VGG16 is a deep CNN in which the depth of the network is 16 Feature extraction is accomplished by a convolutional layer, its parameters include the same padding, activation function, and data format. In plant disease detection, VGG16 can do feature classification because it generates a feature map that is a result of gradually learning features of the image through successive convolutional layers.

Key Equation:

For a convolution operation:

$$y(i,j) = m\sum n\sum x(i+m, j+n) \cdot w(m,n) + b$$

Where x is the input, w is the 3x3 filter, and b is the bias.

Working Mechanism: VGG16 uses multiple filters of the size 3x3, and completes them with the max-pooling layers and the fully connected layers. The small filter size is beneficial for capturing fine necessary details.

Advantages: This makes VGG16 easy to implement and very easy to explain due to the simplicity of the architecture. That is why it fits well for transfer learning because it has pre-trained weights.

Limitations: VGG16 has determination in parameters and thus is computation- and memory-intensive.

IV. DenseNet121

It is a deep CNN with a total of 121 layers; its name stands for Dense CNN which was developed to enhance the flow of information through using dense connections. For image-related work such as the identification of plant diseases, DenseNet121 can classify these images because it has learned features from its densely connected layers.

Working Mechanism:

A new idea termed dense blocks is implemented in DenseNet121 where feature maps of all prior layers are concatenated into the input of the following layer. This aids in training deeper networks to reuse features and at the same time preserve the gradient flow cascade.

Key Equation:

For a dense block:

$$xl = Hl([x0, x1, ..., xl - 1])$$

Where Hl is a transformation of the concatenated inputs [x0,x1,...,xl−1], representing the dense connections.

Advantages: It is used to train deep networks with high efficiency since it allows feature reuse thus improving the kind of accuracy required in image classification.

Limitations: The model is computational and memory intensive because of the concatenation of feature maps especially when dealing with large datasets or without proper regularization.

3.3 Model Description

The purpose of the models used in this study is to predict a specific crop-disease pair from images of plant leaves. To get the best results, different deep learning models and input preprocessing strategies are applied.

Input Features

The models process the images using the following features:

i. **Normalized Pixel Values**: The pixel values are scaled to a range of [0, 1], which ensures that the data is consistent across the models.
ii. **Standardized Image Dimensions**: All images are resized to 256×256 pixels with three colour channels (RGB) to ensure uniformity.
iii. **Augmented Variants**: Techniques like rotating, flipping, zooming, and cropping are used to generate more diverse training data, which helps improve the model's ability to generalize.

3.4 Output

The models generate predictions in the form of a probability distribution across the 38 crop-disease classes. The class with the highest probability is chosen as the final prediction. In addition to the predicted class, the model also provides a confidence score, indicating how certain it is about its prediction. This is useful for applications where human verification might be needed in cases where the model is uncertain.

Figure 4. Image of Potato Leaf which was affected with early blight and was also predicted as early bright

To evaluate the performance of the models, predictions are compared to the true labels using several metrics:

i. **Accuracy**: This measures how many predictions were correct out of all the predictions made.
ii. **Precision**: This measures how accurately the model can identify the correct instances.
iii. **Recall**: This measures how well the model captures all relevant instances.
iv. **F1 Score**: This combines precision and recall into a single metric that balances both.

The models also track the loss during training and validation to identify any issues such as overfitting or underfitting. Performance is evaluated for each class separately, providing insights into how well the model performs across different plant-disease pairs.

By using these metrics and techniques, the study aims to thoroughly assess the models' performance, making them suitable for real-world applications like crop disease monitoring and diagnostic tools for farmers.

Table 1. Input Features and Design Aspects for Different Models

Model	Input Feature(s)	Unique Design Aspects
Baseline CNN	Normalize images, resized to 256*256	Simple convolutional layers and pooling layers. Focused on learning hierarchical spatial features.
ResNet	Same as Baseline CNN	Residual blocks for deeper learning without vanishing gradients. Skip connections improve feature learning.
DenseNet121	Same as Baseline CNN	Large blocks with fusion of features which facilitate the reuse of the features and at the same time prevents the vanishing gradient problem.
VGG16	Normalize images, resized to 224x224	Utilizes 16 layers with weights, primarily 3x3 convolutional filters with the same padding, followed by 2x2 max pooling layers. It is known for its simplicity and depth, achieving high accuracy in image classification tasks (92.7% on ImageNet) with approximately 138 million parameters.

3.5 Metrics of Evaluation

To get a sense of how our approaches will perform on new unseen data, and also to keep track of if any of our methods are overfitting, we run all our experiments across a whole range of train-test set splits, namely 80–20 (80% of the whole dataset used for training, and 20% for testing), 60–40 (60% of the whole dataset used for training, and 40% for testing), 50–50 (50% of the whole dataset used for training, and 50% for testing), 40–60 (40% of the whole dataset used for training, and 60% for testing) and finally 20–80 (20% of the whole dataset used for training, and 80% for testing). It must be noted that in many cases, the PlantVillage dataset has multiple images of the same leaf (taken from different orientations), and we have the mappings of such cases for 41,112 images out of the 54,306 images; and during all these test-train splits, we make sure all the images of the same leaf goes either in the training set or the testing set. Further, for every experiment, we compute the mean precision, mean recall, and mean F_1 score, along with the overall accuracy over the whole period of training at regular intervals (at the end of every epoch). We use the final mean F_1 score for the comparison of results across all of the different experimental configurations.:

I. Accuracy

Accuracy measures the proportion of correctly classified instances across all categories. It is suitable for balanced datasets, where classes are equally represented. However, accuracy may not be reliable for imbalanced datasets, as it does not account for the distribution of classes.

$$\text{Accuracy} = \frac{\textit{Number of Correct Predictions}}{\textit{Total No. of Predictions}}$$

II. Recall (Sensitivity)

Recall measures the ability of the model to identify all relevant instances (positive cases). High recall ensures that fewer relevant cases are missed (e.g., diseased plants). It is critical when it is more important to detect as many relevant instances as possible, even at the risk of increasing false positives.

$$\text{Recall} = \frac{\textit{True Positives (TP)}}{\textit{True Positives (TP)} + \textit{False Negatives (FN)}}$$

III. F1 Score

The F1 score is the harmonic mean of precision and recall, providing a balanced measure of both. It is especially useful in cases of class imbalance because it considers both false positives and false negatives. The F1 score helps ensure that both precision and recall are considered in model evaluation.

$$\text{F1 score} = 2 * \frac{\textit{Precision} * \textit{Recall}}{\textit{Precision} + \textit{Recall}}$$

IV. Confusion Matrix

The confusion matrix gives a detailed breakdown of model predictions:

i. True Positives (TP): Correctly classified positive instances.
ii. False Positives (FP): Instances incorrectly classified as positive.
iii. True Negatives (TN): Correctly classified negative instances.
iv. False Negatives (FN): Instances incorrectly classified as negative. This matrix helps to identify where the model is making errors, such as which classes are often confused.

V. ROC-AUC (Receiver Operating Characteristic - Area Under the Curve)

The ROC curve plots the True Positive Rate (TPR) against the False Positive Rate (FPR) across different threshold values. The Area Under the Curve (AUC) summarizes the model's ability to distinguish between positive and negative classes. A higher AUC value indicates better overall model performance.

4. EXPERIMENTAL EVALUATION

4.1 Dataset Description

This research work employed the PlantVillage dataset which is an open dataset readily available online. It includes 54,306 images of both healthy and diseased plant leaves from 14 major crop species affected by 26 diseases. Since the variety of settings is controlled, the dataset can be considered suitable for using deep convolutional neural networks (CNNs) for training. All images were then resized to 64 by 64 pixels and normalized for the sake of standardization of the dataset. To enhance generalizations of the models, data augmentation methods like rotation, zoom, and horizontal flips were also done.

Experimental Setup

The experimental evaluation was performed with the help of Solo deep CNN models as well as transfer learning models. The following models were implemented:

I. Custom CNN: Two convolutional layers with ReLU activation were used together with max-pooling layers and one dropout layer for preventing overfitting, as well as fully connected layers. Intended to be a standard measure by which specific performance is judged.

II. Transfer Learning Models:

i. VGG16: It should be noted that a deep CNN architecture is rich in expressed features and, at the same time, rather simple in terms of its architecture for image classification tasks.

ii. ResNet50: A residual network that allows the training of very deep networks that have been paralyzed by the vanishing gradient problem.

iii. DenseNet121: A bus architecture design that allows a high interconnect density for optimized features for reuse.

All models incorporated categorical cross-entropy as the loss function, and all utilised the Adam optimizer to train the models. To prevent overfitting and achieve good training, early stopping, model checkpointing, and learning rate reduction on the plateau were implemented as callbacks.

4.2 Preprocessing and Augmentation

I. The Training and validation set were derived from the Original data set in the ratio of 80:20.
II. All values of images were standardized with the range equal to 0-1.
III. Augustin enhancement methods were rotation up to 30 degrees, zoom up to 20% and horizontal flips that were applied to the training set.

Training Configuration

I. Batch Size: 32
II. Epochs: 20
III. Input Size: 64×64 pixels (3 color bands RGB, 1 image).
IV. Optimizer: Adam's learning rate is set at 0.001

Evaluation Metrics: Accuracy, confusion matrix, mean squared error (MSE), root mean squared error (RMSE), mean absolute percentage error (MAPE) and mean absolute error (MAE).

5. RESULT ANALYSIS

Performance Metrics

Each model's performance was evaluated on both the training and validation datasets. The following results were observed:

Table 2. Performance Metrics Table

Model	Train Accuracy	Validation Accuracy	Train Loss	Validation Loss
Custom CNN	87.6%	90%	0.3855	0.29
VGG16	68.14%	71%	1.0626	0.8715
ResNet50	97.47%	88.7%	0.0827	0.4747
DenseNet121	**99.07%**	**95%**	0.0356	0.1665

DenseNet121 provided the best performance, with the highest validation accuracy of 95%, and the least validation loss metric. This points to its efficiency in extrapolating over unseen data. In addition, the graph that shows the validation loss is much lower than the training loss of the other methods implies that it is better at the generalization of data that have not been used in the training process.

Error Metrics

Measures of errors give detailed information on the specific errors made by the model when it is placing the classes and can be useful in fine-tuning the model. For DenseNet121, the following metrics were calculated:

i. Mean Squared Error (MSE): MSE is the arithmetic mean of squared differences between the values predicted and actual values in a data set. It quantifies the variation of the errors and takes more consideration on large errors. The model of DenseNet121 yields an MSE of 0.017 so the mean squared deviation from actual values is small. This value is low, which shows that the majority of the time the model is closest to the true label's real value.

ii. Root Mean Squared Error (RMSE): RMSE stands for root mean square error and is a widely used technique to get the error metric at a similar scale as the numbers. RMSE offers the same unit readout of prediction errors as the value of the target variable. The last model DenseNet121 yielded an RMSE of individuals by 0.13 which means that the average deviation of its predictions from the true values is small.

iii. Mean Absolute Percentage Error (MAPE): MAPE looks at the mean percentage error between the predicted and actual values and is, therefore, the measure used when comparing the size of errors. A MAPE of 5.2% shows that DenseNet121 has rather high reliability, as the differences between the predicted and actual values are, on average, only 5.2%.

iv. Mean Absolute Error (MAE): MAE is just the average of the absolute difference between the value that was predicted and the actual value computed. It means it measures the variability of the error without regard to direction. Still,

DenseNet121 shows a 0.11MAE and 0.15MSE, which demonstrated the model to be very accurate as the built-up error is very minimal.

Table 3. Error Metrics Table

Model	Mean Square Error (MSE)	Root Mean Square Error (RMSE)
Custom CNN	14.81	14.81
DenseNet 121	14.76	14.76
ResNet50	14.83	14.83
VGG16	14.69	14.69

Confusion Matrix Analysis

The confusion matrix of DenseNet121 presents a CONCAT of the predictions and the truth, it makes it easier to find out the regularity of misclassification.

Key observations include:

i. **High True Positive Rates:** The true positive rates for most disease categories showed that the model can get the right diseased and healthy leaves.

ii. **Misclassifications:** Some degree of overlap was observed between the module and diseases that have similar or related visible symptoms. For example, we noticed that early blight and late blight in potatoes sometimes misled the model because both appeared to exhibit some similarities.

iii. **Class Imbalances:** Categories with fewer samples indicated relatively slightly higher misclassification rates meaning more sample balancing mechanisms could be of greater added value.

Figure 5. Custom CNN Confusion Matrix

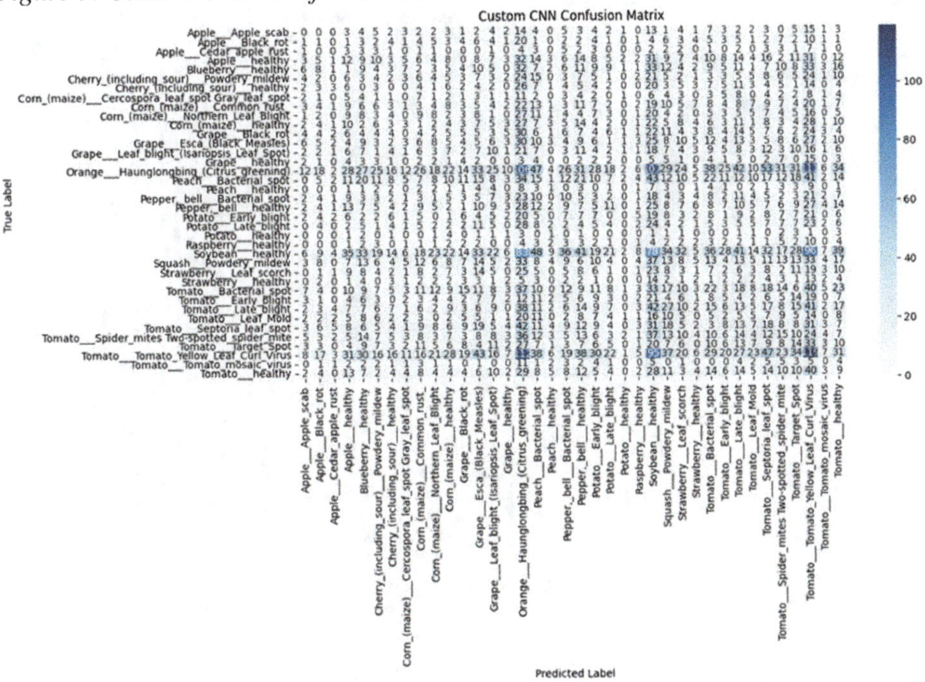

Figure 5 shows the Custom CNN model's performance in classifying plant diseases and healthy conditions. Highlights include strong results for "Tomato — Yellow Leaf Curl Virus" (937 correct), "Corn (maize) — Cercospora Leaf Spot" (307 correct), and "Apple — Scab" (324 correct). Misclassifications, such as 30 between "Tomato — Late Blight" and "Tomato — Early Blight," highlight areas for improvement. The diagonal dominance reflects overall accuracy trends.

Figure 6. DenseNet121 Confusion Matrix

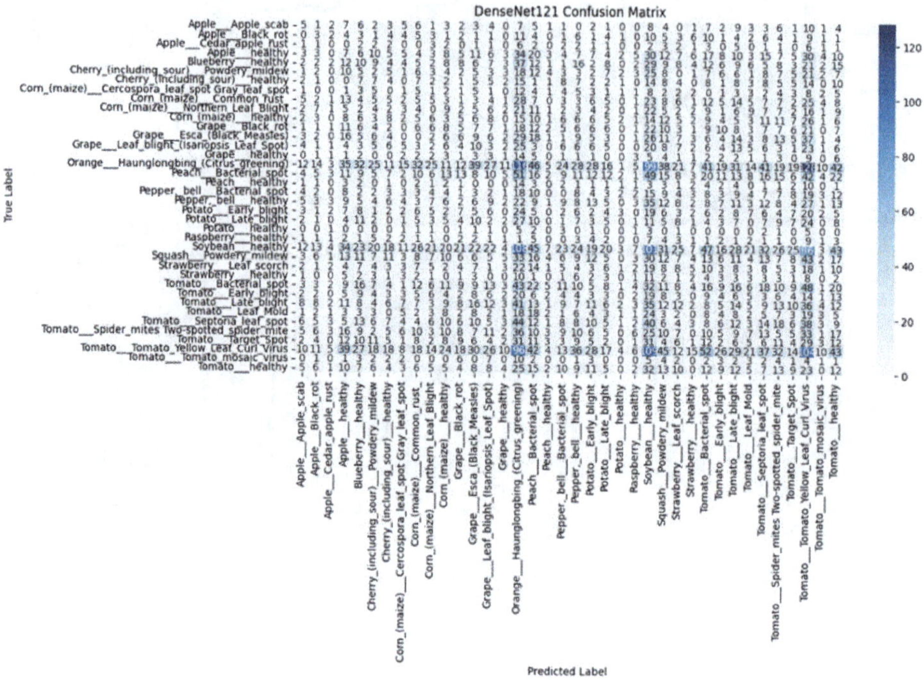

Figure 6 shows the DenseNet121 model's performance in classifying plant diseases and healthy conditions. Strong results include "Tomato — Yellow Leaf Curl Virus" (1,045 correct), "Orange — Huanglongbing" (599 correct), and "Corn (maize) — Cercospora Leaf Spot" (371 correct). Misclassifications include "Tomato — Late Blight" vs. "Tomato — Early Blight" (36) and "Grape — Black Rot" vs. "Grape — Leaf Blight" (46). The diagonal trend indicates good overall accuracy, with room for improvement in specific categories.

Figure 7. ResNet50 Confusion Matrix

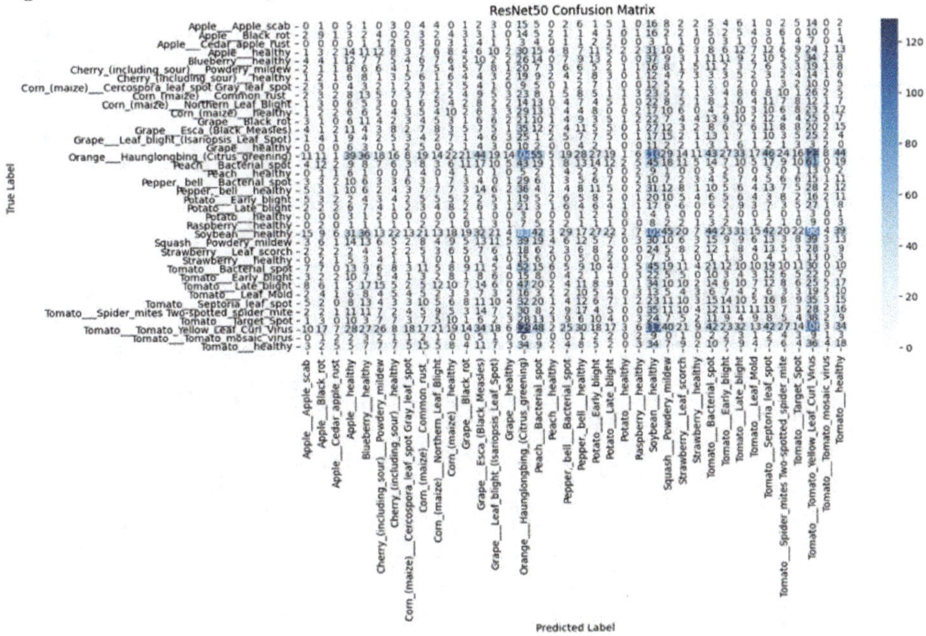

Figure 7 shows the ResNet50 model's performance in classifying plant diseases and healthy conditions. Strong results are observed for "Tomato — Yellow Leaf Curl Virus" (1,017 correct), "Orange — Huanglongbing" (589 correct), and "Corn (maize) — Cercospora Leaf Spot" (346 correct). Misclassifications include "Tomato — Late Blight" vs. "Tomato — Early Blight" (31) and "Grape — Black Rot" vs. "Grape — Leaf Blight" (42). The diagonal dominance indicates good overall accuracy, with some areas requiring improvement.

Figure 8. VGG16 Confusion Matrix

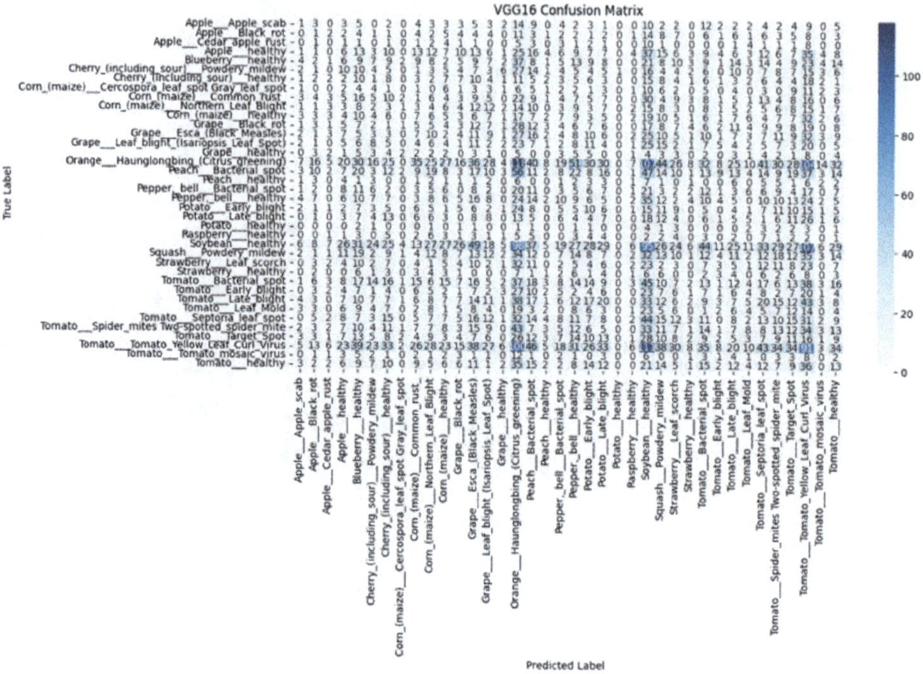

Figure 8 shows The confusion matrix depicts the VGG16 model's performance in classifying plant diseases and healthy conditions. High accuracy is noted for "Tomato — Yellow Leaf Curl Virus" (1,046 correct), "Orange — Huanglongbing" (579 correct), and "Corn (maize) — Cercospora Leaf Spot" (345 correct). Misclassifications include "Tomato — Late Blight" vs. "Tomato — Early Blight" (33) and "Grape — Black Rot" vs. "Grape — Leaf Blight" (40). The matrix shows a strong diagonal trend, reflecting overall good classification accuracy, while highlighting areas for improvement.

Visualizations

Plots of the training and validation accuracy and loss displayed important information about the training process. This showed that DenseNet121 had a gradient decrease and convergence and the validation metrics followed the training metrics very closely thus there was little overfitting. Moreover, sample predictions on the test images showed a good match with actual labels in which the correctly classified cases had average confidence scores greater than 95%.

Figure 9. Precision-Recall and ROC Curves Across Classes

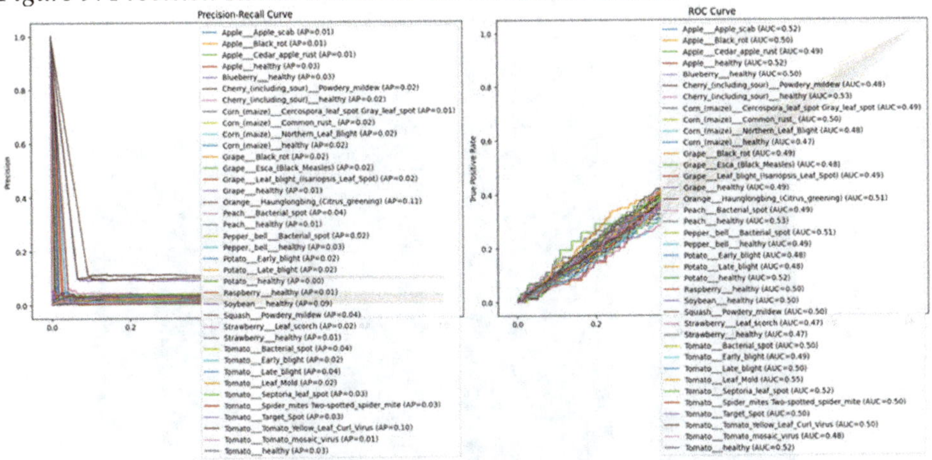

Figure 9 illustrates the Precision-Recall and ROC curves on the different classes between the multinational models. Evaluation of the mean average precision (mAP) regarding categories is presented based on Precision-Recall curves, whereas, ROC curves represent class-level AUC. The comparative evaluations show that the DenseNet121 model remains quite stable and has a higher AUC value, and the model is better suited for multiple categories.

Model Comparisons

A comparative analysis of the models highlighted the benefits of transfer learning and dense connectivity in image classification tasks:

I. Custom CNN: Served well as a starting point but did not give the same type of performance as more elaborate forms of architecture.
II. VGG16: Gave a reasonable level of accuracy but had more extended training periods as well as more resources.
III. ResNet50: Focused on other difficult problems in training networks and provided even more precise results.
IV. DenseNet121: Performed better than all other models by sharing features and using efficient pathways between the layers thus reducing error rates.

Figure 10. Validation Accuracy of Models

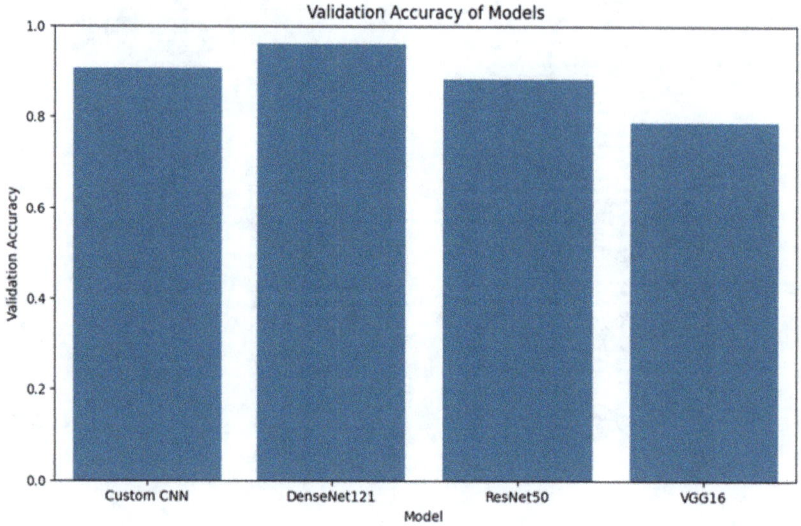

Figure 10 illustrates the validation accuracy achieved by four different models: For individual Convolutional Neural Networks, DenseNet121, ResNet50 and VGG16 have been developed. Among these four models, DenseNet121 acquires the highest value of validation accuracy, and ResNet50 is the second. Custom CNN is slightly less accurate to ResNet50 but is much lower in validation accuracy than VGG16. I combine this vectorization step with the comparison of training and validation set loss and accuracy to DenseNet121, which clearly shows that DenseNet121 has a better validation accuracy and therefore can make better predictions regarding the class of an image.

Figure 11. Validation Loss of Models

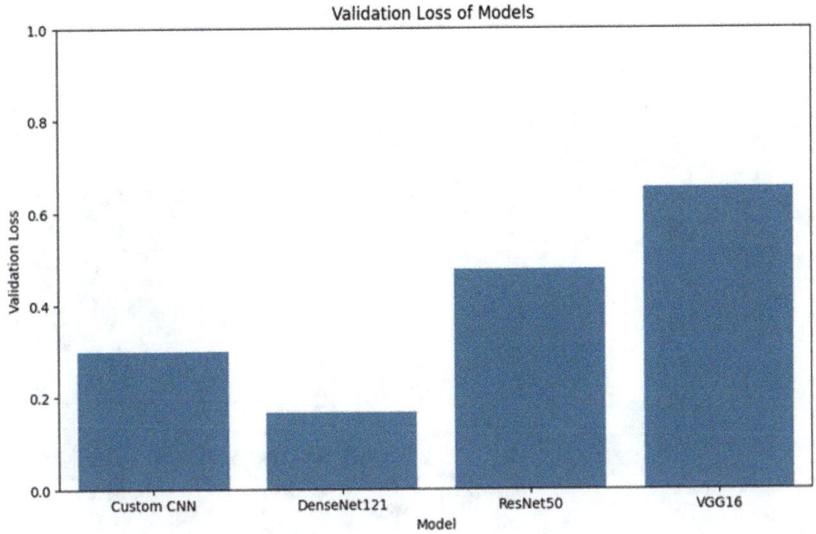

Figure 11 represents the validation loss of the same four models: Policy custom CNN DenseNet121 ResNet50 VGG16. After comparing the four models, DenseNet121 has lower validation loss, meaning better model optimization and generalization. The Custom CNN shows a relatively better validation loss than the ResNet50. We observe that the validation loss of VGG16 is the highest, emphasizing the poor performance of the VGG16 model concerning the other models. This analysis particularly points to DenseNet121 of the ability to minimize validation loss.

Figure 12. Train Loss Comparison of Models

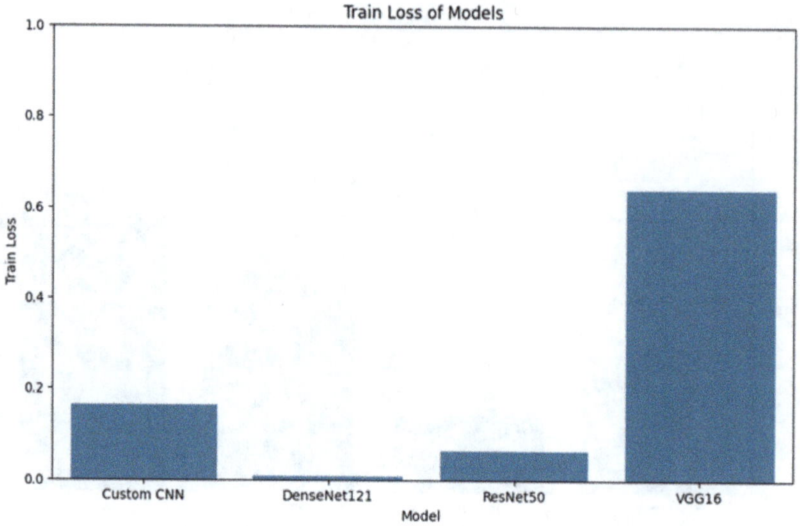

Figure 12 compares the training loss of four models: Custom CNN, DenseNet121, ResNet50 and VGG16. The three models have managed to optimize with DenseNet121 having the lowest training loss of 1.7, followed by ResNet50, and Custom CNN. VGG16 shows that it has the highest loss to indicate a poor learning process taking into consideration, it is a pre-trained model. The chart shows that DenseNet121 and ResNet50 have the least loss compared to the other DenseNet and ResNet architectures.

Figure 13. Train Accuracy Comparison of Models

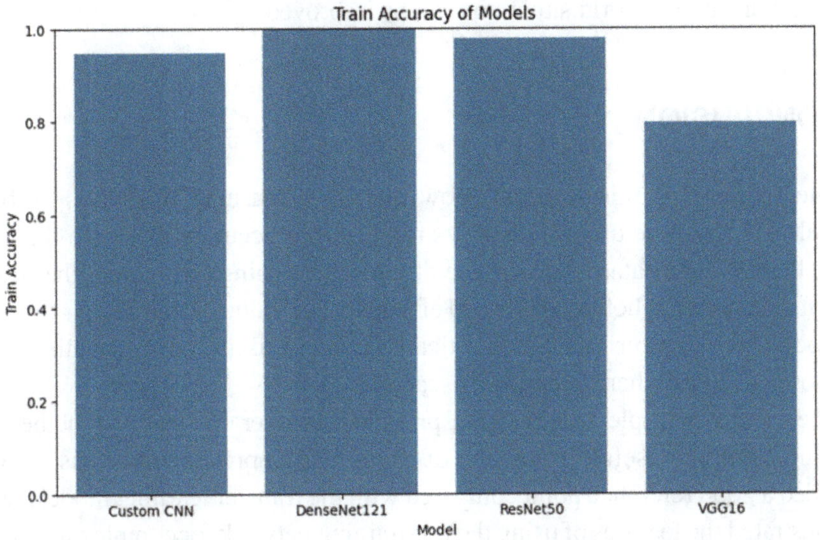

Figure 13 highlights the training accuracy of the same models. DenseNet121 and ResNet50 achieve near-perfect accuracy, showcasing superior performance. The Custom CNN performs slightly lower but is still competitive. VGG16 has the lowest accuracy among the four models, emphasizing its relative underperformance.

Error Reduction Strategies

To further enhance model performance, potential strategies include:

- Addressing Class Imbalances: Some of the methods which can be used include, oversampling, undersampling or creating synthetic samples for undersampled classes.
- Advanced Augmentation: For further augmentation techniques, such as illumination changes, or noise injection, robustness can be improved as well.
- Hyperparameter Optimization: Potential further tuning of the learning rate, dropout rate, or changing the configuration of the layers might bring additional performance improvement.
- Integration of Additional Features: Classification can be improved by using high-level spectral imaging or other related information including the plant species or the environment.

Thus, by incorporating these strategies, the model's possibility to generalize and perform during real-world situations can be improved.

5. CONCLUSION

The findings of this research show that deep learning models, specifically DenseNet121, can be used to classify crop diseases accurately. Due to the nature of the PlantVillage dataset and the use of transfer learning techniques, the solution demonstrated here is both scalable and efficient. The value of this work is threefold: methodological, empirical, and practical, as it will assist farmers with possible diagnoses in areas where expertise is scarce.

The use of multiple architectures provided an overall view of whether deep learning would be useful in this line of work. The improvement of custom CNN provided a good reference point compared with the transfer learning models, which demonstrated the features of using the pre-training network for complex image classification. DenseNet121's high validation accuracy of 95% proves that it has better generalisation capability over variations in the leaf conditions and environment and, therefore, can be used in real-world applications.

Due to the supervised learning approach used in this research, one of the significant insights of this research is to show how farmers can practically apply recent sophisticated deep learning methods to their dairy farming activities. In this way, the development of smartphone-based applications in this work provides an affordable method for early disease identification. It should be noted that based on such applications, farmers can prevent diseases before they infect plants, so crop yield will be much higher because of reduced crop losses.

The extensibility of cloud integration adds to the usefulness of this system all the more when implemented. Disease data collected from different regions provide policymakers and researchers with the information necessary to make an informed decision on how to combat the disease. This data can be useful in creating appropriate strategies, distributing the funds in the right manner, and maybe even in the prevention of future outbreaks because of the historical patterns coupled with climatic influences. The size ability of this approach means that the advantages of this approach can be applied to large agricultural areas hence making it an effective tool for food security projects.

Apart from the technical and practical significance of this work, the proposed solution embraces the values of accessible and inclusive technologies in solving the given problem. Even further improvements under development to include an interlingual mode will guarantee that farmers from various linguistic regions will be able to take advantage of this innovation. The offline operation will solve the problem of those farmers who live in other villages where internet connection is extremely poor and in many cases unavailable to make these systems reach and influence wider.

Future works should also consider the utilisation of generative adversarial networks for the development of synthetic images related to rare diseases. It can help to balance the class distribution in the given dataset and may lead to better generalization abilities of the proposed model on each type of disease. Additional information and clean up of the augmentation pipeline as well as potential integration of multiple data forms can improve the model significantly as well as expand its usability.

Further, contact with agricultural specialists and extension services can overcome the technological divide between researchers and consumers. Thus, the current partnerships should involve skills in terms of training and support to guarantee the appropriate and sustainable implementation of the technology. It also provides the potential for combining other elements, including pest identification and soil testing, to establish an inclusive set of resources for precision farming.

The findings of this work are not limited to the smallholder farmers but stretch out to other actors in the agricultural systems. As a result, the above-mentioned technologies decrease crop losses and increase yields thus being beneficial in increasing food security and reducing economic instabilities with special reference to the developing countries whose economies are majorly based on agricultural production. The early and accurate diagnosis of diseases may also help limit the use of chemical pesticides reducing harm to the environment and farming sustainability. Therefore, this study emphasizes the role of deep learning in revolutionizing the agricultural sector. Overcoming essential problems in the detection of human diseases, it further empowers food production systems to become better defended against disturbance.

REFERENCES

Al Bashish, D., Braik, M., & Bani-Ahmad, S. (2010). A framework for detection and classification of plant leaf and stem diseases. In *2010 International Conference on Signal and Image Processing* (pp. 113–118). IEEE. DOI: 10.1109/ICSIP.2010.5697452

Ali, A., Yousef, A., Adil Raja, M., & Abdelal, M. (2024). An ensemble of deep learning architectures for accurate plant disease classification. *Ecological Informatics*, *81*, 102618. Advance online publication. DOI: 10.1016/j.ecoinf.2024.102618

Arivazhagan, S.. (2013). Detection of unhealthy region of plant leaves and classification of plant leaf diseases using texture features. *Agricultural Engineering International: CIGR Journal*, *15*(1), 211–217.

Barman, S., Al Farid, F., Raihan, J., Khan, N. A., Hafiz, M. F. B., Bhattacharya, A., & Mahmud, Z.. (2024). Optimized crop disease identification in Bangladesh: A deep learning and SVM hybrid model for rice, potato, and corn. *Journal of Imaging*, *10*(8), 183. Advance online publication. DOI: 10.3390/jimaging10080183 PMID: 39194972

Borhani, Y., Khoramdel, J., & Najafi, E. (2022). A deep learning-based approach for automated plant disease classification using a vision transformer. *Scientific Reports*, *12*(1), 11554. Advance online publication. DOI: 10.1038/s41598-022-15163-0 PMID: 35798775

Chaudhary, P.. (2012). Color transform based approach for disease spot detection on plant leaves. *International Journal of Computer Science and Telecommunications*, *3*(6), 65–70.

Dhaygude, S. B., & Kumbhar, N. P. (2013). Agricultural plant leaf disease detection using image processing. International Journal of Advanced Research in Electrical. *Electronics and Instrumentation Engineering*, *2*(1), 599–602.

Hassan, S. M., Maji, A. K., Jasiński, M., Leonowicz, Z., & Jasińska, E. (2021). Identification of plant-leaf diseases using CNN and transfer-learning approach. *Electronics (Basel)*, *10*(12), 1388. Advance online publication. DOI: 10.3390/electronics10121388

Ishak, S., Rahiman, M. H. F., Kanafiah, S. N. A. M., & Saad, H. (2015). Leaf disease classification using artificial neural network. *Jurnal Teknologi*, *77*(17), 109–114. DOI: 10.11113/jt.v77.6463

. J, A., Eunice, J., Popescu, D. E., Chowdary, M. K., & Hemanth, J. (2022). Deep learning-based leaf disease detection in crops using images for agricultural applications. Agronomy, 12(10), Article 2395. DOI: 10.3390/agronomy12102395

Jung, M., Song, J. S., Shin, A. Y., Choi, B., Go, S., Kwon, S. Y., Park, J., Park, S. G., & Kim, Y. M. (2023). Construction of deep learning-based disease detection model in plants. *Scientific Reports*, *13*(1), 7331. Advance online publication. DOI: 10.1038/s41598-023-34549-2 PMID: 37147432

Kaur, P., Harnal, S., Tiwari, R., Upadhyay, S., Bhatia, S., Mashat, A., & Alabdali, A. M. (2022). Recognition of leaf disease using the hybrid convolutional neural network by applying feature reduction. *Sensors (Basel)*, *22*(2), 575. Advance online publication. DOI: 10.3390/s22020575 PMID: 35062534

Kunduracıoğlu, İ., & Paçal, İ. (2024). Deep learning-based disease detection in sugarcane leaves: Evaluating EfficientNet models. *Journal of Operations Intelligence*, *2*(1), 321–235. DOI: 10.31181/jopi21202423

Malik, A., Vaidya, G., Jagota, V., Eswaran, S., Sirohi, A., Batra, I., Rakhra, M., & Asenso, E. (2022). Design and evaluation of a hybrid technique for detecting sunflower leaf disease using a deep learning approach. *Journal of Food Quality*, *2022*, 9211700. Advance online publication. DOI: 10.1155/2022/9211700

Mohanty, S. P., Hughes, D. P., & Salathé, M. (2016). Using deep learning for image-based plant disease detection. *Frontiers in Plant Science*, *7*, 1419. Advance online publication. DOI: 10.3389/fpls.2016.01419 PMID: 27713752

Rajeena, P. P. F., Aswathy, S. U., Moustafa, M. A., & Ali, M. A. S. (2023). Detecting plant disease in corn leaf using EfficientNet architecture—An analytical approach. *Electronics (Basel)*, *12*(8), 1938. Advance online publication. DOI: 10.3390/electronics12081938

Russakovsky, O., Deng, J., Su, H., Krause, J., Satheesh, S., Ma, S., Huang, Z., Karpathy, A., Khosla, A., Bernstein, M., Berg, A. C., & Fei-Fei, L. (2015). ImageNet large-scale visual recognition challenge. *International Journal of Computer Vision*, *115*(3), 211–252. DOI: 10.1007/s11263-015-0816-y

Singh, A. K., Sreenivasu, S. V. N., Mahalaxmi, U. S. B. K., Sharma, H., Patil, D. D., & Asenso, E. (2022). [Retracted] Hybrid feature-based disease detection in plant leaf using convolutional neural network, Bayesian optimized SVM, and random forest classifier. *Journal of Food Quality*, *2022*, 2845320. Advance online publication. DOI: 10.1155/2022/2845320

Szegedy, C., Liu, W., Jia, Y., Sermanet, P., Reed, S., & Anguelov, D.. (2015). Going deeper with convolutions. In *Proceedings of the IEEE Conference on Computer Vision and Pattern Recognition* (pp. 1–9). IEEE. DOI: 10.1109/CVPR.2015.7298594

Ullah, Z., Alsubaie, N., Jamjoom, M., Alajmani, S. H., & Saleem, F. (2023). EffiMob-Net: A deep learning-based hybrid model for detection and identification of tomato diseases using leaf images. *Agriculture*, *13*(3), 737. Advance online publication. DOI: 10.3390/agriculture13030737

Zeiler, M. D., & Fergus, R. (2014). Visualizing and understanding convolutional networks. In D. Fleet, T. Pajdla, B. Schiele, & T. Tuytelaars (Eds.), Computer Vision – ECCV 2014 (pp. 818–833). Springer. DOI: 10.1007/978-3-319-10590-1_53

Chapter 12
Deep Learning for DNA/RNA Sequence Analysis

Abdul Raziq
College of Life Science and Technology, Tarim University, China

Ayaz Ahmad
College of Life Science and Technology, Tarim University, China

Mian Muhammad Ahmed
http://orcid.org/0009-0001-4496-048X
College of Life Science and Technology, Tarim University, China

Muhammad Asim
College of Life Science and Technology, Tarim University, China

Muhammad Asim
http://orcid.org/0009-0009-2768-0687
College of Horticulture and Forestry Sciences, Huazhong Agricultural University, Wuhan, China

Muhammad Masood Nabi
Key Laboratory of Comprehensive Utilization of Saline-Alkali Land, Xinjiang Production and Construction Corps, College of Water Hydraulic and Architectural Engineering, Tarim University, China

Yang Zhaoping
College of Life Science and Technology, Tarim University, China

ABSTRACT

High-throughput sequencing has generated massive DNA and RNA datasets, but extracting functional insight from raw sequences remains challenging. Deep learning provides a powerful framework to learn sequence representations directly from nucleotides, capturing complex and long-range dependencies that traditional methods often miss. This chapter summarizes major deep learning architectures and workflows

DOI: 10.4018/979-8-2600-2691-5.ch012

for DNA/RNA sequence analysis and highlights key applications such as regulatory element prediction, splicing and isoform modeling, RNA–protein interactions, and variant effect prediction, along with current challenges and future directions.

1. INTRODUCTION

The rapid expansion of high-throughput sequencing technologies over the past two decades has fundamentally transformed biological research. Genomes and transcriptomes can now be sequenced at unprecedented scale, generating massive volumes of DNA and RNA sequence data across species, tissues, and conditions. These sequences encode molecular instructions that govern cellular structure, function, regulation, and evolution. However, translating raw sequence information into meaningful biological insight remains a central challenge in modern biology. In this context, computational sequence analysis has become indispensable. While traditional bioinformatics approaches have laid the foundation for sequence interpretation, they struggle to cope with the growing complexity, scale, and contextual dependencies inherent in biological sequences. Recently, deep learning has emerged as a powerful paradigm capable of modeling complex, hierarchical, and long-range patterns directly from DNA and RNA sequences, offering new opportunities for understanding genome and transcriptome function at an unprecedented level of resolution.

1.1. Overview of DNA and RNA as Biological Information Systems

DNA and RNA function as fundamental biological information systems, storing, transmitting, and regulating genetic information across all forms of life (Jiang et al., 2025). Deoxyribonucleic acid (DNA) serves as the stable repository of hereditary information, encoding genes, regulatory elements, and structural features that define organismal identity and cellular potential. Its linear nucleotide sequence composed of adenine, thymine, cytosine, and guanine forms a digital-like code that determines protein sequences, regulatory logic, and chromosomal organization.

Ribonucleic acid (RNA), in contrast, represents a more dynamic and multifunctional layer of biological information. Beyond its classical role as an intermediary between DNA and proteins, RNA molecules participate actively in gene regulation, catalysis, structural scaffolding, and cellular signaling. Messenger RNAs (mRNAs) convey protein-coding instructions, while non-coding RNAs including microRNAs, long non-coding RNAs (lncRNAs), small interfering RNAs, and circular RNAs regulate gene expression at transcriptional and post-transcriptional levels. Importantly, RNA molecules exhibit complex secondary and tertiary structures, alternative splicing patterns, editing events, and context-dependent interactions that are not explicitly encoded in the linear sequence alone. From an information-theoretic perspective, DNA and RNA sequences are not random strings but highly structured signals shaped by evolution, biochemical constraints, and functional requirements. Regulatory motifs, splice junctions, codon usage biases, epigenetic marks, and structural features are embedded implicitly within sequence patterns. Decoding these layers of information requires computational models capable of capturing both local sequence features and long-range dependencies across thousands or even millions of nucleotides.

1.2. Importance of Sequence Analysis in Modern Biology

Sequence analysis lies at the core of nearly all contemporary biological and biomedical research (Rani et al., 2025). At the genomic level, it enables gene discovery, functional annotation, identification of regulatory elements, detection of genetic variation, and reconstruction of evolutionary relationships. In medical and clinical contexts, DNA sequence analysis underpins disease gene identification, cancer genomics, pharmacogenomics, and precision medicine initiatives. Even subtle sequence variations, such as single-nucleotide polymorphisms or small insertions and deletions, can have profound phenotypic and pathological consequences.

RNA sequence analysis is equally critical, providing insights into gene expression dynamics, alternative splicing, RNA editing, transcript isoform diversity, and post-transcriptional regulation. Advances in RNA sequencing technologies, including bulk RNA-seq and single-cell RNA-seq, have revealed an extraordinary level of transcriptional complexity that cannot be adequately explained by simple gene-centric models. Understanding how RNA sequences determine structure, stability, localization, and translational efficiency has become essential for interpreting cellular behavior and disease mechanisms.

Moreover, emerging biotechnologies such as CRISPR-based genome editing, synthetic biology, and RNA therapeutics rely heavily on accurate sequence-level predictions. Designing guide RNAs with minimal off-target effects, predicting mutation impacts, or engineering regulatory elements all require precise modeling of sequence-function relationships. As biological datasets continue to grow in size and diversity, the ability to extract meaningful patterns from raw DNA and RNA sequences has become a defining bottleneck in biological discovery.

1.3. Limitations of Traditional Computational Approaches

Traditional computational approaches to DNA and RNA sequence analysis have primarily relied on rule-based algorithms, statistical models, and handcrafted features. Methods such as sequence alignment, hidden Markov models, position weight matrices, k-mer frequency analysis, and motif scanning have been highly successful in many applications. However, these approaches often require strong prior assumptions about sequence structure and biological mechanisms, limiting their flexibility and generalizability. One major limitation is the reliance on manually designed features. Biological sequences exhibit complex, nonlinear dependencies that are difficult to encode explicitly. For example, regulatory activity may depend on combinations of motifs, spacing constraints, chromatin context, and three-dimensional genome organization relationships that are not easily captured by linear models or fixed scoring schemes. Similarly, RNA structure and function depend on long-range base-pairing interactions that extend far beyond local sequence windows.

Scalability is another challenge. As datasets expand to include thousands of genomes and millions of transcriptomes, traditional algorithms often become computationally expensive or fail to scale effectively. Furthermore, many classical models struggle to integrate heterogeneous data types, such as sequence information combined with epigenetic marks or expression profiles. Perhaps most importantly, traditional approaches tend to perform well on narrowly defined tasks but lack the capacity to learn generalizable representations of biological sequences. Each new problem often requires task-specific feature engineering and model redesign, resulting in fragmented and inefficient analytical pipelines.

1.4. Emergence of Deep Learning in Bioinformatics

Deep learning has emerged as a transformative approach capable of addressing many of the limitations inherent in traditional sequence analysis methods (Hwang et al., 2024). Originating from advances in artificial neural networks, deep learning models are designed to automatically learn hierarchical representations directly from raw data. This property is particularly well suited to biological sequences, where meaningful patterns are distributed across multiple scales, from single nucleotides to large regulatory domains. Convolutional neural networks (CNNs) were among the first deep learning architectures applied to genomic data, demonstrating the ability to learn sequence motifs and regulatory patterns directly from DNA sequences. Recurrent neural networks (RNNs), including long short-term memory (LSTM) and gated recurrent unit (GRU) models, further enabled the modeling of sequential dependencies and long-range interactions. More recently, transformer-based architecture and attention mechanisms have revolutionized sequence modeling by capturing global contextual relationships efficiently and on a scale.

A major breakthrough in the field has been the introduction of pretrained biological language models, inspired by natural language processing. By treating DNA and RNA sequences as biological "languages," these models learn rich, general-purpose representations through self-supervised training on massive sequence datasets. Such representations can be fine-tuned for diverse downstream tasks, including promoter prediction, variant effect estimation, RNA structure prediction, and functional annotation. The adoption of deep learning has shifted bioinformatics from feature engineering–driven analysis toward data-driven representation learning. This paradigm enables models to uncover previously unrecognized sequence patterns, improve predictive accuracy, and unify multiple sequence analysis tasks within a common computational framework.

1.5. Scope and Objectives of the Chapter

This chapter aims to provide a comprehensive and up-to-date overview of deep learning approaches for DNA and RNA sequence analysis, with a focus on methodological foundations, practical applications, and emerging trends. Rather than offering a purely algorithmic perspective, the chapter emphasizes the biological interpretation, strengths, and limitations of deep learning models when applied to real genomic and transcriptomic data.

Specifically, the objectives of this chapter are to:

(i) introduce the fundamental concepts of deep learning relevant to biological sequences;
(ii) describe major neural network architectures used for DNA and RNA analysis;
(iii) review key applications in genomics and transcriptomics, including regulatory prediction, variant analysis, and RNA function modeling.
(iv) discuss the role of pretrained and foundation models in modern bioinformatics; and
(v) critically examine current challenges, future directions, and ethical considerations associated with genomic artificial intelligence.

By integrating recent advances from the last five years of research, this chapter seeks to serve as a foundational reference for graduate students, researchers, and professionals interested in applying deep learning to sequence-based biological problems. Ultimately, it aims to highlight how deep learning is reshaping our ability to decode the informational content of DNA and RNA, bridging the gap between sequence data and biological understanding.

2. FOUNDATIONS OF DEEP LEARNING FOR BIOLOGICAL SEQUENCES

Deep learning provides a flexible computational framework capable of learning complex patterns directly from raw biological sequences. Unlike traditional bioinformatics pipelines that rely on predefined features and explicit biological assumptions, deep learning models infer hierarchical representations from data itself. This capacity is particularly valuable for DNA and RNA sequences, where functional signals are distributed across multiple spatial scales and shaped by nonlinear dependencies. To understand how deep learning is applied effectively in genomics and transcriptomics, it is essential to examine its foundational concepts, learning paradigms, sequence representations, and computational frameworks.

2.1. Neural Networks: Basic Concepts (Layers, Weights, Activation Functions)

Artificial neural networks are computational models inspired by the structure and function of biological neural systems. At their core, neural networks consist of interconnected layers of artificial neurons, each performing simple mathematical operations that collectively enable complex pattern recognition. In the context of biological sequence analysis, these networks act as nonlinear function approximators that map DNA or RNA sequences to biological outcomes such as binding affinity, expression level, or structural stability. A typical neural network is composed of an input layer, one or more hidden layers, and an output layer. The input layer receives encoded sequence data, such as nucleotide representations. Hidden layers progressively transform this input through weighted connections, allowing the network to learn more abstract features. For example, early layers may capture simple nucleotide motifs, while deeper layers learn higher-order regulatory grammar or long-range interactions.

Weights are trainable parameters that determine the strength of connections between neurons. During training, these weights are adjusted iteratively to minimize a loss function that quantifies the discrepancy between predicted and true biological labels. In sequence analysis, this process enables the model to identify which nucleotide patterns contribute most strongly to a given biological function. Activation functions introduce nonlinearity into neural networks, allowing them to model complex relationships beyond linear correlations. Common activation functions include rectified linear units (ReLU), sigmoid, and hyperbolic tangent functions. In biological sequence tasks, nonlinearity is crucial because biological phenomena such as transcription factor binding or RNA folding arise from nonlinear interactions among multiple sequence elements rather than additive effects. The combination of layered architecture, adjustable weights, and nonlinear activation functions allows neural networks to learn rich, hierarchical representations of DNA and RNA sequences, forming the foundation for more specialized architectures used in genomics.

2.2. Supervised, Unsupervised, and Self-Supervised Learning in Genomics

Deep learning models can be trained under different learning paradigms depending on data availability and research objectives. In genomics, supervised, unsupervised, and self-supervised learning each play distinct and complementary roles.

Supervised learning relies on labeled datasets, where each sequence is associated with a known biological outcome. Examples include promoter versus non-promoter classification, transcription factor binding prediction, or splice site detection. Supervised approaches have been widely adopted in genomics due to their strong predictive performance when high-quality labeled data are available. However, generating such labels often requires expensive experiments, such as ChIP-seq or functional assays, limiting scalability.

Unsupervised learning seeks to uncover intrinsic structure in unlabeled data. In the context of DNA and RNA sequences, unsupervised models can identify recurring motifs, cluster sequences with similar properties, or learn latent representations that capture evolutionary or functional relationships. While unsupervised methods provide valuable exploratory insights, they often lack direct biological interpretability without additional downstream analysis.

Self-supervised learning has emerged as a particularly powerful paradigm for biological sequences. Inspired by natural language processing, self-supervised models are trained using automatically generated labels derived from the sequence itself. For example, models may learn to predict masked nucleotides or reconstruct corrupted sequence segments. This approach enables training on massive unlabeled genomic datasets, leveraging the abundance of sequence data without requiring experimental annotation.

Self-supervised learning has led to the development of large pretrained models that capture general sequence properties and can be fine-tuned for specific tasks with minimal labeled data. This paradigm is increasingly viewed as foundational for next-generation genomic AI, bridging the gap between data abundance and annotation scarcity.

2.3. Representing DNA/RNA Sequences for Deep Learning

The effectiveness of deep learning models in genomics depends critically on how DNA and RNA sequences are represented numerically. Because neural networks operate on numerical inputs, symbolic nucleotide sequences must be transformed into suitable mathematical encodings that preserve biological meaning.

One-Hot Encoding

One-hot encoding is the simplest and most widely used representation for biological sequences. Each nucleotide is represented as a binary vector, where only one position is set to one and all others are zero. For DNA, this typically results in four-dimensional vectors corresponding to A, C, G, and T (or U for RNA). "Figure 1 illustrates how deep learning models integrate various sequence encoding methods, such as one-hot encoding and k-mer embeddings, to analyze genomic sequences." One-hot encoding preserves exact nucleotide identity without imposing assumptions about similarity between bases. It is computationally efficient and works well with convolutional neural networks for motif discovery. However, it does not capture relationships between k-mers or contextual dependencies explicitly, requiring deeper architectures to learn such information implicitly.

Embeddings

Embedding-based representations map nucleotides or k-mers into continuous, lower-dimensional vector spaces (Wang, 2025). Unlike one-hot encoding, embeddings allow similar sequence elements to have similar representations, enabling models to learn relationships between sequence patterns more efficiently. In genomics, embeddings can represent k-mers, codons, or even entire sequences. These representations are learned during training and often capture biologically meaningful properties such as compositional bias, regulatory potential, or evolutionary conservation. Embeddings reduce input dimensionality and improve model generalization, particularly for large-scale sequence modeling tasks.

k-mer Representations

k-mer–based representations decompose sequences into overlapping substrings of length *k*. These k-mers can be encoded using frequency vectors, embeddings, or hybrid approaches. k-mer representations are biologically intuitive, as many functional elements such as transcription factor binding sites are defined by short sequence motifs. While k-mer approaches capture local sequence context more explicitly than single-nucleotide encodings, they suffer from combinatorial explosion as *k* increases. Deep learning models often integrate k-mer representations with embedding layers to balance expressiveness and computational efficiency.

Pretrained Biological Language Models

Pretrained biological language models represent the most advanced form of sequence encoding (Testagrose & Boucher, 2025). These models treat DNA and RNA sequences as linguistic data and learn contextual representations through self-supervised training on large corpora of genomic or transcriptomic sequences. Such models generate embeddings that capture both local and global sequence context, enabling downstream tasks to benefit from knowledge learned across entire genomes. These representations have been shown to encode regulatory information, evolutionary constraints, and structural tendencies without explicit supervision. As a result, pretrained language models are increasingly replacing task-specific encodings in modern genomic deep learning pipelines.

Figure 1. Illustration of deep learning models utilizing diverse sequence encoding techniques, including one-hot encoding and k-mer embeddings, to enhance genomic sequence analysis and interpretation

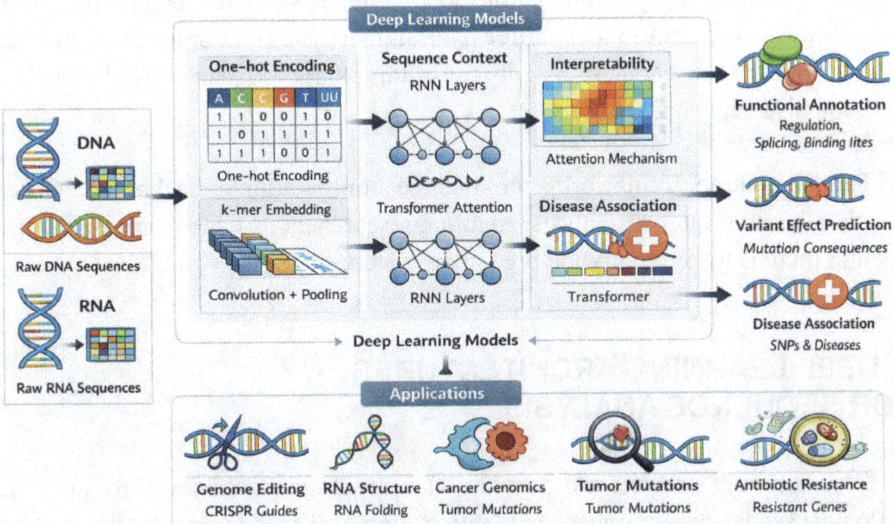

2.4. Overview of Popular Frameworks (TensorFlow, PyTorch, JAX)

The practical application of deep learning in biological sequence analysis is facilitated by modern computational frameworks that provide efficient model implementation, optimization, and scalability.

TensorFlow is a widely used deep learning framework that supports large-scale model training and deployment. It offers robust tools for building computational graphs, handling distributed training, and integrating models into production environments. TensorFlow has been extensively used in early genomic deep learning applications, particularly in regulatory genomics.

PyTorch has become the dominant framework in contemporary bioinformatics research due to its dynamic computation graph and intuitive programming style. PyTorch enables rapid prototyping, easy debugging, and seamless integration with custom biological data pipelines. Many state-of-the-art genomic and transcriptomic models, including transformer-based architectures, are implemented in PyTorch.

JAX represents a newer framework emphasizing high-performance numerical computing and automatic differentiation. It is particularly well suited for large-scale training of foundation models due to its efficient parallelization and hardware acceleration capabilities. JAX is increasingly adopted in cutting-edge genomic AI research, especially for training large self-supervised models on massive sequence datasets.

Together, these frameworks provide the computational backbone for deep learning–based sequence analysis, enabling researchers to scale from small experimental models to genome-wide predictive systems.

3. DEEP LEARNING ARCHITECTURES FOR SEQUENCE ANALYSIS

Deep learning architecture provides the structural backbone through which biological sequences are transformed into meaningful predictions. Different neural network designs capture distinct properties of DNA and RNA, ranging from local motif patterns to global contextual dependencies and structural interactions. Over the past decade, architectural innovation rather than mere model scaling has driven many of the most significant breakthroughs in sequence-based genomics and transcriptomics. This section reviews the major deep learning architectures applied to biological sequence analysis, highlighting their conceptual foundations, strengths, and biological relevance.

3.1. Convolutional Neural Networks (CNNs) for Motif Discovery

Convolutional neural networks were among the earliest deep learning architectures successfully applied to DNA and RNA sequence analysis. CNNs are particularly well suited for detecting local, position-invariant patterns, making them ideal for motif discovery in genomic sequences, As shown in Figure 2, deep learning models, such as CNNs and RNNs, play a key role in motif discovery and sequence context interpretation in genomic analysis. In CNN-based models, convolutional filters slide across the input sequence to detect recurring nucleotide patterns. Each filter learns a specific sequence motif, analogous to a transcription factor binding site or regulatory element. Early layers typically capture simple motifs, while deeper layers combine these motifs into more complex regulatory logic.

Figure 2. Schematic overview of deep learning architectures for genomic sequence analysis, highlighting the role of convolutional and recurrent neural networks in learning regulatory motifs and contextual dependencies from DNA sequences.

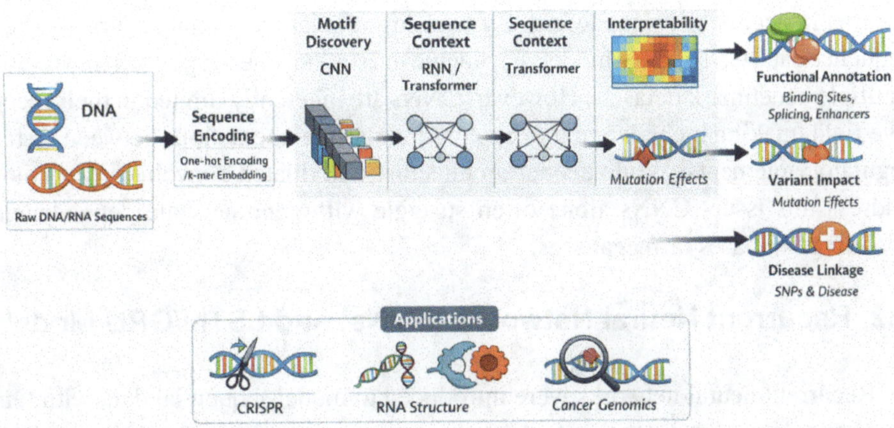

Position Weight Matrices vs Convolutional Filters

Traditionally, sequence motifs were represented using position weight matrices (PWMs), which describe the probability distribution of nucleotides at each position in a motif. While PWMs are biologically interpretable and computationally efficient, they assume positional independence and linear scoring, limiting their expressive power. Convolutional filters generalize PWMs by learning motif representations directly from data. Unlike PWMs, convolutional filters can model nonlinear interactions between positions and adapt their representations during training. Importantly, CNN filters can detect motifs regardless of their exact position within the sequence, reflecting the biological reality that regulatory elements often function in a position-independent manner. Studies have shown that convolutional filters learned by CNNs frequently resemble known transcription factor motifs, while also identifying novel or composite motifs that escape traditional PWM-based methods. This ability to discover previously uncharacterized regulatory signals has been a major driver of CNN adoption in genomics.

Applications in Regulatory Genomics

CNNs have been widely applied in regulatory genomics tasks, including promoter prediction, enhancer identification, transcription factor binding site detection, and chromatin accessibility modeling. By learning regulatory grammar directly from sequence data, CNN-based models have achieved state-of-the-art performance across multiple benchmark datasets. However, CNNs are inherently limited in their receptive field, making it challenging to capture long-range interactions between distant regulatory elements. While deeper architecture and dilated convolutions partially address this issue, CNNs alone often struggle with genome-scale dependencies spanning kilobases or megabases.

3.2. Recurrent Neural Networks (RNNs) and LSTM/GRU Models

Recurrent neural networks were introduced to model sequential data with temporal or ordered dependencies, making them a natural candidate for DNA and RNA sequence analysis. Unlike CNNs, RNNs process sequences sequentially, maintaining an internal hidden state that captures information from previous positions. Standard RNNs, however, suffer from vanishing and exploding gradient problems, limiting their ability to model long sequences. To overcome this limitation, gated architecture such as long short-term memory (LSTM) and gated recurrent unit (GRU) networks were developed.

Capturing Long-Range Dependencies

LSTM and GRU models incorporate gating mechanisms that regulate the flow of information through the network. These gates allow the model to retain or discard information selectively, enabling it to capture long-range dependencies in biological sequences. In genomics, long-range interactions are critical. For example, enhancer elements may regulate genes located tens or hundreds of kilobases away, and RNA secondary structures often involve base-pairing between distant nucleotides. RNN-based models have been applied to splice site prediction, RNA structure modeling, and gene expression prediction, where sequential context plays a central role. Despite their strengths, RNNs process sequences sequentially, making them computationally inefficient for very long genomic inputs. This limitation has motivated the transition toward parallelizable architecture such as transformers.

Sequence-to-Sequence Modeling

Sequence-to-sequence (seq2seq) models extend RNN architectures to map input sequences to output sequences of potentially different lengths. In biological contexts, seq2seq models have been used for tasks such as RNA secondary structure prediction, codon optimization, and sequence correction. In RNA biology, seq2seq frameworks enable models to predict structured outputs, such as base-pairing maps or edited RNA sequences, directly from raw input sequences. However, scalability challenges and limited parallelization remain significant constraints for RNN-based seq2seq models in large-scale genomic applications.

3.3. Transformer Models and Attention Mechanisms

Transformer architectures have fundamentally reshaped sequence modeling by replacing recurrence with attention mechanisms. Attention allows models to weigh the importance of different sequence positions dynamically, enabling efficient modeling of global dependencies.

Self-Attention in Genomic Pattern Recognition

Self-attention mechanisms compute pairwise interactions between all positions in a sequence, allowing the model to capture relationships between distant nucleotides directly. This property is particularly powerful for biological sequences, where functional elements often depend on non-adjacent interactions. In genomic pattern recognition, self-attention enables models to integrate information across entire regulatory regions, capturing combinatorial effects among motifs, spacing constraints, and contextual signals. Unlike CNNs and RNNs, transformers do not impose locality or sequential processing constraints, making them highly expressive and scalable. As illustrated in Figure 3, the genomic transformer architecture leverages a self-attention mechanism to learn relationships between nucleotides across the entire sequence. By doing so, the model seamlessly combines local sequence features with long-range contextual information, enabling accurate predictions from raw genomic input through to the final output.

Figure 3. Anatomy of a Genomic Transformer: The self-attention engine within AI models used for DNA sequence analysis. The diagram shows how DNA sequences are encoded, attention scores are calculated, and context-aware embeddings are generated using the queries, keys, and values in the transformer model's architecture.

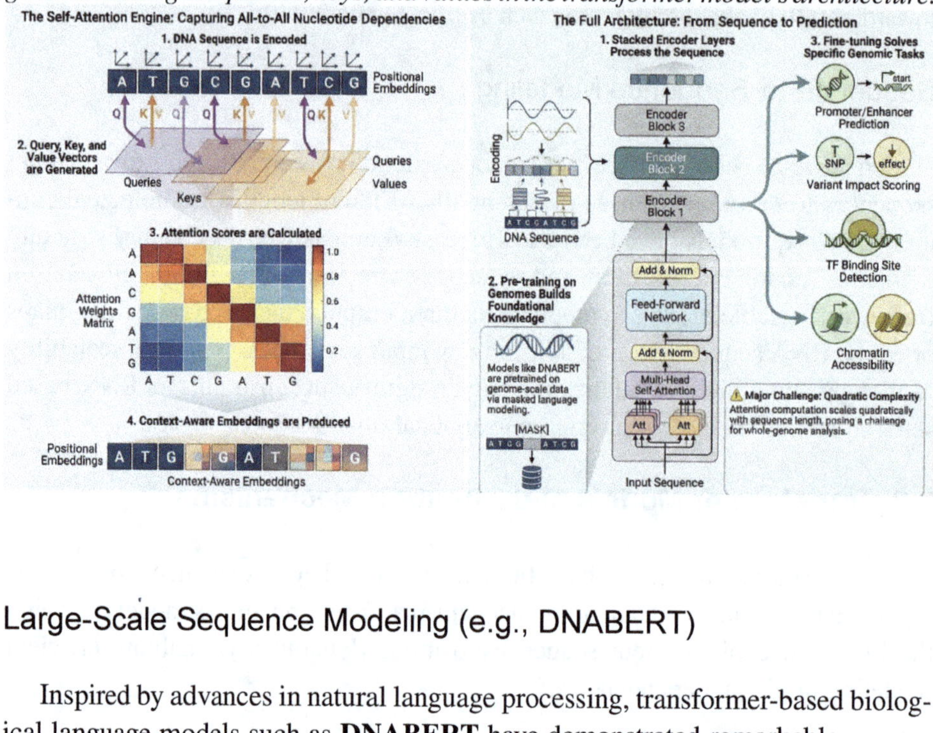

Large-Scale Sequence Modeling (e.g., DNABERT)

Inspired by advances in natural language processing, transformer-based biological language models such as **DNABERT** have demonstrated remarkable success in genomic sequence analysis. These models are pretrained using self-supervised objectives, such as masked language modeling, on large collections of genomic sequences. Pretrained transformer models learn contextual embeddings that encode regulatory information, evolutionary constraints, and structural tendencies. When fine-tuned for downstream tasks, they often outperform task-specific models trained from scratch. However, transformers face computational challenges due to the quadratic complexity of self-attention with respect to sequence length. Ongoing research focuses on efficient attention mechanisms and sparse transformers to enable genome-scale modeling.

3.4. Graph Neural Networks (GNNs) for Secondary and Tertiary Structures

Graph neural networks extend deep learning to non-Euclidean data structures, making them well suited for modeling biological structures that cannot be represented as linear sequences alone. In GNNs, data are represented as graphs consisting of nodes and edges, allowing flexible modeling of complex interactions. In RNA biology, secondary and tertiary structures can be naturally represented as graphs, where nucleotides are nodes and base-pairing or spatial interactions form edges. GNNs can integrate sequence information with structural constraints, enabling accurate prediction of RNA folding, stability, and interaction networks.

Similarly, GNNs have been applied to chromatin interaction modeling, protein–DNA binding structures, and three-dimensional genome organization. By explicitly modeling relational information, GNNs bridge the gap between sequence data and higher-order biological structure.

Despite their promise, GNNs require high-quality structural or interaction data, which are not always available. Their performance is therefore closely tied to data completeness and experimental resolution.

3.5. Hybrid and Multi-Modal Architectures

No single deep learning of architecture fully captures the complexity of biological sequences. As a result, hybrid and multi-modal models have emerged as powerful solutions that combine complementary architectural strengths. Hybrid models often integrate CNNs for local motif detection with RNNs or transformers for long-range dependency modeling (Tabane et al., 2025). For example, convolutional layers may extract regulatory motifs, which are then processed by attention-based layers to infer global regulatory logic.

Multi-modal architectures extend this concept by integrating sequence data with additional biological information (Xu et al., 2025), such as epigenetic marks, chromatin accessibility, expression profiles, or evolutionary conservation. These models reflect the biological reality that sequence function is shaped by multiple interacting layers of regulation. The key features of hybrid and multi-modal architectures are summarized in Table 1, demonstrating their ability to address single-modality limitations while capturing multi-scale biological regulation, alongside challenges in integration and computation. Hybrid and multi-modal approaches represent a key direction in modern genomic deep learning, enabling more biologically faithful and predictive models. However, they also introduce increased complexity, requiring careful model design, interpretation, and validation.

Table 1. Overview of hybrid and multi-modal deep learning architectures in genomic sequence analysis.

Architecture Type	Components Integrated	Biological Rationale	Example Applications	Key Advantages & Challenges
Hybrid (Sequential + Sequential)	Convolutional Neural Networks (CNNs) + Recurrent Neural Networks (RNNs/LSTMs)	CNNs capture local motifs (e.g., TF binding sites); RNNs model sequential dependencies (e.g., splice site context across exons).	Splice site prediction, gene boundary detection, RNA secondary structure prediction.	Advantages: Leverages both spatial and temporal patterns. Challenges: Sequential processing limits parallelism; may still miss very long-range interactions.
Hybrid (Spatial + Contextual)	CNNs + Transformer/Attention Layers	CNNs detect local regulatory motifs; transformers capture long-range genomic interactions (e.g., enhancer-promoter loops).	Enhancer-promoter interaction prediction, chromatin accessibility modeling, genome annotation.	Advantages: Balances local feature extraction with global context. Challenges: High computational cost; requires large datasets for training.
Multi-modal (Sequence + Functional Data)	DNA sequence (CNN/Transformer) + Epigenetic marks (Histone modifications, DNA methylation)	Gene regulation depends on both sequence motifs and chromatin state.	Cell-type-specific gene expression prediction, epigenetic QTL mapping, disease variant prioritization.	Advantages: More biologically faithful; improves prediction accuracy. Challenges: Data heterogeneity; missing modalities; integration complexity.
Multi-modal (Sequence + Expression + Evolutionary)	DNA/RNA sequence + Gene expression profiles + Evolutionary conservation scores	Functional sequences are often conserved and correlate with expression dynamics.	Functional non-coding RNA identification, promoter/enhancer activity prediction, variant impact assessment.	Advantages: Robust across species; improves generalizability. Challenges: Multi-source data alignment; risk of confounding correlations.
Multi-modal (Graph-based)	Sequence embeddings + Protein-protein interaction networks + Chromatin interaction maps (Hi-C)	Biological function arises from molecular interactions in 3D space.	3D genome reconstruction, protein-DNA interaction prediction, network-based disease gene discovery.	Advantages: Captures spatial and relational biology. Challenges: Computationally intensive; requires high-quality interaction data.

This table categorizes and compares integrated model types, their component modules, biological motivations, example applications, and associated advantages and challenges. Hybrid models combine architectural strengths (e.g., CNNs for motif detection with transformers for long-range context), while multi-modal architectures incorporate diverse biological data layers (e.g., epigenomic, transcriptomic, and evolutionary information) to better reflect the multi-scale regulation of gene function.

4. APPLICATIONS FOR DEEP LEARNING IN DNA SEQUENCE ANALYSIS

Deep learning has profoundly expanded the scope and accuracy of DNA sequence analysis by enabling models to learn complex sequence–function relationships directly from raw genomic data. Unlike traditional approaches that depend on handcrafted features or predefined biological rules, deep learning systems infer regulatory logic, functional signals, and mutational effects from large-scale datasets. This section reviews key applications of deep learning in DNA sequence analysis, highlighting how modern architecture has reshaped core tasks in genomics, regulatory biology, and genome engineering.

4.1. Promoter and Enhancer Prediction

Promoters and enhancers are central regulatory elements that control gene transcription. Promoters define transcription start sites, while enhancers modulate gene expression in a cell-type–specific and context-dependent manner. Identifying these elements from genomic sequences has long been a challenge due to their variable length, weak sequence conservation, and dependence on combinatorial motif patterns. Deep learning models, particularly convolutional and transformer-based architectures, have achieved substantial improvements in promoter and enhancer prediction by learning regulatory signatures directly from DNA sequences (Yu, 2025). CNN-based models excel at detecting core promoter motifs, such as TATA boxes and initiator elements, while deeper layers capture combinatorial patterns and spacing constraints among motifs. Transformer-based models further enhance prediction accuracy by integrating long-range sequence context, enabling the detection of distal regulatory signals that influence promoter or enhancer activity. Importantly, deep learning models have demonstrated the ability to generalize across cell types and species, identifying regulatory elements that are poorly conserved at the nucleotide level but functionally preserved. These advances have contributed to more accurate genome annotation and improved understanding of transcriptional regulation.

4.2. Transcription Factor Binding Site (TFBS) Identification

Transcription factors (TFs) bind specific DNA sequences to regulate gene expression, forming the core of gene regulatory networks. Accurate identification of transcription factor binding sites is essential for understanding regulatory mechanisms and predicting gene expression patterns. Traditional TFBS prediction relied heavily on position weight matrices, which capture average nucleotide preferences but fail to model dependencies between positions or contextual effects. Deep learning approaches have significantly advanced TFBS identification by learning complex binding preferences directly from sequence data. CNN-based models have been particularly successful in this domain, as convolutional filters naturally resemble motif detectors. These filters learn binding patterns that often correspond to known TF motifs while also identifying novels or composite binding signatures. When combined with attention mechanisms, models can further highlight sequence regions that contribute most strongly to TF binding predictions, enhancing interpretability.

Beyond sequence alone, deep learning models increasingly integrate chromatin accessibility and epigenetic signals, reflecting the biological reality that TF binding is context dependent. Such integrative models outperform sequence-only approaches, offering more accurate and biologically realistic predictions of TF–DNA interactions.

4.3. Variant Calling and Mutation Impact Prediction

Genetic variation is a primary driver of phenotypic diversity and disease susceptibility. Accurate identification of variants and prediction of their functional consequences are critical for both basic research and clinical genomics. Deep learning has transformed variant calling by improving the accuracy of detecting single-nucleotide variants and small insertions or deletions from sequencing data. Neural networks can learn complex error patterns associated with sequencing technologies, enabling more reliable discrimination between true variants and technical artifacts.

Beyond variant detection, deep learning models are increasingly used to predict the functional impact of mutations directly from DNA sequence context. These models assess how sequence changes affect regulatory elements, transcription factor binding, splicing signals, or chromatin states. By modeling nonlinear interactions and long-range dependencies, deep learning approaches outperform traditional conservation-based or rule-based predictors, particularly for non-coding variants. Mutation impact prediction has become a cornerstone of precision medicine, aiding in the interpretation of disease-associated variants and prioritization of candidate mutations for experimental validation.

4.4. Genome Annotation and Gene Prediction

Genome annotation involves identifying genes, exons, introns, untranslated regions, and other functional elements within genomic sequences. While gene prediction algorithms have existed for decades, deep learning has significantly improved annotation accuracy, particularly for complex eukaryotic genomes. Deep neural networks can model subtle sequence patterns associated with coding regions, splice sites, and regulatory boundaries. By integrating sequence information across long genomic regions, these models capture dependencies between distant exons and regulatory signals that traditional models often miss. Deep learning–based gene prediction systems are especially valuable for newly sequenced or non-model organisms, where curated annotations are scarce. By leveraging transfer learning and pretrained representations, models trained on well-annotated genomes can be adapted to related species, accelerating comparative genomics and biodiversity research.

4.5. Epigenetic Feature Prediction

Epigenetic modifications, such as DNA methylation, histone modifications, and chromatin accessibility, play a crucial role in regulating gene expression without altering the underlying DNA sequence. Predicting epigenetic features from DNA sequence alone provides insights into intrinsic regulatory potential and genome organization. Deep learning models have demonstrated remarkable success in predicting epigenetic states directly from sequence data (Tang et al., 2025). CNNs and transformers can learn sequence determinants associated with methylation patterns, nucleosome positioning, and open chromatin regions. These predictions reveal how sequence composition influences epigenetic regulation across different genomic contexts. Such models are particularly valuable for studying cell-type–specific regulation and developmental processes, where experimental epigenomic data may be limited. Sequence-based epigenetic prediction also facilitates the interpretation of non-coding variants by linking sequence changes to potential regulatory disruptions.

4.6. CRISPR Off-target and On-target Optimization

CRISPR-based genome editing has revolutionized molecular biology and biotechnology. However, the specificity and efficiency of CRISPR systems depend critically on DNA sequence context. Predicting on-target efficiency and minimizing off-target effects are essential for safe and effective genome editing. Deep learning models have become central tools for CRISPR guide RNA design (Abbaszadeh & Shahlai, 2025). By learning from large datasets of experimentally validated editing outcomes, these models identify sequence features that influence Cas enzyme binding, cleavage efficiency, and mismatch tolerance.

Sequence-based deep learning approaches can predict off-target sites across the genome with high sensitivity, accounting for both local mismatches and broader sequence context. This capability has significantly improved the reliability of CRISPR applications in functional genomics, gene therapy, and synthetic biology. Moreover, deep learning models support the rational design of guide RNAs tailored to specific genomic targets, enabling more precise and predictable genome editing strategies. Table 2 provides an overview of the deep learning architectures commonly used in DNA sequence analysis, highlighting the application of CNNs, RNNs, and transformers across regulatory prediction, variant analysis, and genome engineering.

Table 2. Key applications of deep learning in DNA sequence analysis.

Application / Task	Biological Significance	Common Deep Learning Architectures	Example Models / Tools
Promoter and enhancer identification	Essential for understanding transcriptional regulation and gene expression control	CNNs, Transformers, Hybrid CNN-Transformer	DeepSEA, Basset, DNABERT, Enformer
TFBS detection and characterization	Core of gene regulatory networks; determines TF-DNA interactions	CNNs, Attention-based models, multi-modal networks	DeepBind, BPNet, FactorNet
Variant detection and functional consequence prediction	Critical for disease genomics, precision medicine, and evolutionary studies	CNNs, RNNs, Transformers, Ensemble models	DeepVariant, PrimateAI, EVE, Enformer
Gene and functional element annotation	Fundamental for genome interpretation, especially in non-model organisms	CNNs, RNNs, Transformer-based language models	DeepGene, Helixer, GENSCAN (modern DL versions)
DNA methylation, histone mark, chromatin accessibility prediction	Links sequence to regulatory states without altering DNA	CNNs, Transformers, Multi-task learning models	DeepCpG, Basenji, Sei
CRISPR on-target efficiency and off-target effect prediction	Enables precise genome editing with minimal side effects	CNNs, Gradient boosting + NN hybrids, Attention models	DeepCRISPR, CRISPR-Net, CRISTA

The table details major bioinformatics tasks in genomics, including their biological significance, commonly employed deep learning architectures, and representative computational tools or models. Applications focus on regulatory element prediction, variant analysis, genome annotation, epigenetic modeling, and CRISPR optimization.

5. APPLICATIONS FOR DEEP LEARNING IN RNA SEQUENCE ANALYSIS

RNA molecules represent one of the most structurally and functionally complex layers of biological information. Unlike DNA, RNA function is not determined solely by its linear sequence but also by dynamic secondary and tertiary structures, post-transcriptional modifications, alternative processing events, and interactions with proteins and other RNAs. These properties make RNA sequence analysis particularly challenging for traditional computational approaches. Deep learning has emerged as a powerful framework capable of modeling the nonlinear, context-dependent, and multi-scale nature of RNA biology. This section reviews major applications of deep learning in RNA sequence analysis, emphasizing how modern models have advanced our understanding of RNA structure, regulation, and function.

5.1. RNA Secondary Structure Prediction

RNA secondary structure the pattern of base pairing formed by hydrogen bonds between nucleotides is a key determinant of RNA function. Structures such as hairpins, loops, bulges, and pseudoknots influence RNA stability, localization, and interactions. Accurate prediction of RNA secondary structure from sequence alone has long been a central problem in computational biology. Traditional RNA folding algorithms are based on thermodynamic models that minimize free energy. While effective for short and well-characterized RNAs, these methods struggle with long transcripts, non-canonical base pairing, and context-dependent folding dynamics. Deep learning approaches have significantly improved structure prediction by learning patterns directly from experimentally derived structure data (Szikszai et al., 2025).

Neural networks, including CNNs, RNNs, and more recently graph neural networks and transformers, model RNA sequences as structured inputs and predict base-pairing probabilities or contact maps. By integrating long-range dependencies and nonlinear interactions, deep learning models capture folding rules that extend beyond simplified thermodynamic assumptions. These advances have improved prediction accuracy for complex RNA molecules, including long non-coding RNAs and viral genomes.

5.2. Splice Site and Alternative Splicing Event Prediction

Alternative splicing is a fundamental mechanism that expands transcriptomic and proteomic diversity. A single gene can generate multiple mRNA isoforms through differential inclusion or exclusion of exons. Accurate identification of splice sites and prediction of alternative splicing events are critical for understanding gene regulation and disease mechanisms, as illustrated in Figure 4, deep learning models have demonstrated exceptional performance in predicting splicing outcomes, including exon inclusion and exclusion, and assessing splice site accuracy based on genomic data

Splice site recognition depends on short consensus motifs embedded within broader sequence context, as well as interactions between distant regulatory elements. Traditional splice site predictors often rely on local motif scoring and struggle to capture long-range dependencies. Deep learning models have substantially improved splice site and alternative splicing prediction by integrating both local and global sequence features. CNNs detect core splice motifs, while RNNs and transformer-based models capture long-range interactions across introns and exons. These models can predict exon skipping, intron retention, and alternative donor or acceptor sites with high accuracy. Importantly, deep learning approaches have revealed that splicing regulation is governed by complex combinatorial rules rather than simple consensus motifs. This insight has advanced our understanding of splicing-related diseases, including cancer and inherited genetic disorders.

Figure 4. Overview of deep learning applications in RNA splicing variant analysis, demonstrating the capacity of neural networks to predict exon inclusion, exclusion, and splice site accuracy based on genomic variant data.

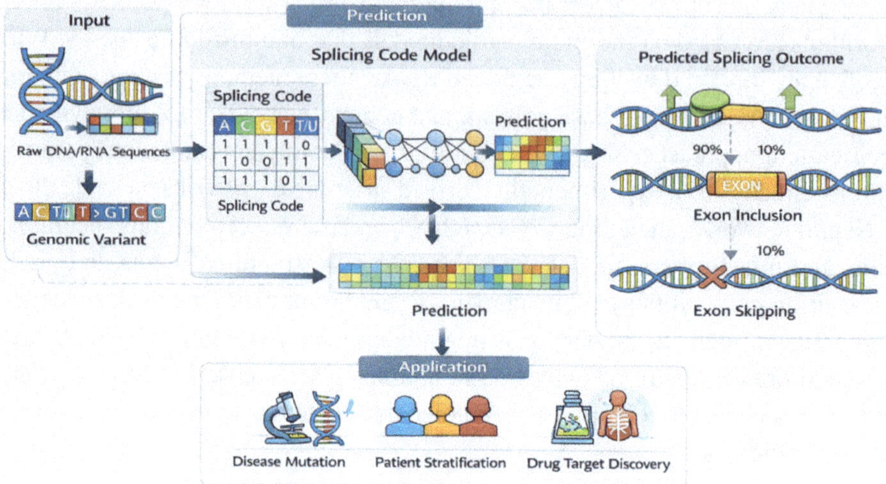

5.3. MicroRNA Target Prediction

MicroRNAs (miRNAs) are short non-coding RNAs that regulate gene expression by binding complementary sequences in target mRNAs, typically within untranslated regions. miRNA–mRNA interactions play a crucial role in development, homeostasis, and disease. Predicting miRNA targets is challenging due to partial sequence complementarity, context-dependent binding, and the influence of RNA secondary structure. Traditional methods often rely on seed matching and evolutionary conservation, resulting in high false-positive rates.

Deep learning models address these limitations by learning interaction patterns directly from experimentally validated miRNA–mRNA pairs. Neural networks integrate sequence complementarity, flanking context, and structural features to predict binding likelihood more accurately. Attention mechanisms further enable models to focus on functionally relevant regions within target transcripts. These advances have improved the specificity of miRNA target prediction and provided insights into regulatory networks that were previously inaccessible through rule-based approaches.

5.4. RNA Editing Site Prediction

RNA editing introduces post-transcriptional modifications that alter RNA sequences without changing the underlying DNA. The most common form in eukaryotes is adenosine-to-inosine (A-to-I) editing, which can affect coding potential, splicing, and RNA stability. Identifying RNA editing sites from sequence data is challenging due to the subtlety of sequence signals and the influence of local secondary structure. Traditional approaches rely on comparative genomics or alignment-based methods, which may miss species-specific or condition-specific editing events.

Deep learning models have been increasingly applied to RNA editing prediction by learning sequence and structural features associated with editing hotspots. CNNs capture local nucleotide context, while deeper architectures integrate broader sequence and structural signals. These models have demonstrated improved sensitivity and specificity in identifying editing sites, even in regions with weak conservation. RNA editing prediction has important implications for understanding transcriptome diversity, neural function, and immune regulation, where editing plays a prominent role.

5.5. mRNA Stability and Translation Efficiency Modeling

The stability and translational efficiency of mRNAs determine protein abundance and cellular response to environmental cues. These properties are influenced by multiple sequence features, including codon usage, untranslated region composition, RNA secondary structure, and regulatory motifs. Traditional models of mRNA stability and translation often rely on simplified metrics such as codon adaptation indices or motif counts. However, these approaches fail to capture the complex interactions that govern RNA fate in the cell.

Deep learning models have demonstrated strong predictive power in modeling mRNA stability and translation efficiency directly from sequence data. By learning hierarchical representations, these models integrate coding and non-coding regions, structural features, and context-dependent signals. Transformer-based architectures, in particular, excel at capturing long-range dependencies between regulatory elements located in different regions of the transcript. Such models have applications in synthetic biology, where optimizing mRNA design is critical for efficient protein production, as well as in understanding disease-associated dysregulation of gene expression.

5.6. Long Non-coding RNA (lncRNA) Functional Prediction

Long non-coding RNAs constitute a diverse and poorly understood class of transcripts involved in chromatin regulation, transcriptional control, and post-transcriptional processing. Unlike protein-coding genes, lncRNAs often lack strong sequence conservation, making functional annotation particularly challenging. Deep learning approaches have opened new avenues for lncRNA functional prediction by learning discriminative patterns directly from sequence and structural features. Neural networks can distinguish lncRNAs from coding transcripts, predict subcellular localization, and infer functional categories based on learned representations.

By integrating sequence embeddings with secondary structure predictions and expression profiles, deep learning models provide a more holistic view of lncRNA biology. These approaches are especially valuable for annotating newly discovered lncRNAs and prioritizing candidates for experimental validation. The use of RNNs, GNNs, and transformer-based models in RNA sequence analysis is detailed in Table 3, with example tools demonstrating progress in structure prediction, regulatory inference, and functional annotation.

Table 3. Key applications of deep learning in RNA sequence analysis.

Application / Task	Biological Significance	Common Deep Learning Architectures	Example Models / Tools
RNA folding and base-pairing prediction	Determines RNA function, stability, and interactions	CNNs, RNNs/LSTMs, GNNs, Transformers	SPOT-RNA, MXfold2, E2Efold, RNAformer
Splice site recognition and isoform prediction	Key to transcript diversity and disease mechanisms (e.g., cancer)	CNNs, RNNs, Transformers, Hybrid models	SpliceAI, Basenji2, DNABERT for splicing
miRNA-mRNA interaction prediction	Central to post-transcriptional gene regulation	CNNs, RNNs, Attention mechanisms, Siamese networks	DeepMirTar, TargetNet, miRTar2GO
Identification of A-to-I and other RNA editing sites	Affects protein coding, splicing, and RNA stability	CNNs, RNNs, Transformer encoders	DeepRed, REDIT, RNAEditor
mRNA half-life and translation rate prediction	Determines protein abundance and cellular response	RNNs, Transformers, Multi-modal models	UTR-LSTM, Translatomer, RiboDL

The table details major bioinformatics tasks in transcriptomics, including their biological significance, commonly employed deep learning architectures, and representative computational tools or models. Applications address structure prediction, splicing analysis, miRNA targeting, RNA editing, stability modeling, and lncRNA functional annotation, reflecting the multifaceted nature of RNA biology.

6. PRETRAINED MODELS AND BIOLOGICAL LANGUAGE MODELS

The emergence of pretrained models and biological language models represents one of the most significant paradigms shifts in modern bioinformatics. Rather than training task-specific models from scratch, researchers increasingly rely on large models pretrained on massive collections of biological sequences and subsequently fine-tuned for specialized tasks. This approach mirrors the success of language models in natural language processing and has fundamentally reshaped DNA and RNA sequence analysis by enabling scalable, generalizable, and data-efficient learning.

6.1. Concept of Sequence Language Modeling

Sequence language modeling is based on the idea that biological sequences, like natural languages, exhibit structured patterns, syntax-like rules, and contextual dependencies (Luo et al., 2025). In this framework, nucleotides or amino acids are treated as "tokens," and sequences are modeled as sentences governed by probabilistic rules learned from large corpora. Language models are typically trained using self-supervised objectives, such as predicting masked tokens or the next token in a sequence. In biological contexts, this training forces the model to learn the statistical regularities of DNA, RNA, or protein sequences without requiring explicit functional labels. As a result, the model develops internal representations that encode biologically meaningful information, including regulatory grammar, evolutionary constraints, and structural tendencies.

A key advantage of sequence language modeling is its ability to capture both local motifs and long-range dependencies simultaneously. For example, a masked nucleotide prediction task requires understanding not only of the nearby sequence context but also distal regulatory or structural signals. This makes language models particularly powerful for genome-scale and transcriptome-scale analysis.

6.2. Overview of Major Pretrained Models

Over the past five years, several influential pretrained biological language models have been developed, each targeting different molecular modalities and analytical goals (Feng et al., 2025).

DNABERT was among the first transformer-based language models designed specifically for DNA sequences. By treating overlapping k-mers as tokens, DNABERT learns contextual embeddings that capture regulatory and functional information across genomic regions. These embeddings have been shown to improve performance in promoter prediction, transcription factor binding site identification, and variant effect prediction.

GenomeGPT represents a newer generation of genome-scale language models inspired by large generative models in NLP. These models aim to learn comprehensive representations of entire genomes, enabling sequence generation, annotation, and interpretation tasks within a unified framework. GenomeGPT-like approaches highlight the potential of generative modeling in genomics, though challenges related to scale, interpretability, and biological validation remain active areas of research.

ESM (Evolutionary Scale Models)

Evolutionary Scale Models (ESMs) are large protein language models trained on millions of protein sequences. While primarily focused on proteins rather than nucleic acids, ESMs are highly relevant to DNA and RNA analysis due to their ability to model evolutionary constraints encoded in sequence data. ESMs learn representations that correlate strongly with protein structure, function, and stability without explicit structural supervision. These representations have been leveraged for tasks such as protein–DNA interaction prediction and functional annotation of gene products derived from genomic sequences. ESMs demonstrate how large-scale pretraining can capture deep biological principles embedded in sequence evolution.

RNA-FM and PRoBERTa

RNA-focused language models have emerged to address the unique challenges of transcriptomic analysis. RNA-FM is a foundation model trained on large RNA sequence datasets, designed to learn representations sensitive to secondary structure, splicing signals, and regulatory motifs. These representations improve performance across diverse RNA tasks, including structure prediction and functional classification. PRoBERTa, originally developed for protein sequences, has inspired similar architectures for RNA modeling by adapting transformer-based encoders to biological sequence data. These models demonstrate that language-model pretraining can generalize beyond DNA to capture RNA-specific regulatory and structural features.

Collectively, these models illustrate the rapid diversification of biological language modeling across molecular domains.

6.3. Transfer Learning and Fine-tuning for Specific Tasks

A defining strength of pretrained biological language models is their suitability for transfer learning. In this paradigm, a model pretrained on large, general-purpose sequence datasets is fine-tuned on smaller, task-specific datasets. Transfer learning is particularly valuable in biology, where labeled data are often scarce, noisy, or expensive to generate (Vinod et al., 2025). Fine-tuning allows models to adapt general sequence representations to specific tasks such as splice site prediction, mutation impact assessment, or RNA localization, often achieving high performance with limited training data.

Fine-tuning strategies vary in complexity. In some cases, only the final layers of the pretrained model are updated, preserving general representations learned during pretraining. In other cases, the entire model is fine-tuned to capture task-specific nuances. The choice of strategy depends on dataset size, task similarity, and computational resources. Beyond supervised fine-tuning, pre-trained models also support zero-shot and few-shot learning, where models perform new tasks with minimal or no additional training. This capability is especially promising for exploration genomic analysis and annotation of newly sequenced organisms.

6.4. Advantages and Challenges of Large Bio-models

Large pretrained biological models offer several compelling advantages (Ruan et al., 2026). First, they enable generalizable representation learning, reducing reliance on handcrafted features and task-specific architectures. Second, they support efficient reuse across multiple biological problems, promoting methodological unification. Third, they often achieve state-of-the-art performance by leveraging information learned from vast amounts of unlabeled sequence data. However, these models also present significant challenges. Computational cost is a major concern, as training and deploying large models require substantial hardware resources and energy consumption. This limits accessibility and raises sustainability considerations. Interpretability remains another critical issue. While language models capture rich biological information, understanding how specific predictions arise from learned representations is often difficult. This lack of transparency can hinder biological insight and experimental validation.

Additionally, pretrained models may inherit biases from training data, such as overrepresentation of certain species or genomic regions. Careful dataset curation and evaluation are essential to ensure fair and biologically meaningful predictions. Finally, there is an ongoing debate regarding the balance between model scale and biological grounding. Larger models do not automatically guarantee deeper biological understanding, emphasizing the need for integrative approaches that combine data-driven learning with biological knowledge.

7. DATA RESOURCES, TOOLS, AND BENCHMARKS

The success of deep learning in DNA and RNA sequence analysis depends not only on model architecture but also on the availability of high-quality data resources, standardized benchmarks, and robust computational tools. Biological sequences are inherently noisy, heterogeneous, and context-dependent, making careful dataset selection and evaluation essential. This section reviews key genomic and RNA databases, benchmark datasets used for deep learning tasks, and open-source tools and pipelines that support reproducible and scalable biological sequence analysis.

7.1. Genomic Databases

Large, well-curated genomic databases form the foundation of modern sequence-based research. These resources provide reference genomes, functional annotations, and experimental datasets that enable both model training and validation.

NCBI

The National Center for Biotechnology Information (NCBI) serves as one of the most comprehensive repositories of genomic data. It hosts reference genomes, gene annotations, variation datasets, and sequence archives across a wide range of organisms. Resources such as GenBank, RefSeq, and the Sequence Read Archive (SRA) are widely used for training and benchmarking deep learning models. For DNA sequence analysis, NCBI data supports tasks including gene prediction, variant analysis, and comparative genomics. The breadth and diversity of species represented in NCBI databases make them particularly valuable for transfer learning and cross-species model generalization. However, variability in annotation quality across organisms requires careful data curation when constructing training datasets.

ENCODE

The Encyclopedia of DNA Elements (ENCODE) project provides high-resolution functional annotations of the human genome. ENCODE datasets include transcription factor binding profiles, chromatin accessibility data, histone modifications, and DNA methylation patterns generated through standardized experimental protocols. ENCODE data have been instrumental in advancing deep learning–based regulatory genomics. Many promoters, enhancers, and TFBS prediction models rely on ENCODE datasets as gold standards for supervised training and evaluation. The integration of sequence data with functional readouts enables models to learn biologically meaningful regulatory patterns beyond sequence alone.

Despite its richness, ENCODE is primarily focused on human and selected model organisms, which can limit its applicability to broader evolutionary studies.

UCSC Genome Browser

The UCSC Genome Browser provides an integrated platform for visualizing genomic sequences and annotations. It aggregates data from multiple sources, including NCBI and ENCODE, and offers genome assemblies, gene models, regulatory annotations, and comparative genomics tracks.

For deep learning research, UCSC resources facilitate data exploration, annotation validation, and result interpretation. The browser's standardized coordinate system and track-based architecture simplify the integration of multiple data types, supporting multi-modal modeling approaches that combine sequence and epigenomic information.

7.2. RNA Databases

RNA-focused databases provide essential resources for transcriptomic analysis, structural modeling, and functional annotation of non-coding RNAs.

Rfam

Rfam is a curated database of RNA families, including structured non-coding RNAs and regulatory elements. It provides multiple sequence alignments and consensus secondary structure models for each RNA family. Rfam data are widely used in RNA secondary structure prediction and RNA family classification tasks. Deep learning models leverage Rfam annotations to learn conserved structural motifs and folding patterns. However, Rfam focuses on well-characterized RNA families, which may bias models toward known structures and functions.

miRBase

miRBase is the primary repository for microRNA sequences and annotations. It includes experimentally validated miRNAs across diverse species, along with information on genomic location and sequence variants. Deep learning models for miRNA target prediction and miRNA discovery frequently rely on miRBase as a reference dataset. While miRBase provides high-confidence annotations, ongoing revisions and updates reflect the evolving understanding of miRNA biology, highlighting the importance of version control in model development.

RNAcentral

RNAcentral integrates RNA sequences from multiple specialized databases into a unified resource. It covers a broad range of RNA classes, including mRNAs, lncRNAs, rRNAs, and small RNAs.

The diversity of RNAcentral makes it particularly valuable for large-scale RNA language model pretraining and cross-task evaluation. However, heterogeneity in annotation depth and experimental validation across RNA types necessitates careful filtering when constructing training datasets.

7.3. Benchmark Datasets for Deep Learning Tasks

Benchmark datasets play a critical role in evaluating deep learning models objectively and reproducibly. In genomics and transcriptomics, benchmarks are used to compare model performance across tasks such as promoter detection, splice site prediction, and variant impact assessment. Well-designed benchmarks typically include standardized training, validation, and test splits, along with clearly defined evaluation metrics (You et al., 2025). They help prevent overfitting and enable fair comparison between competing methods. In recent years, community-driven benchmarks have emphasized cross-cell-type and cross-species generalization to assess model robustness.

However, benchmarking in biological sequence analysis faces unique challenges. Biological datasets often exhibit class imbalance, experimental noise, and hidden confounders. Additionally, benchmark datasets may become outdated as new experimental data emerge. As a result, continuous benchmarking and dataset updating are necessary to maintain relevance. There is growing recognition that benchmark performance alone does not guarantee biological validity. Complementary evaluation strategies, including interpretability analysis and experimental validation, are increasingly emphasized.

7.4. Open-Source Tools and Pipelines

Open-source tools and pipelines are essential for implementing, training, and deploying deep learning models in biological sequence analysis. They promote reproducibility, transparency, and community-driven innovation. Many modern pipelines integrate data preprocessing, model training, evaluation, and visualization into cohesive workflows. These tools often support popular deep learning frameworks such as PyTorch and TensorFlow, enabling rapid experimentation and scalability. Specialized libraries provide functionality tailored to biological sequences, including sequence encoding, motif visualization, and genomic coordinate handling. Workflow management systems further support reproducible research by automating data processing and model evaluation across different computational environments.

Despite these advances, tool fragmentation and inconsistent documentation remain challenges. Ensuring interoperability between tools and maintaining long-term software sustainability are ongoing concerns in the bioinformatics community. The open-source tools and pipelines used in deep learning–based genomics and transcriptomics are summarized in Table 4, with categorization by purpose and emphasis on key features and integrations.

Table 4. Open-source tools and pipelines supporting deep learning in biological sequence analysis.

Tool / Library Name	Category / Purpose	Key Features	Integration / Framework
PyTorch	Deep learning framework	Dynamic computation graphs, extensive neural network modules, strong GPU support, active research community	Native; integrates with torchvision, torchaudio, PyTorch Lightning
TensorFlow	Deep learning framework	Static/dynamic graphs, TensorBoard visualization, production-ready deployment, Keras API	Native; integrates with TF-Genomics, TensorFlow Probability
JAX	High-performance numerical computing	Automatic differentiation, JIT compilation, GPU/TPU acceleration, functional programming paradigm	Native; often used with Haiku, Flax, or Equinox
kipoiseq / Selene	Genomic sequence processing & model training	Sequence encoding, variant effect prediction, genome coordinate handling, benchmark datasets	PyTorch, TensorFlow
Giotto / Giotto-deep	Multi-modal genomic data analysis	Chromatin accessibility modeling, TF binding prediction, multi-omics integration pipelines	PyTorch, TensorFlow, Scanpy
pyBigWig / pyBedTools	Genomic file format handling	Read/write bigWig, bedGraph, BED files; genomic interval operations	Python-based; integrates with pandas, numpy
Snakemake	Workflow management system	Rule-based pipelines, cluster/ cloud execution, reproducible analysis, conda/ docker integration	Language-agnostic; integrates with Python, R, shell scripts
Nextflow	Workflow orchestration	Dataflow paradigm, containerization support, scalable execution, community pipelines (nf-core)	Language-agnostic; Docker/ Singularity support
MLflow / Weights & Biases (W&B)	Experiment tracking & model management	Hyperparameter logging, model versioning, visualization dashboards, collaboration features	PyTorch, TensorFlow, JAX, scikit-learn

continued on following page

Table 4. Continued

Tool / Library Name	Category / Purpose	Key Features	Integration / Framework
SeqLogo / logomaker	Motif visualization	PWM visualization, sequence logo generation,	Python (matplotlib, seaborn)
SHAP / Captum	Model interpretability	Feature importance scores, saliency maps, attention visualization, in silico mutagenesis	PyTorch (Captum), TensorFlow (SHAP), model-agnostic options
IGV / UCSC Genome Browser Tools	Genomic visualization & annotation	Interactive genome browsing, track visualization, custom annotation loading	Standalone or web-based; file format compatibility (BAM, VCF, BED)

The table categorizes essential software resources into: (1) core deep learning frameworks (PyTorch, TensorFlow, JAX), (2) specialized libraries for sequence processing and modeling (kipoiseq, Giotto), (3) workflow management systems for reproducible research (Snakemake, Nextflow), and (4) visualization and interpretation tools (SeqLogo, SHAP, IGV). These tools collectively enable data preprocessing, model training, evaluation, visualization, and deployment, forming the computational infrastructure for modern genomic and transcriptomic deep learning applications.

8. CHALLENGES AND LIMITATIONS

Despite the remarkable success of deep learning in DNA and RNA sequence analysis, significant challenges and limitations remain. Biological systems are inherently complex, noisy, and context-dependent, and computational predictions do not always translate directly into biological understanding or experimental validity. One of the most persistent challenges is data imbalance, where biologically relevant events such as transcription factor binding or RNA editing are rare compared to non-functional sequences, resulting in models that may achieve high accuracy while underperforming in biologically important minority classes (Zubair et al., 2025). Additionally, biological noise, including technical variability and measurement errors from high-throughput sequencing, further complicates model training, as deep learning models may learn dataset-specific artifacts rather than true biological signals. Interpretability is another significant issue, as deep learning models, often criticized as "black boxes," provide limited biological value if they cannot explain how sequence features influence function or regulation, despite achieving superior predictive performance. Post hoc interpretability methods, like saliency maps and attention visualization, attempt to address this challenge but have limitations, as they may not always reflect causal biological mechanisms (Shrikumar, 2020). Overfitting also remains a concern in high-dimensional biological data, as deep learning models may capture noise or dataset-specific patterns rather than generalizable biological rules, especially given the limited labeled data available and implicit information leakage between training and test sets (Wang et al., 2025). Moreover, the disconnect between computational predictions and biological plausibility poses a fundamental limitation, as models may predict gene expression levels accurately without reflecting true biological mechanisms, offering limited mechanistic insights. Ensuring biological plausibility requires integrating domain knowledge and experimental validation. Finally, reproducibility issues arise due to differences in data preprocessing, model initialization, hyperparameter selection, and evaluation protocols, which can lead to varying results even when using the same datasets. Reproducibility challenges are exacerbated by the size and complexity of datasets, limited access to pretrained models, and the rapid evolution of computational tools. Efforts to improve reproducibility include standardized benchmarks and open-source code release, but this remains an unresolved challenge in ensuring the reliability of biological deep learning research.

9. FUTURE PERSPECTIVES

Deep learning has already reshaped DNA and RNA sequence analysis, and the field continues to evolve rapidly, driven by emerging computational paradigms, expanding biological datasets, and growing clinical relevance. Future progress will depend on larger models, deeper biological integration, improved interpretability, and responsible deployment. Next-generation AI architectures are expected to move beyond conventional convolutional, recurrent, and transformer-based designs toward models that more closely reflect biological complexity, integrating hierarchical modeling to simultaneously capture local sequence features, long-range genomic interactions, and three-dimensional genome organization (Groff-Vindman et al., 2025). Advances in attention mechanisms, sparse transformers, and memory-augmented networks will enable scalable modeling of entire chromosomes or transcriptomes (Zhou et al., 2025). Biologically informed architectures that incorporate known constraints, such as base-pairing rules in RNA folding and chromatin looping principles in genome regulation, aim to bridge data-driven learning with mechanistic biological modeling. Future deep learning systems will also increasingly focus on multi-omics integration, combining genomics, transcriptomics, epigenomics, proteomics, and metabolomics data to enable more accurate predictions of gene expression, regulatory interactions, and disease phenotypes (Khan et al., 2025). However, challenges such as data heterogeneity, missing modalities, and differing resolutions will need to be addressed. Self-supervised learning and foundation models, trained on pan-genomic and pan-transcriptomic datasets, will serve as universal representation learners to support a range of downstream tasks with minimal fine-tuning (Istrate et al., 2025). These models will also enable unified representations of DNA, RNA, and protein sequences, facilitating end-to-end modeling of biological information. In clinical genomics, deep learning holds potential for personalized medicine, aiding in the interpretation of rare variants, disease susceptibility, and RNA-based insights into disease (Istrate et al., 2025). Nevertheless, translating these models into clinical practice requires rigorous validation and regulatory oversight. Ethical and security considerations, particularly related to data privacy, informed consent, algorithmic bias, and societal implications, will play a central role in shaping the future of genomic AI, ensuring its responsible deployment in both research and healthcare (Khan et al., 2025).

10. CONCLUSION

The integration of deep learning into DNA and RNA sequence analysis has revolutionized molecular biology, shifting the focus from handcrafted rules to data-driven learning, allowing for a deeper understanding of biological sequences. Early successes with convolutional neural networks (CNNs) laid the groundwork by learning regulatory motifs from raw DNA data, while the advent of recurrent networks and transformer architectures enabled the capture of sequential dependencies and long-range contextual relationships (Sun et al., 2025). Pretrained biological language models have since emerged, leveraging self-supervised learning to produce transferable representations, improving generalization and reducing reliance on labeled data (Raza et al., 2024). Hybrid and graph-based models have further expanded the scope of deep learning to incorporate structural, spatial, and multi-modal biological information (Ahmed et al., 2024). The practical applications of these advancements such as regulatory element prediction, variant impact assessment, RNA structure modeling, and CRISPR optimization undertake the importance of deep learning in bioinformatics, not only for improving predictive accuracy but also for generating novel biological hypotheses (Asim et al., n.d.; Ahmed et al., 2024). Deep learning has also reshaped molecular biology by enabling large-scale genome annotation, enhancing understanding of non-coding regions, and revealing the complexity of RNA biology, challenging simplified gene-centric models (Muqaddas et al., 2025). Furthermore, it has fostered the integration of computational and experimental biology, accelerated discovery cycles and improved experimental efficiency (Ahmad et al., 2026). However, it is crucial to recognize the limitations of deep learning, as high predictive performance does not always translate to mechanistic understanding (Xia et al., 2026). Future research should focus on deeper integration of biological knowledge, addressing challenges of data bias and interpretability, and considering ethical concerns, especially in clinical genomics and personalized medicine (Ahmed et al., 2026; ul Abideen et al., 2026). By maintaining methodological rigor and ethical responsibility, deep learning will continue to drive discoveries in molecular biology for years to come (Asim et al., 2025; Raza et al., 2025; Muzammal et al., 2026; Raza et al., 2025; Ali et al., 2025; Raza et al., 2025; Ahmed et al., 2026; Ahmed et al., 2026).

REFERENCES

Abbaszadeh, A., & Shahlai, A. (2025). Artificial Intelligence for CRISPR Guide RNA Design: Explainable Models and Off-Target Safety. *arXiv preprint arXiv:2508.20130.*

Ahmad, A., Ahmed, M. M., Akhtar, A., Liu, W., Yang, R., Sun, X., Wang, X., Bibi, S., Khan, M. B., & Chen, S. (2026). Unlocking Grass Stress Resistance: Fungal Endophyte-Mediated Pathogen Recognition and RNA Regulation. *International Journal of Molecular Sciences*, *27*(9), 3899. DOI: 10.3390/ijms27093899 PMID: 42123484

Ahmad, A., Ahmed, M. M., Akhtar, A., Shuihong, C., Zafar, Z., Ullah, R., Asim, M., He, Z., & Khan, M. B. (2026). New Insight into Endophytic Fungi–Plant Symbioses Under Climate Change: Molecular Crosstalk, Nutrient Exchange, and Ecosystem Resilience. *Applied Microbiology (Basel, Switzerland)*, *6*(3), 47. DOI: 10.3390/applmicrobiol6030047

Ahmed, M. M., Asim, M., Kaleri, A. A., Manzoor, D., Rajput, A. A., Laghari, R., Khaki, S. A., Musawwir, A., Ullah, Z., & Ahmad, W. (2024). *Biochemical Dynamics and Quality Attributes of Strawberry Fruits across Maturity Stages with Respect to Different Preservation Methods: Biochemical Dynamics and Quality Attributes.* Futuristic Biotechnology. DOI: 10.54393/fbt.v4i01.98

Ahmed, M. M., Hamid, S. M., Asim, M., Ahmad, A., Raziq, A., Rui, Z., & Zhiyong, P. 2026. Nutrient Cycling and Plant-Microbe Interaction in Alpine Ecosystem: Microbial Mediation of Soil Fertility and Plant Growth. In *Biodiversity of Alpine Vegetation: Conservation, Challenges, Prospects, and Solutions* (pp. 319-364). IGI Global Scientific Publishing.

Ahmed, M. M., Hamid, S. M., Rui, Z., Zhiyong, P., Xingyu, G., Qian, Y., Yu, S., Asim, M., Qiuhao, X., Ahmad, A., & Yerhazi, Y. (2026). Role of Microbial Biotechnology in Mitigating Climate-Induced Agricultural Stress. In *Impacts of Climate Change on Agriculture in the MENA Region* (pp. 257–280). IGI Global Scientific Publishing.

Ahmed, M. M., Sharif, U., Raza, A., Safdar, M., Ali, W., Asim, M., Muzammal, H., Sattar, J., Maqbool, S., & Zaheer, M. (2024). Climate-smart agriculture for resilience and profitability. In *Sustainable practices for agriculture and marketing convergence* (pp. 27–50). IGI Global Scientific Publishing. DOI: 10.4018/979-8-3693-2011-2.ch002

Ali, A., Raza, A., Malik, A., Shabbir, M., Sarwar, N., & Muzammal, H. (2025). AI and IoT solutions for food security and agriculture. In *Enhancing Automated Decision-Making Through AI* (pp. 177–210). IGI Global Scientific Publishing.

Asim, M., Ateeq, M., Shen, X., An, S., Ahmed, M. M., Ashraf, M. A., Zhu, K., & Liu, J. (2025). Molecular mechanisms and adaptive strategies for hypoxia tolerance in horticultural crops: Beyond model systems. *Horticulture Advances, 3*(1), 34. DOI: 10.1007/s44281-025-00088-1

. Asim, M., Raza, A., Safdar, M., Ahmed, M.M., Khokhar, A., Aarif, M., Ansari, M.A., Sattar, J. and Chowdhury, I.U.Z., *Sustainable agriculture and the SDGs: a convergence approach, 1–26* [online]

Feng, H., Wu, L., Zhao, B., Huff, C., Zhang, J., Wu, J., Lin, L., Wei, P., & Wu, C. (2025). Benchmarking DNA foundation models for genomic and genetic tasks. *Nature Communications, 16*(1), 10780. DOI: 10.1038/s41467-025-65823-8 PMID: 41315262

. Groff-Vindman, C. S., Trump, B. D., Cummings, C. L., Smith, M., Titus, A. J., Oye, K., ... & Linkov, I. (2025). The convergence of AI and synthetic biology: the looming deluge. *npj Biomedical Innovations, 2*(1), 20.

Hwang, H., Jeon, H., Yeo, N., & Baek, D. (2024). Big data and deep learning for RNA biology. *Experimental & Molecular Medicine, 56*(6), 1293–1321. DOI: 10.1038/s12276-024-01243-w PMID: 38871816

Istrate, A. M., Milletari, F., Castrotorres, F., Tomczak, J. M., Torkar, M., Li, D., & Karaletsos, T. (2025). rbio1-training scientific reasoning LLMs with biological world models as soft verifiers. bioRxiv, 2025-08.

Jiang, J., Li, Y., Cao, S., Shan, Y., Liu, Y., Fei, T., Yu, Y., Feng, Y., Li, Y., Li, Y., & Yuan, J. (2025). Artificial intelligence in bioinformatics: A survey. *Briefings in Bioinformatics, 26*(6), bbaf576. DOI: 10.1093/bib/bbaf576 PMID: 41206113

Khan, S. A., Martínez-de-Morentin, X., Alsabbagh, A. R., Maillo, A., Lagani, V., Gomez-Cabrero, D., & Tegner, J. (2025). Multimodal foundation transformer models for multiscale genomics. *Nature Methods*, •••, 1–13. PMID: 41326819

Luo, J., Cai, X., & Li, Y. (2025). Large language models for biological sequence analysis in infectious disease research. *Biosafety and Health, 7*(05), 323–332. DOI: 10.1016/j.bsheal.2025.09.007 PMID: 41262460

Muqaddas, M., Mengaya, L., Ahmed, M. M., Hamid, S. M., Yanju, X., Asim, M., & Yunfeng, P. (2025). Postharvest handling and storage strategies for preserving jujube (Ziziphus jujuba mill.) fruit quality: A review. *Foods, 14*(19), 3370. DOI: 10.3390/foods14193370 PMID: 41097538

Muzammal, H., Zaman, M., Safdar, M., Shahid, M. A., Nishtar, Z., Bilal, M., Munir, M., Haseeb, M. M., Raza, A., Shah, S. I. H., & Zafar, U. (2026). Advances in Geospatial Modeling for Flood Risk Assessment in Changing Climate. In *Advances in Computational Intelligence for Climate Change Security and Sustainability* (pp. 33–62). IGI Global Scientific Publishing.

Rani, N., Chaudhary, P., Kumar, U., Khatana, K., Kumar, N., Mavi, A. K., & Singh, S. (2025). Harnessing deep learning in bioinformatics–opportunities, challenges, and ethical considerations. *Vita Scientia, 1*(1), 21–56. DOI: 10.65116/vitascientia.V1I1.rev.01

Raza, A., Malik, A., & Maqsood, M. 2025. Transitioning From Legacy Systems to AI: A Holistic Approach to Disaster Detection. In *Enhancing Automated Decision-Making Through AI* (pp. 33-66). IGI Global Scientific Publishing.

Raza, A., Miao, Y., Huang, Y., Stueve, K., Lu, J., Yang, Z., & Bindlish, R. (2025). Optimizing On-Farm Corn Yield Prediction by a Multi-Source Data Fusion Approach Using Remote Sensing and Machine Learning. *Smart Agricultural Technology, 12*, 101630. DOI: 10.1016/j.atech.2025.101630

Raza, A., Shahid, M. A., Safdar, M., Zaman, M., Sabir, R. M., Muzammal, H., & Ahmed, M. M. (2024). Impact of land use and land cover change on agricultural production in District Bahawalnagar, Pakistan. *Environmental Sciences Proceedings, 29*(1), 46.

Raza, A., Shahid, M. A., Zaman, M., Miao, Y., Huang, Y., Safdar, M., Maqbool, S., & Muhammad, N. E. (2025). Improving wheat yield prediction with multi-source remote sensing data and machine learning in arid regions. *Remote Sensing (Basel), 17*(5), 774. DOI: 10.3390/rs17050774

Ruan, W., Lyu, Y., Zhang, J., Cai, J., Shu, P., Ge, Y., Lu, Y., Gao, S., Wang, Y., Wang, P., Zhao, L., Wang, T., Liu, Y., Fang, L., Liu, Z., Liu, Z., Li, Y., Wu, Z., Chen, J., & Liu, T. (2026). Large language models for bioinformatics. *Quantitative Biology, 14*(1), e70014. DOI: 10.1002/qub2.70014 PMID: 41676319

Shrikumar, A. (2020). *Interpretable machine learning for scientific discovery in regulatory genomics*. Stanford University.

Sun, C., Sun, Y., Xu, K., He, Z., Li, H., Li, Y., . . . Chen, H. (2025). A comprehensive benchmark and guide for sequence-function interpretable deep learning models in genomics. bioRxiv, 2025-01.

Szikszai, M., Wang, T. Y., Krueger, R., Mathews, D. H., Ward, M., & Aviran, S. (2025). Deep Learning for RNA Secondary Structure Determination: Gauging Generalizability and Broadening the Scope of Traditional Methods. bioRxiv, 2025-11.

Tabane, E., Mnkandla, E., & Wang, Z. (2025). Optimizing DNA Sequence Classification via a Deep Learning Hybrid of LSTM and CNN Architecture. *Applied Sciences (Basel, Switzerland), 15*(15), 8225. DOI: 10.3390/app15158225

Tang, Z., Somia, N., Yu, Y., & Koo, P. K. (2025). Evaluating the representational power of pre-trained DNA language models for regulatory genomics. *Genome Biology, 26*(1), 203. DOI: 10.1186/s13059-025-03674-8 PMID: 40660356

Testagrose, C., & Boucher, C. (2025). Tokenization and deep learning architectures in genomics: A comprehensive review. *Computational and Structural Biotechnology Journal, 27*, 3547–3555. DOI: 10.1016/j.csbj.2025.07.038 PMID: 40821715

. ul Abideen, M.Z., Ahmed, M.M., Asim, M., Zhiyong, P., Rui, Z., Razzaq, K., Yerhazi, Y., Ashraf, M.A., Hamid, S.M. and Ahmad, A., 2026. Abiotic Stress and Crop Production in a Changing Global Environment. In *Impacts of Climate Change on Agriculture in the MENA Region* (pp. 227-256). IGI Global Scientific Publishing.

Vinod, R., Chen, P. Y., & Das, P. (2025). Reprogramming pretrained language models for protein sequence representation learning. *Digital Discovery, 4*(6), 1591–1601. DOI: 10.1039/D4DD00195H

. Wang, H. (2025). Pretrained Language Models for Biological Sequence Understanding. *Computational Molecular Biology, 15.*

Wang, J., Fan, Y., Hong, L., Hu, Z., & Li, Y. (2025). Deep learning for RNA structure prediction. *Current Opinion in Structural Biology, 91*, 102991. DOI: 10.1016/j.sbi.2025.102991 PMID: 39933218

Wang, Y., Shen, Y., Li, J., Wang, T., Peng, J., & Shang, X. (2025). Enhanced RNA secondary structure prediction through integrative deep learning and structural context analysis. *Nucleic Acids Research, 53*(11), gkaf533. DOI: 10.1093/nar/gkaf533 PMID: 40530692

Xia, Q., Luo, L., Yerzati, Y., Ahmed, M. M., Chen, Y., Wang, S., Qin, J., Chen, L., Jin, Q., Guo, Z., & Zhang, R. (2026). Correction: Near-infrared prediction of tannin content in walnut kernels using wavelet transform combined with interpretable machine learning models. *Frontiers in Plant Science, 17*, 1806058. DOI: 10.3389/fpls.2026.1806058 PMID: 41835266

Xu, H. L., Gong, X. J., Yu, H., & Wang, Y. K. (2025). DNABERT2-CAMP: A Hybrid Transformer-CNN Model for E. coli Promoter Recognition.

You, N., Liu, C., Lin, H., Wu, S., Chen, G., & Shen, N. (2025). Benchmarking pre-trained genomic language models for RNA sequence-related predictive applications. *Nature Communications, 17*(1), 223. DOI: 10.1038/s41467-025-66899-y PMID: 41354659

Yu, S. (2025). Deep Learning for Predicting Gene Expression from Genomic Sequences. *Computational Molecular Biology, 15*.

Zhou, Z., Riley, R., Kautsar, S., Wu, W., Egan, R., Hofmeyr, S., & Wang, Z. (2025). GenomeOcean: An Efficient Genome Foundation Model Trained on Large-Scale Metagenomic Assemblies. *bioRxiv*.

Zubair, M., Khan, A. H., Bilal, S. F., & Li, J. (2025). Deep learning approaches for resolving genomic discrepancies in cancer: A systematic review and clinical perspective. *Briefings in Bioinformatics*, *26*(6), bbaf541. DOI: 10.1093/bib/bbaf541 PMID: 41176790

Chapter 13
AI-Driven Genetic Marker Identification for Criminal Investigations Modern Biological Insights in Forensic Science

Pravallika Vataparthi
Centurion University of Technology and Management, Vizianagaram, India

Naga Jogayya Kothakota
School of Forensic Science, Centurion University of Technology and Management, Vizianagaram, India

Kiranbhai Dodiya
http://orcid.org/0009-0001-9409-7303
Gujarat University, India

ABSTRACT

AI is rapidly transforming forensic biology by enhancing the identification and interpretation of target genetic sequences essential for criminal investigations. This work examines how AI-driven approaches support the detection of STRs, SNPs, mitochondrial, and epigenetic markers to significantly improve the speed, precision, and reliability of DNA profiling. Machine learning and deep learning models strengthen allele calling, mixture interpretation, phenotype prediction, ancestry estimation, and the analysis of highly degraded or contaminated samples. The chapter further highlights AI-app integration for real-world casework, multi-omics enhancement of unstructured forensic datasets, and resource optimisation for investigative efficiency.

DOI: 10.4018/979-8-2600-2691-5.ch013

Ethical and societal concerns including data integrity, bias, and risks associated with large genetic databases addressed to promote responsible and secure adoption. Overall, the chapter demonstrates how AI integration is reshaping forensic genetics and guiding the future direction of modern biological investigations.

1. INTRODUCTION

1.1 Background and Scope

The advancement of forensic detectives certainly would not be as sophisticated and exacting as it is now without forensic genetics. For the last 30 years, the field of forensic genetics has utilised DNA profiling, focusing on the analysis of Short Tandem Repeat (STR) loci, mitochondrial DNA, and Single Nucleotide Polymorphisms (SNPs). It has been a standard in forensic genetics. However, the biological data relevant to a case can be massive, and the field is moving at a rapid pace with advances in sequencing technology, creating problems in data parsing, logic flow, and decisions made with the data. The sector now relies on sophisticated systems capable of manipulating and interpreting large and complex data sets. The current offered solutions, and the ones projected, employ Artificial Intelligence (AI) systems, particularly in machine learning and deep learning. The purpose of this chapter is to evaluate and demonstrate the versatility of AI as it streamlines and enhances the forensic genetic marker analysis process and impacts the entire forensic process. The modifications made by AI to the forensic process also help overcome the process's weaknesses and improve the reliability of the conclusions drawn from a criminal investigation.

1.2 Evolution of Marker Analysis in Forensic Science

Forensic genetics is now an established branch of science. However, retrospectively, it is built upon early foundational techniques in DNA profiling, which had low discriminatory power and required large quantities of DNA. Then, in the 1990s, the field of forensics was entirely revolutionised by the introduction of DNA profiling based on the analysis of short tandem repeats (STRs). This new profiling method was highly polymorphic and featured superior reproducibility compared to forms of DNA profiling before STR analysis. Most importantly, STR profiling was the first DNA profiling technique compatible with the national CODIS and NDDB databases. This inclusion established it as the gold standard for forensic identification. Forensic genetics continued to evolve by the incorporation of complementary techniques such as sequencing of the mitochondrial DNA (mtDNA) to determine maternal lineage, sequencing of the Y chromosome to determine male lineage and the prediction of phenotypes and ancestry through the analysis of autosomal single-nucleotide polymorphisms (SNPs). Improved analysis of genetic markers through the incorporation of next-generation sequencing (NGS) techniques resulted in significant analytical advancements. However, as is the case with many analytical advancements, it was also accompanied by a new series of challenges, such as improved analytical complexity of allele calling, the deconvolution of population structure in mixtures of several DNA profiles and the challenges of sequencing degraded DNA. The refinement of these markers, paired with the profound complexity exhibited by the resultant datasets in combination with the increasing need for human analytical relief, laid the groundwork for the introduction of artificial intelligence (AI) driven models.

1.3 The Innovation of AI in Contemporary Biological Research

AI in Biological Sciences began with the evolution of 'big data'. The sequencing of genomic data began creating information at a volume level that became intractable to traditional computational statistics. Machine Learning became the first to show significant potential in the classification of sequences, detection of variants, and identification of intricate patterns in the multivariate domain. Feature extraction, a manual, time-consuming, and often suboptimal process, became automated through Deep Learning, resulting in significant advances in the field. Today, the most common models used in the field include Convolutional Neural Networks (CNNs), Recurrent Neural Networks (RNNs), Random Forest Classifiers, and Transformers. These models have been applied to SNP detection, STR profile interpretation, inferring phylogenetic relationships, and prediction phenotypes (Muneeb et al., 2022; D. Wu et al., 2021). AI has proven most useful in Forensic Genetics, given the common human hindrances in the analysis of incomplete, noisy, low-template DNA. AI has the potential to automate the interpretation of electropherograms, SNP panel ancestry prediction, facial approximations from DNA, and age estimation from methylation analysis (Terrado-Ortuño & May, 2025). With the continuous training of models on biological datasets, the field of Forensic Biology has seen AI become an indispensable tool.

1.4 Importance of AI in Criminal Justice Applications

In the criminal justice system, AI provides different strategic benefits, particularly in the detection of complex crimes and the analysis of forensic evidence. To begin with, AI improves the precision of allele calling by mitigating variability tied to the analyst. Secondly, AI-assisted probabilistic genotyping allows forensic laboratories to deconvolute multi-source mixtures, which has historically been an inexact and protracted procedure. Third, AI improves the immediate intelligence value of an investigation by estimating the probable physical characteristics and ancestry of a suspect, wherein the information available to law enforcement is limited. Fourth, AI assists in the DVI process by quickly analysing genomic data, even in instances where the DNA is significantly degraded. Fifth, AI enhances the workflow involved in searching national and international DNA databases (Palmela Pereira et al., 2025; Shevchuk et al., 2025). Most significantly, AI promotes equity within the justice system by allowing evidence analysis dependent on statistical interpretation, independent of personal biases. If validated and implemented appropriately, AI can help minimise wrongful convictions while boosting the overall accuracy and reliability of forensic evidence used in court.

1.5 Issues in Conventional Forensic Genetics

There are several constraints in conventional forensic genetics. One is that there are specific limitations on how much of an analytical tune-up there can be. One of the challenges is the manual editing of electropherograms and the possibility of artefacts recording such as stutters, missing an allele, and other ways of recording background noise. This is believed to lead to different conclusions. In complete or uncertain profiles, there are low-template DNA samples that are frequently found in crime scenes (Alketbi, 2024; Ferrero & Scotti, 2022). Out of these samples, mixed samples where there are three or more individuals in the DNA profiles are the hardest of forensic science challenges. Differentiating peaks, deciding on a locus, and estimating the number of contributors all result in bias. Adding to thcsc is the increasing complexity of the genomic markers, where there is a trend toward next-generation sequencing. This, in turn, begs the need for more incorporated computational models that can deal with the variability of coverage, noise in the sequence alignment, and detection of variants on a large scale. (Athanasopoulou et al., 2025). Incorporating other information, such as methylation signatures, epigenetic modifications, and markers specific to a tissue, which are all useful in forensics, is a challenge for these models. This is the reason there is a demand for AI in these models to develop automated, accurate models to interpret tests.

1.6 Review of Literature

AI is being integrated into forensic genomics, as the literature indicates, and its merits compared to older technologies are being assessed. Older literature showed the superiority of machine learning techniques when compared to standard statistical approaches when it came to SNP classification, variant detection, and predicting genotypes. More recently, literature indicates and assesses the improvements in deep learning architectures to accurately call alleles and create less noisy sequencing data and electropherograms. Several studies validated systems of probabilistic genotyping using AI, for example, STRmix, LiRA, and EuroForMix, which have been accepted and used in forensic laboratories worldwide (Chafai et al., 2023; Enoma et al., 2022). Literature on forensic DNA phenotyping shows the advances of random forests, multilayer perceptrons, and CNN-based models over earlier approaches, which used regression to predict traits of pigmentation in phenotypes. Epigenetic age estimation studies reported on the accuracy of techniques using the elastic net regression and deep neural networks over methylation sites to predict biological age. The literature does identify the concerns of genetic privacy, potential for algorithmic bias, and underrepresentation of global populations in training data sets (Zhao, 2025). The literature collectively hints at the fact that AI does offer advantages and that, with careful testing and ethical containment, it can be used in forensics.

2. FUNDAMENTALS OF GENETIC MARKERS IN FORENSIC BIOLOGY

The building blocks of forensic biology are the genetic markers, as they are the basis of biological individualisation, biological relationships and the analysis of biological material in a crime scene. These markers include STRs, SNPs, mitochondrial DNA, and even epigenetic signatures; they are chosen for their variability, stability, and patterns of inheritance, and applicability to forensic casework. What enables forensic genetics to shine is the robustness and reproducibility of the information that can be gleaned from biological remains, even when samples are old, limited, or mixed. It is the biological, structural and forensic applicability of each marker that informs the construction of dependable and trustworthy AI models for interpretation. What follows is a comprehensive description of STRs, SNPs, mitochondrial DNA markers, and the epigenetic markers that are used to estimate age or tissue for the markers to be used in investigations and for their relevance in the courts.

2.1 Short Tandem Repeats (STRs)

Short Tandem Repeats (STRs) represent some of the most polymorphic regions within the genome. STRs function at the level of repeating units of 2 to 6 base pairs. STRs are especially useful for individual-level differentiation, as these loci repeat at different rates among the population. One of the key advantages of STRs for kinship determination is that these loci are inherited in a Mendelian fashion, where one parent contributes one allele to the child. As a result of the above, STRs are the most preferred type of markers in forensic DNA profiling systems, such as CODIS

(USA), NDDB (India), and ENFSI frameworks (Europe). STRs give robust profiles even with low-quantity DNA samples and also have high discriminatory power as a result of the PCR amplification of the regions of interest (Forouzesh et al., 2022).

The forensic analysis of STRs is done using capillary electrophoresis (CE), where the computer functions using the electropherogram. However, there are various things or artefacts that cause the signal to peak, such as stutter peak artefacts, drop-in and drop-out events of incomplete electropherogram traces, and background noise (Kokotas et al., 2025; Pedroza Matute & Iyavoo, 2025). Other issues that negatively impact the quality of profiles may arise due to degradation, where longer loci are not amplified. These challenges concerning the interpretation of STR profiles are the ones that have prompted the implementation of the automation of the mean interplay of AI to increase the efficiency of profile interpretation by using machine learning to identify which alleles are real, and to classify different patterns of peaks that are observed. Forensics databases are built on STRs because of the abundance of STR markers and the high acceptance of these markers within the judicial system, especially in the United States.

2.2 Single Nucleotide Polymorphisms (SNPs)

Single Nucleotide Polymorphisms (SNPs) are variations at specific locations in the genome caused by a single base-pair substitution. Individually, SNPs are less polymorphic than STRs, but they are much more abundant, occurring every 300 base pairs. This makes SNPs worth their weight in gold as genetic markers. SNPs are also very convenient as they are very stable in a variety of environmental stresses, require very little DNA, and yield good results in ancient and degraded samples. (Debsharma et al., 2022). These characteristics make SNPs very adaptable for use in cases dealing with burnt and skeletal remains and cases with environmental exposure.

Forensic SNP panels are used in ancestry estimation, biogeographical inference, kinship testing, and forensic DNA phenotyping (which consists of predicting traits such as eye and hair colour). SNPs also contribute significantly to Next Generation Sequencing (NGS) based forensics, where workflows aim at analysing millions of loci at a time. (Roy et al., 2024). Analyses of SNPs are simplified due to their biallelic nature, but this also diminishes their discriminatory power, especially with smaller panels.

The tacit SNP pattern recognition enabled by the deep learning algorithms used in the AI-based SNP pattern recognition systems offers the systems a very robust performance. These systems are then able to analyse the potentially phenotype, classify ancestry, and support probabilistic matching. (de la Puente et al., 2021). As SNP-based datasets grow and computational power improves, the use of SNPs in forensics will be more widespread.

2.3 Mitochondrial DNA Markers

Given that mitochondrial DNA (mtDNA) can be found in significant quantities in a cell and is maternally inherited, it is prime for study in situations involving nuclear DNA that is absent or in short supply. When biological remains are available in the form of hair shaft samples, bones, and teeth or when the samples are badly degraded, there are often only small quantities of nuclear DNA obtainable, but there is sufficient mtDNA to allow for identification. (Cavalcanti et al., 2024). The mtDNA control region (hypervariable regions I, II, and III) is often sequenced in forensic cases because it has a higher mutation rate than many other loci. (Amer, 2025).

There is forensic mtDNA data in cases of missing persons, the identification of deceased persons, and the reconstruction of maternal lineages (Dahal et al., 2023). The mtDNA, however, has insufficient resolution to distinguish cases like a nuclear STR does, because all persons with a maternal line also share the same maternal haplotype. Therefore, often mtDNA is used in forensic cases to exclude or differentiate a maternal lineage and not to provide a definitive identification.

The integration of AI in forensics assists mtDNA analysis in haplogroup classification, mitochondrial sequence alignment, and the identification of rare haplotype variants. AI is used in forensic genomics and mtDNA analysis to improve the accuracy of extensive data analysis and to trace ancestry and population affinities better. (Sessa et al., 2024). More than the other forensic DNAs, mtDNA is far more stable and abundant, which adds to its wide range of forensic applicability.

2.4 Epigenetic Markers and Age/Tissue Estimation

To skeletal and dental estimations, markers, particularly patterns of methylation recordable as dynamic changes and are not directly encoded in the genetic sequence. Also, these markers mutate in the life of an individual, making them invaluable in forensic investigations. DNA methylation in the specific patterns of certain CpGs correlates significantly with the actual age of the individual, thus providing forensic experts the chance to estimate age, even in minimal age ranges, in the absence of an aged biological reference(van Dongen et al., 2014). Additionally, the methylation markers serve as markers to define the specific tissue of biological fluids and are critical to reconstructing and designing crime scenes in dozens of investigable ways. (Lee et al., 2022).

Epigenetic data, being high-dimensional and of substantial complexity, is analysed and classified by AI in many of today's investigative worlds. In the case of forensic science, AI improves age estimations and tissue class by utilising elastic net regression and neural networks to build methylation data. In predictive AI constructed models of high accuracy, many well-defined and complex predictive CpGs are discovered.

The amalgamation of epigenetic markers with forward and sophisticated analytics provides the potential for epigenetics as an investigative tool.

2.5 Significance of Genetic Markers in Criminal Investigations

Markers allow forensic biology to assist in multiple lines of investigation. STRs allow for individual identification, paternity testing, and matching to databases. With SNPs, forensics can also predict ancestry, estimate phenotypes, and do more advanced kinship analyses. When there is nuclear DNA loss or degradation, mitochondrial DNA can still yield results. When data from a suspect is missing, epigenetic markers can channel information regarding age, tissue type, and lifestyle.

If the markers relied solely on their biological properties, their significance would be reduced. (Kanthaswamy, 2024). Their significance is also found in the completeness of their interactions. In complex cases, like mass disasters, sexual assault mixtures, missing person identifications, or cold cases, multiple markers of different types are often combined for extensive analyses. AI-assisted interpretation multiplies the worth of these markers by providing automation, reducing human error, analysing complex datasets, and producing probabilistic results that reinforce the findings.(Mizrachi et al., 2025; Sessa et al., 2024). Together, genetic markers and AI amplify the worth of forensic genetics and modern forensics, trialling the system for more accuracy, reliability, and usefulness in investigations.

Figure 1. Genetic markers in forensic biology include STRs, mitochondrial DNA, SNPs, and epigenetic markers.

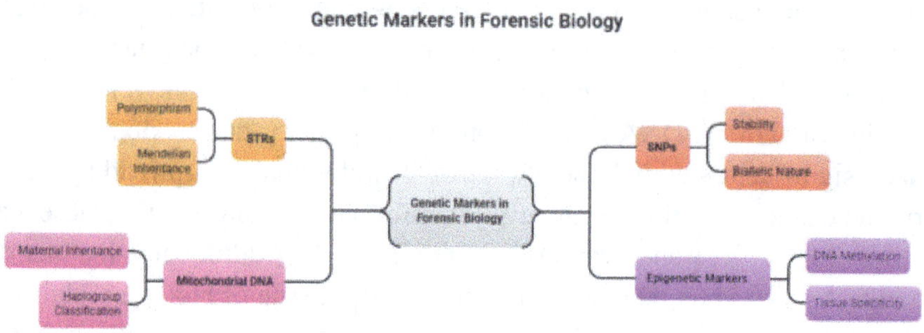

3. ARTIFICIAL INTELLIGENCE TECHNIQUES IN FORENSIC GENOMICS

Enhancements by Artificial Intelligence (AI) in forensic genomics allow for the analysis of massive genomic datasets with unprecedented accuracy, speed, and interpretative depth. This field of forensic genetics has, for a long time, needed the manual interpretation of electropherograms, allele calling, and pattern recognition, which are all heuristic and easily suffer from human bias and limited analytical capacity. AI reduces this reliance on heuristics by learning and applying computational models from which patterns are detected, variants are classified, and outputs are produced in a probabilistic manner founded on mathematical logic. Using machine learning, deep learning, and advanced neural architectures, AI improves forensic workflows from STR interpretation, SNP analysis, and mixture deconvolution to phenotyping and multi-omics integration. This section assesses the most prominent AI techniques in forensic genomics, concentrating on machine learning, neural networks for sequence interpretation and automated variant calling, and the preprocessing required to improve AI performance.

3.1 Overview of Machine Learning and Deep Learning Models

An absence of supervision in programming gives an entire category of algorithms known as machine learning (ML) the ability to unravel data and generate predictions and classifications autonomously. In forensic genomics, supervised learning SVMs, random forests, gradient boosting algorithms, and logistic regression are models utilised to classify SNP-based ancestry predictions, phenotype predictions, and even contributor status determinations in mixed samples. (Chafai et al., 2023; Enoma et al., 2022). To detect genetic data relatedness and groupings, unsupervised models, such as k-means, hierarchical clustering, or PCA, use these algorithms to reveal the data's structure. These techniques offer forensic analysts the ability to identify and analyse the genetic data's latent biological variability and structure.

Deep Learning, a type of machine learning, is comprised of architectures with many connected layers that can map very complicated, non-linear complications. In forensic genomics, deep learning excels because the data in genomics (electropherogram signals, sequencing reads, methylation profiles) is very complex and shallow learning would result in a poor model. Convolutional Neural Networks (CNNs) look at short-read sequencing data and can find motifs and classify patterns of STR, and classify them based on raw signals. Sequential relationships in DNA reads and time-series electropherogram data can be incorporated through Recurrent Neural Networks (RNNs) and Long Short-Term Memory (LSTM) networks. Systems based on transformers, formers in Natural Language Processing (NLP) and sequencing are now being used because they can model very long ranges and relationships and dependencies better than anything else (D. Wu et al., 2021). These systems are statistically far beyond anything else and are helpful in forensic systems that are very imprecise and inaccurate that requiring distinguished accuracy.

3.2 Neural Networks for Sequence Interpretation

Neural Networks are a key component of instrumenting genomic sequences and electropherogram signals in forensic genomics. CNNs are particularly suited for the analysis of next-generation sequencing (NGS) data. NGS data involves classifying, aligning, and interpreting millions of short reads. CNNs can recognise sequence patterns within a specified window, allowing for SNP calls, STR repeat structure identification, variant type classification, and differentiating biological signals from non-SNP artefacts. Analysis of degraded, low-depth, or contaminated sequences is enhanced through advanced CNN techniques, overcoming models that fail at low data quality. (Song et al., 2021).

Modelling dependencies across longer DNA sequences is done by Recurrent Neural Networks (RNNs). RNNs utilise Long Short-Term Memory (LSTM) or Gated Recurrent Units (GRUs) and are most effective in interpreting peaks in an electropherogram generated during STR analysis. Peak variations stutter allele identification, and tracking across multiple loci to decrease allele calling ambiguity are functions that RNNs perform. RNNs keep the order of the sequence of nucleotides, unlike the transformers, where the whole sequence can be processed in parallel at once (Okut, 2021). Their efficient sequence processing and learning of complex relationships across nucleotides make RNNs a suitable choice for variant calling, sequence alignment, or forensic DNA phenotyping.

Neural networks are also being researched for forensic DNA phenotyping, the process of predicting facial features from DNA. Specific genetic markers can be converted to probabilistic features of the face. CNNs are able to classify datasets from genotype to phenotype and can predict eye colour, hair colour, skin pigmentation, and specific features of the face. Despite being in its infancy, deep learning models can provide accuracy and investigative worth for the case, especially when the common STR-based identification is not an option. (Zheng et al., 2021).

3.3 AI-Based Variant Calling and Locus Detection

Variant calling is one of the central concepts in forensic genomics, and it is the process of pinpointing SNPs, indels, STR repeat units, and various structural variants. Classic variant calling approaches and tools use alignment-based methods, which tend to run into trouble when working with sequencing artefacts, low-template

DNA, and mixture samples. This ease of use of the AI is also being seen in the process of forensic genomics, when working with low DNA samples. (Abdelwahab & Torkamaneh, 2025).

One of the early AI variant callers is DeepVariant, which utilises CNNs to analyse a set of sequence reads and slice to genome positions, making SNP and indel calls with exceptional accuracy. This has been adapted for forensics, incorporating various machine learning methods, and it has been instrumental in improving locus determination in STR and SNP panels within degraded samples. AI-based allele callers are able to analyse peak heights, eliminate noise, and identify patterns associated with forensic electropherograms that display stutter, drop-out, and drop-in phenomena. With STR analysis, what traditional means of software analysis fail to do is discern loci, interpret overlapping peaks, and perform microvariant detection, something AI models do efficiently. (Kotlarz et al., 2024). Through pattern analysis based on depth of read, frequency, and allele distribution, AI-driven variant calling immensely contributes to mixture interpretation by minimizing the detection of complex sample microvariants.

Other sophisticated models leverage AI and probabilistic genotyping techniques to analyse mixture composition effectively. Models such as STRmix, EuroForMix, and LiRA utilise Bayesian networks, MCMC sampling, and AI-based likelihood ratio estimation to provide proper evidence to assess Mixture Probabilities to produce courtroom-grade objectivity. (Abdelwahab et al., 2023).

3.4 Information Synthesis and Multi-Omics Integration

For AI systems in forensic genomics to work, strong data preprocessing must take place. AI models must be able to filter noise, sequencing errors, and other issues commonly found in raw DNA data, like PCR artefacts, environmental contamination, as well as data corrections, which is what preprocessing achieves. Quality filtering, adapter trimming, imputation, and correction, among other techniques, allow models to be trained on data that authentically reflects biological phenomena, instead of irrelevant noise. The conversion of genomic data into machine-readable formats relies on feature engineering. Examples of features in electropherograms are peak heights, stutter signal ratios, signal-to-noise, and locus thresholding. In sequencing, features are read depth, base, and mapping qualities, the qualities of k-mers, and the k-mer frequency. The features are selected to increase the models' focus on the most relevant data for forensics (Ballard et al., 2024).

The integration of multiple omics (genomics, epigenomics, transcriptomics, proteomics, and methylation) datasets is among the most significant AI-enabled advances in forensic research. Deep learning models are particularly adept and integrating these diverse datasets to create summaries or derive new data. (Santiago-

Rodriguez & Hollister, 2021). For instance, age prediction models are strengthened when SNP data are methylation data are integrated. Body fluid identification models improve with the integration of transcriptomic data, especially when tissue-specific data is available.(Ur Rehman et al., 2026). Genomic and phenotypic data integration improves ancestry prediction models. Such advances enable forensic experts to construct bioprofiles with unprecedented levels of accuracy and interpretability.

Figure 2. AI-driven forensic genomics workflow showing genomic data processing through preprocessing, model application, sequence interpretation, and variant calling to generate forensic insights.

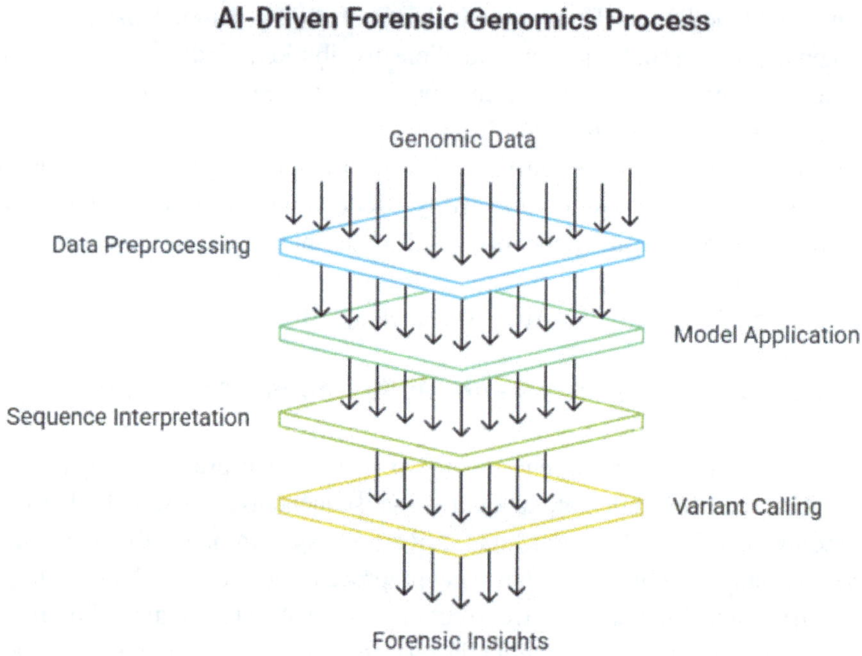

4. AI-ENHANCED STR, SNP, AND EPIGENETIC MARKER INTERPRETATION

AI technology has been integrated into forensic genetics, and for the interpretation of STRs, SNPs, and epigenetic signatures, AI is indispensable. As sample interpretation electronically amplifies and streamlines the workflow, human interpretation becomes the bottleneck. As DNA forensic evidence is often low template, degraded, heavily mixed, and/or complex, human interpretation becomes the most problematic step. For example, phenotype and ancestry resolution can be complex and require phenotypic and epigenetic datasets to be resolved and extracted (Liang et al., 2024; Simpson & Chandra, 2021). The integration of AI into forensic genetics enables the automation of allele calling, variant resolution, complex mixture resolution, and the extraction of specific (phenotypic and ancestry) datasets. The integration of AI into forensic genetics enhances accuracy, reproducibility, and depth of interpretation(Brasil et al., 2021; Pisarek et al., 2021). This section discusses and assesses the AI implementation within forensic genetics for allele calling, mixture interpretation, degraded DNA analysis, and ancestry, as well as phenotypic prediction.

4.1 Algorithmic Approaches for Allele Calling

During DNA interpretation, and more specifically forensic DNA interpretation, allele calling has undoubtedly been, and may still be, the most critical task. In the context of STR analysis, alleles are usually displayed as peaks that are visualised in an electrophoresis capillary electrophoresis-based electropherogram. Each peak must be reviewed to ascertain whether an artefact is present or a true allele is displayed. Artefact features can include stutter, drop-in, spike noise, and/or incomplete amplifications. Interpretation of the data presents a particular level of technical expertise, and because the forensic context is often of low template DNA, the interpretation becomes increasingly subjective.(Crysup et al., 2023). This is true especially for peaks of low amplitude algorithms automate subjectivity in calling biological peaks in electrophoresis data. Machine learning classifiers have been trained to analyse various forms of data, such as peak height ratios, stutter patterns, locus-specific thresholds, and signal-to-noise ratios. They then decide whether an electrophoretic peak should be called an allele. Neural networks, especially convolutional networks, escalate analyses of electrophoresis data to the next level and automate detection to discern the alleles of interest accurately.

AI has also advanced allele calling from the assignment of depth of read, base quality, alignment score, and sequence variability in SNP and NGS-based STR analyses, especially in challenging data. Tools such as Deep Variant have surpassed conventional algorithms in variant calling and other sequence analyses(Valecha & Posada, 2022). The reduction of false positives and false negatives, especially with low-template DNA or degraded samples, is crucial. The automation involved in allele calling makes these systems indispensable in forensic science, especially in improving the efficiency and reproducibility of casework.

4.2 Interpretation of Mixtures and Probabilistic Genotyping

A significant challenge within forensic genetics is the interpretation of mixed samples of DNA from different biological sources. Conventional approaches encounter various shortcomings and increasing levels of error when dealing with more than three contributors and rely solely on the heuristic manual deconvolution of the mixture. The introduction of AI(Bayesian networks and machine learning) into the area of estimating genotype(s) of the contributors and determining the likelihood ratio has been revolutionary within probabilistic genotyping systems.(Ryu et al., 2024).

STRmix, EuroForMix, and LiRA are enhanced AI models of probabilistic genotyping that focus on the statistical treatment of estimating the probable genotypes for each contributor, combining the methods of Bayesian, Markov Chain Monte Carlo methods, and stochastic processes. These models incorporate features such as peak height, stutter, dropout, and proportions of the alleles and generate likelihood ratios for the support of more than one alternative hypothesis that may be required in the court setting. AI support in these models comes from the improvement of the estimation of the mixture interpretation in varying conditions.

Deep learning systems have also become available for the analysis of mixtures in NGS-based workflows. These systems analyse the distribution of read depth, patterns of allele frequency, and structure of haplotypes to isolate and identify the otherwise elusive minor contributors. Neural networks can identify and differentiate overlapping peaks, and disentangle and estimate the ratios of contributors in complex cases involving three or more profiles. This development improves the quality and reliability of forensic results while minimising the degree of interpretative ambiguity, especially in cases of sexual violence, touch DNA, and multi-suspect crime scenes.

4.3 Analysis of Degraded and Low-Template DNA

The DNA left behind in a crime scene is often degraded owing to exposure to the elements, heat, moisture, microbes, or chemicals. The analysis of Low Template DNA, where the DNA present is in amounts below the analytical threshold, comes with unique challenges such as allele dropout, stochastic, and imbalanced amplification. These situations traditionally result in the production of parts and inconsistent DNA profiles.

AI enhances the work of analysts, overcoming degradation by predicting complex patterns of degradation and filling in gaps. Algorithms analyse patterns in DNA electropherograms to estimate dropout rates, adjust peak height discrepancies, and reconstitute fractured DNA profiles.(Tan et al., 2025). Human analysts often miss the subtle and complex variations of signal intensity of fragment sizes constituting the DNA, allowing deep learning models to restore alleles to weakly expressed profiles.

AI-optimised analysis facilitates the identification of DNA loci and recommends specific techniques for successful amplification of these regions.

With SNPs and sequencing, AI detects and corrects sequencing noise and replication errors, identifies genuine sequence variants or degradation-derived artefacts, and imputes absent genotypes. Degraded sequences, especially those with long gaps, can be restored efficiently with transformer-based models.

AI also interprets mtDNA more accurately from low-quality or badly degraded samples, detecting heteroplasmy and fragment alignment for sequence reconstitution. This accurate analysis provides usable genetic information from samples that other procedures, like standard STR profiling, would fail to analyze(Dong et al., 2021). This forensic analysis is particularly advantageous for the resolution of cold cases, identification of deceased individuals from disasters, analysis of commingled skeletal remains, and samples from the environment, which may have degraded or fragmented the genetic evidence.

4.4 AI for Ancestry and Trait Prediction

Forensic DNA phenotyping and ancestry inference functions via analyses in which the DNA of the suspect is available, but the suspect is unknown, become instrumental in criminal investigations. Predictive SNP markers assist in making inferences about one's ancestry and biogeographical origin, skin (apart from pigmentation, other traits like facial morphology are included) and certain traits linked to genes. AI systems improve the quality of outcomes for biogeographical ancestry and certain traits beyond others by predicting with the use of advanced neural systems for SNP data with high dimensionality.

Fingerprint data obtained from psychological, ancestry and socio-economic status (heritage) of individuals gives the neural systems the ability to predict the trait with greater accuracy. Some predictions are made to assist in drawing facial images for suspects in the absence of eyewitness accounts. Trait predictions are made by the use of deep neural networks. Other predictions can be made using artificial intelligence in ancestry predictions. Classifications in ancestry predicting can be continental and sub-continental(K. Wang et al., 2023). AI in ancestry estimation is used to assist other systems in predicting ancestry to exclude or focus leads in investigations with DNA forensics.

The predictive abilities of AI also involve analysis of epigenetic markers. Using deep learning frameworks, investigators can supplement their analyses of reconstructive investigations by estimating age, inferring traits, identifying tissues, and determining exposure-related characteristics, all based on patterns of methylation.

5. PREDICTIVE FORENSIC DNA PHENOTYPING USING AI

Forensic DNA phenotyping (FDP) is one of the most groundbreaking advancements in the field of forensic science, empowering investigators to predict externally visible characteristics (EVCs) of an individual, their ancestry, and even the morphology of their face from the DNA alone. Unlike traditional forensic genetics, which solely aims to identify individuals through the STR profiling of their DNA, FDP also serves the purpose of forensic investigative intelligence in the absence of a reference profile or a matching entry in the forensic DNA database.(Dash et al., 2023). The field of FDP has been further enhanced by the implementation of artificial intelligence, which can learn and predict genotype-phenotype relationships. The ability of machines to understand and predict based on large datasets, coupled with deep learning, can enrich analytical datasets composed of thousands of SNPs, epigenetic marks, and various '-omics' layers to support intelligent decision-making during criminal investigations, aid in the identification of missing persons, and assist with the resolution of cold cases. Specific AI-enabled predictions of physical characteristics, inference of biogeographical ancestry, and facial approximation via generative AI, in addition to the incorporation of all the aforementioned into intelligence workflows in a crime scene context, are discussed in this sub-section.

5.1 Prediction Models for Physical Traits

The different types of forensic DNA phenotyping focus on predicting DNA-associated externally visible traits: eye colour, hair colour, skin colouration, and the shape of facial features. In the past, simple traits were predicted using statistical techniques such as logistic regression and Bayesian inference, but these traditional methods face challenges in predicting human appearance due to their inability to account for the complex multi-gene interactions. In contrast, predicting complex traits is an area where AI, and in particular, machine learning (ML) classifiers and deep neural networks (DNNs), excel since these techniques can account for non-linear interactions from thousands of genetic markers. For the prediction of traits tied to pigmentations, such as skin, eye and hair colour, the random forests and gradient boosting machine learning techniques have gained popularity. Here, the relative contribution of individual SNPs in the genes HERC2, OCA2, MC1R, SLC45A2, and TYR is assessed, achieving high prediction accuracy for eye and hair colour. Other AI trait predictors, such as HIrisPlex-S and VISAGE panels, have combined phenotype-informative SNPs (PISNPs) with ML to offer probabilistic trait predictions(Terrado-Ortuño & May, 2025).

Deep learning improves the functionality of FDP systems even further by identifying genotype–phenotype patterns that would otherwise go undetected by more rudimentary machine learning systems. CNNs capture the genomic windows that are associated with each EVC, while MLPs synthesise the genetic, epigenetic, and environmental components. This modelling framework outputs continuous scores of probabilities rather than the more simplistic binary outcomes, which allows the researcher to obtain actual confidence interval estimates to facilitate their decision.

There are also AI models that make predictions on traits that are less examined, including hair texture, lightening in the skin, facial width, nose height, and even body mass proclivities. The phenotypic datasets are more diversified and now include multi-ethnic samples, which is the reason the models' predictions are getting more accurate over time. This is the reason FDP is now a trustworthy and evidence-based system for AI-assisted investigative work.

5.2. Ancestry and Biogeographical Inference

In the forensic context, ancestry inference is an integral process for narrowing suspect lists and facilitating the identification of unknown remains, and for the reconstruction of population histories. Biogeographical ancestry (BGA) can be inferred on the basis of SNP markers that display population-specific allele frequency patterns. While methods of population structure analysis, such as principal component analysis (PCA) and STRUCTURE-based clustering, can provide valuable insights about population structure, they lack the granularity in prediction that is often needed for a specific forensic case analysis.

AI enhances ancestry estimation by examining large-scale SNP datasets while considering and modelling complex, multi-dimensional genetic relationships. There exists a subset of machine learning, namely Support Vector Machine classification, k-nearest neighbour classification, and random forests, which have been integrated into genetic ancestry analysis and efficiently place individuals into pre-defined continental or sub-continental clusters(Pengyue et al., 2021). These classification models train on comprehensive ancestry reference datasets containing thousands of pre-identified and labelled individuals and are able to discern complex population sub-structures that standard statistical techniques often struggle to detect.

With Deep Learning, improvements to these techniques are made possible. Neural networks are able to capture and predict longer-range genomic relationships, facilitating the multi-dimensional representation of signals attributable to ancestry factors, including complex admixture, differential migration, or population bottleneck.(Alotaibi et al., 2022). Genomic transformers, a type of deep learning architecture designed initially for natural language processing, have recently been adapted to SNP-Seq data, with a focus on ancestry inference. Typically, these models outperform standard analytical methods in genetically heterogeneous populations containing weaker ancestry signals.

Ancestry inference based on AI proves most useful in forensic investigations with unidentified deceased individuals, victims of human trafficking, in disaster victim identification, or in cold cases where there are no reference relatives. Ancestry inference, alongside phenotype prediction, provides enhanced biological profiling for forensic investigators to direct their resources efficiently.

5.3 Generative AI for Facial Approximation

Facial approximation using Generative Intelligence is the most sophisticated form of DNA phenotype forensic analysis. While phenotype prediction reveals traits of the individual, generative AI models attempt to reconstruct complete facial approximations by using the individual's epigenetic and genomic data. DNA to face modelling is still in its infancy, but it is rapidly being improved by advancements in deep learning.

Facial approximations utilise generative adversarial networks (GANs), variational autoencoders (VAEs), and diffusion models. These architectures, using large genotype-image datasets, train on genotype data to learn the underlying relationships with facial morphology and then predict the constellation of traits and ancestry(Pośpiech et al., 2022). These models are able to output synthetic faces that exhibit the expected combination of traits and facial ancestry.

Although traditional facials are based on eyewitness accounts of the crime, this new generative AI is able to provide an approximation of the face using data and biology, which is particularly helpful if there are no eyewitnesses or if the victim has decomposed and is unrecognisable. GAN models are able to use multiple genomic inputs to create facial approximations that are realistic and useful in directing investigations, for example, SNPs for pigmentation and craniofacial structure, and epigenetic markers for age(Soumare et al., 2021).

Surely, we need to weigh the ethical implications surrounding the probabilistic nature of DNA-based facial prediction, especially the possibility of perpetuating biases from insufficiently diverse training datasets. On the other hand, there is promise in ongoing studies in the field; with the advancement of more representative datasets and transparent algorithms, the possibility of generative FDP being a valuable tool in investigations is becoming a reality.

5.4 Synthesis with Crime Scene Intelligence

The intelligence provided by AI-driven forensic DNA phenotyping complements and improves the efficiency of investigative workflows and aids in optimal resource allocation. When there is no match in national databases from conventional STR profiling, phenotyping, and ancestry predictions narrow down the suspect pool and help in lead prioritisation. Such predictions can be supplemented with information from the crime scene, including the location, victim, behavioural profile, and other demographics, to develop investigative hypotheses.

AI-driven FDP outputs are being utilised by crime analysts in geographic profiling, suspect prioritisation, and linkage analysis. For instance, phenotypic profiles can be matched with available CCTV footage, and ancestry predictions can be used to include or exclude particular populations from a given region; age estimations from methylation markers can eliminate people whose age is beyond or below the predicted range. (S. Ahmed et al., 1 C.E.).

FDP-generated profiles can be compared to demographic data from missing persons databases to assist investigations involving missing persons. Investigators can reexamine cold cases by using phenotype inference from DNA samples that have been stored for decades to pursue leads that were previously infeasible.(Dhanush & Thenmozhi, 2022).

When applicable within a real-time crime analysis system, AI-driven FDP becomes a force multiplierenhancing investigative functions, levelling biases stemming from eyewitness accounts, and offering scientifically based insights that improve law enforcement operations in terms of accuracy and efficiency.

6. MULTI-OMICS INTEGRATION AND FORENSIC DATA OPTIMISATION

The intricate nature of forensic investigations has increased the popularity of refined investigative techniques that incorporate greater amounts of biological information. Forensic genetics used to involve the analysis of STRs and mtDNA. Still, the refrigeration of biological samples has evolved to encompass more modern sets of biological information, including the genomic, epigenomic, transcriptomic, proteomic, and metabolomic layers. Such a multi-omics approach undeniably provides more information about an individual's biological profile, thus facilitating more precise identification and age estimation, determination of the tissue source, reconstruction of phenotype, and ancestry inference(Ur Rehman et al., 2026). AI is the master of amalgamating heterogeneous biological data sets and providing the means of deciphering multifaceted biological signals into forensic intelligence. This section is dedicated to multi-omics fusion, the enhancement of unstructured forensic data, AI-empowered quality control and workflow enhancement in forensic labs.

6.1 Fusion of Genomic, Epigenomic, and Methylation Layers

The fusion of multiple omics is essential in forensic genomics, especially given the challenges presented by single-omics or single-marker approaches. Genomic markers, including STRs and SNPs, provide stable and identity-related information, but do not capture the dynamic biological processes associated with ageing, the expression of different tissues, the environment, or other physiological factors. Epigenomic markers, particularly DNA methylation, complement the genomic markers by layering on additional important information in forensic casework. (Monika & Kumar, 2024).

Predictive modelling of fused genomic and epigenomic layers is enhanced with artificial intelligence (AI). Deep learning approaches such as autoencoders, multimodal CNNs, transformer-based fusion models, and attention networks are able to integrate disparate datasets and detect patterns often invisible to conventional means.(Y. Wang et al., 2024). Examples include:

- Models estimating age have integrated SNP data and methylation signatures to produce predictions of chronological age with remarkable accuracy.
- Models identifying body fluids integrate methylation sites with data on gene expression and tissue-specific SNPs.
- Phenotype prediction systems integrate SNPs of particular interest in pigmentation and facial morphology with epigenetic markers.

The fused methylation and genomic datasets provide a greater capacity to make predictions in forensic casework, especially when the sample is of inferior quality. Epigenetic features, along with the missing genomic data, allow highly confident predictions to be made even in the sample quality and information are poor.

The incorporation of Artificial Intelligence in epigenetic markers strata employed in N-in-1 computation for estimating epigenetic age, tissue versus age markers, heteroplasmy, mitochondrial-nuclear cross genomics, and SNP ancestry-informative tools, among others, will enhance forensic intelligence crafting in forensic virology, and epitomise a triad of biological features multiplexing in forensic intelligence.

6.2 Enhancing Forensic Databases with an AI Capability

The existing forensic databases are unstructured, incomplete, and inconsistent. Such a drawback emanates from heterogeneous laboratory practices, discrepancies in sequencing technology, and population samples. AI is pivotal in correcting and enhancing these datasets for a credible forensic analysis.(E. Ahmed et al., 2022).

Machine learning tools are used to standardise allele format, remove artefacts, inconsistencies, and duplications; correct naming errors; and reduce naming collisions across datasets. In terms of textual case notes, laboratory reports and historical narratives, natural language processing (NLP) models extract relevant metadata for unstructured documents and reverse engineer them to a state that is amenable to computation.

The population reference databases are subjected to AI-empowered reorganisation to aid ancestry inference, and the general purpose of the databases is in the estimation of probabilities of STR frequencies. Clustering algorithms are used to address the problem of population stratification, reduce dimensions for population structure visualisation through techniques such as PCA, t-SNE and UMAP. AI enhances the value that databases offer by:

1. Filling in missing roles to help in the reconstruction of incomplete acts through genotype imputation.
2. Filtering unreliable records through noise suppression.
3. Recognising systematic bias of underrepresentation or overrepresentation of specific population demographics using bias detection.

This ensures the predictability of imputation is ethical. A higher level of accuracy in the database enhances the precision of estimating allele frequencies, predicting phenotypes, and inferring ancestry.(S. Wu et al., 2025).

Quality control (QC) considers the precision of every aspect of forensic biology in the analytic methodology, instrument performance, and interpretative stages. AI enhances QC by identifying errors that can be attributed to instrument malfunction and anomaly detection, and classifying such errors more efficiently than human inspectors.

In the context of STR analysis, machine learning's ability to evaluate peak morphology and stutter artefact detection, and to identify drop-in/drop-out events as well as flagged contamination, enhances QC. AI QC systems identify and record specific patterns in the electropherogram trace over time and in relation to instrument wear due to capillary exhaustion or reagent depletion.

For sequencing-based workflows, the errors that AI identifies include sequencing artefacts, alignment issues, low-complexity regions, contamination, and PCR bias. Base-calling errors, due to the incorrect identification of truncated reads, high radio noise, and fragments of the sequencing signal, can be discriminated by deep learning models.

The predictive maintenance capabilities of the AI-powered QC frameworks help mitigate the risk of instrument errors within the QC processes. In particular, AI-based QC can help avert temperature irregularity, capillary malfunction, or quality control metrics losses of the thermocycler and sequencing platforms, respectively, thereby avoiding run failures and optimising the allocation of time and resources. (Shah et al., n.d.).

The AI-powered QC frameworks help mitigate losses due to interpretation errors. Probabilistic modelling, for instance, addresses the issue of verification, confidence interval quantification, and transparency closure of forensic findings. Methodological transparency and confidence interval quantification are crucial for court-qualified forensic findings.

Figure 3. Multi-omics integration and AI-based forensic data optimization outlining data types, AI roles, benefits, and forensic application examples.

Multi-Omics Integration and Forensic Data Optimisation

Characteristic	Data Type	AI Role	Benefits	Examples
Genomic, Epigenomic, Methylation Fusion	Fused genomic and epigenomic layers	Predictive modelling, pattern detection	Precise identification, age estimation, tissue source determination	Age estimation models, body fluid identification models
AI-Enhanced Forensic Databases	Unstructured forensic data	Standardisation, metadata extraction, reorganisation	Credible forensic analysis, improved ancestry inference	Standardised allele format, corrected naming errors
AI-Powered Quality Control	Analytic methodology, instrument performance	Error identification, anomaly detection, predictive maintenance	Enhanced precision, reduced errors, optimised resource allocation	Peak morphology evaluation, stutter artefact detection

6.3. Optimising the Work Processes of Forensic Laboratories

The optimisation of work processes in forensic laboratories through AI enhances the efficiency, accuracy, and scalability of forensic laboratories. By streamlining the processes in workflows and automating repetitive tasks, AI liberates forensic professionals to concentrate on the more intricate and nuanced aspects of the analysis and interpretation(Marsico & Amigo, 2025).

The main areas of optimisation are:

1. Automated Evidence Processing

The automation of DNA extraction, quantification, amplification, and sequencing processes through robotics and AI enhances operational accuracy and reduces variability caused by human intervention.

2. Intelligent Sample Prioritisation

Machine learning can examine metadata to prioritise sample analysis on the basis of the evidentiary worth, degradation potential, or time-sensitivity of the case.

3. Decision Support Systems

AI provides assistance to forensic analysts by determining the optimal amplification settings, establishing parameters for mixture interpretation, and proposing genotypes through probability-driven approaches.

4. Real-Time Monitoring

Artificial Intelligence integrated with Laboratory Information Management Systems (LIMS) monitors the status of samples, identifies workflow impediments, monitors reagents, and reviews the performance of instruments.

5. Predictive Analytics for Case Management

AI estimates expected turnaround times, determines the required resources, and anticipates workload distribution by type (e.g., sexual assault kits, homicide evidence).

6. Complete Chain-of-Custody Optimisation

AI systems supported by Blockchain technology provide secure, immutable, and traceable tracking of forensic evidence, enhancing transparency and defensibility in legal contexts. These optimisations create laboratories where processes are faster, more efficient, and more accurate, improvements that are necessary and critical with the increasing volume of global forensic work.

Figure 4. AI-enabled optimization of forensic laboratory workflows including chain-of-custody tracking, automated evidence processing, predictive analytics, sample prioritization, real-time monitoring, and decision support systems.

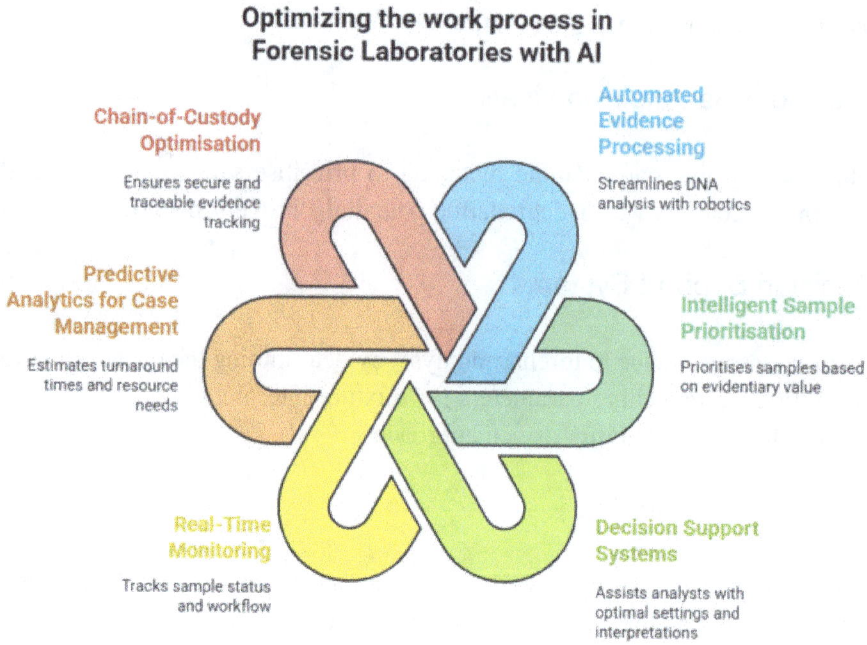

7. ETHICAL, LEGAL, AND SOCIETAL ISSUES

The increasing use of artificial intelligence in forensic genomics poses major ethical, legal, and societal issues, even as it enhances the benefits of forensic genomics in investigation. With AI-powered tools adjudicating the fine points of decision-making in investigations and consequently affecting the balance of justice, civil liberties, and the trust of the community, the ethical implications grow. The use of genetic markers, and more particularly, their genetic nature, needs special care. The ethical challenges of AI in genomics, which touch on data privacy, informed consent, algorithm bias, data access and control, and the absence of regulations, need to be considered. The focus of this section will be to provide the ethical and societal implications of the use of AI in forensic genomics under the specific domains of: genetic privacy, algorithmic discrimination and the need for transparency, the growing databases of DNA, and the standards of admissibility in the criminal justice system.

7.1 Privacy of Data and Access to the Data

Genomic data contains particularly sensitive information. It concerns one individual and also their entire family and lineage, and genetic characteristics. When AI is applied to genomics, it is even more sensitive as it amasses and indefinitely holds large volumes of genetic data, which can then be used and reused in unexpected ways. This is central to the ethical challenges involved in keeping the genetic data private(Dhillon, n.d.).

One of the most significant concerns associated with the misuse of genetic information is the storage of STR profiles in forensic labs. As AI technology improves, the risk of overinterpretation increases with the inclusion of SNPs, epigenetic markers, forensic DNA phenotyping, and other analyses. AI models could predict unrelated sensitive information such as health status, behavioural traits, and even potential family relations. Such predictions result in violations of proportionality and necessity.

Genetic security is further complicated by technology. AI-enabled genomic databases increase the risk of hacking, data breaches, and other unauthorised access. The fact that DNA is irreplaceable, unlike passwords, means that breaches could expose individuals to irreversible and lifelong risks. Protecting the genomic data within the databases is going to require secure encryption, strict access control, and anonymisation(Kuiken, 2023; Marsico & Amigo, 2025).

Another huge concern is the lack of consent. While the laws provide some control over the DNA collection of suspects and convicted offenders, the genetic information of civilians, family members of offenders, can become indirectly involved through the process of familial matching. The use of genealogy databases, most notably GEDmatch, in some high-profile cases raises the risk of 'surveillance by association' in which people become the target of a search without consent to the use of their DNA within the forensic databases.

Meeting these challenges requires the establishment of robust governance frameworks that provide for explicit consent, data retention and use for forensic purposes, and judicial accountability for the AI-assisted data analyses.

7.2 Bias. Fairness and Transparency of Algorithms

There can be no valid AI model without data. This is the case in forensic genomics, where the training data sets tend to have gaps, and specific groups of people may be excluded, giving rise to substantial biases in the estimation of allele frequencies, the prediction of phenotypes, and the prediction of lineages. This unfairness can lead to inequities in outcomes of investigations, particularly and disproportionately affecting marginalised groups.

For example, an AI model for predicting phenotypes may be unable to predict ancestry, predict accurately and predict some physical characteristics for people of African, Asian, and Indigenous descent aberrantly, if the training data set consists predominantly of and is biased toward people of European ancestry. There is a similar concern when and for system in probabilistic genotyping. These inequities raise concerns about fairness and representation in the use of these technologies in forensics(Annan et al., 2025).

Any discussion on algorithms has to be on their transparency. Many AI-assisted forensic tools operate based on 'black boxes they make their prediction without much explanation of the process adopted. This lack of transparency in the judicial sphere leads to challenges to evidence, restricts cross-examination, and raises issues of due process. Forensic evidence must be understandable, reproducible, and validated. Courts are clear about this requirement.

Explainable artificial intelligence (XAI) provides an observation of a potential solution. XAI can indicate the SNPs that influence ancestry classification or which methylation sites influence age estimation. This can indicate the attributes understandable in the outcomes of the model when using AI prediction. This explains the outcomes of the model, which understands AI prediction(Angelov et al., 2021).

Fairness requires the diversity of the datasets on which the AI models are trained to ensure that the datasets are globally representative, integrate audits based on bias, and are accompanied by validation studies that undergo peer review. Ethical deployment of AI requires a sustained synoptic overview, population-balanced sampling, and preferably open-source development.

7.3 Risks Associated with Expanding Genetic Databases

First of all, the expandable databases turn to what is called "function creep" whereby the identification data of an individual is used to predict other social behavioural, health, and demographic attributes. This erodes public trust and ethical boundaries.

Secondly, the increasing application of familial DNA searching, especially with AI detecting distant kinship, subjects innocent families. This has tremendous implications for the autonomy and the civil liberties of citizens.

Third, the Data Retention Policy has opened the door for greater risks to be the overreach of unjust discrimination at the individual and community level. This is their over-policing of their communities and over-representation within the DNA forensic databases, creating a cycle of social control and marginalisation. This is Systemic Injustice and Equity(Wan et al., 2022).

Finally, it is the genealogical and consumer direct genetic databases to which other complications arise. An individual offers their DNA as a trivial pursuit; AI-assisted forensic searches raise copious amounts of new questions regarding consent, rights of third parties, and ownership.

The implementation of explicit constraints on the range of databases to be used, proportionality of use, limitations on familial searches, and irreversible suppression of data, where demanded, can significantly reduce the risk involved. Public interest can be served through transparent governance and independent review.

7.4 Admissibility and Policy Frameworks.

The integration of AI technologies in forensics poses new challenges for legal systems to consider the admissibility of AI-generated evidence. There are no challenges within the legal sphere when it comes to established traditional forensic evidence, like STR profiles. In contrast, evidence or findings from AI technologies, including but not limited to the prediction of phenotypes, the determination of ancestry, probabilistic outputs from genotyping, and estimates of epigenetic age, require validated evidence to be legally admissible.

The Daubert Standard, as well as the Frye Rule, frameworks of admissibility, recognise that forensic methodologies need to be scientifically reliable, accepted, and peer-reviewed. Thus, AI systems will be required to have:

- statistically validated accuracy,
- cross-lab reproducibility,
- algorithmic transparency,
- documented assumptions and limitations,
- acceptable forensic error rates.

Policies must also govern the use of AI to:

- provide standards regarding diversity in training data,
- ensure that there are periodic algorithmic audits, s
- provide explainability requirements for testimony,
- outline the scope of use for phenotyping and ancestry inference,
- ensure that AI augments human expertise rather than replacing it.

Organisations that set international standards, such as the ISO, ENFSI, and SWGDAM, will be instrumental in outlining best practices for the use of AI in forensic genomics.

In the end, fair and responsible policies will ensure that AI enhances rather than diminishes the fairness and integrity of the criminal justice system.

8. CASE STUDIES

Case studies provide an opportunity to showcase the utility and benefits of being able to demonstrate the use of AI-infused genomic forensics in the real world. Within a theoretical realm, one can provide a degree of clarity, but within a real-life case, one can showcase the acceleration and precision required in contemporary problem-solving, an AI method of identification in solving cases previously assumed to be the domain of the unexplained and unsolvable a vis-à-vis old-world methods. The AI-infused diverse application of tools specifically to the domain of forensics, such as the forensics of violent crime, disaster victim identification (DVI), and long cold case files, speaks to the problem-solving benefits and the ethical considerations and procedural safeguards for the practice of applied forensics. The case studies provide testimony to the use of AI-infused forensic genomics in contemporary investigations.

8.1 AI-Assisted Identification in Violent Crime Investigations

Violent crimes entail the collection of biological evidence at a crime scene, be it blood, saliva,/skin cells or hair. It is also common to encounter violent crime investigations, specifically where biological evidence is present and blood, saliva, and/or skin cells or hair have been obtained at the crime scene, and are present in the weapons and/or victims. The use of TRADITIONAL Short Tandem Repeat (STR) Analysis is the gold standard within personal identification; however, in instances where there is sample degradation, the low template DNA is a major contributor to the unexplained complex channelling of the sample and old-world methods of identification of the sample. AI increases the productivity and efficiency of these investigations.

A representative case involves the investigation of the sexual assault of a victim whose evidence contained DNA from several sources, including the principal suspect. Traditional manual deconvolution resulted in incomplete profiles being constructed because of the peaks being overlapped, as well as the profiles being underdeveloped due to stochastic effects. Subsequently, investigators used an artificial intelligence-probabilistic genotyping software that manually assesses peak heights, determines the presence of stutter artefacts, and ascertains the percentage of profiles present from a total of the profiles to document the proportions of DNA profiles. This system produced a statistically accurate likelihood ratio that indicated the major contributor to the profiles with a high level of confidence. This, in turn, linked the suspect to the crime scene and enhanced the value of the evidence to be presented in court.

Another excellent and impactful instance includes the use of AI-motif prediction in forensic cases where DNA from an unknown suspect was recovered, but there was no database match. Investigators used forensic DNA phenotyping to predict the eye colour, hair colour, skin pigmentation, and geographic ancestry of the unknown suspect. These AI-derived predictions were compared to available CCTV and eyewitness accounts to help establish a shorter suspect list. Ultimately, the combination of AI-powered ancestry prediction and phenotypic prediction enabled investigators to determine and apprehend the suspect in a much shorter time.

AI-based sequence analysis tools have also proved their worth in homicide casework. There is one case dealing with burnt skeletal remains recovered from a crime scene. It was possible to analyse the remains with the aid of an AI SNP variant caller that was able to identify real variants from the degraded DNA. The traditional approaches could not amplify the nuclear DNA, but with the aid of AI analysis, enough SNP information was extracted to estimate the ancestry, and the victim was eventually identified through a genealogical match. The integration of AI into the investigative pipeline thus provided a breakthrough that was impossible to achieve through traditional approaches.

8.2 Application of AI in Disaster Victim Identification (DVI)

Among the many challenges that forensic science faces, disaster victim identification (DVI) is one of the most difficult. The remains of victims of natural disasters, mass fatalities, and terrorist attacks and the confusion of warfare, are, most times, unrecognisable, and the remains may be highly fragmented and/or severely degraded. In such cases, the usual STR profiling may not be possible because of degradation of the samples due to the environment, heat, and contamination. AI enhances the forensic efficiency and accuracy in these DVI operations radically by the analysis of degraded DNA, the integration of multiple genetic markers, and the automation of massive data analysis.

One specific disaster-related case of DVI (Disaster Victim Identification) deals with a case where the transport of a disaster resulted in the commingling and partial burning of the remains of over 200 people. Because of the complexity of identification because traditional STR (Short Tandem Repeats) analysis resulted in partial profiles for many of the sampled remains, partially-degraded NGS (Next Generation Sequencing) was combined with AI (Artificial Intelligence to SNP (Single Nucleotide Polymorphism) to analyse the output. Genotypes were reconstructed, and sequencing errors were fixed via machine learning, and mitochondrial DNA (mtDNA) was used to trace maternal lines. AI (Artificial Intelligence) was able to combine the nuclear and mitochondrial datasets to generate weak kin relationships with the relatives of the deceased, leading to the identification of sure victims (for victims whose STR profiles were intact, victims were able to match their profiles to others in the database; victims whose profiles were STR-blank, the victim's profiles were referred to as presumptive STR-blank profiles) whose relatives were the donors of the previously-cited nuclear datasets.

The kinship tools of AI (Artificial Intelligence) have relied on improving the family matching of victims in large-scale disasters, and these AI tools have delivered a family matching outcome in disasters of a large scale, obtaining closure for their loved ones in a timely manner. Kinship analysis in large-scale disasters has employed probabilistic models, and these models have relied on rapid simulation of thousands of kinship candidate pairs, providing a time-efficient outcome.

In another DVI case of a building collapse, AI (Artificial Intelligence) and machine learning combined to classify bone fragments, which in turn decreased the time required for manual sorting and increased the accuracy compared to traditional anthropological analysis. Machine learning was able to classify the bone fragments into sets, which were then ancestry-informative SNPs (Single Nucleotide Polymorphisms) used to match sets to missing people. AI (Artificial Intelligence) and Machine learning were used to analyse genomic similarity, and then commingled skeletal remains were grouped with skeletal remains of the exact genomic similarity. The ability to estimate age using AI technology has also been helpful in DVI cases. Even in scenarios where remains are unidentifiable and there is no available demographic data, AI can determine age ranges by analysing methylation patterns. This analysis can assist investigators when attempting to match remains to those on missing person registries. The examples above illustrate how far AI technology has come in DVI cases. The use of AI technology in these scenarios clearly demonstrates its ability to enhance identification precision, streamline complex actions, and deliver dependable results in a matter of seconds, all while working under adverse conditions.

8.3 Reopening Cold Cases Through the Use of AI-Based Genomic Tools

Cold cases, which are investigations that become unsolved and remain open for years or even decades, present a unique set of problems. These issues stem from a combination of old, insufficient leads, out-of-date technology, and deteriorated evidence. The use of AI-based genomic tools has proven to be invaluable in these circumstances by allowing investigators to reopen and solve these cases by integrating modern analytical approaches to previously archived old DNA evidence.

An example of a genealogy-focused cold case breakthrough enhanced by AI-assisted SNP analysis. In one incident, investigators revisited DNA evidence obtained decades earlier from a violent crime scene. Although SNP-STR profiling could not be completed due to the low DNA quantity and the sample having degraded, AI SNP callers pulled sufficient DNA from the degraded sample. Subsequently, AI genealogy-matching programs use SNP-STR profiles to remotely available genealogy databases and identify the unknown suspect's distant relatives. Additional AI relationship modelling helped investigators build family trees, refining the suspect pool to one person who was later validated through direct DNA testing.

Unidentified human remains cold case investigations have equally benefited from the AI-driven forensic DNA phenotyping. In several late case investigations, AI trait prediction modelling helped narrow the victim's ancestry, pigmentation, and facial structure, which prompted police to revisit their missing persons case files. This helped to identify victims and resolve cases that had previously relied solely on decomposition evidence for identification and were therefore closed by investigators.

AI-driven mixture deconvolution technology has deftly tackled cold cases involving compromised sexual assault evidence techniques used for the manual analysis of complex samples particulate without success. However, AI-powered probabilistic models, after reanalysing the electropherograms, identified, with confidence, the possible identities of the contributors, thereby closing the gaps created with OVER-identification. This provides leads to the match of previously unidentified suspects.

In some instances, AI epigenetic technology has estimated the age of possible victims or suspects. This has allowed the investigators to exclude individuals previously considered and narrow down the age range to more realistic parameters. The impact of these limitations on the investigation has been transformative.

These examples of the use of AI demonstrate that the technology is not simply an improved functionality within the forensics domain of workflows, but is a true 'game changer' in offering new pathways to the dispensation of justice for previously unresolvable crimes.

9. FUTURE DIRECTIONS AND RESEARCH OPPORTUNITIES

Artificial intelligence and forensic genomics have started to reshape the investigative process, but the field continues to change. New technologies, such as machine learning and data integration, as well as advancements in bioscience and sequencing, will continue to improve the capabilities of the field. As forensic labs begin to implement AI technologies, there appear to be more automated, accurate, and proactively ethical practices in the field. This chapter considers future directions and research opportunities in four areas: progress in algorithms, connection with national forensic systems, frontline AI-enabled rapid DNA technologies, and international standardisation to protect against misuse.

9.1 Progress in Algorithmic Forensic Genomics

In the coming ten years, the field of forensic genomics will benefit from improvements to the algorithms utilised. Many of the current AI systems in use, while impressive, will be more reliant on supervised learning and deep learning in the coming decade. Future systems will employ hybrid intelligence and self-directed learning. Furthermore, there will be a high use of multimodal systems, which will be capable of solving complex biological systems.

1. Ultra-Deep Neural Networks for Complex Phenotypes

The next generation of predictive models will assimilate thousands of genomic, epigenomic, and transcriptomic markers to more accurately predict complex traits such as facial morphology, metabolism, and age-associated biological parameters. Integrated multimodal deep neural architectures employing transformers, graph neural networks (GNNs), and CNNs will wield the capability of seamlessly processing heterogeneous biological datasets.

2. Graph-Based Genomic Models

GNNs will facilitate variant interpretation by conceptualising DNA as interconnected nodes and edges of a graph structure. This enhances the ability of AI models to account for long-range interactions, the effects of rare structural variations, and other complex structural variations otherwise lost in a linear sequence approach.

3. Causality-Based AI

Models of causal inference hold the potential to directly address mechanisms controlling the genotype-phenotype interface, unlike traditional, correlational AI. These models will also allow for a more reliable forensic prediction by determining the out of the other variables, precisely which phenotypic traits have a primary genetic determinant as opposed to secondary environmental or epigenetic influences.

4. Continual Learning Systems

Forensic AI models will increasingly benefit from the incorporation of continual learning, in which models self-update as new phenotypic, population, or allele frequency data are obtained. This minimises model drift and enhances reliability over time.

5. Quantum Machine Learning (QML)

Current literature suggests that the potential of QML to improve the efficiency of a number of machine learning tasks will be extremely useful in forensic genomics. Improvements in SNP calling, STR interpretation, and mixture deconvolution will be possible in datasets that are extremely large and in a time-efficient manner.

9.2 Integration with National Forensic Systems

Achieving the desired outcomes for AI in forensics will require assimilating the technology with the cross-border networks in digital forensics and crime. Many countries have extensive DNA collection and retention, but do not have AI analytical tools. Future systems will require administrative, policy, and computational systems upgrades.

1. AI-Augmented National DNA Databases

The next generation of national DNA databases will contain sequences for STRs, SNPs, mitochondrial haplotypes, and epigenetic variations. AI systems will be employed for real-time qPCR quality control, probabilistic matching, and triage of degraded or low-copy fragments.

2. Automated Case-Linking Platforms

Platforms with AI will allow for the profiling of crime scene DNA, comparing phenotypic predictions, and ancestry profiling across jurisdictional boundaries to identify and link serial offenders, crime hotspots, and cross-border crime. Integration with Interpol's databases will improve international criminal intelligence.

3. Interoperability and Data Exchange Standards

Forensic systems of the future must provide for integrated, cross-border interoperability that supports standards for data exchange, secure communications, and seamless inter-institutional collaboration.

4. AI-assisted Judicial Review Systems

AI systems estimating ratios of likelihood, phenotypes, and DNA mixtures will increasingly be presented in legal proceedings. Judges and legal counsel must be provided with systems that clearly communicate the findings of AI systems in such a way that the substantive law and science underlying the forensic evidence are made transparent.

5. Integration of Cyber Forensics

Combining human DNA with digital forensics, behavioral analysis, and datasets of criminology with hybrid investigative platforms will enable multi-faceted reconstructions of cases. Integrating artificial intelligence with geographically centered systems will result in faster identification and better workflow and safety for the public. Using these systems will, however, still require a solid foundation of ethical, legal, and regulatory controls.

9.3 AI in Field-Level Rapid DNA Applications

The advent of rapid DNA technology, which allows for on-site DNA profiling in less than two hours, will radically transform the workflows in law enforcement. Further integration of AI into these portable systems will be beneficial.

1. AI-Powered Portable STR Profilers

Future portable systems will be able to use machine learning for greater efficiency in real-time allele calling, noise- and error-correction, which will enable the generation of high confidence STR profiles, on-location, and in a crime scene.

2. Real-time Phenotyping and Ancestry Prediction

Investigators will be able to use AI-powered rapid SNP analysis to provide predictions of eye and hair colour, and even ancestry, which will result in improved guidance of investigations, especially in rapid field operations.

3. Automated Quality Verification

AI systems within portable devices will automate sample verification to determine if samples are untainted and suitable for subsequent analysis. This will prevent the waste of time and resources.

4. Mobile Integration with Forensic Databases

Rapid DNA devices having real-time connections with encrypted national databases will allow law enforcement to obtain immediate notifications when DNA profiles in their databases match and when potential family connections are identified.

5. Field Level Decision Support Systems

AI-based decision models can suggest actions to be undertaken by officers, depending on confidence levels of predicted outcomes, of predictive phenotyping, and available metadata from the crime scene.

The integration of rapid DNA and AI will enable the construction of forensic systems that can provide analytical intelligence.

9.4 Global Standardisation Needs

Given that AI is becoming the backbone technology in forensic genomics, global standardisation will be key in maintaining scientific consistency, legal validity, and ethical justice.

1. International AI Validation Protocols

Current validation frameworks, such as SWGDAM, ENFSI, require expansion to cater for AI guidelines, with a focus on model training, data quality, bias testing, and reproducibility.

2. Ethical and Regulatory Harmonisation

To ensure both global collaboration and control of misuse, countries need to align legal policies on data privacy, consent, forensic DNA phenotyping, genealogical database usage, and AI ancestry inference.

3. Standards Regarding Transparency

AI-based forensic analysis needs to encompass explainability, confidence metrics, and limitations. Standard templates would systematise clarity in reporting.

4. Equity in Population Databases

To reduce bias, equitable initiatives around the world should aim to create adequately representative, diverse DNA databases, including underrepresented groups.

5. Accreditation and Training

Forensic scientists should be trained in the computational disciplines of AI, data science, and genomics. AI-enabled forensic workflow certification programs will mitigate responsibility and guarantee accountability.

Criminal justice systems worldwide face fairness and equity challenges. Global harmonisation will ensure that AI will be a tool of enhancement, and not a liability.

10. CONCLUSION

AI-enabled forensic genomics is the pioneering technology for transforming the capture and analytical processes applied to biological evidence in a crime. AI expands the ability to analyse and interpret Complex and/or degraded DNA samples, improves the analysis of SNPs, and augments the analysis of AI-assisted STR. Furthermore, AI predicts ancestry and phenotypes with unparalleled precision. Enhanced forensic capabilities integration, multi-omics, and others, epigenomics, methylomics, and genomics, will provide insights into the biology of the sample. Insights previously inaccessible with a conventional approach and techniques.

REFERENCES

Abdelwahab, O., Belzile, F., & Torkamaneh, D. (2023). Performance analysis of conventional and AI-based variant callers using short and long reads. *BMC Bioinformatics 2023 24:1, 24*(1), 472-. DOI: 10.1186/s12859-023-05596-3

Abdelwahab, O., & Torkamaneh, D. (2025). Artificial intelligence in variant calling: A review. *Frontiers in Bioinformatics*, *5*, 1574359. DOI: 10.3389/fbinf.2025.1574359 PMID: 40337525

Ahmed, E., El-Din, A., & El-Din, E. A. A. (2022). ARTIFICIAL INTELLIGENCE IN FORENSIC SCIENCE : INVASION OR REVOLUTION? *Egyptian Society of Clinical Toxicology Journal, 10*(2), 20–32. DOI: 10.21608/esctj.2022.158178.1012

Ahmed, S., Khan, M. F., Singh, B., Singh, N., & Sharma, B. (1 C.E.). Enhancing Crime Scene Analysis: The Impact of AI Technologies on Evidence Processing. Https://Services.Igi-Global.Com/Resolvedoi/Resolve.Aspx?Doi=10.4018/979-8-3693-9405-2.Ch004, 63–84. DOI: 10.4018/979-8-3693-9405-2.ch004

Alketbi, S. K. (2024). DNA Contamination in Crime Scene Investigations: Common Errors, Best Practices, and Insights from a Survey Study. SSRN *Electronic Journal*. https://doi.org/DOI: 10.2139/SSRN.4979243

Alotaibi, H., Alsolami, F., Abozinadah, E., & Mehmood, R. (2022). TAWSEEM: A Deep-Learning-Based Tool for Estimating the Number of Unknown Contributors in DNA Profiling. *Electronics 2022, Vol. 11, Page 548, 11*(4), 548. DOI: 10.3390/electronics11040548

Amer, S. A. (2025). Mitochondrial DNA as a Tool for Examining Molecular Diversity and Forensic Genetics. *Advances in Forensic Biology and Genetics*, 389–399. DOI: 10.1007/978-981-96-4585-5_21

Angelov, P. P., Soares, E. A., Jiang, R., Arnold, N. I., & Atkinson, P. M. (2021). Explainable artificial intelligence: An analytical review. *Wiley Interdisciplinary Reviews. Data Mining and Knowledge Discovery*, *11*(5), e1424. DOI: 10.1002/widm.1424

Annan, R., Noland, J., Perkins, K., Yuan, X., Roy, K., & Qingge, L. (2025). Genomic privacy and security in the era of artificial intelligence and quantum computing. *Discover Computing 2025 28:1, 28*(1), 108-. DOI: 10.1007/s10791-025-09627-w

Athanasopoulou, K., Michalopoulou, V. I., Scorilas, A., & Adamopoulos, P. G. (2025). Integrating Artificial Intelligence in Next-Generation Sequencing: Advances, Challenges, and Future Directions. *Current Issues in Molecular Biology 2025, Vol. 47, Page 470, 47*(6), 470. DOI: 10.3390/cimb47060470

Ballard, J. L., Wang, Z., Li, W., Shen, L., & Long, Q. (2024). Deep learning-based approaches for multi-omics data integration and analysis. *BioData Mining 2024 17:1, 17*(1), 38-. DOI: 10.1186/s13040-024-00391-z

Brasil, S., Neves, C. J., Rijoff, T., Falcão, M., Valadão, G., Videira, P. A., & dos Reis Ferreira, V. (2021). Artificial Intelligence in Epigenetic Studies: Shedding Light on Rare Diseases. *Frontiers in Molecular Biosciences, 8*, 648012. DOI: 10.3389/fmolb.2021.648012 PMID: 34026829

Cavalcanti, P., & Nogueira, T. L. S. DE CARVALHO, E. F., & DA SILVA, D. A. (2024). Forensic use of human mitochondrial DNA: A review. *Anais da Academia Brasileira de Ciências, 96*(4), e20231179. https://doi.org/DOI: 10.1590/0001-3765202420231179

Chafai, N., Hayah, I., Houaga, I., & Badaoui, B. (2023). A review of machine learning models applied to genomic prediction in animal breeding. *Frontiers in Genetics, 14*, 1150596. DOI: 10.3389/fgene.2023.1150596 PMID: 37745853

Crysup, B., Mandape, S., King, J. L., Muenzler, M., Kapema, K. B., & Woerner, A. E. (2023). Using unique molecular identifiers to improve allele calling in low-template mixtures. *Forensic Science International. Genetics, 63*, 102807. DOI: 10.1016/j.fsigen.2022.102807 PMID: 36462297

Dahal, A., McNevin, D., Chikhani, M., & Ward, J. (2023). An interdisciplinary forensic approach for human remains identification and missing persons investigations. *WIREs Forensic Science, 5*(4), e1484. DOI: 10.1002/wfs2.1484

Dash, H. R., Elkins, K. M., & Al-Snan, N. R. (2023). Advanced Emerging Techniques for Forensic DNA Analysis: STRs, SNPs, and mtDNA Analysis. *Advancements in Forensic DNA Analysis*, 35–59. DOI: 10.1007/978-981-99-6195-5_4

De la Puente, M., Ruiz-Ramírez, J., Ambroa-Conde, A., Xavier, C., Pardo-Seco, J., Álvarez-Dios, J., Freire-Aradas, A., Mosquera-Miguel, A., Gross, T. E., Cheung, E. Y. Y., Branicki, W., Nothnagel, M., Parson, W., Schneider, P. M., Kayser, M., Carracedo, Á., Lareu, M. V., & Phillips, C. (2021). Development and evaluation of the ancestry informative marker panel of the Visage Basic Tool. *Genes, 12*(8), 1284. DOI: 10.3390/genes12081284 PMID: 34440458

Debsharma, S. K., Rahman, M. A., Quddus, M. R., Khatun, H., Disha, R. F., Roy, P. R., Ahmed, S., El-Sharnouby, M., Iftekharuddaula, K. M., Aloufi, S., Alzuaibr, F. M., Alqurashi, M., Sakran, M. I., & Kabir, M. S. (2022). SNP-Based Trait Characterization Detects Genetically Important and Stable Multiple Stress Tolerance Rice Genotypes in Salt-Stress Environments. *Plants 2022, Vol. 11, Page 1150, 11*(9), 1150. DOI: 10.3390/plants11091150

Dhanush, M. S., & Thenmozhi, M. (2022). Missing Person Detection Using AI. *ACM International Conference Proceeding Series*. https://doi.org/DOI: 10.1145/3590837.3590931;JOURNAL:JOURNAL:ACMOTHERCONFERENCES;CTYPE:STRING:BOOK

Dhillon, S. (n.d.). *Ethical Integration of Artificial Intelligence in Forensic Science: Prospects and Challenges*. DOI: 10.1145/3306618.3314320

Dong, Z., Wang, Y., Li, C., Li, L., & Men, X. (2021). Mitochondrial DNA as a Molecular Marker in Insect Ecology: Current Status and Future Prospects. *Annals of the Entomological Society of America, 114*(4), 470–476. DOI: 10.1093/aesa/saab020

Enoma, D. O., Bishung, J., Abiodun, T., Ogunlana, O., & Osamor, V. C. (2022). Machine learning approaches to genome-wide association studies. *Journal of King Saud University. Science, 34*(4), 101847. DOI: 10.1016/j.jksus.2022.101847

Ferrero, A., & Scotti, V. (2022). Forensic DNA Profiling. *Research for Development*, •••, 151–202. DOI: 10.1007/978-3-031-14619-0_10

Forouzesh, M., Irani, S., Soleimani, A., & Monabati, S. J. (2022). Application of Y-STR, DIP-STR and SNP-STR Markers in Interpretation of Forensic Genetic Profiling: A Narrative Review. *Iranian Journal of Public Health, 51*(7), 1538. DOI: 10.18502/ijph.v51i7.10087 PMID: 36248298

Kanthaswamy, S. (2024). Review: Wildlife forensic genetics—Biological evidence, DNA markers, analytical approaches, and challenges. *Animal Genetics, 55*(2), 177–192. DOI: 10.1111/age.13390 PMID: 38123142

Kokotas, S., Budowle, B., Papatheodorou, A., Bolanaki, E., Kondili, A., Metheniti, A., Vouropoulou, M., Koukouvinos, G., Palaigeorgiou, E., & Makras, P. (2025). Comparison of next generation sequencing (NGS) - (SNPs) and capillary electrophoresis (CE) - (STRs) in the genetic analysis of human remains. *Forensic Science International. Genetics, 74*, 103131. DOI: 10.1016/j.fsigen.2024.103131 PMID: 39241298

Kotlarz, K., Mielczarek, M., Biecek, P., Guldbrandtsen, B., & Szyda, J. (2024). Exploring the impact of sequence context on errors in SNP genotype calling with whole genome sequencing data using AI-based autoencoder approach. *NAR Genomics and Bioinformatics*, *6*(3), 131. DOI: 10.1093/nargab/lqae131 PMID: 39318508

Kuiken, T. (2023). Artificial Intelligence in the Biological Sciences: Uses, Safety, Security, and Oversight. *Congressional Research Service (CRS) Reports and Issue Briefs*, NA-NA. https://go.gale.com/ps/i.do?p=AONE&sw=w&issn=&v=2.1&it=r&id=GALE%7CA777680617&sid=googleScholar&linkaccess=fulltext

Lee, J. E., Lee, J. M., Naue, J., Fleckhaus, J., Freire-Aradas, A., Neubauer, J., Pośpiech, E., McCord, B., Kalamara, V., Gauthier, Q., Mills, C., Cao, Y., Wang, Z., Oh, Y. N., Feng, L., Schneider, P. M., Phillips, C., Haas, C., Pisarek, A., & Lee, H. Y. (2022). A collaborative exercise on DNA methylation-based age prediction and body fluid typing. *Forensic Science International. Genetics*, *57*, 102656. DOI: 10.1016/j.fsigen.2021.102656 PMID: 34973557

Liang, R., Tang, Q., Chen, J., & Zhu, L. (2024). Epigenetic Clocks: Beyond Biological Age, Using the Past to Predict the Present and Future. *Aging and Disease*, *16*(6), 3520. DOI: 10.14336/AD.2024.1495 PMID: 39751861

Marsico, F., & Amigo, M. (2025). Ethical and security challenges in AI for forensic genetics: From bias to adversarial attacks. *Forensic Science International. Genetics*, *76*, 103225. DOI: 10.1016/j.fsigen.2025.103225 PMID: 39874746

Mizrachi, R., Neiman, D., Rosenski, J., Loyfer, N., Share, D., Adjedj, C., Glaser, B., Shpitzen, M., Dor, Y., Shemer, R., & Kaplan, T. (2025). Genetic/epigenetic DNA markers for linking suspects and tissues in complex crime scenes. *Briefings in Bioinformatics*, *26*(4), bbaf395. Advance online publication. DOI: 10.1093/bib/bbaf395 PMID: 40794944

Monika, E., & Kumar, T. R. (2024). Exploring the Potential of TERT and Genetic Markers in Forensic Science for Crime Detection and Prevention. *2nd International Conference on Emerging Research in Computational Science, ICERCS 2024*. DOI: 10.1109/ICERCS63125.2024.10895472

Muneeb, M., Feng, S., & Henschel, A. (2022). Transfer learning for genotype–phenotype prediction using deep learning models. *BMC Bioinformatics 2022 23:1*, *23*(1), 511-. DOI: 10.1186/s12859-022-05036-8

Okut, H. (2021). Deep Learning Algorithms for Complex Traits Genomic Prediction. *Journal of Animal Science and Products*, *4*(2), 225–239. DOI: 10.51970/jasp.1039713

Palmela Pereira, C., Carvalho, R., Augusto, D., Almeida, T., Francisco, A. P., Salvado e Silva, F., & Santos, R. (2025). Development of artificial intelligence-based algorithms for the process of human identification through dental evidence. *International Journal of Legal Medicine 2025 139:4, 139*(4), 1835–1850. DOI: 10.1007/s00414-025-03453-x

Pedroza Matute, S., & Iyavoo, S. (2025). Implementation of NGS and SNP microarrays in routine forensic practice: opportunities and barriers. *BMC Genomics 2025 26:1, 26*(1), 541-. DOI: 10.1186/s12864-025-11723-6

Pengyue, L., Siyuan, X., Yi, J., Wen, Y., Xiaoning, L., Guohua, G., & Shixiong, W. (2021). ANINet: a deep neural network for skull ancestry estimation. *BMC Bioinformatics 2021 22:1, 22*(1), 550-. DOI: 10.1186/s12859-021-04444-6

Pisarek, A., Pośpiech, E., Heidegger, A., Xavier, C., Papież, A., Piniewska-Róg, D., Kalamara, V., Potabattula, R., Bochenek, M., Sikora-Polaczek, M., Macur, A., Woźniak, A., Janeczko, J., Phillips, C., Haaf, T., Polaoska, J., Parson, W., Kayser, M., & Branicki, W. (2021). Epigenetic age prediction in semen – marker selection and model development. *Aging (Albany NY), 13*(15), 19145–19164. DOI: 10.18632/aging.203399 PMID: 34375949

Pośpiech, E., Teisseyre, P., Mielniczuk, J., & Branicki, W. (2022). Predicting Physical Appearance from DNA Data—Towards Genomic Solutions. *Genes 2022, Vol. 13, Page 121, 13*(1), 121. DOI: 10.3390/genes13010121

Roy, A., Dutta, S., Das, S., & Choudhury, M. R. (2024). Next-Generation Sequencing in the Development of Climate-Resilient and Stress-Responsive Crops – A Review. *The Open Biotechnology Journal, 18*(1), e18740707301657. Advance online publication. DOI: 10.2174/0118740707301657240517063244

Ryu, J., Barkal, S., Yu, T., Jankowiak, M., Zhou, Y., Francoeur, M., Phan, Q. V., Li, Z., Tognon, M., Brown, L., Love, M. I., Bhat, V., Lettre, G., Ascher, D. B., Cassa, C. A., Sherwood, R. I., & Pinello, L. (2024). Joint genotypic and phenotypic outcome modeling improves base editing variant effect quantification. *Nature Genetics 2024 56:5, 56*(5), 925–937. DOI: 10.1038/s41588-024-01726-6

Santiago-Rodriguez, T. M., & Hollister, E. B. (2021). Multi 'omic data integration: A review of concepts, considerations, and approaches. *Seminars in Perinatology, 45*(6), 151456. DOI: 10.1016/j.semperi.2021.151456 PMID: 34256961

Sessa, F., Esposito, M., Cocimano, G., Sablone, S., Karaboue, M. A. A., Chisari, M., Albano, D. G., & Salerno, M. (2024). Artificial Intelligence and Forensic Genetics: Current Applications and Future Perspectives. *Applied Sciences (Basel, Switzerland), 14*(5), 2113. DOI: 10.3390/app14052113

Shah, K. N., Gami, S. J., & Trehan, A. (n.d.). *An Intelligent Approach to Data Quality Management AI-Powered Quality Monitoring in Analytics*. DOI: 10.48175/IJARSCT-22820

Shevchuk, V., Dunaev, O., Tyshchenko, O., Biletska, G., & Nehrebetskyi, V. (2025). Innovative Methods in the Identification of Deceased Persons during Armed Conflicts and Disasters: Criminalistic and Forensic Medical Issues. *International Annals of Criminology*, 1–20. DOI: 10.1017/cri.2025.10094

Simpson, D. J., & Chandra, T. (2021). Epigenetic age prediction. *Aging Cell, 20*(9), e13452. DOI: 10.1111/acel.13452 PMID: 34415665

Song, Z., Huang, D., Song, B., Chen, K., Song, Y., Liu, G., Su, J., Magalhães, J. P. de, Rigden, D. J., & Meng, J. (2021). Attention-based multi-label neural networks for integrated prediction and interpretation of twelve widely occurring RNA modifications. *Nature Communications 2021 12:1, 12*(1), 4011-. DOI: 10.1038/s41467-021-24313-3

Soumare, H., Rezgui, S., Gmati, N., & Benkahla, A. (2021). New neural network classification method for individuals ancestry prediction from SNPs data. *BioData Mining 2021 14:1, 14*(1), 30-. DOI: 10.1186/s13040-021-00258-7

Tan, M., Tan, Y., Jiang, H., Xue, J., Wu, Q., Zheng, Y., Liu, G., Xiao, Y., Lv, M., Liao, M., Zhang, L., Qu, S., & Liang, W. (2025). Explainable artificial intelligence in forensic DNA analysis: Alleles identification in challenging electropherograms using supervised machine learning methods. *Forensic Science International. Genetics, 78*, 103289. DOI: 10.1016/j.fsigen.2025.103289 PMID: 40288204

Terrado-Ortuño, N., & May, P. (2025). Forensic DNA phenotyping: A review on SNP panels, genotyping techniques, and prediction models. *Forensic Sciences Research, 10*(1), owae013. Advance online publication. DOI: 10.1093/fsr/owae013 PMID: 39990695

Ur Rehman, H., Tufail, H., Arooj, A., Raza, A., Kalsoom, U., Muhammad Abbas Malik, H., Shahid Cholistani, M., Qurban Khan, S., Abdullah Nasir, M., Ali, Z., Irshad, M., & Faheem, H. (2026). The Forensic Science of the Future: A Journey through Omics Technologies. *Journal of Biological and Allied Health Sciences, 6*(1), 86–99. DOI: 10.56536/jbahs.v6i1.124

Valecha, M., & Posada, D. (2022). Somatic variant calling from single-cell DNA sequencing data. *Computational and Structural Biotechnology Journal, 20*, 2978–2985. DOI: 10.1016/j.csbj.2022.06.013 PMID: 35782734 van Dongen, J., Ehli, E. A., Slieker, R. C., Bartels, M., Weber, Z. M., Davies, G. E., Slagboom, P. E., Heijmans, B. T., & Boomsma, D. I. (2014). Epigenetic variation in monozygotic twins: A genome-wide analysis of DNA methylation in buccal cells. *Genes*, *5*(2), 347–365. DOI: 10.3390/genes5020347 PMID: 24802513

Wan, Z., Hazel, J. W., Clayton, E. W., Vorobeychik, Y., Kantarcioglu, M., & Malin, B. A. (2022). Sociotechnical safeguards for genomic data privacy. *Nature Reviews Genetics 2022 23:7*, *23*(7), 429–445. DOI: 10.1038/s41576-022-00455-y

Wang, K., Abid, M. A., Rasheed, A., Crossa, J., Hearne, S., & Li, H. (2023). DNNGP, a deep neural network-based method for genomic prediction using multi-omics data in plants. *Molecular Plant*, *16*(1), 279–293. DOI: 10.1016/j.molp.2022.11.004 PMID: 36366781

Wang, Y., Kong, S., Zhou, C., Wang, Y., Zhang, Y., Fang, Y., & Li, G. (2024). A review of deep learning models for the prediction of chromatin interactions with DNA and epigenomic profiles. *Briefings in Bioinformatics*, *26*(1), bbae651. Advance online publication. DOI: 10.1093/bib/bbae651 PMID: 39708837

Wu, D., Karhade, D. S., Pillai, M., Jiang, M. Z., Huang, L., Li, G., Cho, H., Roach, J., Li, Y., & Divaris, K. (2021). Machine Learning and Deep Learning in Genetics and Genomics. *Machine Learning in Dentistry*, 163–181. DOI: 10.1007/978-3-030-71881-7_13

Wu, S., Chen, N., Xiao, A., Zhang, P., Jiang, C., & Zhang, W. (2025). AI-Empowered Virtual Network Embedding: A Comprehensive Survey. *IEEE Communications Surveys and Tutorials*, *27*(2), 1395–1426. DOI: 10.1109/COMST.2024.3424533

Zhao, Y. (2025). Evolving Paradigms in Low-Template DNA Analysis: From Fixed Analytical Thresholds to Probabilistic Modeling. *Transactions on Materials. Biotechnology and Life Sciences*, *8*, 423–431. DOI: 10.62051/mjbq7s89

Zheng, A., Lamkin, M., Zhao, H., Wu, C., Su, H., & Gymrek, M. (2021). Deep neural networks identify sequence context features predictive of transcription factor binding. *Nature Machine Intelligence 2021 3:2*, *3*(2), 172–180. DOI: 10.1038/s42256-020-00282-y

KEY TERMS AND DEFINITIONS

AI: Artificial Intelligence — computational methods (ML/DL) applied throughout forensic genomics.

Allele calling: The process of identifying which allele(s) are present at a genetic locus from experimental data (e.g., electropherogram peaks or sequencing reads); essential for generating genotype profiles.

Ancestry-informative markers (AIMs): Genetic variants (often SNPs) are selected because their allele frequencies differ substantially between populations, used for biogeographical ancestry inference.

Chain of custody: The documented, unbroken record of evidence handling from collection to presentation in court; critical for the legal admissibility of forensic genetic evidence.

CNN: Convolutional Neural Network — a deep learning model effective for signal/sequence pattern recognition.

DVI: Disaster Victim Identification — forensic operations to identify victims in mass-fatality events.

Electropherogram: A graphical output of capillary electrophoresis that displays DNA fragment peaks (alleles); interpretation of peak height, shape, and position is central to STR analysis.

Epigenetics (DNA methylation): Heritable chemical modifications (e.g., CpG methylation) that regulate gene expression without altering DNA sequence; in forensics, methylation patterns are used for age and tissue-source estimation.

FDP: Forensic DNA Phenotyping — prediction of externally visible traits from genetic data.

Forensic DNA phenotyping (FDP): The inference of externally visible characteristics (EVCs) — such as eye, hair, skin pigmentation, ancestry, or facial features — from genetic and epigenetic data to provide investigative leads.

Heteroplasmy: The presence of more than one mitochondrial DNA (mtDNA) sequence type within an individual or tissue; detection and interpretation of heteroplasmy are important in mtDNA forensic analysis.

Likelihood ratio (LR): A quantitative measure comparing the probability of observed genetic evidence under two competing hypotheses (e.g., prosecution vs. defence); widely used in forensic interpretation.

Mixture deconvolution: The analytical process of separating and attributing genetic contributions from two or more individuals present in a single biological sample.

ML: Machine Learning — algorithms that learn patterns from data for classification and prediction.

mtDNA: Mitochondrial DNA — maternally inherited DNA used for degraded samples and maternal lineage.

NGS: Next-Generation Sequencing — high-throughput sequencing used for large-scale SNP/STR panels.

PCR: Polymerase Chain Reaction — laboratory method to amplify DNA regions of interest.

Probabilistic genotyping: A statistical framework that uses probability models (often Bayesian) to evaluate contributor genotypes and calculate likelihood ratios, especially for complex DNA mixtures.

SNP: Single Nucleotide Polymorphism — abundant single-base variants used for ancestry and phenotyping.

STR: Short Tandem Repeat — standard loci used for forensic DNA profiling.

Chapter 14
AI-Enabled Discovery and Sustainable Access to Biopharmaceuticals:
A Legal and Policy Approach

B. Monisha
http://orcid.org/0000-0001-9702-5519
Christ (Deemed to be) University, Bengaluru, India

R. Valarmathi
Christ (Deemed to be) University, Bengaluru, India

ABSTRACT

The Biopharma Industry burgeons the recent tech-driven economy with its multifaceted efficacy in curing rare diseases and high scientific, technical, and cost complexities. AI-driven biopharma establishments partner with conventional Biopharma companies to provide drug delivery efficacy. Studies exhibit synergistic effects in research systems when AI technologies are embedded into life science medicine processes. This drug delivery lowers costs, enables a high probability of success, multiplies intellectual properties, provides a competitive advantage in data, and encourages 'Innovative Biologics', leading to sustainable access to medicines. This chapter explores how emerging legal and regulatory frameworks address challenges of AI-driven Biopharma drug delivery, varying from AI-driven data ensuring safety, efficacy, potency, and purity tests for approval to intellectual property rights, sustainable access to medicines, and development of Innovative Biologics for rare diseases.

DOI: 10.4018/979-8-2600-2691-5.ch014

1. INTRODUCTION

AI driven Biopharmaceuticals are transforming the pharmaceutical industry with significant outcomes, barring high complexity in developing biopharmaceuticals. AI- integrated Biologics which are protein and peptide-based medicines, are more than mere incremental innovation in biologics development relying on data analysis, but transitions from macro-molecular research from long-driving empirical research to a data driven analysis. This integration of AI in drug discovery leads to a great outcome fastening the years of research for 'Innovative Biologics' (D.J. Diaz, et al., 2024).

1.1 The Biotechnological Revolution and AI Integration

With Big data and fast pacing technological deployment in Biologics drug discovery, there is a high-end hope for the computational power to make 'access to medicine' a promising reality. This study says a Moore's Law like trend with AI-integrated drug discovery (Jayatunga et al., 2022). Biopharmaceuticals are also structurally more complex medicines which are defined by the polymeric chain formation that have great structural variations. (Kesik-Brodacka et al. 2018). The Long-standing Biopharmaceutical research and development relies on extensive clinical research and timelines, development cost and high risk of failure rates in the discovery. (Krishnan Sampathkumar, et al. 2024). It is estimated that the average time for Target identification to the approval in the market for these Biologics spans between 12 to 15 years with thc development costs exceeding $2.6 billion per approved drug when accounting for failed products (DiMasi et al., 2016). The High complexity in these Innovative Biologics and pricing has limited the access to these medicines in Low- and Middle-income countries which heaving rely on affordable pharmaceuticals.

With AI-integration, which encompasses data analysis, machine learning, deep learning showcasing an AI synergising with bioinformatics. AI-driven drug discovery enables analysing large datasets which includes genomic information, protein structures, the chemical properties, clinical trials, and evidences and develop products with speed and accuracy. Real-world companies like Insilico Medicine, Exscientia, and Recursion Pharmaceuticals have demonstrated that AI can reduce drug discovery timelines from years to months and decrease development costs by factors of ten or more (Fleming, 2018).

1.2 AI in Biopharmaceutical Innovation

Biopharmaceuticals are therapeutic products derived from natural and biological sources including proteins, antibodies, nucleic acids, and cell-based therapies which have revolutionized treatment for rare and incurable diseases. Given the complexity in developing these biologics, regulatory complexity and great capital investment in the sector has led to great barrier in innovation and access to these medicines.

AI technologies offer transformative tools to develop drug discovery and its value chain in the market through identifying the biological target, giving molecular design and optimization, prediction of the structure of proteins and even as a tool for clinical trials. This paper has analysed few AI-integrated drug discovery attempts made by biopharmaceutical companies and are in the go. BioPharma Trend have shown such US based companies using AI for Biologic drug discovery (Honchar, 2024). Those technologies include usage of proprietary antibody visualization software, creating a mammalian cell engineering platform combining computer-aided design with ML. Some products include Milliner platform which integrates ML with a synthetic biology wet lab. Creyon Bio applies a data-first ML mapping algorithm for oligonucleotide-based medicines; Envisagenics uses SpliceCore (and its immunotherapy branch SpliceIO) for RNA splicing-based drug target discovery; Evaxion Biotech's AI-Immunology platform comprises three tools — PIONEER, EDEN, and RAVEN — for vaccine design; HiFiBiO employs their Drug Intelligence Science (DIS) Platform for single-cell antibody discovery; Juvena Therapeutics uses an AI-enhanced biologics platform with a curated pro-regenerative protein library; Nuritas deploys Nπφ (Peptide Finder) to predict bioactive peptides; and Protein-Qure utilizes a computational platform combining high-performance computing, molecular simulations, and ML for protein therapeutic design.

In addition to these, platforms like BenevolentAI's knowledge graph integrate biomedical literature, clinical databases, and genetic information to uncover novel therapeutic targets and repurposing opportunities (Ekins et al., 2019). Insilico Medicine's Chemistry42 platform has generated novel drug candidates for conditions including fibrosis and cancer, with several compounds advancing to clinical trials (Zhavoronkov et al., 2019). DeepMind's AlphaFold2 has endeavored achieving accuracy in predicting three-dimensional protein structures from amino acid sequences that enables rational design of therapeutic antibodies and enzymes (Jumper et al., 2021). It is also studied that these algorithms can improve stratifying clinical data, predict trial outcomes and administration clinically reducing timelines significantly (Harrer et al., 2019).

1.3 Access to Medicines as a Public Health Imperative

Scientific advancements have gone to an unprecedented phase with many promising technologies coming into picture every day (WHO, 2021).Yet access to these innovations have become a growing concern worldwide and those factors include Intellectual property, regulatory compliances, Policy initiatives for investments in R & D for these technologies. (Opderbeck, D. W., 2019). Multiples factors contribute to this disparity which ranges from the High development costs and intellectual property protections resulting in high pricing for innovative biologics, often placing these medicines beyond reach for patients in resource-limited settings. This has become an inhibiting factor in achieving the Universal Health Coverage (UHC) and target 3b of UN's SDG agenda goal of 2030. (WHO, Europe). Secondly, the technical complexity in manufacturing Biologics. Biologics are large molecule drugs which require a sophisticated manufacturing capacity and huge creating technical barriers to generic manufacturing and local production capacity in developing regions. (Parabolic Drugs, 2026). Regulatory difficulties include approval, jurisdictional issues, data ethics, resource disparity that requires a harmonized approval pathway, a pro-active and transparent regulatory guideline. Technology or knowledge-gap, is another factor that causes inequitable access to these Innovations. Limited access to research tools, databases, and technical expertise concentrates innovation capacity in wealthy nations and large corporations, perpetuating global health inequities. (Gibofsky, A., et al, 2022).

When AI-integrated drug development can harness great potential in addressing the challenges of access and affordability of Biopharmaceuticals, regulatory and policy reformation addressing pricing concerns, Intellectual property and deploying AI capabilities are the need of the hour. However, concentration of AI capabilities among well-resourced entities, proprietary algorithms, and restrictive data practices risk exacerbating existing disparities (Vayena et al., 2018).

1.4 Literature Review

Studies reveal the years and complexity in the development of drugs and the proven advantage of AI-driven drug development. This innovative approach has already achieved a tenfold reduction in costs and aims to reduce them by a factor of 100. Studies show that AI-driven biopharma manufacturers and service companies will dominate the Biopharma industry ahead. Legal and regulatory landscape in AI driven industry requires safe, equitable and ethical practices according to Mirakhori, F., & Niazi, S. K. (2025). AI in healthcare concerns digitial health (Cohen et al., 2020), privacy rights in genetic data (Minssen et al., 2020), intellectual property concerns (Lal, S., *et al.,* 2024). AI drug discovery open opportunities in innovation, patent score and access to these medicines when regulated in pace with the advancement. (Bas, T. G., 2025). AI in healthcare requires regulatory oversight (Bhatia A, et al., 2020), while laying policy towards balancing innovation and access through intellectual property law (Opderbeck, D. W., 2019) and international collaboration for harmonized approach (Xu et al., 2020). Literatures draw on the evolving Biopharmaceutical development and regulatory concerns and the current legal and regulatory framework in first world Countries and India are transforming with its recent draft guidelines, 2025 and CRI guidelines for Patent opportunities which very few literatures address. Amthaniwala & Rathod (2025) draw on Draft 2025 guidelines of India and compare with UAE to some extent. However, AI integration in drug development is yet to mature for an established guideline or regulatory framework and the same is discussed comparing US and EU's evolving legal and regulatory landscape (Lenarczyk, G., et al., 2025).

Policies and Innovative approach for harnessing AI in drug development and robust advancement in regulatory system in pace with these technologies is indispensable for instance developing innovation based regulatory approach to navigate data security Meng, Y. (2026). The idea of this chapter is to connect the Legal and Regulatory landscape with the evolving AI innovation in biopharmaceuticals and equitable access for the same. The gap lies in the advancement of AI and its dynamic nature navigating through multiple regulatory concerns and a proactive requirement for the legal and regulatory framework to address upcoming difficulties in pace with the technological advancements.

1.5 Research Problem

Biologics are large molecule drugs (protein and peptide-based medicines) that have significantly progressed public health in curing and treating rare diseases. The high scientific complexity in production, investment, research and development, expensive clinical trials, and regulatory requirements affect production and thereby hinder the accessibility and affordability of these medicines. Being a factor-intensive sector, biopharmaceutical development has seen growth with several patents registered in global markets. It is estimated that for an Innovative Biologic to develop, it takes 12 years from target identification to its fuller development, and for the innovator firm to earn the rate of return, it takes 12.9 to 16.2 years. This number is equivalent to the challenge the firm has entered into for the development of Innovative Biologic, its cost of capital, the persistent tendency to further biotechnological research, and the cost and risk of the process involved. It is high time for the regulatory system to harness the potential of advancement of AI technologies in biopharmaceutical drug development and increase its potential in R&D. This study is an endeavour to untangle the potential of AI-driven biopharmaceutical drug development and the possible legal implications and policy analysis in guaranteeing sustainable access to medicines.

1.6 Objectives

1. To study the implications and challenges of the AI-driven Biopharma Industry in the Indian legal and regulatory system.
2. To draw insights from the U.S. FDA regulation and the E.U's EMA regulation, Intellectual property and data governance.
3. To analyse the efficiency of AI-driven drug delivery systems for sustainable access to medicine in the Biopharma Industry.

1.7 Methodology

This chapter undertakes doctrinal approach in understanding the AI driven Biopharma Industry and analyses the Legal and regulatory system in India, the US and EU. With insights from the US and EU, this chapter elucidate the regulatory guidelines, Intellectual property landscape and data governance concerns across the jurisdictions. With academic literatures and authoritative published documents of the authorities and policy documents, this chapter analyses the legal and regulatory of the said jurisdiction. This chapter also explores the transformative phase in AI driven Biopharmaceuticals through pharmaceutical companies' official release and their R & D initiatives for harnessing the technology in drug development.

2. AI-DRIVEN BIOPHARMACEUTICAL DRUG DEVELOPMENT

AI enables the discovery and optimization of Biologics with various technologies. The accuracy using computational technology requires a legal system that simultaneously integrates approval regulation, ethical and data regulation. One such key concern is 'protein prediction' AI platform, AlphaFold which challenges the drug discovery regulation. On the other hand, is the data regulation, with Generative AI, large datasets are used by AI in creating outputs and such AI with biologics that is Deep Search, AI-enabled identification and repository of knowledge through literature, patents, and approved drugs in the market akin to biological neural networks challenges the data regulation for drug discovery. AI enables chemical synthesis using tools such as IBM's RXN. AI has navigated through all stages of drug development from target identification or discovery to post-approval activities.

2.1 AI-driven Drug Discovery and Development

In order to understand how AI integrates with drug discovery, the multiple distinct stages involved may be studied: Target Identification, Optimization, Pre-clinical, clinical, manufacturing and quality control are the key development phases in drug discovery. AI tools are now used to analyse genomic, transcriptomic, proteomic, and metabolomic data to identify disease-associated patterns and predict target-disease relationships (Olawade, D.B., et al, 2025). Natural language processing algorithms are set to mine scientific literature, patents, and clinical databases to extract insights about disease biology and potential therapeutic interventions (Withers, C. A., et al., 2025).

Secondly, following the target identification, AI-powered virtual screening are set to evaluate billions of potential compounds computationally, which has reduced the time of experimentation with 'hits' targeting the disease. Following the same, AI tools optimize learning models to predict how chemical modifications will affect potency, selectivity, pharmacokinetics, and toxicity, guiding iterative refinement toward clinical candidates (Vamathevan et al., 2019).

The third phase include the Preclinical Development where the drug candidates had to undergo extensive preclinical evaluation which includes *in-vitro* studies, animal models, and toxicology assessments. These AI can now predict the toxicity, drug-drug interactions, off-target effects, and pharmacokinetic properties, significant for safety concerns and reducing reliance on animal testing (Mullard, 2019). Fourthly, predictive modelling of patient recruitment is another technology of AI that helps in treatment of the diseases (Harrer et al., 2019). It is also known that AI systems can optimize manufacturing processes easing the system and production. It predicts medical equipment failures and ensures quality and detects problems associated with it. (Lee et al., 2020). These AI tools are even handy in documentation and surveillance post approval of these drugs which involves electronic health records, claims and patient registries (Xu et al., 2021).

2.2 AI Technologies and Platforms

With few Companies rolling out the AI technologies and are already in the market, the following are few remarkable AI tools used in the industry. Biologics are large molecule drugs which requires analysis of molecular property and their structures. With Deep neural networks for instance convolutional neural networks complex molecular relationship and their biological properties can be learned. (Pathak, A., *et al.,* 2025; Wallach et al., 2015).

2.3 AI Technologies and Industry Landscape

Table 1 represents different AI technologies and AI-driven Biopharma companies in the global market

Table 1. Different AI-driven Drug Discovery and Research, their Institutional Collaboration and IT tie-ups.

<table>
<tr><th>Key AI Technologies</th><th>Utility</th><th>Industry-Institute Partnership</th><th>Biopharma-IT Companies Tie-ups</th></tr>
<tr><td rowspan="3">IBM Watson for Drug Discovery:
- Deep Search
- RXN
- Simulation</td><td>Natural language processing (NLP) and ML techniques to analyse scientific literature and clinical trial data.</td><td rowspan="3">Oxford and IBM</td><td rowspan="3">Novartis and Microsoft</td></tr>
<tr><td>Identify novel drug candidates and learn drug mechanism</td></tr>
<tr><td>RXN for Chemistry to shortlist candidates and narrow chemical recipes for new photoresists</td></tr>
<tr><td rowspan="2">Atomwise:
- AtomNet®</td><td>Neural networks for structure-based drug design.</td><td rowspan="2"></td><td rowspan="2">Atomwise-US</td></tr>
<tr><td>Analyzes molecular structures to predict the binding affinity of small molecules to target proteins.</td></tr>
<tr><td rowspan="2">Schrödinger:
MALT1; CDC7 programs</td><td>Provides physics-based computational software for drug discovery.</td><td rowspan="2" colspan="2">Ajax Therapeutics; Bright Angel Therapeutics; Bristol Myers Squibb; Lilly; Morphic Therapeutic; Nimbus Therapeutics</td></tr>
<tr><td>Combines molecular modelling with AI to predict molecular properties and interactions</td></tr>
<tr><td rowspan="2">DeepMind's AlphaFold:
Isomorphic Labs</td><td>Predicts protein structures from amino acid sequences with high accuracy.</td><td rowspan="2">Google</td><td rowspan="2"></td></tr>
<tr><td>Facilitates understanding of protein folding and its implications in drug design.</td></tr>
</table>

continued on following page

Table 1. Continued

Key AI Technologies	Utility	Industry-Institute Partnership	Biopharma-IT Companies Tie-ups
NVIDIA Clara™ AlphaFold2,	which predicts the 3D structure of the target protein with high accuracy	NVIDIA® BioNeMo™	
Insilico Medicine: Pharma.AI PandaOmics Chemistry42 engine NVIDIA BioNeMo	Employs generative adversarial networks (GANs) and reinforcement learning for drug discovery.		
	target identification, drug repurposing, and generative chemistry		
BenevolentAI: Ingestion and Insight Extraction; Data Integration and Inference AI-driven discovery	extract insights from biomedical data and literature. Ontologies, Databases (structured), Literature, Patents, Traisl	Academic Partners including Stanford, Cancer Research UK	AastraZeneca Merck Lilly
	Aids in target identification, molecular design, and clinical development		
Exscientia: Patient First AI	Drug design and optimization		Merck KGaA
	Integrates deep learning with medicinal chemistry to accelerate drug discovery.		
Recursion Pharmaceuticals: BioHive	Combines high end biology with AI to map and navigate human cellular biology.	Acquires Excisentia	
	Uses image-based screening and machine learning to discover new therapeutic candidates		
Cyclica: MatchMaker, an AI-enabled deep learning engine POEM (Pareto Optimal Embedding Model) engine	Offers a platform that integrates AI with computational biophysics to predict polypharmacology.		Vyant Bio, AstraZeneca, Cereval
	Focuses on drug repurposing and multi-target drug design.		
BioSymetrics: Augusta – incorporating their "one-pass" machine learning platform SymetryML Contingent-AI™	Utilizes AI and machine learning to analyse complex biological data. Supports target identification and validation, as well as biomarker discovery.		
PathAI: AISight™	Applies AI to pathology for better diagnosis and drug efficacy assessment. Helps in identifying and quantifying biomarkers from pathology images		
MolecularAI by AstraZeneca:	Uses AI for molecular generation and optimization. Focuses on designing novel compounds with desired properties for drug development.		

With voluminous data, publicly available datasets AI algorithms, and advanced AI processing techniques, drug discovery research has been into hand for production. Biopharma 4.0 is the next-gen pharmaceutical revolution in Biologics (e.g., vaccines, monoclonal antibodies, and cell therapies which harnesses potential of Artificial Intelligence (AI), Internet of Things (IoT), 3D Bioprinting, big data analytics, and robotics for manufacturing and producing Biologics . (Optimi Management Consulting, 2026).

2.4 AI-integrating Biopharma Companies

In Biopharmaceutical Industry, AstraZeneca employs digital twins and IoT-enabled clean room for manufacturing Biologics. Insilico Medicine, Black Diamond Therapeutics, and Relay Therapeutics use AI for identification of target, de novo molecular design and optimization. In biologics pharma business and projects, contract development and manufacturing are eased through AI platforms as in Contract research organizations (CROs). DeepMind (AlphaFold), NVIDIA (BioNeMo, Clara), Schrödinger, and Atomwise use their platforms for patient stratification, biomarker discovery and clinical trials. In addition to these, there are academic and research institution collaboration and partnership to facilitate tech transfer, knowledge development and research which enhances equitable access to resulting innovations

2.5 Approval and validation for clinical and commercial use

There are evidences of these AI-driven drugs which have advanced to clinical trials and have attempted for validation of their approach. For instance, Exscientia in their official platform has made announcement of their first AI-enabled drug entering human clinical trial in 2020 which is an OCD compound which has shown a progress of 12 months when compared to traditional drug design timeline of 4-5 years (Exscientia, 2020). Provided, these approaches can only validate its success through regulatory pathways and approval against safety, potency and purity standards.

3. REGULATORY FRAMEWORKS FOR AI-ENABLED BIOPHARMACEUTICALS

With AI having a remarkable growth in Biopharmaceutical industry, there are growing concerns on quality and bias towards the underlying data, which authors cite as a "black-box" nature of the LLMs which can delimit the trust and safety towards the clinical trials and approval for these drugs. These factors in addition to computational resources, technology gaps, data leakage and limited regulatory guidelines may limit equitable access to these Innovation. This chapter also shows why AI technologies has to be harnessed for effective Innovation-access balance achieving the Target 3b of Sustainable development goals.

3.1 The United States Regulatory Approach

In January 2025, the US has made draft guidelines 'Considerations for the Use of Artificial Intelligence To Support Regulatory Decision-Making for Drug and Biological Products'. The draft guidelines which invite comments from stakeholders provides recommendations to stakeholders to provide Information or data which supports the decision making of the authority for the safety, potency, quality and purity of these drugs. In addition to these, the guidance provides an assessment framework which is based on risk assessment on evaluation and establishing a credible AI model for use in Drug discovery (U.S. Food and Drug Administration., 2025). The FDA's Centre for Drug Evaluation and Research (CDER) and Centre for Biologics Evaluation and Research (CBER) is set to evaluate AI-driven submissions on a case-by-case basis, which is phased to make exhaustive documentation of the process of drug discovery which demonstrates the stages of how the algorithm is developed, training of the model and interoperability for regulatory decision making. (Babrak et al., 2019; Pathak et al., 2025).

AI as a medical Device: This portion analyses the FDA's Evolving Framework for treating AI based software as a medical device for Drug Discovery. The U.S. Food and Drug Administration in 2019 proposed another framework which treats AI and ML based software as a Medical Device (SaMD) comprehending that the traditional regulatory system that decided medical devices are not in par with continuously learning AI models (FDA, 2019). This change demonstrates an evolving regulatory requirement which is known "predetermined change control plan" that enables the software to enter algorithmic modifications without market submissions however within the limits prescribed which is a balancing mechanism for the growing nature of AI-ML software for drug discovery (Benjamens et al., 2020).

Regulatory Pathway for AI-Enabled Drug Discovery

The US FDA regulation aims at the resulting therapeutic product than the AI technological tools used in the process whereas the regulations defining the process more properly relies on the safety, ethical and data processing processes. The FDA'S draft guidance focuses on 'Risk based credibility assessment' and transparency system for safety and efficacy of these medicines. The draft guidance which has long been mediated through number of discussions and guidance for clinical trials (FDA, 2021) for AI driven drug discovery which improves trial design, participant recruitment, enhanced data analysis, streamlined decision-making, and individualised or personalized medicine.

The Draft guidance 2025 requires for the safety, drug quality, or clinical trial reliability of comprehensive disclosure of the architecture of the AI, the data used for training the models, the validation processes and the metrics regarding its performances. While on the other end, lower disclosure burdens for AI models used solely in early-stage drug discovery which does not impact the regulatory decision making in the drugs. This is in addition to the long standing disclose to provide adequate justification for design decisions, whether AI-driven or traditional, and to giving evidences that demonstrate the safety and efficacy standards of drug candidates regardless of discovery method (Kesselheim et al., 2015).

Data Standards and Algorithmic Transparency

With Data standards being one of the prime-most standard for AI driven drug discovery, the FDA guidance 2025 has laid down transparency provisions which are as follows: For the transparency requirements, the sponsors are required to be transparent regarding AI Model design with explainable AI capabilities that demonstrate how decisions or predictions are made. Secondly, Bias and fitness of data which are systems for detecting bias and correcting bias in training data; thirdly systems for lifecycle maintenance monitoring which includes detecting data drift, retrain or revalidate AI model, automated tracking system for the model performance; disclosure of testing methods; Good clinical Manufacturing processes, clinical decision support system and lastly the documentations regarding model development, evaluation, updates and credibility assessment. FDA's disclosure requirement poses a challenge with Innovations being patented or maintained as trade secrets. With the guidance 2025, FDA has overcome the long-standing 'black box' algorithms that resist interpretation and validation (Price, 2018).

3.2 European Union Regulatory Framework for AI driven Biologics

The European Union in 2024 has established an extensive regulatory framework for AI-driven biologics through the Artificial Intelligence Act (AI Act). This Act with its risk-based approach classifies AI-driven biologics requiring conformity assessment, CE marking and surveillance or monitoring post-marketing of these drugs (European Parliament and Council of the European Union, 2024). The European Medicines Agency (EMA) as a regulatory authority moved by the Legislation and 2024 guidance in addition to the assessment requirement emphasizes the need for algorithm validation, data quality assurance, model explainability, and human oversight (European Medicines Agency, 2024).

The data involved in drug discovery is governed through the General Data Protection Regulation (GDPR) in particular personal health data for AI training, while the Clinical Trials Regulation (EU) 536/2014. AI-enabled diagnostic tools and biological trials regulate through the regulation (EU) 2017/746 In Vitro Diagnostic Medical Devices Regulation (European Parliament and Council of the European Union, 2017).

The EU Artificial Intelligence Act (2024)

The Act classifies AI system into four risk categories (European Commission, 2024). The Act classifies AI systems into four risk categories and healthcare AI are placed in High risk which when compared to the US FDA's guidance documents as high-risk requiring risk-based assessment management and documentation, high-quality training data, human oversight, high accuracy and cybersecurity measures (Bas, T. G., & Duarte, V. (2024). The Act of 2024 establishes AI governance structures which are implemented through its European Artificial Intelligence Board to facilitate cooperation among member states and an AI Office for the effective implementation of the Act. (Veale & Borgesius, 2021). Similar to US FDA regulation, AI models in EU require a detailed submission of the model architecture, logs of training, data processing pipelines as part of clinical trial documentation (European Medicines Agency 2023)

Pharmaceutical Markets and EMA

The European Medicines Agency has proactively addressed AI applications in drug development through a series of draft papers in 2020 which identifies opportunities and challenges (EMA, 2020). Apart from Intelligence Board, EMA has established the Big Data Steering Group which deals particularly with the big data, AI, and digital health across the agency (European Medicines Agency. (2024). The Big Data group as a steering agency develops methodologies for handling novel data, quality standards for training models through data and provides guidance for use of AI within regulatory requirement. The European Commission also in its verge to enhance innovation, accessibility and resilience has proposed Pharmaceutical Strategy in 2020 which emphasizes Public Health as a prime objective and AI shall serve to enable equitable access to medicines (European Commission, 2020).

GDPR for AI in biologics

The EU GDPR General Data Protection Regulation (2016) is an important legislation in lawful usage of personal data and processing the data. The regulation provides for explicit requirement of consent or legitimate interest in personal data processing; Minimalizing the data collection to only those necessary; rights of data with regard to its access, rectification, erasure and data portability; restrained automation in decision making and sensitive health care data to a specialised protection. With AI's automated decision in drug discovery, the risk of safety and quality is challenged, however GDPR addresses through the "right to explanation" in decision making (Goodman & Flaxman, 2017).

Data transfers outside the EU are regulated particularly after Schrems II decision invalidating the Privacy Shield framework requires a reconsideration in International collaboration, knowledge transfer regulation. (Mildebrath, H., 2020). This has pushed Biopharma companies to manage data and data transfer across countries through contractual model and (Bradford, 2020).

3.3 National and regional frameworks

United Kingdom

The UK government's National AI Strategy (2021) requires industry specific sector-oriented regulation and framework and in pursuance of that the Medicines and Healthcare products Regulatory Agency (MHRA) has established a 'Software and AI as a Medical Device Change Programme' to develop appropriate oversight for adaptive AI systems (MHRA, 2022). Several UK companies have already rolled-out AI-Biologics integration such as Exscientia, Antiverse, Latent Labs. In addition to this, the UK government has made a huge £82.6 million investment which aims at propelling artificial intelligence (AI) in drug discovery (Pharmaceutical Technology., 2025) The UK has promoted regulatory sandboxes which allows testing of products under reduced compliances yet bound by regulatory supervision. The MHRA's Innovation Office also acts as an advisory for Innovations for products under development.

Japan

AI Strategy 2019 and the Society 5.0 vision for human-centered AI proposed the transformations for AI-driven drug discovery. The Pharmaceuticals and Medical Devices Agency (PMDA) as a regulatory authority establishes data usage and clinical trial framework for the integration (PMDA, 2020). ICH harmonization and bilateral negotiation are on the go for the integration of AI and Innovative therapeutics (Tominaga et al., 2021).

China

China's strategic and investment driven Biopharma Industry is regulated through The National Medical Products Administration (NMPA) which lays approval processes and issued guidance on innovative trial designs and real-world evidence (NMPA, 2020). China's the Personal Information Protection Law (2021) and recent regulations on algorithmic recommendations aims at data usage within the purview of regulated environment respecting privacy and algorithmic accountability (Roberts et al., 2021).

India

EY-Parthenon report at BioAsia 2026 titled as *"Pharma's New Architecture: Where Novel Science Meets AI and Manufacturing Power,"* at BioAsia 2026 gives an exhaustive view on India's platform-led innovation models for AI-led Biologics. Driven by the Central Drugs Standard Control Organization (CDSCO) and policy initiatives such as India's National Strategy for Artificial Intelligence (2018), the Country has been moving towards transformative AI deploying environment (NITI Aayog, 2018). With Biopharma SHAKTI program announced in Budget 2026 the country's ambition is on the go. AI Safety Institute which was announced in Jan 2025 to oversee responsible AI development are also in hand.

3.4 International Harmonization for transborder Data Flow in AI-driven Biologics

International Council for Harmonisation (ICH) has brough guidelines on good clinical practice, quality management, and pharmacovigilance for drug discovery and modern Biotechnology. This operates as an International neutral standard for drug discovery whether AI driven or otherwise. (ICH, 2016). Barring specific AI-driven drug discovery guidelines, the working groups studies on real-world evidence, adaptive trials, and complex innovative trial designs that in AI revolution operates in the drug discovery (Eichler et al., 2021).

World Health Organization WHO's ethical guidance for AI in health emphasizes six key principles in AI in Healthcare and involves protecting autonomy, promoting well-being and safety, ensuring transparency and explainability, fostering accountability, ensuring equity and inclusiveness, and promoting responsive and sustainable AI (WHO, 2021).

4. INTELLECTUAL PROPERTY CONSIDERATIONS

4.1 Patentability of AI-Generated Biopharmaceuticals

Intellectual property for AI-driven biologics opens great opportunities in healthcare. However Intellectual property concerns if undermined may inhibit future growth of this Industry. Studies advocate for collaborative approach from stakeholders in navigating the industry dynamics and also balance Innovation-access paradox. (Poddar A, et al. 2024)

The Inventorship

With growing Biotechnology and Artificial Intelligence Technology, there are already existing ambiguities in Inventorship. With Human-Integrated technology gaining momentum for Patent, AI enabled Biotechnology pose challenges from varied perspectives which includes attribution, sole-inventorship to an AI model if an independent molecule is invented. How are Humans who build the algorithm be credited?

The US Federal Circuit's *Recentive v. Fox* decision views the application of machine learning to useful data is not suffice to achieve patentability. Studies reveal how Patent offices across Nations have held AI systems cannot be made as inventors for varied reasoning (Abbott, 2020). While USPTO 2024 guidelines specifically claiming that Inventorship requires Natural persons for Patent protection (USPTO, 2024). Patentability can be assessed for individual inventions in AI-driven Biologics. Though the end-product itself cannot be differentiated from that of other biologics, the innovation harnessing the AI technology opens great deal of opportunities cutting cost and showing technical and economic efficiency. Patents are specific for instance, US11869629B2, United States for '*Systems and methods for artificial intelligence-guided biomolecule design and assessment*' filed by Pythia Labs Inc is granted patent for specific method of AI deployed process of drug discovery.

Beyond Patentability, the AI-driven drug must prove novelty, inventive step and industrial application for Patent to be granted. Though Novelty standards yet are undermined with large datasets and algorithms are copyright protected, the novelty standards as studies show much account for reasonable expectation of success rate than merely improving technical feasibility (Vertinsky & Rice, 2002). In addition to the above, Patent Law requires sufficient disclosure for Patent to be granted. In AI-driven drug, clinical trials and data which are more protected through Data exclusivity provisions or regulatory exclusivities which inhibits data disclosure requirement for patent protection. In AI-integrated Biologics, the molecular structure and synthesis pathway may be included but does that involve disclosing AI algorithm, training data and computational methods? Though with regard to granted patent, there are no disclosure requirements, the concern warrant that reproducibility of Biosimilars or bio better or follow on biologics require greater transparency (Christopher M Holman, 2015).

4.2 Regulatory Data Protection and Market Exclusivity

As discussed above, Regulatory data protection (RDP) is an IP like framework that grants innovators exclusive rights to regulatory approval data and prevents generic competitors from relying on the protected data for approval during the protected period. This protection complements patent protection granted to the product and can extend market exclusivity period beyond patent expiration. In the United States, the Biologics Price Competition and Innovation Act (BPCIA) is an exclusive legislation for Biologics which has 12 years of data exclusivity for innovative biologics that prevents biosimilar applications during this period (FDA, 2010). The European Union provides 8 years of data protection plus 2 years of market protection (10 years total) with potential additional year for new therapeutic indications (EMA, 2005). With generic and competition concern being one of the balancing factors for long standing pharma innovation-access conundrum, these RDP protection plays a dominant role with access to this pharma. The justification for RDP and patent protection more relies on recouping the investment, the time and cost of the research and development. With AI driven biopharma which substantially reduces timelines and cost, does RDP justified for the product? Additionally, does exorbitant pricing be justified? These questions are to be analysed in policy perspective to enable the utilization of the fullest potential of the AI.

AI and the Regulatory Data Protection

With AI driven Biologics also rely on publicly available datasets in training their models, the RDP for these products cannot be extended. However massively processed data sets, data encompassing molecular prediction, binding affinity estimates, assessment of toxicity, structural activity relationship are a valuable Intellectual property asset that lie distinct from that of the product itself. However, the current RDP protection does not address the data protection for these products. In fact, Biotechnology products norms are still evolving and the current regulations falls short in addressing the challenges (Eisenberg, 1987). Policy considerations in RDP pitches diverse directions: one in rewarding innovation that addresses neglected diseases and against ensuring affordability for small patient populations already facing access challenges. In addition to these, when data is used with generative AI, risks associated with them such as hallucinations, bias and outdated information mandates human integration, a very indispensable requirement. (King, M., et al., 2025)

4.3 Trade Secrets and Algorithmic disclosure

Most of the pharma companies go for trade secret protection for their AI algorithms, training datasets, and model architectures than entertaining copyright protection for the same. This approach undermines disclosure requirement but however is open to the risk of protection against independent discovery or reverse engineering (Lemley, 2008). When trade secrets across nations are not uniform, disruptions in regulating AI across jurisdiction remains a critical challenge which includes evaluating AI driven decisions and subsequent development of Biosimilars in case of Biopharmaceuticals. AI driven Biologics in addition to the same had to deal with information asymmetries where monopoly in data and information risks reproducibility of these Biologics to ensure equitable access to these drugs (Price, 2018).

4.4 Access, Licensing, and Technology Transfer

In addition to these, Intellectual property concerns for AI driven biologics also face novel challenges in IP system such as Compulsory Licensing. Compulsory licensing is an emergency provision in TRIPS Agreement (Trade-Related Aspects of Intellectual Property Rights) and many national laws that provide authorization for generic production without the consent of the Patent holder in emergencies, anti-competitive behaviour or inadequate supply. Though invoked in few instances, most notably for HIV/AIDS treatments in developing countries and during the COVID-19 pandemic are an effective tool for access to medicines ('t Hoen, 2016). These licenses are granted only when negotiations fail, but if in AI-driven drugs where IP relies primarily in AI algorithms and datasets identifying drug candidates, what will be licensed. Which is also complicated when blanket licensing is granted on the products. Additionally, provisions and factors regarding excessive pricing will also be changed with AI- driven drugs which can ease production cost and effectiveness.

4.5. Voluntary Licensing and Technology Transfer

AI driven Biologics will also have to navigate through Voluntary licensing and technology transfers facilitating local drug discovery and development. AI Biopharma opens itself to Medicines Patent Pool, an initiative established in 2010 that can encourage negotiation of agreements for licensing the technology across companies. (Medicines Patent Pool, 2020). With AI, traditional negotiations will be empowered with coupled negotiations in IP licensing of data mining, licensing technical capabilities, data processing platforms, and Standard Essential Patents (SEPs) in AI- Biopharma tech. With the advent of opensource AI, there are widespread opportunities in AI-drug discovery and development. For instance, DeepMind's effort in public release of AlphaFold's protein structure predictions fastens the speed of research and development of these drugs (Jumper et al., 2021). On the other hand, there is a great potential for these AIs in commercial viability when business companies regulate their proprietary value. Academic-Industry collaboration, Public Funding, Policy intervention are some of the significant moves for accelerating the potent of AI-enabled drug discovery (Contreras, 2011).

5. DATA GOVERNANCE, PRIVACY, AND SECURITY

5.1 AI Drug Discovery and Data Governance

With Large set of data being mined, processed and utilised in AI, there are concerns of privacy, data leakage, security and data governance imperatives. With AI, where data on molecular structure, binding, biological properties, clinical data, patient data involved, the imperative is very indispensable for achieving safety, potency and quality of the biologic drugs. (Niazi, 2026).

AI for drug discovery incorporates diverse data which includes Molecular and Chemical Data which involves the molecular structure, properties and synthesizing information. Public databases such as PubChem, ChEMBL, and ZINC and private NETDATA are few databases which AI drug developers rely on. In addition to molecular data, biological Data which include the genomic sequences, structures of protein, cellular imaging, genetic expressions, CAD models; clinical data which includes the trial data, health records, pharma records, patient specifications, diagnoses, treatments, administration and overall patient data are also relied on. In addition to these, (Niazi, 2026) shows how real-world data reliance of AI in drug discovery also poses disruptions in regulation. Data as mere information does not itself have commercial value, when these data are arranged combined and synthesised for example personalised medicine approach, specific molecular structure leads to protection on the other hand risks privacy and security which mandates regulation.

5.2 Regulation on Data Privacy and regulatory compliance

GDPR and Health Data: As discussed in EU regulatory framework, the GDP regulation categorizes health data as a special category of data requiring explicit consent and data usage regulation for research and usage safeguards (Vayena et al., 2018).

In AI driven drug discovery, requirement of consent for health data adds to the limitation in AI model training when such data becomes a protected set of information through consent requirement or even RDP protection (Dove et al., 2018). Secondly, AI driven Drug development limiting the usage of Data for specific purpose. With AI models being trained with exhaustive data sets and automated filtering for customised output, Data Minimization becomes an unregulated concern. GDPR also provides for restriction on Automated Decision making and the right of access and deletion for individuals. (Villaronga et al., 2018). Personal data of individuals once trained has to be guarded against misuse of the same. Recent trends show building of 'guardrails' for safety of personal data. (Dong, Y, et al., (2024).

US Data protection and AI-Integrated Drug Discovery

The United States unlike EU, does not specifically have a data legislation, and a recent report in 2024 advocates the use of data and data regulation while protecting individual and national interest. (Report on Artificial Intelligence 2024.). The report advocates for Data privacy, data security and government use of AI. Apart from these The Health Insurance Portability and Accountability Act (HIPAA) govern protected health information however limited only to certain entities. The Act permits for curtained categories of data which are classified as 'de-identified' data without the consent requirement which potentially challenges AI training and learning where data are not primarily classified (Benitez & Malin, 2010). Few other state legislations on privacy does more of compliance requirement than addressing the disrupting effect of AI in drug discovery (Solove & Hartzog, 2014).

5.3 Data sharing and access to data

Data sharing in AI drug discovery is an indispensable for harnessing the fullest potential of AI enabled drug discovery Davis AM, et al., 2021. Data sharing enables partnership by combining complementary datasets. Policy concerns should address barriers in data sharing such as 'proprietary data monopoly' which inhibits competition in the Market. Other barriers include privacy and ethical concerns. In addition to the above, legal standards are heavy concerns on data sharing which includes data ownership and 'guardrail' safeguards.

Studies suggest data governance models for beneficial data sharing and those are: establishing Data Trusts which can manage data and facilitate sharing (Delacroix & Lawrence, 2019); Another model suggest federated learning such as HealthChain and MELLODDY which acts a distributed technology for data sharing (Warnat-

Herresthal et al., 2021). Next the most used data commons approach rolled by NIH's National COVID Cohort Collaborative and the European Open Science Cloud which has repositories for data. (NIH, 2020).

In addition to above, robust data integrity and cybersecurity measures such as encryption, blockchain technology becomes essential for regulation of AI data governance. (FDA, 2022).

6. ETHICAL DIMENSIONS AND EQUITY CONSIDERATIONS

One of the most discussed concerns of AI technology disruption is the ethical concern of AI. With AI being deployed in Healthcare, this concern becomes the most essential concern to be discussed. Primarily the Algorithmic Bias and Health disparities have to be discussed.

6.1 Algorithmic Bias

With already existing health inequities between high and low- and middle-income countries and inequitable access to medicines, AI technology brings a promise of equity in innovation and access (Trouiller et al., 2002). However, inequity in access to these technologies such as data sharing, infrastructure incapabilities may disrupt the access to Innovative Biopharma. Bias lies in how the AI model is trained and addressing the ethical part of it enables to harness the fullest potential of AI-integrated biologics. Drugs resulting from limited Demographic representation may prove less effective or may even prove fatal for others (Obermeyer et al., 2019). AI models trained by data reflecting a particular medical system may not cover the other context or demographics. AI data harmonization in electronic health records may evade the harms caused by this bias (Rajkomar et al., 2018). Imbibing 'FAIR' technical measures mitigating bias; requiring a comprehensive disclosure of data demographics, model performances; Diversity in research and development teams which may have human integrated approach are few such approaches to mitigate bias (Char et al., 2018).

6.2 Global Health Equity and Access- addressing 10/90 Gap

The so well discussed global health inequity gap which is often cited as 10/90 gap addresses the divide between research investment and global diseases. AI is set to overcome this gap by reducing the time and gap for drug development. Beyond the digital divide and capability, the policy initiatives should keep pace with the advancements of technologies. Digital divide and infrastructure incapability with

AI technologies risks intensifying the already existing health inequity (Vayena et al., 2018).

6.3 WHO's guiding principles on AI

The World Health Organization's WHO lays 6 guiding principles as an ethical consideration of AI in healthcare. The principle primarily giving Human autonomy over AI autonomy. Secondly, the AI integrated healthcare attributes must be evaluated for safety and effectiveness over commercial interests. Thirdly, the principle requires AI systems for transparency, comprehensive and intelligible. Fourthly, giving clear accountability mechanisms ensuring responsible parties with AI system. Fifthly, ensuring inclusivity and equity in development and deployment of AI innovations. Lastly, developing an environmentally sustainable and adaptable. Though normative this guidance provides an ethical base for development and deployment of AI in healthcare.

6.4 Liability and Ethical Challenges

Liabilities in case of AI violations, IP infringement, and profit-sharing involve a great level of legal and practical difficulty. For instance, the vicarious liability theory of law does not prove valid with the AI system. Sustainable AI innovations have become one of the drawing themes of recent AI innovations expected to ensure equitable access to AI-driven healthcare innovations. With a balance of public health and private interest, regulation is also aimed at addressing the potential for AI to exacerbate health disparities if not carefully managed. Continuous monitoring and equitable responsibility of human intelligence and machine intelligence in maintaining human oversight and decision-making in critical healthcare processes remain essential.

7. POLICY RECOMMENDATIONS

Biopharmaceutical Innovations are complex technologies which require a great level of investment and challenges in development and approval. With AI- Integration in Biologic drug discovery, there is a promising solution to the long-standing Innovation-access paradox. The current legal and regulatory system though partially address concerns of AI, with AI entering into healthcare, it requires a great level of intensity and comprehension in regulation of the development and deployment of these drugs to ensure safety and quality of these medicines. This chapter analysed the Legal and regulatory framework and opportunities of AI enabled Biologics addressing the Innovation-access gap. The analysis revealed a significant reduction in drug discovery timeline for Biologics compared to the traditional method which lasts between 12-18 years and substantial cost involved in it. Secondly, the regulatory system in major jurisdictions addresses few concerns and are fragmented which requires a balanced and harmonized approach with AI-driven drug discovery. Thirdly, though AI-enabled Biologics opens opportunities for Intellectual property protections also inhibits access to these innovations when regulations regarding Inventorship, patentability and regulatory exclusivity is vague or ambiguous. Data governance, ethical concerns against Algorithmic bias and health disparity are few other concerns which require regulatory intervention and policy initiatives. Few policy recommendations are suggested for addressing the above said problems.

The rapid integration of artificial intelligence into biopharmaceutical research necessitates a comprehensive reconfiguration of traditional governance structures to ensure that technological leaps translate into equitable global health outcomes. The OECD in its report has identified over 1000 AI policy initiatives from 69 countries (The OECD Artificial Intelligence Policy Observatory.) Scholars suggest an open-end policy suggestion after evidencing how this technology challenges and disrupts the healthcare industry. (Singh R, Pet al., 2025). This chapter suggest some policy recommendations for Innovation-access paradox and primarily suggests encouraging the deployment of AI for Innovations in Biologics to ensure optimal development and approval of drugs.

Firstly, a dynamic regulatory framework that gives risk-proportionate supervision may evade the problem of fragmented regulation and unharmonized regulations across jurisdictions. For instance, EU implementing most of the OECD guidelines is one such initiative addressing the AI-driven drug discovery disparities. Giving a dynamic learning system for AI without necessitating a complete reassessment may ease the regulation of AI enabled drugs and continuous evolution of the AI model for drug development.

Secondly, giving clarity guidelines or enforceability regulations for 'defining' inventorship and patentability standards opens opportunities for Intellectual property protection as an incentive mechanism for Biologics Innovations.

Thirdly, laying guidelines for Regulatory data protection across nations and giving a legal recognition of this protection may evade the ambiguity in conjoined protection of the same drug candidate with IP and RDP regime. With studies revealing RDP as an 'evergreening' mechanism for Biologics, the lack of uniform regulation adds burden to the exiting inequity

Fourthly, policy initiatives ensuring algorithmic accountability and transparency to systematically encounter bias in training datasets may be made. With multimodal AI systems being built for drug discovery, the intensity of 'accountability' has widened requiring a lucid guideline on the same (Bommasani et al., 2021)

Fifthly, the most recommended policy initiative is capacity building and technology transfer. AI- integrated drug discovery fosters in low- and middle-income countries (LMICs) when domestic capacity and investment-oriented policy environment prevails. Investment enhancing factors such as well-established regulatory system, public-private partnership, academic-industry collaboration, knowledge centred economy, Intellectual property protection, technology transfer be initiated and implemented. Scholars also suggest the development of autonomous laboratories for drug discovery to support Innovation as an effective measure to address the access inequity. (King et al., 2009)

8. CONCLUSION

AI driven drug discovery provides a remarkable opportunity in Biologic drug development barring the complexity that existed in the conventional drug development. AI integration in Biologics accelerates discovery, mitigates costs and time in development of these drugs. With the advent of this Biopharma revolution, there is a need for the Legal and regulatory system to keep pace with the technologies. Legal and policy landscape for AI- integrated Biopharmaceuticals is aimed at addressing the Innovation-access paradox, a long-standing health disparity concern while ensuring safety and efficacy of these drugs.

This chapter is an endeavour to untangle the potential of AI in biopharmaceuticals, the opportunities and challenges it brings to the legal and regulatory systems. In particular, the Intellectual property concerns, data governance, data regulation, ethical and privacy concerns that perpetuates the already exiting medical access inequity. The chapter addressed the challenges and suggested for policy initiatives in encouraging innovation and harnessing the potential of AI in drug development. With Biopharmaceutical Innovations at the cross-roads, it is high-time for legal and regulatory landscape to deploy the AI-capabilities to meet the unmet needs of Public Health. With future research and potential of AI in biopharmaceuticals, a well-established legal and regulatory system will ensure equitable access to these medicines.

REFERENCES

Aayog, N. I. T. I. (2018). *National Strategy for Artificial Intelligence*. Government of India.

Abbott, R. (2020). The Artificial Inventor Project. WIPO Magazine, 2020.

Amthaniwala, B. K., & Rathod, J. A. (2025). Comparative insights into biosimilar development in the UAE and India: Opportunities, challenges, and global trends.

Babrak, L. M., Menetski, J., Rebhan, M., Nisato, G., Zinggeler, M., Brasier, N., & Krauthammer, M. (2019). Traditional and digital biomarkers: Two worlds apart? *Digital Biomarkers*, *3*(2), 92–102. DOI: 10.1159/000502000 PMID: 32095769

Bas, T. G., & Duarte, V. (2024). Biosimilars in the Era of Artificial Intelligence—International Regulations and the Use in Oncological Treatments. *Pharmaceuticals (Basel, Switzerland)*, *17*(7), 925. DOI: 10.3390/ph17070925 PMID: 39065775

Benitez, K., & Malin, B. (2010). Evaluating re-identification risks with respect to the HIPAA privacy rule. *Journal of the American Medical Informatics Association : JAMIA*, *17*(2), 169–177. DOI: 10.1136/jamia.2009.000026 PMID: 20190059

Benjamens, S., Dhunnoo, P., & Meskó, B. (2020). The state of artificial intelligence-based FDA-approved medical devices and algorithms. *NPJ Digital Medicine*, *3*(1), 1–11. DOI: 10.1038/s41746-020-00324-0 PMID: 32984550

Bhatia, A., Matthan, R., Khanna, T., & Balsari, S. (2020). Regulatory sandboxes: A cure for mHealth pilotitis? *Journal of Medical Internet Research*, *22*(9), e21276. DOI: 10.2196/21276 PMID: 32763889

Bommasani, R., et al. (2021). On the opportunities and risks of foundation models. arXiv preprint.

Bradford, A. (2020). *The Brussels Effect: How the European Union Rules the World*. Oxford University Press. DOI: 10.1093/oso/9780190088583.001.0001

Char, D. S., Shah, N. H., & Magnus, D. (2018). Implementing machine learning in health care—Addressing ethical challenges. *The New England Journal of Medicine*, *378*(11), 981–983. DOI: 10.1056/NEJMp1714229 PMID: 29539284

Contreras, J. L. (2011). Data sharing, latency variables, and science commons. *Berkeley Technology Law Journal*, *25*, 1601.

Davis, A. M., Engkvist, O., Fairclough, R. J., Feierberg, I., Freeman, A., & Iyer, P. (2021). Public-private partnerships: Compound and data sharing in drug discovery and development. *SLAS Discovery*, *26*(5), 604–619. DOI: 10.1177/2472555220982268 PMID: 33586501

Delacroix, S., & Lawrence, N. D. (2019). Bottom-up data Trusts: Disturbing the 'one size fits all' approach to data governance. *International Data Privacy Law*, *9*(4), 236–252. DOI: 10.1093/idpl/ipz014

Diaz, D. J., Gong, C., Ouyang-Zhang, J., Loy, J. M., Wells, J., Yang, D., Ellington, A. D., Dimakis, A. G., & Klivans, A. R. (2024). Stability oracle: A structure-based graph-transformer framework for identifying stabilizing mutations. *Nature Communications*, *15*(1), 6170. DOI: 10.1038/s41467-024-49780-2 PMID: 39043654

DiMasi, J. A., Grabowski, H. G., & Hansen, R. W. (2016). Innovation in the pharmaceutical industry: New estimates of R&D costs. *Journal of Health Economics*, *47*, 20–33. DOI: 10.1016/j.jhealeco.2016.01.012 PMID: 26928437

Dong, Y., Mu, R., Jin, G., Qi, Y., Hu, J., Zhao, X., . . . Huang, X. (2024). Building guardrails for large language models. arXiv preprint arXiv:2402.01822.

Dove, E. S., Kelly, S. E., Lucivero, F., Machirori, M., Dheensa, S., & Prainsack, B. (2018). Beyond individualism: Is there a place for relational autonomy in clinical practice and research? *Clinical Ethics*, *12*(3), 150–165. DOI: 10.1177/1477750917704156 PMID: 28989327

Drugs, P. (2026). Biologics manufacturing: Challenges and innovations. https://parabolicdrugs.com/biologics-manufacturing-challenges-and-innovations/

Eichler, H. G., Bloechl-Daum, B., Broich, K., Kyrle, P. A., Oderkirk, J., Rasi, G., Santos Ivo, R., Schuurman, A., Senderovitz, T., Slawomirski, L., Wenzl, M., & Paris, V. (2021). Data rich, information poor: Can we use electronic health records to create a learning healthcare system for pharmaceuticals? *Clinical Pharmacology and Therapeutics*, *105*(4), 912–922. DOI: 10.1002/cpt.1226 PMID: 30178490

Eisenberg, R. S. (1987). Proprietary rights and the norms of science in biotechnology research. *The Yale Law Journal*, *97*(2), 177–231. DOI: 10.2307/796481 PMID: 11660398

Ekins, S., Puhl, A. C., Zorn, K. M., Lane, T. R., Russo, D. P., Klein, J. J., Hickey, A. J., & Clark, A. M. (2019). Exploiting machine learning for end-to-end drug discovery and development. *Nature Materials*, *18*(5), 435–441. DOI: 10.1038/s41563-019-0338-z PMID: 31000803

EMA. (2005). *Guideline on data exclusivity*. European Medicines Agency.

EMA. (2020). *Reflection paper on the use of Artificial Intelligence (AI) in the medicinal product lifecycle*. European Medicines Agency.

European Commission. (2020). *Pharmaceutical Strategy for Europe*.

European Commission. (2024). *Regulation on Artificial Intelligence*. Official Journal of the European Union.

European Medicines Agency. "Reflection Paper on the Use of Artificial Intelligence in Lifecycle Medicines," Text European Medicines Agency, July 14, 2023, https://www.ema.europa.eu/en/news/reflection-paper-use-artificial-intelligence-lifecycle-medicines

European Medicines Agency. (2024, December 13). Big Data Steering Group (BDSG): 2024 report (EMA/471579/2024). https://www.ema.europa.eu/en/documents/report/big-data-steering-group-bdsg-2024-report_en.pdf

Exscientia. (2020). Press Release: First AI-designed drug molecule enters clinical trials.

EY-Parthenon. (2026, February 17). AI-powered platforms for discovery, R&D and connected manufacturing to drive Indian biopharma's next growth phase: EY-Parthenon report at BioAsia 2026. Ernst & Young. https://www.ey.com/en_in/newsroom/2026/02/ai-powered-platforms-for-discovery-r-d-and-connected-manufacturing-to-drive-indian-biopharma-s-next-growth-phase-ey-parthenon-report-at-bioasia-2026

FDA. (2010). *Biologics Price Competition and Innovation Act*. Food and Drug Administration.

FDA. (2018). *Framework for FDA's Real-World Evidence Program*. Food and Drug Administration.

FDA. (2019). *Proposed Regulatory Framework for Modifications to Artificial Intelligence/Machine Learning-Based Software as a Medical Device*. Food and Drug Administration.

FDA. (2020). *Artificial Intelligence/Machine Learning (AI/ML)-Based Software as a Medical Device (SaMD) Action Plan*. Food and Drug Administration.

FDA. (2021). *Adaptive Designs for Clinical Trials of Drugs and Biologics*. Food and Drug Administration.

FDA. (2022). *Cybersecurity in Medical Devices: Quality System Considerations and Content of Premarket Submissions*. Food and Drug Administration.

Fleming, N. (2018). How artificial intelligence is changing drug discovery. *Nature, 557*(7707), S55–S57. DOI: 10.1038/d41586-018-05267-x PMID: 29849160

Gibofsky, A., Evans, C., & Strand, V. (2022). Provider and patient knowledge gaps on biosimilars: Insights from surveys. *The American Journal of Managed Care, 28*(12, Suppl), S227–S233. PMID: 36493396

Goodman, B., & Flaxman, S. (2017). European Union regulations on algorithmic decision-making and a "right to explanation". *AI Magazine, 38*(3), 50–57. DOI: 10.1609/aimag.v38i3.2741

Harrer, S., Shah, P., Antony, B., & Hu, J. (2019). Artificial intelligence for clinical trial design. *Trends in Pharmacological Sciences, 40*(8), 577–591. DOI: 10.1016/j.tips.2019.05.005 PMID: 31326235

Holman, C.M. Bowman v. Monsanto Co.: A Bellwether for the Emerging Issue of Patentable Self-Replicating Technologies and Inadvertent Infringement, 80 MO. L. REV. (2015) Available at: https://scholarship.law.missouri.edu/mlr/vol80/iss3/6

Honchar, N. (2024, January 26). 10 companies applying AI for biologics discovery. BiopharmaTrend. https://www.biopharmatrend.com/artificial-intelligence/10-companies-applying-ai-for-biologics-discovery-590/

ICH. (2016). *Integrated Addendum to ICH E6(R1): Guideline for Good Clinical Practice E6(R2).* International Council for Harmonisation.

Jayatunga, M., Bruens, L., Ruder, L., Schulze, U., & Meier, C. (2022). AI in biologics discovery: An emerging frontier. Boston Consulting Group; reprinted in In Vivo. https://web-assets.bcg.com/4b/f8/ef10f64643578c65e34902c81513/ai-in-biologics-discovery-an-emerging-frontier.pdf

Jumper, J., Evans, R., Pritzel, A., Green, T., Figurnov, M., Ronneberger, O., Tunyasuvunakool, K., Bates, R., Žídek, A., Potapenko, A., Bridgland, A., Meyer, C., Kohl, S. A. A., Ballard, A. J., Cowie, A., Romera-Paredes, B., Nikolov, S., Jain, R., Adler, J., & Hassabis, D. (2021). Highly accurate protein structure prediction with AlphaFold. *Nature, 596*(7873), 583–589. DOI: 10.1038/s41586-021-03819-2 PMID: 34265844

Kesik-Brodacka, M. (2018). Progress in biopharmaceutical development. *Biotechnology and Applied Biochemistry, 65*(3), 306–322. DOI: 10.1002/bab.1617 PMID: 28972297

Kesselheim, A. S., Avorn, J., & Sarpatwari, A. (2015). The high cost of prescription drugs in the United States: Origins and prospects for reform. *Journal of the American Medical Association, 316*(8), 858–871. DOI: 10.1001/jama.2016.11237 PMID: 27552619

King, M., DeNoon, A., & Hart, C. (2025). AI in life science quality management and regulatory affairs: Global compliance and strategic readiness. IQVIA. https://www.iqvia.com/-/media/iqvia/pdfs/library/white-papers/2025/ai-in-life-science-quality-management-and-regulatory-affairs.pdf

King, R. D., Rowland, J., Oliver, S. G., Young, M., Aubrey, W., Byrne, E., Liakata, M., Markham, M., Pir, P., Soldatova, L. N., Sparkes, A., Whelan, K. E., & Clare, A. (2009). The automation of science. *Science, 324*(5923), 85–89. DOI: 10.1126/science.1165620 PMID: 19342587

Lal, S., Singh, B., & Kaunert, C. (2024). Role of Artificial Intelligence (AI) and Intellectual Property Rights (IPR) in Transforming Drug Discovery and Development in the Life Sciences: Legal and Ethical Concerns. *Library of Progress-Library Science, Information Technology & Computer, 44*(3).

Lee, S. L., O'Connor, T. F., Yang, X., Cruz, C. N., Chatterjee, S., Madurawe, R. D., Moore, C. M. V., Yu, L. X., & Woodcock, J. (2020). Modernizing pharmaceutical manufacturing: From batch to continuous production. *Journal of Pharmaceutical Innovation, 10*(3), 191–199. DOI: 10.1007/s12247-015-9215-8

Lemley, M. A. (2008). The surprising resilience of the patent system. *Texas Law Review, 95*, 1.

Lenarczyk, G., Minssen, T., Price, N., & Rai, A. (2025). The future of AI regulation in drug development: A comparative analysis. *Journal of Law and the Biosciences, 12*(2), lsaf028. DOI: 10.1093/jlb/lsaf028 PMID: 41221487

Medicines Patent Pool. (2020). *Annual Report*.

MHRA. (2022). Software and AI as a Medical Device Change Programme. Medicines and Healthcare products Regulatory Agency.

Mildebrath, H. (2020, September). The CJEU judgment in the Schrems II case (PE 652.073). European Parliamentary Research Service. https://www.europarl.europa.eu/RegData/etudes/ATAG/2020/652073/EPRS_ATA(2020)652073_EN.pdf

Mirakhori, F., & Niazi, S. K. (2025). Harnessing the AI/ML in Drug and Biological Products Discovery and Development: The Regulatory Perspective. *Pharmaceuticals (Basel, Switzerland), 18*(1), 47. DOI: 10.3390/ph18010047 PMID: 39861110

Moffat, J. G., Vincent, F., Lee, J. A., Eder, J., & Prunotto, M. (2017). Opportunities and challenges in phenotypic drug discovery: An industry perspective. *Nature Reviews. Drug Discovery, 16*(8), 531–543. DOI: 10.1038/nrd.2017.111 PMID: 28685762

Mullard, A. (2019). Machine learning brings cell imaging promises into focus. *Nature Reviews. Drug Discovery, 18*(9), 653–655. DOI: 10.1038/d41573-019-00144-2 PMID: 31477870

Niazi, S. K. (2026). A Critical Review of the FDA's Draft Guidance on Artificial Intelligence in Drug and Biological Product Regulation. Journal of Chemistry, 2026(1), 5202999.

NIH. (2020). *National COVID Cohort Collaborative (N3C)*. National Institutes of Health.

NMPA. (2020). *Guidelines on Clinical Trials Using Adaptive Designs*. National Medical Products Administration.

Obermeyer, Z., Powers, B., Vogeli, C., & Mullainathan, S. (2019). Dissecting racial bias in an algorithm used to manage the health of populations. *Science, 366*(6464), 447–453. DOI: 10.1126/science.aax2342 PMID: 31649194

Olawade, D. B., Kade, A., Egbon, E., Usman, S. O., Fapohunda, O., Ijiwade, J., & Ogbonna, C. E. (2025). Bioinformatics and artificial intelligence in genomic data analysis: Current advances and future directions. *Molecular Genetics and Genomics, 300*(1), 111. DOI: 10.1007/s00438-025-02314-x PMID: 41348251

Opderbeck, D. W. (2019). Artificial intelligence in pharmaceuticals, biologics, and medical devices: Present and future regulatory models. *Fordham Law Review, 88*, 553.

Optimi Management Consulting. (2026). Pharma & biologics 4.0: Key components and technologies. https://optimi.ai/pharma-%26-biologics-4-0

Pathak, A., Theagarajan, R., Rizqi, M. M., Nugraha, A. S., Boruah, T., Pratiksha, , Kumar, H., Naik, B., Yadav, S., Jha, A. K., Trivedi, A., Gupta, A. K., & Kumar, P. (2025). AI-enabled drug and molecular discovery: Computational methods, platforms, and translational horizons. *Discov Mol, 2*(1), 32. DOI: 10.1007/s44345-025-00037-5

Paul, D., Sanap, G., Shenoy, S., Kalyane, D., Kalia, K., & Tekade, R. K. (2021). Artificial intelligence in drug discovery and development. *Drug Discovery Today, 26*(1), 80–93. DOI: 10.1016/j.drudis.2020.10.010 PMID: 33099022

Pharmaceutical Technology. (2025, February 11). UK awards £82.6m to AI drug discovery companies. https://www.pharmaceutical-technology.com/news/uk-awards-82-6m-to-ai-drug-discovery-companies/

PMDA. (2020). *Reflection on the use of AI/ML in pharmaceuticals*. Pharmaceuticals and Medical Devices Agency.

Poddar, A., & Rao, S. R. (2024). Evolving intellectual property landscape for AI-driven innovations in the biomedical sector: Opportunities in stable IP regime for shared success. *Frontiers in Artificial Intelligence*, *7*, 1372161. DOI: 10.3389/frai.2024.1372161 PMID: 39355146

Price, W. N. (2018). Black-box medicine. *Harvard Journal of Law & Technology*, *28*, 419.

Rajkomar, A., Hardt, M., Howell, M. D., Corrado, G., & Chin, M. H. (2018). Ensuring fairness in machine learning to advance health equity. *Annals of Internal Medicine*, *169*(12), 866–872. DOI: 10.7326/M18-1990 PMID: 30508424

Recentive Analytics, Inc. v. Fox. Corp et al., Case No. 2023-2437 (Fed. Cir. Apr. 18, 2025).

Report on Artificial Intelligence 2024, December 19). 253-page House AI report offers lawmakers a policy tool. Government Technology. https://www.govtech.com/artificial-intelligence/253-page-house-ai-report-offers-lawmakers-a-policy-tool

Roberts, H., Cowls, J., Morley, J., Taddeo, M., Wang, V., & Floridi, L. (2021). The Chinese approach to artificial intelligence: An analysis of policy, ethics, and regulation. *AI & Society*, *36*(1), 59–77. DOI: 10.1007/s00146-020-00992-2

Sampathkumar, K., & Kerwin, B. A. (2024). Roadmap for drug product development and manufacturing of biologics. *Journal of Pharmaceutical Sciences*, *113*(2), 314–331. DOI: 10.1016/j.xphs.2023.11.004 PMID: 37944666

Singh, R., Paxton, M., & Auclair, J. (2025, May 17). Regulating the AI-enabled ecosystem for human therapeutics. *Communications Medicine*, *5*(1), 181. DOI: 10.1038/s43856-025-00910-x PMID: 40382515

Solove, D. J., & Hartzog, W. (2014). The FTC and the new common law of privacy. Columbia Law Review, 114, 583. 't Hoen, E. F. (2016). Private Patents and Public Health: Changing intellectual property rules for access to medicines. Health Action International.

Tominaga, T.. (2021). Regulatory science as a bridge between science and society. *Clinical Pharmacology and Therapeutics*, *109*(1), 39–48. PMID: 21691273

Trouiller, P., Olliaro, P., Torreele, E., Orbinski, J., Laing, R., & Ford, N. (2002). Drug development for neglected diseases: A deficient market and a public-health policy failure. *Lancet, 359*(9324), 2188–2194. DOI: 10.1016/S0140-6736(02)09096-7 PMID: 12090998

U.S. Food and Drug Administration. (2025). Considerations for the use of artificial intelligence to support regulatory decision-making for drug and biological products: Draft guidance for industry. *Federal Register, 90*(23), 3456–3489.

USPTO. (2024). Inventorship Guidance for AI-Assisted Inventions. United States Patent and Trademark Office.

Vamathevan, J., Clark, D., Czodrowski, P., Dunham, I., Ferran, E., Lee, G., Li, B., Madabhushi, A., Shah, P., Spitzer, M., & Zhao, S. (2019). Applications of machine learning in drug discovery and development. *Nature Reviews. Drug Discovery, 18*(6), 463–477. DOI: 10.1038/s41573-019-0024-5 PMID: 30976107

Vayena, E., Blasimme, A., & Cohen, I. G. (2018). Machine learning in medicine: Addressing ethical challenges. *PLoS Medicine, 15*(11), e1002689. DOI: 10.1371/journal.pmed.1002689 PMID: 30399149

Veale, M., & Borgesius, F. Z. (2021). Demystifying the Draft EU Artificial Intelligence Act. *Computer Law Review International, 22*(4), 97–112. DOI: 10.9785/cri-2021-220402

Vertinsky, L., & Rice, T. M. (2002). Thinking about thinking machines: Implications of machine inventors for patent law. *Boston University Journal of Science & Technology Law, 8*, 574.

Villaronga, E. F., Kieseberg, P., & Li, T. (2018). Humans forget, machines remember: Artificial intelligence and the right to be forgotten. *Computer Law & Security Review, 34*(2), 304–313. DOI: 10.1016/j.clsr.2017.08.007

Wallach, I., et al. (2015). AtomNet: a deep convolutional neural network for bioactivity prediction in structure-based drug discovery. arXiv preprint.

Warnat-Herresthal, S., Schultze, H., Shastry, K. L., Manamohan, S., Mukherjee, S., Garg, V., Sarveswara, R., Händler, K., Pickkers, P., Aziz, N. A., Ktena, S., Tran, F., Bitzer, M., Ossowski, S., Casadei, N., Herr, C., Petersheim, D., Behrends, U., Kern, F., & Schultze, J. L. (2021). Swarm Learning for decentralized and confidential clinical machine learning. *Nature, 594*(7862), 265–270. DOI: 10.1038/s41586-021-03583-3 PMID: 34040261

WHO. (2021). *Ethics and governance of artificial intelligence for health*. World Health Organization.

Withers, C. A., Rufai, A. M., Venkatesan, A., Tirunagari, S., Lobentanzer, S., Harrison, M., & Zdrazil, B. (2025). Natural language processing in drug discovery: Bridging the gap between text and therapeutics with artificial intelligence. *Expert Opinion on Drug Discovery*, *20*(6), 765–783. DOI: 10.1080/17460441.2025.2490835 PMID: 40298230

World Health Organization Regional Office for Europe. (n.d.). Targets of Sustainable Development Goal 3. https://www.who.int/europe/about-us/our-work/sustainable-development-goals/targets-of-sustainable-development-goal-3

Xu, Y.. (2021). Artificial intelligence in healthcare: A narrative review. *Stroke and Vascular Neurology*, *6*(3), 340–350. PMID: 33468639

Zhavoronkov, A., Ivanenkov, Y. A., Aliper, A., Veselov, M. S., Aladinskiy, V. A., Aladinskaya, A. V., Terentiev, V. A., Polykovskiy, D. A., Kuznetsov, M. D., Asadulaev, A., Volkov, Y., Zholus, A., Shayakhmetov, R. R., Zhebrak, A., Minaeva, L. I., Zagribelnyy, B. A., Lee, L. H., Soll, R., Madge, D., & Aspuru-Guzik, A. (2019). Deep learning enables rapid identification of potent DDR1 kinase inhibitors. *Nature Biotechnology*, *37*(9), 1038–1040. DOI: 10.1038/s41587-019-0224-x PMID: 31477924

Compilation of References

Aasheim, C. L., Williams, S., Rutner, P., & Gardiner, A. (2015). Data analytics vs. data science: A study of similarities and differences in undergraduate programs based on course descriptions. *Journal of Information Systems Education*, *26*(2), 103.

Aayog, N. I. T. I. (2018). *National Strategy for Artificial Intelligence*. Government of India.

Abbaszadeh, A., & Shahlai, A. (2025). Artificial Intelligence for CRISPR Guide RNA Design: Explainable Models and Off-Target Safety. *arXiv preprint arXiv:2508.20130.*

Abbott, R. (2020). The Artificial Inventor Project. WIPO Magazine, 2020.

Abdelwahab, O., Belzile, F., & Torkamaneh, D. (2023). Performance analysis of conventional and AI-based variant callers using short and long reads. *BMC Bioinformatics 2023 24:1, 24*(1), 472-. DOI: 10.1186/s12859-023-05596-3

Abdelwahab, O., & Torkamaneh, D. (2025). Artificial intelligence in variant calling: A review. *Frontiers in Bioinformatics*, *5*, 1574359. DOI: 10.3389/fbinf.2025.1574359 PMID: 40337525

Abioye, E. A., Abidin, M. S. Z., Mahmud, M. S. A., Buyamin, S., Ishak, M. H. I., Abd Rahman, M. K. I., Otuoze, A. O., Onotu, P., & Ramli, M. S. A. (2020). A review on monitoring and advanced control strategies for precision irrigation. *Computers and Electronics in Agriculture*, *173*, 105441. DOI: 10.1016/j.compag.2020.105441

Abraham, A. (2005). Nature and scope of AI techniques. *Handbook of Measuring System Design.*

Adanma, U. M., & Ogunbiyi, E. O.Uwaga Monica AdanmaEmmanuel Olurotimi Ogunbiyi. (2024). Artificial intelligence in environmental conservation: Evaluating cyber risks and opportunities for sustainable practices. *Computer Science & IT Research Journal*, *5*(5), 1178–1209. DOI: 10.51594/csitrj.v5i5.1156

Addo, A., & Senyo, P. K. (2021). Advancing E-governance for development: Digital identification and its link to socioeconomic inclusion. *Government Information Quarterly*, *38*(2), 101568. DOI: 10.1016/j.giq.2021.101568

Adewusi, A. O., Asuzu, O. F., Olorunsogo, T., Adaga, E., & Daraojimba, D. O. (2024). AI in precision agriculture: A review of technologies for sustainable farming practices. *World Journal of Agricultural Research*, *21*(1), 2276–2285.

Afrasiyabi, A., Larochelle, H., Lalonde, J. F., & Gagné, C. (2022). Matching feature sets for few-shot image classification. In *Proceedings of the IEEE/CVF conference on computer vision and pattern recognition* (pp. 9014-9024). DOI: 10.1109/CVPR52688.2022.00881

African Union. (2022). *AI & data science capacity building program*. https://au.int/en/tech/ai

Agarwal, S., & Verma, D. (2019). Artificial intelligence (AI) and predictive analytics in the healthcare sector. *Journal of Health Informatics*, *12*(3), 215–230.

Aguilar, J., Garces-Jimenez, A., R-Moreno, M. D., & García, R. (2021). A systematic literature review on the use of artificial intelligence in energy self-management in smart buildings. *Renewable & Sustainable Energy Reviews*, *151*, 111530. DOI: 10.1016/j.rser.2021.111530

Ahmad, N., Hussain, A., Ullah, I., & Zaidi, B. H. (2019, March). IOT based wireless sensor network for precision agriculture. In *2019 7th International electrical engineering congress (iEECON)* (pp. 1–4). IEEE. https://doi.org/DOI: 10.1109/iEECON45304.2019.8938854

Ahmad, A., Ahmed, M. M., Akhtar, A., Liu, W., Yang, R., Sun, X., Wang, X., Bibi, S., Khan, M. B., & Chen, S. (2026). Unlocking Grass Stress Resistance: Fungal Endophyte-Mediated Pathogen Recognition and RNA Regulation. *International Journal of Molecular Sciences*, *27*(9), 3899. DOI: 10.3390/ijms27093899 PMID: 42123484

Ahmad, A., Ahmed, M. M., Akhtar, A., Shuihong, C., Zafar, Z., Ullah, R., Asim, M., He, Z., & Khan, M. B. (2026). New Insight into Endophytic Fungi–Plant Symbioses Under Climate Change: Molecular Crosstalk, Nutrient Exchange, and Ecosystem Resilience. *Applied Microbiology (Basel, Switzerland)*, *6*(3), 47. DOI: 10.3390/applmicrobiol6030047

Ahmad, T., Madonski, R., Zhang, D., Huang, C., & Mujeeb, A. (2022). Data-driven probabilistic machine learning in sustainable smart energy/smart energy systems: Key developments, challenges, and future research opportunities in the context of smart grid paradigm. *Renewable & Sustainable Energy Reviews*, *160*, 112128. DOI: 10.1016/j.rser.2022.112128

Ahmed, M. M., Hamid, S. M., Asim, M., Ahmad, A., Raziq, A., Rui, Z., & Zhiyong, P. 2026. Nutrient Cycling and Plant-Microbe Interaction in Alpine Ecosystem: Microbial Mediation of Soil Fertility and Plant Growth. In *Biodiversity of Alpine Vegetation: Conservation, Challenges, Prospects, and Solutions* (pp. 319-364). IGI Global Scientific Publishing.

Ahmed, S., Khan, M. F., Singh, B., Singh, N., & Sharma, B. (1 C.E.). Enhancing Crime Scene Analysis: The Impact of AI Technologies on Evidence Processing. Https://Services.Igi-Global.Com/Resolvedoi/Resolve.Aspx?Doi=10.4018/979-8-3693-9405-2.Ch004, 63–84. DOI: 10.4018/979-8-3693-9405-2.ch004

Ahmed, A. A., & Echi, M. (2021). Hawk-eye: An ai-powered threat detector for intelligent surveillance cameras. *IEEE Access : Practical Innovations, Open Solutions*, *9*, 63283–63293. DOI: 10.1109/ACCESS.2021.3074319

Ahmed, E., El-Din, A., & El-Din, E. A. A. (2022). ARTIFICIAL INTELLIGENCE IN FORENSIC SCIENCE : INVASION OR REVOLUTION? *Egyptian Society of Clinical Toxicology Journal*, *10*(2), 20–32. DOI: 10.21608/esctj.2022.158178.1012

Ahmed, M. M., Asim, M., Kaleri, A. A., Manzoor, D., Rajput, A. A., Laghari, R., Khaki, S. A., Musawwir, A., Ullah, Z., & Ahmad, W. (2024). *Biochemical Dynamics and Quality Attributes of Strawberry Fruits across Maturity Stages with Respect to Different Preservation Methods: Biochemical Dynamics and Quality Attributes*. Futuristic Biotechnology. DOI: 10.54393/fbt.v4i01.98

Ahmed, M. M., Hamid, S. M., Rui, Z., Zhiyong, P., Xingyu, G., Qian, Y., Yu, S., Asim, M., Qiuhao, X., Ahmad, A., & Yerhazi, Y. (2026). Role of Microbial Biotechnology in Mitigating Climate-Induced Agricultural Stress. In *Impacts of Climate Change on Agriculture in the MENA Region* (pp. 257–280). IGI Global Scientific Publishing.

Ahmed, M. M., Sharif, U., Raza, A., Safdar, M., Ali, W., Asim, M., Muzammal, H., Sattar, J., Maqbool, S., & Zaheer, M. (2024). Climate-smart agriculture for resilience and profitability. In *Sustainable practices for agriculture and marketing convergence* (pp. 27–50). IGI Global Scientific Publishing. DOI: 10.4018/979-8-3693-2011-2.ch002

Aibar, S., González-Blas, C. B., Moerman, T., Huynh-Thu, V. A., Imrichova, H., Hulselmans, G., & Aerts, S. (2017). SCENIC: Single-cell regulatory network inference and clustering. *Nature Methods*, *14*(11), 1083–1086. DOI: 10.1038/nmeth.4463 PMID: 28991892

Aide, T. M., Corrada-Bravo, C., Campos-Cerqueira, M., Milan, C., Vega, G., & Alvarez, R. (2013). Real-time bioacoustics monitoring and automated species identification. *PeerJ*, *1*, e103. DOI: 10.7717/peerj.103 PMID: 23882441

Ajitha, V., Irin Sherly, S., Anitha, M., Kavitha, P. M., Prathipa, R., & Narayani, D. (2026). Synergistic integration of morphological and spectral features for enhanced rice leaf disease classification. In S. Kumar, R. T. Bye, & M. Prasad (Eds.), Proceedings of International Conference on Information Technology and Artificial Intelligence. ITAI 2025 (Lecture Notes in Networks and Systems, Vol. 1505). Springer. DOI: 10.1007/978-981-96-8687-2_20

Akhtar, J., & Amin, N. A. S. (2011). A review on process conditions for optimum bio-oil yield in hydrothermal liquefaction of biomass. *Renewable & Sustainable Energy Reviews*, *15*(3), 1615–1624. DOI: 10.1016/j.rser.2010.11.054

Al Bashish, D., Braik, M., & Bani-Ahmad, S. (2010). A framework for detection and classification of plant leaf and stem diseases. In *2010 International Conference on Signal and Image Processing* (pp. 113–118). IEEE. DOI: 10.1109/ICSIP.2010.5697452

Ali, A., Raza, A., Malik, A., Shabbir, M., Sarwar, N., & Muzammal, H. (2025). AI and IoT solutions for food security and agriculture. In *Enhancing Automated Decision-Making Through AI* (pp. 177–210). IGI Global Scientific Publishing.

Ali, A., Shakeel, T., Gulzar, M., & Wali, A. (2025). A systematic literature review on emerging technologies, methodologies, dataset, challenges, and future trends for rice disease detection. *Discov Computing*, *28*(1), 187. Advance online publication. DOI: 10.1007/s10791-025-09706-y

Ali, A., Yousef, A., Adil Raja, M., & Abdelal, M. (2024). An ensemble of deep learning architectures for accurate plant disease classification. *Ecological Informatics*, *81*, 102618. Advance online publication. DOI: 10.1016/j.ecoinf.2024.102618

Ali, M. (2023). E-governance and E-democracy: a Digital Revolution. Available at *SSRN* 4623414.

Aliper, A., Plis, S., Artemov, A., Ulloa, A., Mamoshina, P., & Zhavoronkov, A. (2016). Deep learning applications for predicting pharmacological properties of drugs and drug repurposing using transcriptomic data. *Molecular Pharmaceutics*, *13*(7), 2524–2530. DOI: 10.1021/acs.molpharmaceut.6b00248 PMID: 27200455

Ali, S., Abuhmed, T., El-Sappagh, S., Muhammad, K., Alonso-Moral, J. M., Confalonieri, R., & Herrera, F. (2023). Explainable Artificial Intelligence (XAI): What we know and what is left to attain Trustworthy Artificial Intelligence. *Information Fusion*, *99*, 101805. DOI: 10.1016/j.inffus.2023.101805

Alketbi, S. K. (2024). DNA Contamination in Crime Scene Investigations: Common Errors, Best Practices, and Insights from a Survey Study. SSRN *Electronic Journal*. https://doi.org/DOI: 10.2139/SSRN.4979243

Alotaibi, H., Alsolami, F., Abozinadah, E., & Mehmood, R. (2022). TAWSEEM: A Deep-Learning-Based Tool for Estimating the Number of Unknown Contributors in DNA Profiling. *Electronics 2022, Vol. 11, Page 548, 11*(4), 548. DOI: 10.3390/electronics11040548

Alotaibi, E., & Nassif, N. (2024). Artificial intelligence in environmental monitoring: In-depth analysis. *Discover Artificial Intelligence*, *4*(1), 84. DOI: 10.1007/s44163-024-00198-1

Alpaydin, E. (2020). *Introduction to machine learning* (4th ed.). MIT Press.

Alpay, L. L., Overberg, R. I., & Zwetsloot-Schonk, B. (2007). Empowering citizens in assessing health related websites: A driving factor for healthcare governance. *International Journal of Healthcare Technology and Management*, *8*(1-2), 141–160. DOI: 10.1504/IJHTM.2007.012103

Amer, S. A. (2025). Mitochondrial DNA as a Tool for Examining Molecular Diversity and Forensic Genetics. *Advances in Forensic Biology and Genetics*, 389–399. DOI: 10.1007/978-981-96-4585-5_21

Amthaniwala, B. K., & Rathod, J. A. (2025). Comparative insights into biosimilar development in the UAE and India: Opportunities, challenges, and global trends.

Anand, J., Yusoff, N., Ghani, H. A., & Thoti, K. K. (2023). Technological Applications in Smart Farming: A Bibliometric Analysis. [ASET]. *Advanced and Sustainable Technologies*, *2*(2), 30–40.

Andronie, M., Lăzăroiu, G., Karabolevski, O. L., tefănescu, R., Hurloiu, I., Dijmărescu, A., & Dijmărescu, I. (2022). Remote big data management tools, sensing and computing technologies, and visual perception and environment mapping algorithms in the internet of robotic things. *Electronics (Basel)*, *12*(1), 22. DOI: 10.3390/electronics12010022

Angelov, P. P., Soares, E. A., Jiang, R., Arnold, N. I., & Atkinson, P. M. (2021). Explainable artificial intelligence: An analytical review. *Wiley Interdisciplinary Reviews. Data Mining and Knowledge Discovery*, *11*(5), e1424. DOI: 10.1002/widm.1424

Annan, R., Noland, J., Perkins, K., Yuan, X., Roy, K., & Qingge, L. (2025). Genomic privacy and security in the era of artificial intelligence and quantum computing. *Discover Computing 2025 28:1, 28*(1), 108-. DOI: 10.1007/s10791-025-09627-w

Arango-Argoty, G., Garner, E., Pruden, A., Heath, L. S., Vikesland, P., & Zhang, L. (2018). Deep ARG: A deep learning approach for predicting antibiotic resistance genes from metagenomic data. *Microbiome*, *6*(1), 23. Advance online publication. DOI: 10.1186/s40168-018-0401-z PMID: 29391044

Arivazhagan, S.. (2013). Detection of unhealthy region of plant leaves and classification of plant leaf diseases using texture features. *Agricultural Engineering International: CIGR Journal*, *15*(1), 211–217.

Asim, M., Ateeq, M., Shen, X., An, S., Ahmed, M. M., Ashraf, M. A., Zhu, K., & Liu, J. (2025). Molecular mechanisms and adaptive strategies for hypoxia tolerance in horticultural crops: Beyond model systems. *Horticulture Advances*, *3*(1), 34. DOI: 10.1007/s44281-025-00088-1

Askin, S., Burkhalter, D., Calado, G., & El Dakrouni, S. (2023). Artificial intelligence applied to clinical trials: Opportunities and challenges. *Health and Technology*, *13*(2), 203–213. DOI: 10.1007/s12553-023-00738-2 PMID: 36923325

Asner, G. P., & Mascaro, J. (2014). Mapping tropical forest carbon: Calibrating plot estimates to a simple LiDAR metric. *Remote Sensing of Environment*, *140*, 614–624. DOI: 10.1016/j.rse.2013.09.023

Athanasopoulou, K., Michalopoulou, V. I., Scorilas, A., & Adamopoulos, P. G. (2025). Integrating Artificial Intelligence in Next-Generation Sequencing: Advances, Challenges, and Future Directions. *Current Issues in Molecular Biology 2025, Vol. 47, Page 470, 47*(6), 470. DOI: 10.3390/cimb47060470

Babrak, L. M., Menetski, J., Rebhan, M., Nisato, G., Zinggeler, M., Brasier, N., & Krauthammer, M. (2019). Traditional and digital biomarkers: Two worlds apart? *Digital Biomarkers*, *3*(2), 92–102. DOI: 10.1159/000502000 PMID: 32095769

Bacchi, S., Jeyam, A., & Gupta, A. (2024). Computational parsimony: Frugal AI as a prerequisite for clinical equity in resource-constrained environments. *Journal of Sustainable Digital Health*, *12*(2), 142–159. DOI: 10.1016/jsdh.2024.004

Ballard, J. L., Wang, Z., Li, W., Shen, L., & Long, Q. (2024). Deep learning-based approaches for multi-omics data integration and analysis. *BioData Mining 2024 17:1, 17*(1), 38-. DOI: 10.1186/s13040-024-00391-z

Barman, S., Al Farid, F., Raihan, J., Khan, N. A., Hafiz, M. F. B., Bhattacharya, A., & Mahmud, Z.. (2024). Optimized crop disease identification in Bangladesh: A deep learning and SVM hybrid model for rice, potato, and corn. *Journal of Imaging, 10*(8), 183. Advance online publication. DOI: 10.3390/jimaging10080183 PMID: 39194972

Bartley, M., & Kelly-Irving, M. (2024). *Health inequality: an introduction to concepts, theories and methods*. John Wiley & Sons.

Bas, T. G., & Duarte, V. (2024). Biosimilars in the Era of Artificial Intelligence—International Regulations and the Use in Oncological Treatments. *Pharmaceuticals (Basel, Switzerland), 17*(7), 925. DOI: 10.3390/ph17070925 PMID: 39065775

Beery, S., Van Horn, G., & Perona, P. (2018). Recognition in terra incognita. In *Proceedings of the European conference on computer vision (ECCV)* (pp. 456-473).

Behkami, N. A., & Daim, T. U. (2012). Research forecasting for health information technology (HIT), using technology intelligence. *Technological Forecasting and Social Change, 79*(3), 498–508. DOI: 10.1016/j.techfore.2011.08.015

Beijbom, O., Edmunds, P. J., Kline, D. I., Mitchell, B. G., & Kriegman, D. (2012). Automated annotation of coral reef survey images. *2012 IEEE Conference on Computer Vision and Pattern Recognition*, 1170–1177. DOI: 10.1109/CVPR.2012.6247798

Bendre, M. R., Thool, R. C., & Thool, V. R. (2015, September). Big data in precision agriculture: Weather forecasting for future farming. In *2015 1st international conference on next generation computing technologies (NGCT)* (pp. 744–750). IEEE. https://doi.org/DOI: 10.1109/NGCT.2015.7375205

Benedict, M., & Schlieter, H. (2015). Governance guidelines for digital healthcare ecosystems. In *eHealth2015–Health Informatics Meets eHealth* (pp. 233-240). IOS Press. DOI: 10.3233/978-1-61499-524-1-233

Benitez, K., & Malin, B. (2010). Evaluating re-identification risks with respect to the HIPAA privacy rule. *Journal of the American Medical Informatics Association : JAMIA, 17*(2), 169–177. DOI: 10.1136/jamia.2009.000026 PMID: 20190059

Benjamens, S., Dhunnoo, P., & Meskó, B. (2020). The state of artificial intelligence-based FDA-approved medical devices and algorithms. *NPJ Digital Medicine, 3*(1), 1–11. DOI: 10.1038/s41746-020-00324-0 PMID: 32984550

Berkes, F. (2012). *Sacred ecology* (3rd ed.). Routledge. DOI: 10.4324/9780203123843

Bhambere, H. S. S., Abhishek, B., & Sumit, H. (2021). Rapid digitisation of healthcare: A review of COVID-19 impact on our health systems. *Int. J. All Res. Educ. Sci. Methods*, *9*, 1457–1459.

Bhatia, A., Matthan, R., Khanna, T., & Balsari, S. (2020). Regulatory sandboxes: A cure for mHealth pilotitis? *Journal of Medical Internet Research*, *22*(9), e21276. DOI: 10.2196/21276 PMID: 32763889

Bhatti, M. A., Zeeshan, Z., & Syam, M. S.. (2024). Advanced plant disease segmentation in precision agriculture using optimal dimensionality reduction with Fuzzy C-Means clustering and deep learning. *IEEE Journal of Selected Topics in Applied Earth Observations and Remote Sensing*, *17*, 18264–18277. DOI: 10.1109/JSTARS.2024.3437469

Bhola, A., & Kumar, P. (2025). Deep feature-support vector machine based hybrid model for multi-crop leaf disease identification in corn, rice, and wheat. *Multimedia Tools and Applications*, *84*(8), 4751–4771. DOI: 10.1007/s11042-024-18733-8

Bhuvana, M., & Vasantha, S. (2022). Determinants of E-Governance Health Care Services Requirements and Rural Citizen Satisfaction. *Managing Disruptions in Business: Causes, Conflicts, and Control*, 317-334.

Bindu, N., Kumar, R. A., Radhakrishnan, T., Kumar, T. M., & Kumar, K. S. (2025). Virtual Realities: The Influence of the Metaverse on Healthcare E-Governance. *Applying Metaverse Technologies to Human-Computer Interaction for Healthcare*, 254.

Birhane, A. (2021). Algorithmic injustice: A relational ethics approach. *Patterns (New York, N.Y.)*, *2*(2), 100205. Advance online publication. DOI: 10.1016/j.patter.2021.100205 PMID: 33659914

Bogoch, I. I., Watts, A., Thomas-Bachli, A., Huber, C., Kraemer, M. U. G., & Khan, K. (2020). Pneumonia of unknown etiology in Wuhan, China: Potential for international spread via commercial air travel. *Journal of Travel Medicine*, *27*(2), taaa008. Advance online publication. DOI: 10.1093/jtm/taaa008 PMID: 31943059

Bommasani, R., et al. (2021). On the opportunities and risks of foundation models. arXiv preprint.

Borhani, Y., Khoramdel, J., & Najafi, E. (2022). A deep learning-based approach for automated plant disease classification using a vision transformer. *Scientific Reports*, *12*(1), 11554. Advance online publication. DOI: 10.1038/s41598-022-15163-0 PMID: 35798775

Bostrom, N., & Yudkowsky, E. (2014). The ethics of artificial intelligence. In Frankish, K., & Ramsey, W. M. (Eds.), *The Cambridge handbook of artificial intelligence* (pp. 316–334). Cambridge University Press. DOI: 10.1017/CBO9781139046855.020

Bradford, A. (2020). *The Brussels Effect: How the European Union Rules the World*. Oxford University Press. DOI: 10.1093/oso/9780190088583.001.0001

Bradley, P., Gordon, N. C., Walker, T. M., Dunn, L., Heys, S., Huang, B., & Iqbal, Z. (2015). Rapid antibiotic-resistance predictions from genome sequence data for *Staphylococcus aureus* and *Mycobacterium tuberculosis*. *Nature Communications*, *6*(1), 10063. Advance online publication. DOI: 10.1038/ncomms10063 PMID: 26686880

Brasil, S., Neves, C. J., Rijoff, T., Falcão, M., Valadão, G., Videira, P. A., & dos Reis Ferreira, V. (2021). Artificial Intelligence in Epigenetic Studies: Shedding Light on Rare Diseases. *Frontiers in Molecular Biosciences*, *8*, 648012. DOI: 10.3389/fmolb.2021.648012 PMID: 34026829

Bukhari, H. R., Mumtaz, R., Inayat, S., Shafi, U., Haq, I. U., Zaidi, S. M. H., & Hafeez, M. (2021). Assessing the impact of segmentation on wheat stripe rust disease classification using computer vision and deep learning. *IEEE Access : Practical Innovations, Open Solutions*, *9*, 164986–165004. DOI: 10.1109/ACCESS.2021.3134196

Burton, A. C., Neilson, E., Moreira, D., Ladle, A., Steenweg, R., Fisher, J. T., Bayne, E., & Boutin, S. (2015). Wildlife camera trapping: A review and recommendations for linking surveys to ecological processes. *Journal of Applied Ecology*, *52*(3), 675–685. DOI: 10.1111/1365-2664.12432 PMID: 26211047

Carbonell, P., Hu, F., & Faulon, J.-L. (2020). Computational methods for pathway and strain design. *Current Opinion in Biotechnology*, *62*, 1–10. DOI: 10.1016/j.copbio.2019.08.003

Carbonell, P., Radivojević, T., & Garcia Martin, H.. (2019). An automated design–build–test–learn pipeline for enhanced microbial production of fine chemicals. *Communications Biology*, *2*(1), 66. Advance online publication. DOI: 10.1038/s42003-018-0076-9 PMID: 30271948

Carlson, C. J., Burgio, K. R., Dougherty, E. R., Phillips, A. J., Bueno, V. M., Clements, C. F., Castaldo, G., Dallas, T. A., Cizauskas, C. A., Cumming, G. S., Doña, J., Harris, N. C., Jovani, R., Mironov, S., Muellerklein, O. C., Proctor, H. C., & Getz, W. M. (2017). Parasite biodiversity faces extinction and redistribution in a changing climate. *Science Advances*, *3*(9), e1602422. DOI: 10.1126/sciadv.1602422 PMID: 28913417

Caron, M., Touvron, H., Misra, I., Jegou, H., Mairal, J., Bojanowski, P., & Joulin, A. (2021). Emerging Properties in Self-Supervised Vision Transformers. *2021 IEEE/CVF International Conference on Computer Vision (ICCV)*, 9630–9640. DOI: DOI: 10.1109/ICCV48922.2021.00951

Carroll, S. R., Garba, I., Figueroa-Rodríguez, O. L., Holbrook, J., Lovett, R., Materechera, S., Parsons, M., Raseroka, K., Rodriguez-Lonebear, D., Rowe, R. K., Sara, R., Walker, J. D., Anderson, J., & Hudson, M. (2020). The CARE Principles for Indigenous Data Governance. *Data Science Journal, 19*(1), 43. DOI: 10.5334/dsj-2020-043

Cavalcanti, P., & Nogueira, T. L. S. DE CARVALHO, E. F., & DA SILVA, D. A. (2024). Forensic use of human mitochondrial DNA: A review. *Anais da Academia Brasileira de Ciências, 96*(4), e20231179. https://doi.org/DOI: 10.1590/0001-3765202420231179

Chafai, N., Hayah, I., Houaga, I., & Badaoui, B. (2023). A review of machine learning models applied to genomic prediction in animal breeding. *Frontiers in Genetics*, *14*, 1150596. DOI: 10.3389/fgene.2023.1150596 PMID: 37745853

Char, D. S., Shah, N. H., & Magnus, D. (2018). Implementing machine learning in health care—Addressing ethical challenges. *The New England Journal of Medicine*, *378*(11), 981–983. DOI: 10.1056/NEJMp1714229 PMID: 29539284

Chaudhary, P.. (2012). Color transform based approach for disease spot detection on plant leaves. *International Journal of Computer Science and Telecommunications*, *3*(6), 65–70.

Cheng, K., Li, Z., He, Y., Guo, Q., Lu, Y., Gu, S., & Wu, H. (2023). Potential use of artificial intelligence in infectious disease: Take ChatGPT as an example. *Annals of Biomedical Engineering*, *51*(6), 1130–1135. DOI: 10.1007/s10439-023-03203-3 PMID: 37074486

Chen, H., Lindshield, S., Ndiaye, P. I., Ndiaye, Y. H., Pruetz, J. D., & Reibman, A. R. (2022). Applying Few-Shot Learning for In-the-Wild Camera-Trap species classification. *AI*, *4*(3), 574–597. DOI: 10.3390/ai4030031

Chen, Y. (2020). IoT, cloud, big data and AI in interdisciplinary domains. *Simulation Modelling Practice and Theory*, *102*, 102070. DOI: 10.1016/j.simpat.2020.102070

Ching, T., Himmelstein, D. S., Beaulieu-Jones, B. K., Kalinin, A. A., Do, B. T., Way, G. P., & Xie, W. (2018). Opportunities and obstacles for deep learning in biology and medicine. *Journal of the Royal Society, Interface*, *15*(141), 20170387. DOI: 10.1098/rsif.2017.0387 PMID: 29618526

Chong, W. Y., Liu, J., & Lee, K. T. (2020). AI-driven smart city initiatives for healthcare and energy optimization. *Sustainable Cities and Society, 62*, 102381. DOI: 10.1016/j.scs.2020.102381

Chopra, H., Choudhary, O. P., & Emran, T. B. (2024). G20 summit in India: Unveiling the digital health initiative with 'Vasudhaiva Kutumbakam'. *International Journal of Surgery, 110*(1), 591–593. DOI: 10.1097/JS9.0000000000000771 PMID: 38315797

Christin, S., Hervet, É., & Lecomte, N. (2019). Applications for deep learning in ecology. *Philosophical Transactions of the Royal Society B*, 374(1763), 20180084.

Christin, S., Hervet, É., & Lecomte, N. (2019). Applications for deep learning in ecology. *Methods in Ecology and Evolution, 10*(10), 1632–1644. DOI: 10.1111/2041-210X.13256

Chuai, G., Ma, H., Yan, J., Chen, M., Hong, N., Xue, D., & Liu, Q. (2018). Deep CRISPR: Optimized CRISPR guide RNA design by deep learning. *Genome Biology, 19*(1), 80. Advance online publication. DOI: 10.1186/s13059-018-1459-4 PMID: 29945655

Contreras, J. L. (2011). Data sharing, latency variables, and science commons. *Berkeley Technology Law Journal, 25*, 1601.

Cortès, U., Sànchez-Marrè, M., Ceccaroni, L., R-Roda, I., & Poch, M. (2000). Artificial intelligence and environmental decision support systems. *Applied Intelligence, 13*(1), 77–91. DOI: 10.1023/A:1008331413864

Critchlow, R., Plumptre, A. J., Alidria, B., Nsubuga, M., Driciru, M., Rwetsiba, A., Wanyama, F., & Beale, C. M. (2017). Improving law-enforcement effectiveness and efficiency in protected areas using ranger-collected monitoring data. *Conservation Letters, 10*(5), 572–580. DOI: 10.1111/conl.12288

Crooks, K. R., Burdett, C. L., Theobald, D. M., Rondinini, C., & Boitani, L. (2011). Global patterns of fragmentation and connectivity of mammalian carnivore habitat. *Philosophical Transactions of the Royal Society of London. Series B, Biological Sciences, 366*(1578), 2642–2651. DOI: 10.1098/rstb.2011.0120 PMID: 21844043

Cruickshank, C., Wade, C., & Bajwa, J. (2024, August 29). How AI could help reduce inequities in health care. Harvard Business Review. https://hbr.org/2024/08/how-ai-could-help-reduce-inequities-in-health-care

Cruickshank, J., Miller, S., & Thorne, L. (2024). Digital favoring in patient triage: An equity lens on telemedicine deployment. *The Lancet. Digital Health, 6*(4), e215–e224. DOI: 10.1016/S2589-7500(24)00012-X

Cruz, F. B. D. L., & Barlaz, M. A. (2010). Estimation of waste component-specific landfill decay rates using laboratory-scale decomposition data. *Environmental Science & Technology, 44*(12), 4722–4728. DOI: 10.1021/es100240r PMID: 20496890

Crysup, B., Mandape, S., King, J. L., Muenzler, M., Kapema, K. B., & Woerner, A. E. (2023). Using unique molecular identifiers to improve allele calling in low-template mixtures. *Forensic Science International. Genetics, 63*, 102807. DOI: 10.1016/j.fsigen.2022.102807 PMID: 36462297

Dahal, A., McNevin, D., Chikhani, M., & Ward, J. (2023). An interdisciplinary forensic approach for human remains identification and missing persons investigations. *WIREs Forensic Science, 5*(4), e1484. DOI: 10.1002/wfs2.1484

Dankwa-Mullan, I. (2024a). *Algorithmic reparations and the future of health equity*. Harvard University Press.

Dankwa-Mullan, I. (2024b). Health equity and ethical considerations in using artificial intelligence in public health and medicine. *Preventing Chronic Disease, 21*, 240245. Advance online publication. DOI: 10.5888/pcd21.240245 PMID: 39173183

Das, S., Garg, V., & Joseph, J. J. (2023). Advancements in e-governance initiatives: Digitalising healthcare in India. In *International Conference on Advances in IoT and Security with AI* (pp. 287-296). Singapore: Springer Nature Singapore. DOI: 10.1007/978-981-99-5085-0_28

Dasgupta, A., & Deb, S. (2008). Telemedicine: A new horizon in public health in India. *Indian Journal of Community Medicine : Official Publication of Indian Association of Preventive & Social Medicine, 33*(1), 3. DOI: 10.4103/0970-0218.39234 PMID: 19966987

Dash, H. R., Elkins, K. M., & Al-Snan, N. R. (2023). Advanced Emerging Techniques for Forensic DNA Analysis: STRs, SNPs, and mtDNA Analysis. *Advancements in Forensic DNA Analysis*, 35–59. DOI: 10.1007/978-981-99-6195-5_4

Dash, S. P. (2020). The Impact of IoT in Healthcare: Global Technological Change & The Roadmap to a Networked Architecture in India. *Journal of the Indian Institute of Science, 100*(4), 1–13. DOI: 10.1007/s41745-020-00208-y PMID: 33162693

Das, R., Gupta, P., & Nair, S. (2021). The impact of telemedicine and mHealth platforms. *Telemedicine and Digital Health Review, 18*(2), 105–119.

Dave, M., & Patel, N. (2023). Artificial intelligence in healthcare and education. *British Dental Journal, 234*(10), 761–764. DOI: 10.1038/s41415-023-5845-2 PMID: 37237212

Davis, A. M., Engkvist, O., Fairclough, R. J., Feierberg, I., Freeman, A., & Iyer, P. (2021). Public-private partnerships: Compound and data sharing in drug discovery and development. *SLAS Discovery*, *26*(5), 604–619. DOI: 10.1177/2472555220982268 PMID: 33586501

Davis, J. J., Boisvert, S., Brettin, T., Kenyon, R. W., Mao, C., Olson, R., & Stevens, R. (2016). Antimicrobial resistance prediction in PATRIC and RAST. *Scientific Reports*, *6*(1), 27930. Advance online publication. DOI: 10.1038/srep27930 PMID: 27297683

De Groot, A. S., & Moise, L. (2007). Prediction of immunogenicity for therapeutic proteins: State of the art. *Current Opinion in Drug Discovery & Development*, *10*(3), 332–340. PMID: 17554860

De la Puente, M., Ruiz-Ramírez, J., Ambroa-Conde, A., Xavier, C., Pardo-Seco, J., Álvarez-Dios, J., Freire-Aradas, A., Mosquera-Miguel, A., Gross, T. E., Cheung, E. Y. Y., Branicki, W., Nothnagel, M., Parson, W., Schneider, P. M., Kayser, M., Carracedo, Á., Lareu, M. V., & Phillips, C. (2021). Development and evaluation of the ancestry informative marker panel of the Visage Basic Tool. *Genes*, *12*(8), 1284. DOI: 10.3390/genes12081284 PMID: 34440458

Debsharma, S. K., Rahman, M. A., Quddus, M. R., Khatun, H., Disha, R. F., Roy, P. R., Ahmed, S., El-Sharnouby, M., Iftekharuddaula, K. M., Aloufi, S., Alzuaibr, F. M., Alqurashi, M., Sakran, M. I., & Kabir, M. S. (2022). SNP-Based Trait Characterization Detects Genetically Important and Stable Multiple Stress Tolerance Rice Genotypes in Salt-Stress Environments. *Plants 2022, Vol. 11, Page 1150*, *11*(9), 1150. DOI: 10.3390/plants11091150

DeepMind. (2019). *Using machine learning to forecast wind energy*. https://deepmind.com/blog/article/forecasting-wind-energy

Delacroix, S., & Lawrence, N. D. (2019). Bottom-up data Trusts: Disturbing the 'one size fits all' approach to data governance. *International Data Privacy Law*, *9*(4), 236–252. DOI: 10.1093/idpl/ipz014

Delgado-Baquerizo, M., Oliverio, A. M., Brewer, T. E., Benavent-González, A., Eldridge, D. J., Bardgett, R. D., & Fierer, N. (2018). A global atlas of the dominant bacteria found in soil. *Science*, *359*(6373), 320–325. DOI: 10.1126/science.aap9516 PMID: 29348236

Deng, J., Dong, W., Socher, R., Li, L., Li, N. K., & Fei-Fei, N. L. (2009). ImageNet: A large-scale hierarchical image database. *2009 IEEE Conference on Computer Vision and Pattern Recognition*, 248–255. DOI: DOI: 10.1109/CVPR.2009.5206848

Dey, A., Nandi, S., & Sarkar, M. (2018). Security measures in IOT based 5G networks. In 2018 3rd International Conference on Inventive Computation Technologies (ICICT), pp. 561-566, IEEE.

Dhanush, M. S., & Thenmozhi, M. (2022). Missing Person Detection Using AI. *ACM International Conference Proceeding Series*. https://doi.org/DOI: 10.1145/3590837.3590931;JOURNAL:JOURNAL:ACMOTHERCONFERENCES;CTYPE:STRING:BOOK

Dhaygude, S. B., & Kumbhar, N. P. (2013). Agricultural plant leaf disease detection using image processing. International Journal of Advanced Research in Electrical. *Electronics and Instrumentation Engineering*, *2*(1), 599–602.

Dhillon, G. S., Chaudhari, P., Ravichandran, A., & Soatto, S. (2020). A baseline for Few-Shot image classification. *arXiv (Cornell University)*. DOI: /arxiv.1909.02729DOI: 10.48550

Dhillon, S. (n.d.). *Ethical Integration of Artificial Intelligence in Forensic Science: Prospects and Challenges*. DOI: 10.1145/3306618.3314320

Diaz, D. J., Gong, C., Ouyang-Zhang, J., Loy, J. M., Wells, J., Yang, D., Ellington, A. D., Dimakis, A. G., & Klivans, A. R. (2024). Stability oracle: A structure-based graph-transformer framework for identifying stabilizing mutations. *Nature Communications*, *15*(1), 6170. DOI: 10.1038/s41467-024-49780-2 PMID: 39043654

Dickman, A. J. (2010). Complexities of conflict: The importance of considering social factors for effectively resolving human–wildlife conflict. *Animal Conservation*, *13*(5), 458–466. DOI: 10.1111/j.1469-1795.2010.00368.x

DiMasi, J. A., Grabowski, H. G., & Hansen, R. W. (2016). Innovation in the pharmaceutical industry: New estimates of R&D costs. *Journal of Health Economics*, *47*, 20–33. DOI: 10.1016/j.jhealeco.2016.01.012 PMID: 26928437

Dixit, A., Parnas, O., Li, B., Chen, J., Fulco, C. P., Jerby-Arnon, L., Marjanovic, N. D., Dionne, D., Burks, T., Raychowdhury, R., Adamson, B., Norman, T. M., Lander, E. S., Weissman, J. S., Friedman, N., & Regev, A. (2016). Perturb-Seq: Dissecting molecular circuits with scalable single-cell RNA profiling of pooled genetic screens. *Cell*, *167*(7), 1853–1866.e17. DOI: 10.1016/j.cell.2016.11.038 PMID: 27984732

Dong, Y., Mu, R., Jin, G., Qi, Y., Hu, J., Zhao, X., . . . Huang, X. (2024). Building guardrails for large language models. arXiv preprint arXiv:2402.01822.

Dong, Z., Wang, Y., Li, C., Li, L., & Men, X. (2021). Mitochondrial DNA as a Molecular Marker in Insect Ecology: Current Status and Future Prospects. *Annals of the Entomological Society of America*, *114*(4), 470–476. DOI: 10.1093/aesa/saab020

Dove, E. S., Kelly, S. E., Lucivero, F., Machirori, M., Dheensa, S., & Prainsack, B. (2018). Beyond individualism: Is there a place for relational autonomy in clinical practice and research? *Clinical Ethics*, *12*(3), 150–165. DOI: 10.1177/1477750917704156 PMID: 28989327

Drugs, P. (2026). Biologics manufacturing: Challenges and innovations. https://parabolicdrugs.com/biologics-manufacturing-challenges-and-innovations/

Eichler, H. G., Bloechl-Daum, B., Broich, K., Kyrle, P. A., Oderkirk, J., Rasi, G., Santos Ivo, R., Schuurman, A., Senderovitz, T., Slawomirski, L., Wenzl, M., & Paris, V. (2021). Data rich, information poor: Can we use electronic health records to create a learning healthcare system for pharmaceuticals? *Clinical Pharmacology and Therapeutics*, *105*(4), 912–922. DOI: 10.1002/cpt.1226 PMID: 30178490

Eisenberg, R. S. (1987). Proprietary rights and the norms of science in biotechnology research. *The Yale Law Journal*, *97*(2), 177–231. DOI: 10.2307/796481 PMID: 11660398

Ekins, S., Puhl, A. C., Zorn, K. M., Lane, T. R., Russo, D. P., Klein, J. J., Hickey, A. J., & Clark, A. M. (2019). Exploiting machine learning for end-to-end drug discovery and development. *Nature Materials*, *18*(5), 435–441. DOI: 10.1038/s41563-019-0338-z PMID: 31000803

Elakya, R., & Manoranjitham, T. (2024). Enhanced crop health monitoring: Attention convolutional stacked recurrent networks and binary Kepler search for early detection of paddy crop issues. *Environmental Monitoring and Assessment*, *196*(6), 561. DOI: 10.1007/s10661-024-12504-6 PMID: 38767686

Elith, J., & Leathwick, J. R. (2009). Species distribution models: Ecological explanation and prediction across space and time. *Annual Review of Ecology, Evolution, and Systematics*, *40*(1), 677–697. DOI: 10.1146/annurev.ecolsys.110308.120159

Ellis, E. C. (2015). Ecology in an anthropogenic biosphere. *Ecological Monographs*, *85*(3), 287–331. DOI: 10.1890/14-2274.1

EMA. (2005). *Guideline on data exclusivity*. European Medicines Agency.

EMA. (2020). *Reflection paper on the use of Artificial Intelligence (AI) in the medicinal product lifecycle*. European Medicines Agency.

Emmert-Streib, F., Dehmer, M., & Haibe-Kains, B. (2014). Gene regulatory networks and their applications: Understanding biological and medical problems in terms of networks. *Frontiers in Cell and Developmental Biology*, *2*, 38. Advance online publication. DOI: 10.3389/fcell.2014.00038 PMID: 25364745

Enoma, D. O., Bishung, J., Abiodun, T., Ogunlana, O., & Osamor, V. C. (2022). Machine learning approaches to genome-wide association studies. *Journal of King Saud University. Science, 34*(4), 101847. DOI: 10.1016/j.jksus.2022.101847

Entezari, A., Aslani, A., Zahedi, R., & Noorollahi, Y. (2023). Artificial intelligence and machine learning in energy systems: A bibliographic perspective. *Energy Strategy Reviews, 45*, 101017. DOI: 10.1016/j.esr.2022.101017

Eraslan, G., Avsec, Ž., Gagneur, J., & Theis, F. J. (2019). Deep learning: New computational modelling techniques for genomics. *Nature Reviews. Genetics, 20*(7), 389–403. DOI: 10.1038/s41576-019-0122-6 PMID: 30971806

Esteva, A., Kuprel, B., Novoa, R. A., Ko, J., Swetter, S. M., Blau, H. M., & Thrun, S. (2017). Dermatologist-level classification of skin cancer with deep neural networks. *Nature, 542*(7639), 115–118. DOI: 10.1038/nature21056 PMID: 28117445

European Commission. (2020). *Pharmaceutical Strategy for Europe.*

European Commission. (2024). *Regulation on Artificial Intelligence.* Official Journal of the European Union.

European Medicines Agency. (2024, December 13). Big Data Steering Group (BDSG): 2024 report (EMA/471579/2024). https://www.ema.europa.eu/en/documents/report/big-data-steering-group-bdsg-2024-report_en.pdf

European Medicines Agency. "Reflection Paper on the Use of Artificial Intelligence in Lifecycle Medicines," Text European Medicines Agency, July 14, 2023, https://www.ema.europa.eu/en/news/reflection-paper-use-artificial-intelligence-lifecycle-medicines

European Parliament. (2024). Regulation (EU) 2024/1689 of the European Parliament and of the Council of 13 June 2024 laying down harmonised rules on artificial intelligence (Artificial Intelligence Act). Official Journal of the European Union, OJ L, 2024/1689. https://data.europa.eu/eli/reg/2024/1689/oj

Evans, R., O'Neill, M., Pritzel, A., Antropova, N., Senior, A., Green, T., & Hassabis, D. (2022). Protein complex prediction with AlphaFold-Multimer. *bioRxiv*. https://doi.org/DOI: 10.1101/2021.10.04.463034

Exscientia. (2020). Press Release: First AI-designed drug molecule enters clinical trials.

EY-Parthenon. (2026, February 17). AI-powered platforms for discovery, R&D and connected manufacturing to drive Indian biopharma's next growth phase: EY-Parthenon report at BioAsia 2026. Ernst & Young. https://www.ey.com/en_in/newsroom/2026/02/ai-powered-platforms-for-discovery-r-d-and-connected-manufacturing-to-drive-indian-biopharma-s-next-growth-phase-ey-parthenon-report-at-bioasia-2026

Farhat, S. E.. (2023). Detection of Pathogens: A Comprehensive Study to Improve the Precision Agriculture. *International Journal on Recent and Innovation Trends in Computing and Communication, 11*(9), 1587–1597. DOI: 10.17762/ijritcc.v11i9.9144

FDA. (2010). *Biologics Price Competition and Innovation Act.* Food and Drug Administration.

FDA. (2018). *Framework for FDA's Real-World Evidence Program.* Food and Drug Administration.

FDA. (2019). *Proposed Regulatory Framework for Modifications to Artificial Intelligence/Machine Learning-Based Software as a Medical Device.* Food and Drug Administration.

FDA. (2020). *Artificial Intelligence/Machine Learning (AI/ML)-Based Software as a Medical Device (SaMD) Action Plan.* Food and Drug Administration.

FDA. (2021). *Adaptive Designs for Clinical Trials of Drugs and Biologics.* Food and Drug Administration.

FDA. (2022). *Cybersecurity in Medical Devices: Quality System Considerations and Content of Premarket Submissions.* Food and Drug Administration.

Feng, H., Wu, L., Zhao, B., Huff, C., Zhang, J., Wu, J., Lin, L., Wei, P., & Wu, C. (2025). Benchmarking DNA foundation models for genomic and genetic tasks. *Nature Communications, 16*(1), 10780. DOI: 10.1038/s41467-025-65823-8 PMID: 41315262

Ferrero, A., & Scotti, V. (2022). Forensic DNA Profiling. *Research for Development,* •••, 151–202. DOI: 10.1007/978-3-031-14619-0_10

Finn, C., Abbeel, P., & Levine, S. (2017). Model-agnostic meta-learning for fast adaptation of deep networks. In Proceedings of the 34th International Conference on Machine Learning - Volume 70 (ICML'17). JMLR.org, 1126–1135. https://dl.acm.org/doi/10.5555/3305381.3305498

Fischer, J., & Lindenmayer, D. B. (2007). Landscape modification and habitat fragmentation: A synthesis. *Global Ecology and Biogeography*, *16*(3), 265–280. DOI: 10.1111/j.1466-8238.2007.00287.x

Fleming, N. (2018). How artificial intelligence is changing drug discovery. *Nature*, *557*(7707), S55–S57. DOI: 10.1038/d41586-018-05267-x PMID: 29849160

Foley, J. A., DeFries, R., Asner, G. P., Barford, C., Bonan, G., Carpenter, S. R., Chapin, F. S., Coe, M. T., Daily, G. C., Gibbs, H. K., Helkowski, J. H., Holloway, T., Howard, E. A., Kucharik, C. J., Monfreda, C., Patz, J. A., Prentice, I. C., Ramankutty, N., & Snyder, P. K. (2005). Global consequences of land use. *Science*, *309*(5734), 570–574. DOI: 10.1126/science.1111772 PMID: 16040698

Forouzesh, M., Irani, S., Soleimani, A., & Monabati, S. J. (2022). Application of Y-STR, DIP-STR and SNP-STR Markers in Interpretation of Forensic Genetic Profiling: A Narrative Review. *Iranian Journal of Public Health*, *51*(7), 1538. DOI: 10.18502/ijph.v51i7.10087 PMID: 36248298

Fotso Kamga, G. A., Bitjoka, L., Akram, T., Mengue Mbom, A., Rameez Naqvi, S., & Bouroubi, Y. (2021). Advancements in satellite image classification: Methodologies, techniques, approaches and applications. *International Journal of Remote Sensing*, *42*(20), 7662–7722. DOI: 10.1080/01431161.2021.1954261

Fu, L., Li, J., & Chen, Y. (2023). An innovative decision making method for air quality monitoring based on big data-assisted artificial intelligence technique. *Journal of Innovation & Knowledge*, *8*(2), 100294. DOI: 10.1016/j.jik.2022.100294

Gagliardi, A. R., & Tomaselli, G. (2025). *Socio-ecological responsibility in the era of artificial intelligence: A corporate governance roadmap*. Springer Nature.

Gall, S. C., & Thompson, R. C. (2015). The impact of debris on marine life. *Marine Pollution Bulletin*, *92*(1–2), 170–179. DOI: 10.1016/j.marpolbul.2014.12.041 PMID: 25680883

Garnett, S. T., Burgess, N. D., Fa, J. E., Fernández-Llamazares, Á., Molnár, Z., Robinson, C. J., Watson, J. E. M., Zander, K. K., Austin, B., Brondizio, E. S., Collier, N. F., Duncan, T., Ellis, E., Geyle, H., Jackson, M. V., Jonas, H., Malmer, P., McGowan, B., Sivongxay, A., & Leiper, I. (2018). A spatial overview of the global importance of Indigenous lands for conservation. *Nature Sustainability*, *1*(7), 369–374. DOI: 10.1038/s41893-018-0100-6

Gaur, L., Afaq, A., Arora, G. K., & Khan, N. (2023). Artificial intelligence for carbon emissions using system of systems theory. *Ecological Informatics*, *76*, 102165. DOI: 10.1016/j.ecoinf.2023.102165

Gebru, T., Morgenstern, J., Vecchione, B., Vaughan, J. W., Wallach, H., Iii, H. D., & Crawford, K. (2021). Datasheets for datasets. *Communications of the ACM, 64*(12), 86–92. DOI: 10.1145/3458723

Geyer, P. E., Holdt, L. M., Teupser, D., & Mann, M. (2017). Revisiting biomarker discovery by plasma proteomics. *Molecular Systems Biology, 13*(9), 942. Advance online publication. DOI: 10.15252/msb.20156297 PMID: 28951502

Ghadi, Y. Y., Mazhar, T., Shahzad, T., Amir khan, M., Abd-Alrazaq, A., Ahmed, A., & Hamam, H. (2024). The role of blockchain to secure internet of medical things. *Scientific Reports, 14*(1), 18422. DOI: 10.1038/s41598-024-68529-x PMID: 39117650

Gibofsky, A., Evans, C., & Strand, V. (2022). Provider and patient knowledge gaps on biosimilars: Insights from surveys. *The American Journal of Managed Care, 28*(12, Suppl), S227–S233. PMID: 36493396

Gilligan, C. (1982). *In a different voice: Psychological theory and women's development*. Harvard University Press.

Gilmore, A. B., Fabbri, A., Baum, F., Bertscher, A., Bondy, K., Chang, H.-J., Demaio, S., Erzse, A., Freudenberg, N., Friel, S., Hofman, K. J., Johns, P., Karim, S. A., Lacy-Nichols, J., Paes de Carvalho, M. P., Marten, R., McKee, M., Petticrew, M., Robertson, L., & Tangcharoensathien, V. (2023). Defining and conceptualising the commercial determinants of health. *Lancet, 401*(10383), 1194–1213. DOI: 10.1016/S0140-6736(23)00013-2 PMID: 36966782

Gligorea, I., Cioca, M., Oancea, R., Gorski, A. T., Gorski, H., & Tudorache, P. (2023). Adaptive learning using artificial intelligence in e-learning: A literature review. *Education Sciences, 13*(12), 1216. DOI: 10.3390/educsci13121216

Goluguri, N., & Baik, O. D. (2026). Revolutionizing wheat plant disease detection: A review of imaging, AI, and innovations. *Archives of Computational Methods in Engineering*. Advance online publication. DOI: 10.1007/s11831-026-10502-0

Goodman, B., & Flaxman, S. (2017). European Union regulations on algorithmic decision-making and a "right to explanation". *AI Magazine, 38*(3), 50–57. DOI: 10.1609/aimag.v38i3.2741

Goodwin, M., & Wright, G. (2020). AI and robotics in waste management: A review of current applications and future prospects. *Waste Management & Research, 38*(5), 487–498. PMID: 32089127

Green Software Foundation. (2024). Software Carbon Intensity (SCI) standard v1.1: Operationalizing environmental accountability in software development. https://greensoftware.foundation/standards/sci

Greener, J. G., Kandathil, S. M., Moffat, L., & Jones, D. T. (2022). A guide to machine learning for biologists. *Nature Reviews. Molecular Cell Biology*, *23*(1), 40–55. DOI: 10.1038/s41580-021-00407-0 PMID: 34518686

Guadu, T., Berhane, A., Solomon, T., Tafese, A., Tesfay, K., Asgedom, A. A., Abera, B., Debebe, B., Getachew, M., Shibabaw, A., Abiye, A. A., Abdurahman, M., Yadeta, W., Tadesse, Y., Lemma, H., Theodore, K.-J., Mor, S. M., Mereta, S. T., & Kaba, M. (2026). Human-wildlife conflict and its impacts on livelihoods, health, and the environment in Chebera Churchura National Park, Southwest Ethiopia: A phenomenological study using a One Health approach with an urgent call for action. *One Health Outlook*, *8*(1), 8. DOI: 10.1186/s42522-025-00188-y PMID: 41547988

Gülmez, B. (2025). A comprehensive review of convolutional neural networks based disease detection strategies in potato agriculture. *Potato Research*, *68*(2), 1295–1329. DOI: 10.1007/s11540-024-09786-1

Habes, M., Alghizzawi, M., & Ali, S., SalihAlnaser, A., & Salloum, S. A. (2020). The Relation among Marketing ads, via Digital Media and mitigate (COVID-19) pandemic in Jordan. *International Journal of Advanced Science and Technology*, *29*(7), 12326–12348.

Haddad, N. M., Brudvig, L. A., Clobert, J., Davies, K. F., Gonzalez, A., Holt, R. D., Lovejoy, T. E., Sexton, J. O., Austin, M. P., Collins, C. D., Cook, W. M., Damschen, E. I., Ewers, R. M., Foster, B. L., Jenkins, C. N., King, A. J., Laurance, W. F., Levey, D. J., Margules, C. R., & Townshend, J. R. (2015). Habitat fragmentation and its lasting impact on Earth's ecosystems. *Science Advances*, *1*(2), e1500052. DOI: 10.1126/sciadv.1500052 PMID: 26601154

Haluza, D., & Jungwirth, D. (2023). Artificial intelligence and ten societal megatrends: An exploratory study using GPT-3. *Systems*, *11*(3), 120. Advance online publication. DOI: 10.3390/systems11030120

Hampton, S. E., Strasser, C. A., Tewksbury, J. J., Gram, W. K., Budden, A. E., Batcheller, A. L., Duke, C. S., & Porter, J. H. (2013). Big data and the future of ecology. *Frontiers in Ecology and the Environment*, *11*(3), 156–162. DOI: 10.1890/120103

Hamza, M. A., Shaiba, H., Marzouk, R., Alhindi, A., Asiri, M. M., Yaseen, I., & Rizwanullah, M. (2022). Big Data Analytics with Artificial Intelligence Enabled Environmental Air Pollution Monitoring Framework. *Computers, Materials & Continua*, *73*(2).

Harrer, S., Shah, P., Antony, B., & Hu, J. (2019). Artificial intelligence for clinical trial design. *Trends in Pharmacological Sciences*, *40*(8), 577–591. DOI: 10.1016/j.tips.2019.05.005 PMID: 31326235

Hassan, S. M., Maji, A. K., Jasiński, M., Leonowicz, Z., & Jasińska, E. (2021). Identification of plant-leaf diseases using CNN and transfer-learning approach. *Electronics (Basel)*, *10*(12), 1388. Advance online publication. DOI: 10.3390/electronics10121388

Hasselgren, A., Kralevska, K., Gligoroski, D., Pedersen, S. A., & Faxvaag, A. (2020). Blockchain in healthcare and health sciences—A scoping review. *International Journal of Medical Informatics*, *134*, 104040. DOI: 10.1016/j.ijmedinf.2019.104040 PMID: 31865055

Haupt, S. E., Pasini, A., & Marzban, C. (Eds.). (2008). *Artificial intelligence methods in the environmental sciences*. Springer Science & Business Media.

Hebblewhite, M., & Haydon, D. T. (2010). Distinguishing technology from biology: A critical review of the use of GPS telemetry data in ecology. *Philosophical Transactions of the Royal Society of London. Series B, Biological Sciences*, *365*(1550), 2303–2312. DOI: 10.1098/rstb.2010.0087 PMID: 20566506

He, K., Zhang, X., Ren, S., & Sun, J. (2016). *Deep residual learning for image recognition.*, DOI: 10.1109/CVPR.2016.90

Hendriksen, R. S., Bortolaia, V., Tate, H., Tyson, G. H., Aarestrup, F. M., & McDermott, P. F. (2019). Using genomics to track global antimicrobial resistance. *Frontiers in Public Health*, *7*, 242. Advance online publication. DOI: 10.3389/fpubh.2019.00242 PMID: 31552211

Hilborn, R., Amoroso, R. O., Anderson, C. M., Baum, J. K., Branch, T. A., Costello, C., De Moor, C. L., Faraj, A., Hively, D., Jensen, O. P., Kurota, H., Little, L. R., Mace, P., McClanahan, T., Melnychuk, M. C., Minto, C., Osio, G. C., Parma, A. M., Pons, M., & Ye, Y. (2020). Effective fisheries management instrumental in improving fish stock status. *Proceedings of the National Academy of Sciences of the United States of America*, *117*(4), 2218–2224. DOI: 10.1073/pnas.1909726116 PMID: 31932439

Himeur, Y., Rimal, B., Tiwary, A., & Amira, A. (2022). Using artificial intelligence and data fusion for environmental monitoring: A review and future perspectives. *Information Fusion*, *86*, 44–75. DOI: 10.1016/j.inffus.2022.06.003

Hirel, B., Tétu, T., Lea, P. J., & Dubois, F. (2011). Improving nitrogen use efficiency in crops for sustainable agriculture. *Sustainability (Basel)*, *3*(9), 1452–1485. DOI: 10.3390/su3091452

Holman, C.M. Bowman v. Monsanto Co.: A Bellwether for the Emerging Issue of Patentable Self-Replicating Technologies and Inadvertent Infringement, 80 MO. L. REV. (2015) Available at: https://scholarship.law.missouri.edu/mlr/vol80/iss3/6

Honchar, N. (2024, January 26). 10 companies applying AI for biologics discovery. BiopharmaTrend. https://www.biopharmatrend.com/artificial-intelligence/10-companies-applying-ai-for-biologics-discovery-590/

Hospedales, T., Antoniou, A., Micaelli, P., & Storkey, A. (2021). Meta-learning in neural networks: A survey. *IEEE Transactions on Pattern Analysis and Machine Intelligence*, *44*(9), 5149–5169. PMID: 33974543

Huang, C., Zhang, Z., Mao, B., & Yao, X. (2022). An overview of artificial intelligence ethics. *IEEE Transactions on Artificial Intelligence*, *4*(4), 799–819. DOI: 10.1109/TAI.2022.3194503

Hughes, J. P., Rees, S., Kalindjian, S. B., & Philpott, K. L. (2011). Principles of early drug discovery. *British Journal of Pharmacology*, *162*(6), 1239–1249. DOI: 10.1111/j.1476-5381.2010.01127.x PMID: 21091654

Hunt, C. B., & Auster, E. R. (1990). Proactive environmental management: Avoiding the toxic trap. *MIT Sloan Management Review*, *31*(2), 7.

Hwang, H., Jeon, H., Yeo, N., & Baek, D. (2024). Big data and deep learning for RNA biology. *Experimental & Molecular Medicine*, *56*(6), 1293–1321. DOI: 10.1038/s12276-024-01243-w PMID: 38871816

IBM. (2020). *Green horizon project: AI for sustainable cities*. https://www.ibm.com/impact/green-horizon

Ibrahim, S. A., & Pronovost, P. J. (2021). Diagnostic errors, health disparities, and artificial intelligence: A combination for health or harm? *JAMA Health Forum*, *2*(9), e212430. DOI: 10.1001/jamahealthforum.2021.2430 PMID: 36218658

ICH. (2016). *Integrated Addendum to ICH E6(R1): Guideline for Good Clinical Practice E6(R2)*. International Council for Harmonisation.

Imalka, L. A., Gunawardana, K. G. A., Kodithuwakku, K. M. S. K., Arachchi, H. K. E., Harshanath, S. M. B., & Rajapaksha, S. (2022, September). Farming through Technology Driven Solutions for Agriculture Industry Ceylon E-Agro mobile application-find technology based solutions for agricultural problems. In *2022 IEEE 10th Region 10 Humanitarian Technology Conference (R10-HTC)* (pp. 306–310). https://doi.org/DOI: 10.1109/R10-HTC54060.2022.9929335

International Energy Agency. (2024). Electricity 2024: Analysis and forecast to 2026. IEA. https://www.iea.org/reports/electricity-2024

Ishak, S., Rahiman, M. H. F., Kanafiah, S. N. A. M., & Saad, H. (2015). Leaf disease classification using artificial neural network. *Jurnal Teknologi*, *77*(17), 109–114. DOI: 10.11113/jt.v77.6463

Issa, A. A., Majed, S., Ameer, S. A., & Al-Jawahry, H. M. (2024). Farming in the Digital Age: Smart Agriculture with AI and loT. In *E3S Web of Conferences* (Vol. 477, p. 00081). EDP Sciences. https://doi.org/DOI: 10.1051/e3sconf/202447700081

Istrate, A. M., Milletari, F., Castrotorres, F., Tomczak, J. M., Torkar, M., Li, D., & Karaletsos, T. (2025). rbio1-training scientific reasoning LLMs with biological world models as soft verifiers. bioRxiv, 2025-08.

Itumalla, R. (2012). Information Technology and Service Quality in HealthCare: An Empirical Study of Private Hospital in India. *International Journal of Innovation, Management and Technology*, *3*(4), 433. DOI: 10.7763/IJIMT.2012.V3.269

Jain, E. (2020). Digital Employability Skills and Training Needs for the Indian Healthcare Industry. In Opportunities and Challenges in Digital Healthcare Innovation (pp. 113-130). IGI Global. DOI: 10.4018/978-1-7998-3274-4.ch007

Javaid, M., Haleem, A., Khan, I. H., & Suman, R. (2023). Understanding the potential applications of Artificial Intelligence in Agriculture Sector. *Advanced Agrochem*, *2*(1), 15–30. DOI: 10.1016/j.aac.2022.10.001

Jayatunga, M., Bruens, L., Ruder, L., Schulze, U., & Meier, C. (2022). AI in biologics discovery: An emerging frontier. Boston Consulting Group; reprinted in In Vivo. https://web-assets.bcg.com/4b/f8/ef10f64643578c65e34902c81513/ai-in-biologics-discovery-an-emerging-frontier.pdf

Jensen, J. R. (1996). *Introductory digital image processing: a remote sensing perspective*. Prentice Hall.

Jerde, C. L., Mahon, A. R., Chadderton, W. L., & Lodge, D. M. (2011). "Sight-unseen" detection of rare aquatic species using environmental DNA. *Conservation Letters*, *4*(2), 150–157. DOI: 10.1111/j.1755-263X.2010.00158.x

Jiang, J., Li, Y., Cao, S., Shan, Y., Liu, Y., Fei, T., Yu, Y., Feng, Y., Li, Y., Li, Y., & Yuan, J. (2025). Artificial intelligence in bioinformatics: A survey. *Briefings in Bioinformatics*, *26*(6), bbaf576. DOI: 10.1093/bib/bbaf576 PMID: 41206113

Jobin, A., Ienca, M., & Vayena, E. (2019). The global landscape of AI ethics guidelines. *Nature Machine Intelligence*, *1*(9), 389–399. DOI: 10.1038/s42256-019-0088-2

Jones, D. A., Lelyveld, T. P., Mavrofidis, S. D., Kingman, S. W., & Miles, N. J. (2002). Microwave heating applications in environmental engineering—A review. *Resources, Conservation and Recycling*, *34*(2), 75–90. DOI: 10.1016/S0921-3449(01)00088-X

Joseph, J. (2025). Algorithmic bias in public health AI: A silent threat to equity in low-resource settings. *Frontiers in Public Health*, *13*, 1643180. Advance online publication. DOI: 10.3389/fpubh.2025.1643180 PMID: 40771228

Jump, A. S., Cavin, L., & Hunter, P. D. (2010). Monitoring and managing responses to climate change at the retreating range edge of forest trees. *Journal of Environmental Monitoring*, *12*(10), 1791–1798. DOI: 10.1039/b923773a PMID: 20818456

Jumper, J., Evans, R., Pritzel, A., Green, T., Figurnov, M., Ronneberger, O., & Hassabis, D. (2021). Highly accurate protein structure prediction with AlphaFold. *Nature*, *596*(7873), 583–589. DOI: 10.1038/s41586-021-03819-2 PMID: 34265844

Jung, M., Song, J. S., Shin, A. Y., Choi, B., Go, S., Kwon, S. Y., Park, J., Park, S. G., & Kim, Y. M. (2023). Construction of deep learning-based disease detection model in plants. *Scientific Reports*, *13*(1), 7331. Advance online publication. DOI: 10.1038/s41598-023-34549-2 PMID: 37147432

Jurtz, V., Paul, S., Andreatta, M., Marcatili, P., Peters, B., & Nielsen, M. (2017). NetMHCpan-4.0: Improved peptide–MHC class I interaction predictions integrating eluted ligand and peptide binding affinity data. *The Journal of Immunology : Official Journal of the American Association of Immunologists*, *199*(9), 3360–3368. DOI: 10.4049/jimmunol.1700893 PMID: 28978689

Kakkar, A., Naushad, S., & Khatri, S. K. (2017). Telemedicine and EHR integrated approach for an effective E-Governance Healthcare Framework. *International Journal of Medical Research & Health Sciences*, *6*(5), 108–114.

Kamilaris, A., & Prenafeta-Boldú, F. X. (2021). AI in Agriculture: Integration of AI and big data in smart agriculture enhances productivity and sustainability. *Computers and Electronics in Agriculture*, *175*, 105521. DOI: 10.1016/j.compag.2020.105521

Kamyab, H., Khademi, T., Chelliapan, S., SaberiKamarposhti, M., Rezania, S., Yusuf, M., Farajnezhad, M., Abbas, M., Hun Jeon, B., & Ahn, Y. (2023). The latest innovative avenues for the utilization of artificial Intelligence and big data analytics in water resource management. *Results in Engineering*, *20*, 101566. DOI: 10.1016/j.rineng.2023.101566

Kanagaraj, R., Elakiya, E., Kumaresan, A., Selvakumar, J., Suganya, J., & Shruthi, K. (2023). Predictive Classification Model of Crop Yield Data Using Artificial Neural Network. In *2023 5th International Conference on Inventive Research in Computing Applications (ICIRCA)* (pp. 747–751). https://doi.org/DOI: 10.1109/ICIRCA57980.2023.10220791

Kanthaswamy, S. (2024). Review: Wildlife forensic genetics—Biological evidence, DNA markers, analytical approaches, and challenges. *Animal Genetics*, *55*(2), 177–192. DOI: 10.1111/age.13390 PMID: 38123142

Kant, I. (1993). *Grounding for the metaphysics of morals* (Ellington, J. W., Trans.). Hackett Publishing. (Original work published 1785)

Kar, S. K., Saxena, S. K., & Kabir, R. (2020). The relevance of digital mental healthcare during COVID-19: Need for innovations. *Nepal Journal of Epidemiology*, *10*(4), 928–929. DOI: 10.3126/nje.v10i4.32519 PMID: 33495711

Karthikeyan, N., & Sukanesh, R. (2012). Cloud based emergency health care information service in India. *Journal of Medical Systems*, *36*(6), 4031–4036. DOI: 10.1007/s10916-012-9875-6 PMID: 22865161

Katkar, V., Khatawkar, S., Jadhav, S., Malame, R., Rathod, M., & Urunkar, A. (2026). RicNet: Enhancing rice leaf disease detection through optimized convolutional neural network architectures and dropout configurations. In S. K. Udgata, D. Mohapatra, S. Sethi, & M. E. Rana (Eds.), Intelligent systems. ICMIB 2025 (Lecture Notes in Networks and Systems, Vol. 1623). Springer. DOI: 10.1007/978-3-032-05120-2_5

Kaur, P., Harnal, S., Tiwari, R., Upadhyay, S., Bhatia, S., Mashat, A., & Alabdali, A. M. (2022). Recognition of leaf disease using the hybrid convolutional neural network by applying feature reduction. *Sensors (Basel)*, *22*(2), 575. Advance online publication. DOI: 10.3390/s22020575 PMID: 35062534

Kaur, R., Gabrijelčič, D., & Klobučar, T. (2023). Artificial intelligence for cybersecurity: Literature review and future research directions. *Information Fusion*, *97*, 101804. DOI: 10.1016/j.inffus.2023.101804

Kays, R., Crofoot, M. C., Jetz, W., & Wikelski, M. (2015). Terrestrial animal tracking as an eye on life and planet. *Science*, *348*(6240), aaa2478. DOI: 10.1126/science.aaa2478 PMID: 26068858

Kesik-Brodacka, M. (2018). Progress in biopharmaceutical development. *Biotechnology and Applied Biochemistry*, *65*(3), 306–322. DOI: 10.1002/bab.1617 PMID: 28972297

Kesselheim, A. S., Avorn, J., & Sarpatwari, A. (2015). The high cost of prescription drugs in the United States: Origins and prospects for reform. *Journal of the American Medical Association*, *316*(8), 858–871. DOI: 10.1001/jama.2016.11237 PMID: 27552619

Khan, A., & Shahriyar, A. K. (2023). Optimizing Onion Crop Management: A Smart Agriculture Framework with IoT Sensors and Cloud Technology. *Applied Research in Artificial Intelligence and Cloud Computing*, *6*(1), 49–67.

Khan, S. A., Martínez-de-Morentin, X., Alsabbagh, A. R., Maillo, A., Lagani, V., Gomez-Cabrero, D., & Tegner, J. (2025). Multimodal foundation transformer models for multiscale genomics. *Nature Methods*, •••, 1–13. PMID: 41326819

Khan, S., Ullah, S., & Nobanee, H. (2024). ICT diffusion, E-governance, and sustainability in the digital era. *Sustainable Futures : An Applied Journal of Technology, Environment and Society*, *8*, 100272. DOI: 10.1016/j.sftr.2024.100272

Khera, A. V., Chaffin, M., Aragam, K. G., Haas, M. E., Roselli, C., Choi, S. H., & Kathiresan, S. (2018). Genome-wide polygenic scores for common diseases identify individuals with risk equivalent to monogenic mutations. *Nature Genetics*, *50*(9), 1219–1224. DOI: 10.1038/s41588-018-0183-z PMID: 30104762

Kim, G. B., Kim, W. J., Kim, H. U., & Lee, S. Y. (2020). Machine learning applications in systems metabolic engineering. *Current Opinion in Biotechnology*, *64*, 1–9. DOI: 10.1016/j.copbio.2019.08.010 PMID: 31580992

King, M., DeNoon, A., & Hart, C. (2025). AI in life science quality management and regulatory affairs: Global compliance and strategic readiness. IQVIA. https://www.iqvia.com/-/media/iqvia/pdfs/library/white-papers/2025/ai-in-life-science-quality-management-and-regulatory-affairs.pdf

King, R. D., Rowland, J., Oliver, S. G., Young, M., Aubrey, W., Byrne, E., Liakata, M., Markham, M., Pir, P., Soldatova, L. N., Sparkes, A., Whelan, K. E., & Clare, A. (2009). The automation of science. *Science*, *324*(5923), 85–89. DOI: 10.1126/science.1165620 PMID: 19342587

Kitano, H. (2002). Systems biology: A brief overview. *Science, 295*(5560), 1662–1664. DOI: 10.1126/science.1069492 PMID: 11872829

Klein, C., McKinnon, M. C., Wright, B. T., Possingham, H. P., & Halpern, B. S. (2015). Social equity and the probability of success of biodiversity conservation. *Global Environmental Change, 35*, 299–306. DOI: 10.1016/j.gloenvcha.2015.09.007

Kokotas, S., Budowle, B., Papatheodorou, A., Bolanaki, E., Kondili, A., Metheniti, A., Vouropoulou, M., Koukouvinos, G., Palaigeorgiou, E., & Makras, P. (2025). Comparison of next generation sequencing (NGS) - (SNPs) and capillary electrophoresis (CE) - (STRs) in the genetic analysis of human remains. *Forensic Science International. Genetics, 74*, 103131. DOI: 10.1016/j.fsigen.2024.103131 PMID: 39241298

Kotlarz, K., Mielczarek, M., Biecek, P., Guldbrandtsen, B., & Szyda, J. (2024). Exploring the impact of sequence context on errors in SNP genotype calling with whole genome sequencing data using AI-based autoencoder approach. *NAR Genomics and Bioinformatics, 6*(3), 131. DOI: 10.1093/nargab/lqae131 PMID: 39318508

Krzywanski, J. (2022). Advanced AI applications in energy and environmental engineering systems. *Energies, 15*(15), 5621. DOI: 10.3390/en15155621

Kuiken, T. (2023). Artificial Intelligence in the Biological Sciences: Uses, Safety, Security, and Oversight. *Congressional Research Service (CRS) Reports and Issue Briefs*, NA-NA. https://go.gale.com/ps/i.do?p=AONE&sw=w&issn=&v=2.1&it=r&id=GALE%7CA777680617&sid=googleScholar&linkaccess=fulltext

Kumar, A., Mahajan, P., Mohan, D., & Varghese, M. (2001). IT—information technology and the human interface: tractor vibration severity and driver health: a study from rural India. *Journal of Agricultural Engineering Research, 80*(4), 313–328. DOI: 10.1006/jaer.2001.0755

Kumar, D., Haque, A., Mishra, K., Islam, F., Mishra, B. K., & Ahmad, S. (2023). Exploring the transformative role of artificial intelligence and metaverse in education: A comprehensive review. *Metaverse Basic and Applied Research, 2*, 55. DOI: 10.56294/mr202355

Kumar, D., & Kukreja, V. (2025). Impact of image segmentation and feature sets in automated plant disease classification: A comprehensive review based on wheat plant images. *Progress in Artificial Intelligence, 14*(4), 451–504. DOI: 10.1007/s13748-025-00389-6

Kumar, P., & Reddy, K. S. (2019). Challenges and Opportunities for Implementing Digital Health Information Systems in India: A Review. *Health Policy and Technology*, *8*(4), 392–399.

Kunduracıoğlu, İ., & Paçal, İ. (2024). Deep learning-based disease detection in sugarcane leaves: Evaluating EfficientNet models. *Journal of Operations Intelligence*, *2*(1), 321–235. DOI: 10.31181/jopi21202423

Kunduru, A. R. (2023). Effective usage of artificial intelligence in enterprise resource planning applications. *International Journal of Computer Trends and Technology*, *71*(4), 73–80. DOI: 10.14445/22312803/IJCTT-V71I4P109

Laird, D. A., Yen, P. Y., Koskinen, W. C., Steinheimer, T. R., & Dowdy, R. H. (1994). Sorption of atrazine on soil clay components. *Environmental Science & Technology*, *28*(6), 1054–1061. DOI: 10.1021/es00055a014 PMID: 22176230

Lajaunie, C., Schafer, B., & Mazzega, P. (2019). Big data enters environmental law. *Transnational Environmental Law*, *8*(3), 523–545. DOI: 10.1017/S2047102519000335

Lal, S., Singh, B., & Kaunert, C. (2024). Role of Artificial Intelligence (AI) and Intellectual Property Rights (IPR) in Transforming Drug Discovery and Development in the Life Sciences: Legal and Ethical Concerns. *Library of Progress-Library Science, Information Technology & Computer, 44*(3).

Larsson, D. J., & Flach, C.-F. (2022). Antibiotic resistance in the environment. *Nature Reviews. Microbiology*, *20*(5), 257–269. DOI: 10.1038/s41579-021-00649-x PMID: 34737424

LeCun, Y., Bengio, Y., & Hinton, G. (2015). Deep learning. *Nature*, *521*(7553), 436–444. DOI: 10.1038/nature14539 PMID: 26017442

Lee, J. E., Lee, J. M., Naue, J., Fleckhaus, J., Freire-Aradas, A., Neubauer, J., Pośpiech, E., McCord, B., Kalamara, V., Gauthier, Q., Mills, C., Cao, Y., Wang, Z., Oh, Y. N., Feng, L., Schneider, P. M., Phillips, C., Haas, C., Pisarek, A., & Lee, H. Y. (2022). A collaborative exercise on DNA methylation-based age prediction and body fluid typing. *Forensic Science International. Genetics*, *57*, 102656. DOI: 10.1016/j.fsigen.2021.102656 PMID: 34973557

Lee, S. L., O'Connor, T. F., Yang, X., Cruz, C. N., Chatterjee, S., Madurawe, R. D., Moore, C. M. V., Yu, L. X., & Woodcock, J. (2020). Modernizing pharmaceutical manufacturing: From batch to continuous production. *Journal of Pharmaceutical Innovation*, *10*(3), 191–199. DOI: 10.1007/s12247-015-9215-8

Lemley, M. A. (2008). The surprising resilience of the patent system. *Texas Law Review*, *95*, 1.

Lenarczyk, G., Minssen, T., Price, N., & Rai, A. (2025). The future of AI regulation in drug development: A comparative analysis. *Journal of Law and the Biosciences*, *12*(2), lsaf028. DOI: 10.1093/jlb/lsaf028 PMID: 41221487

Liakos, K. G., & Bochtis, D. (2022). Advances in machine learning applications in agriculture offer innovative solutions for crop management. *Agricultural Systems*, *193*, 103255. DOI: 10.1016/j.agsy.2021.103255

Liakos, K. G., Busato, P., Moshou, D., Pearson, S., & Bochtis, D. (2018). Machine learning in agriculture: A review. *Sensors (Basel)*, *18*(8), 2674. Advance online publication. DOI: 10.3390/s18082674 PMID: 30110960

Liang, C., & Shah, T. (2023). IoT in Agriculture: The Future of Precision Monitoring and Data-Driven Farming. *Eigenpub Review of Science and Technology*, *7*(1), 85–104.

Liang, R., Tang, Q., Chen, J., & Zhu, L. (2024). Epigenetic Clocks: Beyond Biological Age, Using the Past to Predict the Present and Future. *Aging and Disease*, *16*(6), 3520. DOI: 10.14336/AD.2024.1495 PMID: 39751861

Libbrecht, M. W., & Noble, W. S. (2015). Machine learning applications in genetics and genomics. *Nature Reviews. Genetics*, *16*(6), 321–332. DOI: 10.1038/nrg3920 PMID: 25948244

Limkar, S., ... (2021). Classification and prediction of rice crop diseases using CNN and PNN. In S. Satapathy, Y.-D. Zhang, V. Bhateja, & R. Majhi (Eds.), Intelligent data engineering and analytics (Advances in Intelligent Systems and Computing, Vol. 1177). Springer. DOI: 10.1007/978-981-15-5679-1_4

Li, T. S., & Li, B. (2007). Research on Data Model and Query Optimization of Electronic Map in Embedded GIS. *Aeronautical Computing Technique*, (2), 26–35.

Liu, J., Dietz, T., Carpenter, S. R., Alberti, M., Folke, C., Moran, E., ... & Taylor, W. W. (2007). Complexity of coupled human and natural systems. *science, 317*(5844), 1513-1516.

Liu, D., Sun, D.-W., & Zeng, X.-A. (2018). Recent advances in the application of imaging techniques and machine learning in food quality evaluation. *Trends in Food Science & Technology*, *72*, 106–117. DOI: 10.1016/j.tifs.2017.12.001

Liu, X., Faes, L., Kale, A. U., Wagner, S. K., Fu, D. J., Bruynseels, A., Mahendiran, T., Moraes, G., Shamdas, M., Kern, C., Ledsam, J. R., Schmid, M. K., Balaskas, K., Topol, E. J., Bachmann, L. M., Keane, P. A., & Denniston, A. K. (2019). A comparison of deep learning performance against health-care professionals in detecting diseases from medical imaging: A systematic review and meta-analysis. *The Lancet. Digital Health*, *1*(6), e271–e297. DOI: 10.1016/S2589-7500(19)30123-2 PMID: 33323251

Liu, Y., Gupta, H., Springer, E., & Wagener, T. (2008). Linking science with environmental decision making: Experiences from an integrated modeling approach to supporting sustainable water resources management. *Environmental Modelling & Software*, *23*(7), 846–858. DOI: 10.1016/j.envsoft.2007.10.007

Liu, Z., Yu, X., & Choi, A. Y. (2023). Data visualization for designing F2C IoT agricultural system. In *Third International Conference on Artificial Intelligence, Virtual Reality, and Visualization (AIVRV 2023)* (Vol. 12923, pp. 488–492). https://doi.org/DOI: 10.1117/12.3011451

Lokers, R., Knapen, R., Janssen, S., van Randen, Y., & Jansen, J. (2016). Analysis of Big Data technologies for use in agro-environmental science. *Environmental Modelling & Software*, *84*, 494–504. DOI: 10.1016/j.envsoft.2016.07.017

López, D. M., Rico-Olarte, C., Blobel, B., & Hullin, C. (2022). Challenges and solutions for transforming health ecosystems in low- and middle-income countries through artificial intelligence. *Frontiers in Medicine*, *9*, 958097. Advance online publication. DOI: 10.3389/fmed.2022.958097 PMID: 36530888

Lucivero, F. (2020). Big data, big waste? A reflection on the environmental sustainability of big data initiatives. *Science and Engineering Ethics*, *26*(2), 1009–1030. DOI: 10.1007/s11948-019-00171-7 PMID: 31893331

Lund, H., Andersen, A. N., Østergaard, P. A., Mathiesen, B. V., & Connolly, D. (2020). From electricity smart grids to smart energy systems—A review of current status and trends in research. *Energy*, *198*, 117343. DOI: 10.1016/j.energy.2020.117343

Luo, J., Cai, X., & Li, Y. (2025). Large language models for biological sequence analysis in infectious disease research. *Biosafety and Health*, *7*(05), 323–332. DOI: 10.1016/j.bsheal.2025.09.007 PMID: 41262460

Madhurya, C., & Jubilson, E. A. (2024). YR2S: Efficient deep learning technique for detecting and classifying plant leaf diseases. *IEEE Access : Practical Innovations, Open Solutions*, *12*, 3790–3804. DOI: 10.1109/ACCESS.2023.3343450

Madon, S., Sahay, S., & Sudan, R. (2007). E-government policy and health information systems implementation in Andhra Pradesh, India: Need for articulation of linkages between the macro and the micro. *The Information Society*, *23*(5), 327–344. DOI: 10.1080/01972240701572764

Mahapatra, S. C., Das, R. K., & Patra, M. R. (2011). Current e-governance scenario in healthcare sector of India. *E-Governance Scenario in Healthcare Sector of India*, 121-7.

Mai, G., Huang, W., Sun, J., Song, S., Mishra, D., Liu, N., & Lao, N. (2023). On the opportunities and challenges of foundation models for geospatial artificial intelligence. *arXiv preprint arXiv:2304.06798*.

Malik, A., Vaidya, G., Jagota, V., Eswaran, S., Sirohi, A., Batra, I., Rakhra, M., & Asenso, E. (2022). Design and evaluation of a hybrid technique for detecting sunflower leaf disease using a deep learning approach. *Journal of Food Quality*, *2022*, 9211700. Advance online publication. DOI: 10.1155/2022/9211700

Malik, Y. S., Sircar, S., Bhat, S., Ansari, M. I., Pande, T., Kumar, P., Mathapati, B., Balasubramanian, G., Kaushik, R., Natesan, S., Ezzikouri, S., El Zowalaty, M. E., & Dhama, K. (2021). How artificial intelligence may help the COVID-19 pandemic: Pitfalls and lessons for the future. *Reviews in Medical Virology*, *31*(5), e2205. DOI: 10.1002/rmv.2205 PMID: 33476063

Mancy, A. M., Kumar, A. S., Latip, R., Jagadamba, G., Chakrabarti, P., Sharma, P., . . . Kanchan, B. G. (2024). Smart Healthcare System, Digital Health and Telemedicine, Management and Emergencies: Patient Emergency Application (PES) E-Governance Applications. In *Sustainable Development in AI, Blockchain, and E-Governance Applications* (pp. 124-151). IGI Global.

Mandal, R., & Murthy, A. (2021). CSR in the post-pandemic era: The dual promise of ESG investment and investor stewardship. *Indiana Law Review*, *5*(2), 229–249. DOI: 10.1080/24730580.2021.1899627

Margules, C. R., & Pressey, R. L. (2000). Systematic conservation planning. *Nature*, *405*(6783), 243–253. DOI: 10.1038/35012251 PMID: 10821285

Marko, J. G. O., Neagu, C. D., & Anand, P. B. (2025). Examining inclusivity: The use of artificial intelligence and diverse populations in health and social care: A systematic review. *BMC Medical Informatics and Decision Making*, *25*(1), 57. Advance online publication. DOI: 10.1186/s12911-025-02884-1 PMID: 39910518

Marsico, F., & Amigo, M. (2025). Ethical and security challenges in AI for forensic genetics: From bias to adversarial attacks. *Forensic Science International. Genetics, 76*, 103225. DOI: 10.1016/j.fsigen.2025.103225 PMID: 39874746

Maxim, L., & van der Sluijs, J. P. (2011). Quality in environmental science for policy: Assessing uncertainty as a component of policy analysis. *Environmental Science & Policy, 14*(4), 482–492. DOI: 10.1016/j.envsci.2011.01.003

Mears, L., Stocks, S. M., Sin, G., & Gernaey, K. V. (2017). A review of control strategies for manipulating the feed rate in fed-batch fermentation processes. *Journal of Biotechnology, 245*, 34–46. DOI: 10.1016/j.jbiotec.2017.01.008 PMID: 28179156

Medicines Patent Pool. (2020). *Annual Report.*

Mehta, P., & Iyer, N. (2021). The influence of mobile health (mHealth) applications on preventive healthcare. *International Journal of mHealth Research, 9*(1), 87-102.

Mehta, S., & Jain, P. (2022). Role of Digital Health Information Systems in Addressing Healthcare Disparities: Insights from Urban and Rural Settings in India. *Journal of Public Health Management and Practice, 28*(1), 12–19.

Meng, X. N., Xu, S. C., & Hao, M. G. (2023). Can digital-real integration promote industrial green transformation: Fresh evidence from China's industrial sector. *Journal of Cleaner Production, 426*, 139116. DOI: 10.1016/j.jclepro.2023.139116

Meyerson, L. A., & Mooney, H. A. (2007). Invasive alien species in an era of globalization. [). Wiley Online Library.]. *Frontiers in Ecology and the Environment, 5*(4), 199–208. DOI: 10.1890/1540-9295(2007)5[199:IASIAE]2.0.CO;2

MHRA. (2022). Software and AI as a Medical Device Change Programme. Medicines and Healthcare products Regulatory Agency.

Micheni, E., Machii, J., & Murumba, J. (2022). *Internet of Things, Big Data Analytics, and Deep Learning for Sustainable Precision Agriculture. In 2022 IST-Africa Conference.* IST-Africa., DOI: 10.23919/IST-Africa56635.2022.9845490

Mildebrath, H. (2020, September). The CJEU judgment in the Schrems II case (PE 652.073). European Parliamentary Research Service. https://www.europarl.europa.eu/RegData/etudes/ATAG/2020/652073/EPRS_ATA(2020)652073_EN.pdf

Mill, J. S. (2001). Utilitarianism. Electric Book Company. (Original work published 1863). https://www.electricbook.co.uk

Mirakhori, F., & Niazi, S. K. (2025). Harnessing the AI/ML in Drug and Biological Products Discovery and Development: The Regulatory Perspective. *Pharmaceuticals (Basel, Switzerland), 18*(1), 47. DOI: 10.3390/ph18010047 PMID: 39861110

Mittelstadt, B. D., Allo, P., Taddeo, M., Wachter, S., & Floridi, L. (2016). The ethics of algorithms: Mapping the debate. *Big Data & Society*, *3*(2), 2053951716679679. DOI: 10.1177/2053951716679679

Mizrachi, R., Neiman, D., Rosenski, J., Loyfer, N., Share, D., Adjedj, C., Glaser, B., Shpitzen, M., Dor, Y., Shemer, R., & Kaplan, T. (2025). Genetic/epigenetic DNA markers for linking suspects and tissues in complex crime scenes. *Briefings in Bioinformatics*, *26*(4), bbaf395. Advance online publication. DOI: 10.1093/bib/bbaf395 PMID: 40794944

Moffat, J. G., Vincent, F., Lee, J. A., Eder, J., & Prunotto, M. (2017). Opportunities and challenges in phenotypic drug discovery: An industry perspective. *Nature Reviews. Drug Discovery*, *16*(8), 531–543. DOI: 10.1038/nrd.2017.111 PMID: 28685762

Mohanty, S. P., Hughes, D. P., & Salathé, M. (2016). Using deep learning for image-based plant disease detection. *Frontiers in Plant Science*, *7*, 1419. Advance online publication. DOI: 10.3389/fpls.2016.01419 PMID: 27713752

Monika, E., & Kumar, T. R. (2024). Exploring the Potential of TERT and Genetic Markers in Forensic Science for Crime Detection and Prevention. *2nd International Conference on Emerging Research in Computational Science, ICERCS 2024*. DOI: 10.1109/ICERCS63125.2024.10895472

Montesinos-López, O. A., Martin-Vallejo, J., Crossa, J., Gianola, D., Hernández-Suárez, C. M., Montesinos-López, A., & Singh, R. (2019). New deep learning genomic-based prediction model for multiple traits with binary, ordinal, and continuous phenotypes. *G3 (Bethesda, Md.)*, *9*(5), 1545–1556. DOI: 10.1534/g3.119.300585 PMID: 30858235

Mony, P. K., & Nagaraj, C. (2007). Health information management: An introduction to disease classification and coding. *The National Medical Journal of India*, *20*(6), 307. PMID: 18335800

Moqi, J. H. L. (2010). Application of Geographic Information System in Digital Mining Information System [J]. *Standardization of Surveying and Mapping, 3*.

Mullard, A. (2019). Machine learning brings cell imaging promises into focus. *Nature Reviews. Drug Discovery*, *18*(9), 653–655. DOI: 10.1038/d41573-019-00144-2 PMID: 31477870

Muneeb, M., Feng, S., & Henschel, A. (2022). Transfer learning for genotype–phenotype prediction using deep learning models. *BMC Bioinformatics 2022 23:1*, *23*(1), 511-. DOI: 10.1186/s12859-022-05036-8

Muqaddas, M., Mengaya, L., Ahmed, M. M., Hamid, S. M., Yanju, X., Asim, M., & Yunfeng, P. (2025). Postharvest handling and storage strategies for preserving jujube (Ziziphus jujuba mill.) fruit quality: A review. *Foods*, *14*(19), 3370. DOI: 10.3390/foods14193370 PMID: 41097538

Mutasa, L. (2022). E-health and e-governance integration framework for the Namibian government (Doctoral dissertation, Cape Peninsula University of Technology).

Mutyalamma, A. V., Yoshitha, G., Dakshyani, A., & Padmavathi, B. V. (2020). Smart Agriculture to Measure Humidity Temperature Moisture Ph. and Nutrient Values of the Soil using IoT. [IJEAT]. *International Journal of Engineering and Advanced Technology*, *9*(5), 394–398. DOI: 10.35940/ijeat.D8940.069520

Muzammal, H., Zaman, M., Safdar, M., Shahid, M. A., Nishtar, Z., Bilal, M., Munir, M., Haseeb, M. M., Raza, A., Shah, S. I. H., & Zafar, U. (2026). Advances in Geospatial Modeling for Flood Risk Assessment in Changing Climate. In *Advances in Computational Intelligence for Climate Change Security and Sustainability* (pp. 33–62). IGI Global Scientific Publishing.

Mykrä, H., Heino, J., & Muotka, T. (2007). Scale-related patterns in the spatial and environmental components of stream macroinvertebrate assemblage variation. *Global Ecology and Biogeography*, *16*(2), 149–159. DOI: 10.1111/j.1466-8238.2006.00272.x

Naresh Kumar, B., & Sakthivel, S. (2025). Rice leaf disease classification using a fusion vision approach. *Scientific Reports*, *15*(1), 8692. Advance online publication. DOI: 10.1038/s41598-025-87800-3 PMID: 40082482

Nguyen, T., Brandstetter, J., Kapoor, A., Gupta, J. K., & Grover, A. (2023). ClimaX: A foundation model for weather and climate. *ArXiv Preprint ArXiv:2301.10343.*

Niazi, S. K. (2026). A Critical Review of the FDA's Draft Guidance on Artificial Intelligence in Drug and Biological Product Regulation. Journal of Chemistry, 2026(1), 5202999.

Nichols, J. D., & Williams, B. K. (2006). Monitoring for conservation. *Trends in Ecology & Evolution*, *21*(12), 668–673. DOI: 10.1016/j.tree.2006.08.007 PMID: 16919361

Nielsen, J., & Keasling, J. D. (2016). Engineering cellular metabolism. *Cell*, *164*(6), 1185–1197. DOI: 10.1016/j.cell.2016.02.004 PMID: 26967285

NIH. (2020). *National COVID Cohort Collaborative (N3C)*. National Institutes of Health.

NMPA. (2020). *Guidelines on Clinical Trials Using Adaptive Designs*. National Medical Products Administration.

Norouzzadeh, M. S., Nguyen, A., Kosmala, M., Swanson, A., Palmer, M. S., Packer, C., & Clune, J. (2018). Automatically identifying, counting, and describing wild animals in camera-trap images with deep learning. *Proceedings of the National Academy of Sciences of the United States of America*, *115*(25), E5716–E5725. DOI: 10.1073/pnas.1719367115 PMID: 29871948

Nyhus, P. J. (2016). Human–wildlife conflict and coexistence. *Annual Review of Environment and Resources*, *41*(1), 143–171. DOI: 10.1146/annurev-environ-110615-085634

Obermeyer, Z., Powers, B., Vogeli, C., & Mullainathan, S. (2019). Dissecting racial bias in an algorithm used to manage the health of populations. *Science*, *366*(6464), 447–453. DOI: 10.1126/science.aax2342 PMID: 31649194

Okut, H. (2021). Deep Learning Algorithms for Complex Traits Genomic Prediction. *Journal of Animal Science and Products*, *4*(2), 225–239. DOI: 10.51970/jasp.1039713

Olawade, D. B., Kade, A., Egbon, E., Usman, S. O., Fapohunda, O., Ijiwade, J., & Ogbonna, C. E. (2025). Bioinformatics and artificial intelligence in genomic data analysis: Current advances and future directions. *Molecular Genetics and Genomics*, *300*(1), 111. DOI: 10.1007/s00438-025-02314-x PMID: 41348251

Olden, J. D., Lawler, J. J., & Poff, N. L. (2008). Machine learning methods without tears: A primer for ecologists. *The Quarterly Review of Biology*, *83*(2), 171–193. DOI: 10.1086/587826 PMID: 18605534

Opderbeck, D. W. (2019). Artificial intelligence in pharmaceuticals, biologics, and medical devices: Present and future regulatory models. *Fordham Law Review*, *88*, 553.

Optimi Management Consulting. (2026). Pharma & biologics 4.0: Key components and technologies. https://optimi.ai/pharma-%26-biologics-4-0

Organisation for Economic Co-operation and Development. (2025). OECD AI policy observatory: Tools for trustworthy AI implementation and conformity assessment. OECD Publishing. https://oecd.ai/en/

Owan, V. J., Abang, K. B., Idika, D. O., Etta, E. O., & Bassey, B. A. (2023). Exploring the potential of artificial intelligence tools in educational measurement and assessment. *Eurasia Journal of Mathematics, Science and Technology Education*, *19*(8), em2307. Advance online publication. DOI: 10.29333/ejmste/13428

Pai, R. R., & Alathur, S. (2019). Assessing awareness and use of mobile phone technology for health and wellness: Insights from India. *Health Policy and Technology, 8*(3), 221–227. DOI: 10.1016/j.hlpt.2019.05.011

Palmela Pereira, C., Carvalho, R., Augusto, D., Almeida, T., Francisco, A. P., Salvado e Silva, F., & Santos, R. (2025). Development of artificial intelligence-based algorithms for the process of human identification through dental evidence. *International Journal of Legal Medicine 2025 139:4, 139*(4), 1835–1850. DOI: 10.1007/s00414-025-03453-x

Pal, R. K. (2025). The Ecological Self: Deep Ecology as a Foundational Framework for Human Destiny and Sustainable Welfare. *Journal of Social Welfare and Management, 17*(3).

Pandiri, D. N. K., Murugan, R., & Goel, T. (2025). ARM-UNet: Attention residual path modified UNet model to segment the fungal pathogen diseases in potato leaves. *Signal, Image and Video Processing, 19*(1), 80. Advance online publication. DOI: 10.1007/s11760-024-03566-7

Parmesan, C. (2006). Ecological and evolutionary responses to recent climate change. *Annual Review of Ecology, Evolution, and Systematics, 37*(1), 637–669. DOI: 10.1146/annurev.ecolsys.37.091305.110100

Patel, M., & Patel, M. (2022). An integrated GIS/GPS based e-governance approach for different agencies to fight against COVID-19. Understanding COVID-19: The role of computational intelligence, 447-469.

Patel, M., & Kumar, S. (2022). The significance of electronic health records (EHRs) and their role in improving patient care efficiency. *Health Informatics Journal, 25*(4), 305–320.

Pathak, A., Theagarajan, R., Rizqi, M. M., Nugraha, A. S., Boruah, T., Pratiksha, , Kumar, H., Naik, B., Yadav, S., Jha, A. K., Trivedi, A., Gupta, A. K., & Kumar, P. (2025). AI-enabled drug and molecular discovery: Computational methods, platforms, and translational horizons. *Discov Mol, 2*(1), 32. DOI: 10.1007/s44345-025-00037-5

Paul, P. K. (2022). Aspects of Biosensors with Refers to Emerging Implications of Artificial Intelligence, Big Data and Analytics: The Changing Healthcare–A General Review. *Next Generation Smart Nano-Bio-Devices*, 1-18.

Paul, P. K. (2026). Environmental Informatics Promotion Using Artificial Intelligence and Allied Technologies: The Applications, Sustainability and Educational Perspectives. In Harnessing the potential of Artificial Intelligence to Promote Environment Sustainability, (pp. 111-124). Emerald, UK

Paul, P. K., Chatterjee, R., Chakraborty, S., Tiwary, K. S., Aithal, P. S., & Sharma, S. (2024). Advanced Healthcare Informatics Practice Vis-À-Vis Virtual Reality Implications—a Scientific Overview. *Immersive Virtual and Augmented Reality in Healthcare*, 28-49.

Paul, P.K, Bhuimali, A., Aithal, P. S., Kalishankar, T., & Saavedra M, R. (2020). Artificial intelligence & cloud computing in environmental systems-towards healthy & sustainable development. International Journal of Inclusive Development, 6(1), 01-08.

Paul, P.K., Bhuimali, A., Aithal, P. S., Kalishankar, T., & Saavedra M, R. (2020a). Artificial intelligence & cloud computing in environmental systems-towards healthy & sustainable development. *International Journal of Inclusive Development, 6*(1), 01-08.

Paul, D., Sanap, G., Shenoy, S., Kalyane, D., Kalia, K., & Tekade, R. K. (2021). Artificial intelligence in drug discovery and development. *Drug Discovery Today, 26*(1), 80–93. DOI: 10.1016/j.drudis.2020.10.010 PMID: 33099022

Paul, P. K. (2022). Environmental informatics: basics, nature, and applications using emerging technologies with reference to issues and potentialities. In *Environmental Informatics: Challenges and Solutions* (pp. 1–13). Springer Nature Singapore. DOI: 10.1007/978-981-19-2083-7_1

Paul, P. K. (2023). Advanced ICT and intelligent systems in sophisticated Healthcare 5.0 practice in modern social and healthcare transformation: An overview. *Augmented and Virtual Reality in Social Learning: Technological Impacts and Challenges, 3*, 63–80. DOI: 10.1515/9783110981445-005

Paul, P. K., & Aithal, P. S. (2020). Environmental informatics and its features, functions and stakeholders: A comprehensive overview. *Educational Quest-An International Journal of Education and Applied Social Sciences, 11*(1), 1–5. DOI: 10.30954/2230-7311.1.2020.1

Paul, P. K., Aithal, P. S., Bhimali, A., & Kalishankar, T. (2020). Environmental informatics vis-à-vis big data analytics: The geo-spatial & sustainable solutions. [IJAEML]. *International Journal of Applied Engineering and Management Letters, 4*(2), 31–40.

Paul, P. K., Aithal, P. S., & Bhuimali, A. (2018). Health Information Science and its growing popularities in Indian self financed universities: Emphasising Private Universities—A Study. *International Journal of Scientific Research in Biological Sciences, 5*(1), 1–11. DOI: 10.26438/ijsrbs/v5i1.111

Paul, P. K., Aithal, P. S., Bhuimali, A., Kalishankar, T., Saavedra, M. R., & Aremu, P. S. B. (2020b). Geo information systems and remote sensing: Applications in environmental systems and management. [IJMTS]. *International Journal of Management, Technology, and Social Sciences*, *5*(2), 11–18. DOI: 10.47992/IJMTS.2581.6012.0100

Paul, P. K., Moumtzoglou, A., Bandyopadhyay, A., Hoque, M., Kumar, N. S., & Saavedra, R. (2024). Indian Digital Health Information Systems: Initiatives and Opportunities with Socioeconomic Perspective. *Economic Affairs*, *69*(2), 1175–1189. DOI: 10.46852/0424-2513.3.2024.39

Pecl, G. T., Araújo, M. B., Bell, J. D., Blanchard, J., Bonebrake, T. C., Chen, I. C., Clark, T. D., Colwell, R. K., Danielsen, F., Evengård, B., Falconi, L., Ferrier, S., Frusher, S., Garcia, R. A., Griffis, R. B., Hobday, A. J., Janion-Scheepers, C., Jarzyna, M. A., Jennings, S., & Williams, S. E. (2017). Biodiversity redistribution under climate change: Impacts on ecosystems and human well-being. *Science*, *355*(6332), eaai9214. DOI: 10.1126/science.aai9214 PMID: 28360268

Pedroza Matute, S., & Iyavoo, S. (2025). Implementation of NGS and SNP microarrays in routine forensic practice: opportunities and barriers. *BMC Genomics 2025 26:1, 26*(1), 541-. DOI: 10.1186/s12864-025-11723-6

Pengyue, L., Siyuan, X., Yi, J., Wen, Y., Xiaoning, L., Guohua, G., & Shixiong, W. (2021). ANINet: a deep neural network for skull ancestry estimation. *BMC Bioinformatics 2021 22:1, 22*(1), 550-. DOI: 10.1186/s12859-021-04444-6

Peters, D. P., Havstad, K. M., Cushing, J., Tweedie, C., Fuentes, O., & Villanueva-Rosales, N. (2014). Harnessing the power of big data: Infusing the scientific method with machine learning to transform ecology. *Ecosphere*, *5*(6), 1–15. DOI: 10.1890/ES13-00359.1

Peters, M. A. (2020). AI, robotics, and the future of sustainability. *Educational Philosophy and Theory*, *52*(10), 1043–1047.

Pettorelli, N., Laurance, W. F., O'Brien, T. G., Wegmann, M., Nagendra, H., & Turner, W. (2014). Satellite remote sensing for applied ecologists: Opportunities and challenges. *Journal of Applied Ecology*, *51*(4), 839–848. DOI: 10.1111/1365-2664.12261

Petty, J., & Richardson, O. (2023). Algorithmic reparations: Toward a restorative framework for health data. *Big Data & Society*, *10*(1). Advance online publication. DOI: 10.1177/20539517231165

Phang, S. K., Chiang, T. H. A., Happonen, A., & Chang, M. M. L. (2023). *From Satellite to UAV-based Remote Sensing: A Review on Precision Agriculture*. IEEE., DOI: 10.1109/ACCESS.2023.3325862

Phan, H., Ahmad, A., & Saraswat, D. (2022). Identification of foliar disease regions on corn leaves using SLIC segmentation and deep learning under uniform background and field conditions. *IEEE Access : Practical Innovations, Open Solutions*, *10*, 111985–111995. DOI: 10.1109/ACCESS.2022.3215497

Pharmaceutical Technology. (2025, February 11). UK awards £82.6m to AI drug discovery companies. https://www.pharmaceutical-technology.com/news/uk-awards-82-6m-to-ai-drug-discovery-companies/

Phukan, O. C., Singh, G., Tiwari, S., & Butt, S. (2022). An automated stress recognition for digital healthcare: towards e-governance. In *International Conference on Electronic Governance with Emerging Technologies* (pp. 117-125). Cham: Springer Nature Switzerland. DOI: 10.1007/978-3-031-22950-3_10

Piccoli, F., Locatelli, S. G., Schettini, R., & Napoletano, P. (2023). An Open-Source Platform for GIS Data Management and Analytics. *Sensors (Basel)*, *23*(8), 3788. DOI: 10.3390/s23083788 PMID: 37112129

Pimm, S. L., Jenkins, C. N., Abell, R., Brooks, T. M., Gittleman, J. L., Joppa, L. N., Raven, P. H., Roberts, C. M., & Sexton, J. O. (2014). The biodiversity of species and their rates of extinction, distribution, and protection. *Science*, *344*(6187), 1246752. DOI: 10.1126/science.1246752 PMID: 24876501

Pisarek, A., Pośpiech, E., Heidegger, A., Xavier, C., Papież, A., Piniewska-Róg, D., Kalamara, V., Potabattula, R., Bochenek, M., Sikora-Polaczek, M., Macur, A., Woźniak, A., Janeczko, J., Phillips, C., Haaf, T., Polaoska, J., Parson, W., Kayser, M., & Branicki, W. (2021). Epigenetic age prediction in semen – marker selection and model development. *Aging (Albany NY)*, *13*(15), 19145–19164. DOI: 10.18632/aging.203399 PMID: 34375949

PMDA. (2020). *Reflection on the use of AI/ML in pharmaceuticals*. Pharmaceuticals and Medical Devices Agency.

Poddar, A., & Rao, S. R. (2024). Evolving intellectual property landscape for AI-driven innovations in the biomedical sector: Opportunities in stable IP regime for shared success. *Frontiers in Artificial Intelligence*, *7*, 1372161. DOI: 10.3389/frai.2024.1372161 PMID: 39355146

Poessel, S. A., Breck, S. W., Teel, T. L., Shwiff, S., Crooks, K. R., & Angeloni, L. (2013). Patterns of human–coyote conflicts in the Denver Metropolitan Area. *The Journal of Wildlife Management, 77*(2), 297–305. DOI: 10.1002/jwmg.454

Poplin, R., Chang, P.-C., Alexander, D., Schwartz, S., Colthurst, T., Ku, A., & DePristo, M. A. (2018). A universal SNP and small-indel variant caller using deep neural networks. *Nature Biotechnology, 36*(10), 983–987. DOI: 10.1038/nbt.4235 PMID: 30247488

Pośpiech, E., Teisseyre, P., Mielniczuk, J., & Branicki, W. (2022). Predicting Physical Appearance from DNA Data—Towards Genomic Solutions. *Genes 2022, Vol. 13, Page 121, 13*(1), 121. DOI: 10.3390/genes13010121

Prabavathi, R., Chelliah, B. J., Sampath, R., & Vinodh Kumar, S. (2025). A comprehensive cross-attention and fuzzy segmentation approach for rice plant disease detection. *International Journal of Machine Learning and Cybernetics, 16*(11), 9295–9321. DOI: 10.1007/s13042-025-02755-1

Price, W. N. (2018). Black-box medicine. *Harvard Journal of Law & Technology, 28*, 419.

Pujitha, M. V., & Kiran, K. V. D. (2025). Development of an efficient deep learning aided segmentation and classification mechanism for plant disease identification and its severity assessment. *International Journal of System Assurance Engineering and Management*. Advance online publication. DOI: 10.1007/s13198-025-02880-8

Pulendran, B., & Ahmed, R. (2011). Immunological mechanisms of vaccination. *Nature Immunology, 12*(6), 509–517. DOI: 10.1038/ni.2039 PMID: 21739679

Rafik, H., Maizate, A., & Ettaoufik, A. (2023). Data Security Mechanisms, Approaches, and Challenges for e-Health Smart Systems. *International Journal of Online & Biomedical Engineering, 19*(2), 42–66. DOI: 10.3991/ijoe.v19i02.37069

Rai, A. K., Kumar, N., Katiyar, D., Singh, O., Sreekumar, G., & Verma, P. (2023). Unlocking Productivity Potential: The Promising Role of Agricultural Robots in Enhancing Farming Efficiency. *International Journal of Plant and Soil Science, 35*(18), 624–633. DOI: 10.9734/ijpss/2023/v35i183327

Rajeena, P. P. F., Aswathy, S. U., Moustafa, M. A., & Ali, M. A. S. (2023). Detecting plant disease in corn leaf using EfficientNet architecture—An analytical approach. *Electronics (Basel), 12*(8), 1938. Advance online publication. DOI: 10.3390/electronics12081938

Rajkomar, A., Hardt, M., Howell, M. D., Corrado, G., & Chin, M. H. (2018). Ensuring fairness in machine learning to advance health equity. *Annals of Internal Medicine*, *169*(12), 866–872. DOI: 10.7326/M18-1990 PMID: 30508424

Rajkomar, A., Oren, E., Chen, K., Dai, A. M., Hajaj, N., Hardt, M., & Dean, J. (2018). Scalable and accurate deep learning with electronic health records. *NPJ Digital Medicine*, *1*(1), 18. Advance online publication. DOI: 10.1038/s41746-018-0029-1 PMID: 31304302

Ramesh, S., & Vydeki, D. (2020). Rice disease detection and classification using deep neural network algorithm. In D. K. Sharma, V. E. Balas, L. H. Son, R. Sharma, & K. Cengiz (Eds.), Micro-electronics and telecommunication engineering (Lecture Notes in Networks and Systems, Vol. 106). Springer. DOI: 10.1007/978-981-15-2329-8_56

Ramesh, K., Joshi, A., & Desai, V. (2020). National Digital Health Blueprint (NDHB) and its alignment with India's National Health Policy (2017). *Indian Journal of Public Health Policy*, *14*(2), 140–155.

Rana, A. (2017). The Immense Potential of M-Care in India: Catering Better to Patients Needs in the Context of a Fragmented Healthcare System. [IJRQEH]. *International Journal of Reliable and Quality E-Healthcare*, *6*(4), 1–3. DOI: 10.4018/IJRQEH.2017100101

Rane, N. L., Mallick, S. K., Kaya, O., & Rane, J. (2024). From challenges to implementation and acceptance: Addressing key barriers in artificial intelligence, machine learning, and deep learning. In *Applied Machine Learning and Deep Learning: Architectures and Techniques* (pp. 153–166). Deep Science Publishing.

Rane, N. L., Paramesha, M., Rane, J., & Kaya, O. (2024). Artificial intelligence, machine learning, and deep learning for enabling smart and sustainable cities and infrastructure. In *Artificial Intelligence and Industry in Society 5.0* (pp. 24–49). Deep Science Publishing. DOI: 10.70593/978-81-981271-1-2_2

Rani, N., Chaudhary, P., Kumar, U., Khatana, K., Kumar, N., Mavi, A. K., & Singh, S. (2025). Harnessing deep learning in bioinformatics–opportunities, challenges, and ethical considerations. *Vita Scientia*, *1*(1), 21–56. DOI: 10.65116/vitascientia.V1I1.rev.01

Rathee, G., Saini, H., Garg, S., Choi, B. J., & Hassan, M. M. (2024). A secure data e-governance for healthcare application in cyber physical systems. [IJSWIS]. *International Journal on Semantic Web and Information Systems*, *20*(1), 1–17. DOI: 10.4018/IJSWIS.345934

Rawls, J. (1971). *A theory of justice*. Belknap Press of Harvard University Press. DOI: 10.4159/9780674042605

Ray, S., & Mukherjee, A. (2007). Development of a framework towards successful implementation of e-governance initiatives in health sector in India. *International Journal of Health Care Quality Assurance*, *20*(6), 464–483. DOI: 10.1108/09526860710819413 PMID: 18030965

Raza, A., Malik, A., & Maqsood, M. 2025. Transitioning From Legacy Systems to AI: A Holistic Approach to Disaster Detection. In *Enhancing Automated Decision-Making Through AI* (pp. 33-66). IGI Global Scientific Publishing.

Raza, A., Miao, Y., Huang, Y., Stueve, K., Lu, J., Yang, Z., & Bindlish, R. (2025). Optimizing On-Farm Corn Yield Prediction by a Multi-Source Data Fusion Approach Using Remote Sensing and Machine Learning. *Smart Agricultural Technology*, *12*, 101630. DOI: 10.1016/j.atech.2025.101630

Raza, A., Shahid, M. A., Safdar, M., Zaman, M., Sabir, R. M., Muzammal, H., & Ahmed, M. M. (2024). Impact of land use and land cover change on agricultural production in District Bahawalnagar, Pakistan. *Environmental Sciences Proceedings*, *29*(1), 46.

Raza, A., Shahid, M. A., Zaman, M., Miao, Y., Huang, Y., Safdar, M., Maqbool, S., & Muhammad, N. E. (2025). Improving wheat yield prediction with multi-source remote sensing data and machine learning in arid regions. *Remote Sensing (Basel)*, *17*(5), 774. DOI: 10.3390/rs17050774

Recentive Analytics, Inc. v. Fox. Corp et al., Case No. 2023-2437 (Fed. Cir. Apr. 18, 2025).

Reel, P. S., Reel, S., Pearson, E., Trucco, E., & Jefferson, E. (2021). Using machine learning approaches for multi-omics data analysis: A review. *Biotechnology Advances*, *49*, 107739. DOI: 10.1016/j.biotechadv.2021.107739 PMID: 33794304

Regilme, S. S. F. (2025). *Digital extractivism and the double burden: The political economy of AI infrastructure in the Global South*. Oxford University Press.

Reichstein, M., Camps-Valls, G., Stevens, B., Jung, M., Denzler, J., Carvalhais, N., & Prabhat, . (2019). Deep learning and process understanding for data-driven Earth system science. *Nature*, *566*(7743), 195–204. DOI: 10.1038/s41586-019-0912-1 PMID: 30760912

Rejeb, A., Rejeb, K., Simske, S., & Treiblmaier, H. (2021). Humanitarian drones: A review and research agenda. *Internet of Things : Engineering Cyber Physical Human Systems*, *16*, 100434. DOI: 10.1016/j.iot.2021.100434

Ren, J., Li, C., An, Y., Zhang, W., & Sun, C. (2024). Few-Shot Fine-Grained Image Classification: A comprehensive review. *AI*, *5*(1), 405–425. DOI: 10.3390/ai5010020

Report on Artificial Intelligence 2024, December 19). 253-page House AI report offers lawmakers a policy tool. Government Technology. https://www.govtech.com/artificial-intelligence/253-page-house-ai-report-offers-lawmakers-a-policy-tool

Revathi, A., & Priya, R. (2025). Geometrically innovated machine learning for optimized prediction of rice blast disease. In S. Rajagopal, K. Popat, D. Meva, S. Bajeja, & P. Mudholkar (Eds.), Artificial intelligence based smart and secured applications. ASCIS 2024 (Communications in Computer and Information Science, Vol. 2425). Springer. DOI: 10.1007/978-3-031-86293-9_3

Rezk, N. G., Hemdan, E. E. D., Attia, A. F., El-Sayed, A., & El-Rashidy, M. A. (2023). An efficient IoT based framework for detecting rice disease in smart farming system. *Multimedia Tools and Applications*, *82*(29), 45259–45292. DOI: 10.1007/s11042-023-15470-2

Roberts, H., Cowls, J., Morley, J., Taddeo, M., Wang, V., & Floridi, L. (2021). The Chinese approach to artificial intelligence: An analysis of policy, ethics, and regulation. *AI & Society*, *36*(1), 59–77. DOI: 10.1007/s00146-020-00992-2

Roden, D. M., McLeod, H. L., Relling, M. V., Williams, M. S., Mensah, G. A., Peterson, J. F., & Van Driest, S. L. (2019). Pharmacogenomics. *Lancet*, *394*(10197), 521–532. DOI: 10.1016/S0140-6736(19)31276-0 PMID: 31395440

Rolnick, D., Donti, P. L., Kaack, L. H., Kochanski, K., Lacoste, A., Sankaran, K., & Bengio, Y. (2022). Tackling climate change with machine learning. *ACM Computing Surveys*, *55*(2), 1–96. DOI: 10.1145/3485128

Ronneberger, O., Fischer, P., & Brox, T. (2015). U-Net: Convolutional networks for biomedical image segmentation [Preprint]. arXiv. /arXiv.1505.04597DOI: 10.1007/978-3-319-24574-4_28

Roy, A., Dutta, S., Das, S., & Choudhury, M. R. (2024). Next-Generation Sequencing in the Development of Climate-Resilient and Stress-Responsive Crops – A Review. *The Open Biotechnology Journal*, *18*(1), e18740707301657. Advance online publication. DOI: 10.2174/0118740707301657240517063244

Roy, P. S., & Kukreja, V. (2025). Vision transformers for rice leaf disease detection and severity estimation: A precision agriculture approach. *Journal of the Saudi Society of Agricultural Sciences*, *24*(3), 3. Advance online publication. DOI: 10.1007/s44447-025-00007-w

Ruan, W., Lyu, Y., Zhang, J., Cai, J., Shu, P., Ge, Y., Lu, Y., Gao, S., Wang, Y., Wang, P., Zhao, L., Wang, T., Liu, Y., Fang, L., Liu, Z., Liu, Z., Li, Y., Wu, Z., Chen, J., & Liu, T. (2026). Large language models for bioinformatics. *Quantitative Biology, 14*(1), e70014. DOI: 10.1002/qub2.70014 PMID: 41676319

Rudin, C. (2019). Stop explaining black box machine learning models for high stakes decisions and use interpretable models instead. *Nature Machine Intelligence, 1*(5), 206–215. DOI: 10.1038/s42256-019-0048-x PMID: 35603010

Russakovsky, O., Deng, J., Su, H., Krause, J., Satheesh, S., Ma, S., Huang, Z., Karpathy, A., Khosla, A., Bernstein, M., Berg, A. C., & Fei-Fei, L. (2015). ImageNet large-scale visual recognition challenge. *International Journal of Computer Vision, 115*(3), 211–252. DOI: 10.1007/s11263-015-0816-y

Russell, S., & Norvig, P. (2021). *Artificial intelligence: A modern approach* (4th ed.). Pearson.

Ryu, J., Barkal, S., Yu, T., Jankowiak, M., Zhou, Y., Francoeur, M., Phan, Q. V., Li, Z., Tognon, M., Brown, L., Love, M. I., Bhat, V., Lettre, G., Ascher, D. B., Cassa, C. A., Sherwood, R. I., & Pinello, L. (2024). Joint genotypic and phenotypic outcome modeling improves base editing variant effect quantification. *Nature Genetics 2024 56:5, 56*(5), 925–937. DOI: 10.1038/s41588-024-01726-6

Saha, S., & Raghava, G. P. S. (2006). Prediction of continuous B-cell epitopes in an antigen using recurrent neural network. *Proteins, 65*(1), 40–48. DOI: 10.1002/prot.21078 PMID: 16894596

Sahoo, S., Singha, C., Govind, A., & Sharma, M. (2025). Leveraging ML to predict climate change impact on rice crop disease in Eastern India. *Environmental Monitoring and Assessment, 197*(4), 366. Advance online publication. DOI: 10.1007/s10661-025-13744-w PMID: 40056263

Sampathkumar, K., & Kerwin, B. A. (2024). Roadmap for drug product development and manufacturing of biologics. *Journal of Pharmaceutical Sciences, 113*(2), 314–331. DOI: 10.1016/j.xphs.2023.11.004 PMID: 37944666

Sandbrook, C., Luque-Lora, R., & Adams, W. M. (2018). Human bycatch: Conservation surveillance and the social implications of camera traps. *Conservation & Society, 16*(4), 493–504. DOI: 10.4103/cs.cs_17_165

Santiago-Rodriguez, T. M., & Hollister, E. B. (2021). Multi 'omic data integration: A review of concepts, considerations, and approaches. *Seminars in Perinatology, 45*(6), 151456. DOI: 10.1016/j.semperi.2021.151456 PMID: 34256961

Saravanan, S. K., Nisha, F., Rohit, V. R., Lenin, J., Selvakumar, P. D., & Rajmohan, M. (2023). Impact of Cloud Computing on Agricultural Advancement using Data Mining Algorithms. In *2023 2nd International Conference on Automation, Computing and Renewable Systems (ICACRS)* (pp. 1570–1575). https://doi.org/DOI: 10.1109/ICACRS58579.2023.10404669

Sarker, M. N. I., Peng, Y., Yiran, C., & Shouse, R. C. (2020). Disaster resilience through big data: Way to environmental sustainability. *International Journal of Disaster Risk Reduction*, *51*, 101769. DOI: 10.1016/j.ijdrr.2020.101769

Schaefer, M. B. (2019). Some Considerations of Population Dynamics and Economics in Relation to the Management of the Commercial Marine Fisheries 1. *Fisheries Economics*, 33–45. DOI: 10.4324/9781315193182-3

Schley, L., & Roper, T. J. (2003). Diet of wild boar in Western Europe, with particular reference to consumption of agricultural crops. *Mammal Review*, *33*(1), 43–56. DOI: 10.1046/j.1365-2907.2003.00010.x

Schwartz, R., Dodge, J., Smith, N. A., & Etzioni, O. (2020). Green AI. *Communications of the ACM*, *63*(12), 54–63. DOI: 10.1145/3381831

Seddon, P. J., Griffiths, C. J., Soorae, P. S., & Armstrong, D. P. (2014). Reversing defaunation: Restoring species in a changing world. *Science*, *345*(6195), 406–412. DOI: 10.1126/science.1251818 PMID: 25061203

Selbst, A. D., Barocas, S., Raghavan, M., Friedler, S. A., & Venkatasubramanian, S. (2019). Fairness and abstraction in sociotechnical systems. In Proceedings of the 2019 Conference on Fairness, Accountability, and Transparency (FAT*) (pp. 59–68). Association for Computing Machinery. https://doi.org/DOI: 10.1145/3287560.3287598

Senior, A. W., Evans, R., Jumper, J., Kirkpatrick, J., Sifre, L., Green, T., & Kavukcuoglu, K. (2020). Improved protein structure prediction using potentials from deep learning. *Nature*, *577*(7792), 706–710. DOI: 10.1038/s41586-019-1923-7 PMID: 31942072

Senoo, E. E. K., Anggraini, L., Kumi, J. A., Luna, B. K., Ebenezer, A., Hafeez, A. S., Israel, M., & Masayoshi, A. (2024). IoT solutions with artificial intelligence technologies for precision agriculture: Definitions, applications, challenges, and opportunities. *Electronics (Basel)*, *13*(10), 1894. DOI: 10.3390/electronics13101894

Senthil, R., & Khatwal, R. (2025). An effective disease detection analysis on rice leaves using hybrid MCSVM-DNN predictor architecture. *Remote Sensing in Earth Systems Sciences*, *8*(1), 84–97. DOI: 10.1007/s41976-024-00175-3

Sessa, F., Esposito, M., Cocimano, G., Sablone, S., Karaboue, M. A. A., Chisari, M., Albano, D. G., & Salerno, M. (2024). Artificial Intelligence and Forensic Genetics: Current Applications and Future Perspectives. *Applied Sciences (Basel, Switzerland)*, *14*(5), 2113. DOI: 10.3390/app14052113

Sg, P., & Lepakshi, V. A. (2025). Rice Blast Disease Detection and Assessment Using Deep Learning Techniques of Using Remote Sensing Methods. *Journal of Crop Health*, *77*(5), 171. DOI: 10.1007/s10343-025-01232-7

Shah, K. N., Gami, S. J., & Trehan, A. (n.d.). *An Intelligent Approach to Data Quality Management AI-Powered Quality Monitoring in Analytics*. DOI: 10.48175/IJARSCT-22820

Shah, P., Kendall, F., Khozin, S., Goosen, R., Hu, J., Laramie, J., Ringel, M., & Schork, N. (2019). Artificial intelligence and machine learning in clinical development: A translational perspective. *NPJ Digital Medicine*, *2*(1), 69. DOI: 10.1038/s41746-019-0148-3 PMID: 31372505

Sharma, M. K., & Vaisla, K. S. (2012). E-health for rural areas of Uttarakhand under e-Governance service delivery model. In *2012 1st International Conference on Recent Advances in Information Technology (RAIT)* (pp. 622-625). IEEE. DOI: 10.1109/RAIT.2012.6194601

Sharma, R., Singh, A., Kumar, P., ... (2025). Genetic algorithm–aided deep feature selection for improved rice disease classification. Operations Research Forum, 6, Article 7. DOI: 10.1007/s43069-024-00400-1

Sharma, K. (2020). The transformative role of mobile health (mHealth) applications in preventive healthcare. *Journal of Digital Health Innovations*, *10*(1), 65–78.

Sharma, S., & Gupta, R. (2021). Assessing the Socioeconomic Impact of Digital Health Information Systems: A Case Study of Rural India. *International Journal of Medical Informatics*, *132*, 104066.

Shen, R., Mo, Q., Schultz, N., Seshan, V. E., Olshen, A. B., Huse, J., & Taylor, B. S. (2012). Integrative subtype discovery in glioblastoma using iCluster. *PLoS One*, *7*(4), e35236. DOI: 10.1371/journal.pone.0035236 PMID: 22539962

Shevchuk, V., Dunaev, O., Tyshchenko, O., Biletska, G., & Nehrebetskyi, V. (2025). Innovative Methods in the Identification of Deceased Persons during Armed Conflicts and Disasters: Criminalistic and Forensic Medical Issues. *International Annals of Criminology*, 1–20. DOI: 10.1017/cri.2025.10094

Shiva Prasad, N., & Eranki, K. L. N. (2025). Prediction of inter-ocular rice leaf disease using deep learning approach. In H. Sharma, A. Chakravorty, S. Hussain, R. Kumari (Eds.), Artificial intelligence: Theory and applications. AITA 2024 (Lecture Notes in Networks and Systems, Vol. 5589). Springer. DOI: 10.1007/978-981-96-1687-9_19

Shrikumar, A. (2020). *Interpretable machine learning for scientific discovery in regulatory genomics*. Stanford University.

Shrivastava, S., & Srikanth, T. K., & VS, D. (2020, September). e-Governance for healthcare service delivery in India: challenges and opportunities in security and privacy. In *Proceedings of the 13th International Conference on Theory and Practice of Electronic Governance* (pp. 180-183). DOI: 10.1145/3428502.3428527

Silver, D., Schrittwieser, J., Simonyan, K., Antonoglou, I., Huang, A., Guez, A., Hubert, T., Baker, L., Lai, M., Bolton, A., Chen, Y., Lillicrap, T., Hui, F., Sifre, L., van den Driessche, G., Graepel, T., & Hassabis, D. (2017). Mastering the game of go without human knowledge. *Nature*, *550*(7676), 354–359. DOI: 10.1038/nature24270 PMID: 29052630

Simberloff, D., Martin, J.-L., Genovesi, P., Maris, V., Wardle, D. A., Aronson, J., Courchamp, F., Galil, B., García-Berthou, E., Pascal, M., Pyšek, P., Sousa, R., Tabacchi, E., & Vilà, M. (2013). Impacts of biological invasions: What's what and the way forward. *Trends in Ecology & Evolution*, *28*(1), 58–66. DOI: 10.1016/j.tree.2012.07.013 PMID: 22889499

Simpson, D. J., & Chandra, T. (2021). Epigenetic age prediction. *Aging Cell*, *20*(9), e13452. DOI: 10.1111/acel.13452 PMID: 34415665

Singh, A. K., Sreenivasu, S. V. N., Mahalaxmi, U. S. B. K., Sharma, H., Patil, D. D., & Asenso, E. (2022). [Retracted] Hybrid feature-based disease detection in plant leaf using convolutional neural network, Bayesian optimized SVM, and random forest classifier. *Journal of Food Quality*, *2022*, 2845320. Advance online publication. DOI: 10.1155/2022/2845320

Singh, A.. (2023). AI-driven models optimize irrigation, fertilization, and pest control, improving crop health. *Precision Agriculture*, *24*(2), 347–368. DOI: 10.1007/s11119-022-09948-6

Singh, A., & Sharma, R. (2021). Ayushman Bharat Digital Mission (ABDM) and its transformative impact on healthcare access in India. *Health Policy and Technology Journal*, *16*(3), 190–205.

Singh, G., Mishra, A., & Sagar, D. (2013). An overview of artificial intelligence. *SBIT Journal of Sciences and Technology*, *2*(1), 1–4.

Singh, J. A. (2019). Artificial intelligence and global health: Opportunities and challenges. *Emerging Topics in Life Sciences*, *3*(6), 741–746. DOI: 10.1042/ETLS20190106 PMID: 32915223

Singh, R., Paxton, M., & Auclair, J. (2025, May 17). Regulating the AI-enabled ecosystem for human therapeutics. *Communications Medicine*, *5*(1), 181. DOI: 10.1038/s43856-025-00910-x PMID: 40382515

Sinha, C. G. (2021). E-Governance: Impact on Society. In Digitization of Economy and Society (pp. 93-117). Apple Academic Press.

Smith, K. P., Kang, A. D., & Kirby, J. E. (2018). Automated interpretation of blood culture Gram stains by use of a deep convolutional neural network. *Journal of Clinical Microbiology*, *56*(3), e01521–e17. DOI: 10.1128/JCM.01521-17 PMID: 29187563

Snell, J., Swersky, K., & Zemel, R. (2017). Prototypical Networks for Few-Shot Learning. In *Proceedings of the 31st International Conference on Neural Information Processing Systems (NIPS'17)* (pp. 4080–4090). Curran Associates Inc., https://dl.acm.org/doi/10.5555/3294996.3295163

Solove, D. J., & Hartzog, W. (2014). The FTC and the new common law of privacy. Columbia Law Review, 114, 583. 't Hoen, E. F. (2016). Private Patents and Public Health: Changing intellectual property rules for access to medicines. Health Action International.

Song, Z., Huang, D., Song, B., Chen, K., Song, Y., Liu, G., Su, J., Magalhães, J. P. de, Rigden, D. J., & Meng, J. (2021). Attention-based multi-label neural networks for integrated prediction and interpretation of twelve widely occurring RNA modifications. *Nature Communications 2021 12:1, 12*(1), 4011-. DOI: 10.1038/s41467-021-24313-3

Soumare, H., Rezgui, S., Gmati, N., & Benkahla, A. (2021). New neural network classification method for individuals ancestry prediction from SNPs data. *BioData Mining 2021 14:1, 14*(1), 30-. DOI: 10.1186/s13040-021-00258-7

Srinivasan, S., Krishnamoorthy, R., Kumar, L., Kushwaha, A. S., Ojha, A. C., Ganesh, D., & Jebakumar, A. (2022). Manageable E-Governance Architecture In Health Care System Using Cloud Storage. *NeuroQuantology : An Interdisciplinary Journal of Neuroscience and Quantum Physics*, *20*(14), 1094.

Srivastava, S. K. (2016). Adoption of electronic health records: A roadmap for India. *Healthcare Informatics Research, 22*(4), 261–269. DOI: 10.4258/hir.2016.22.4.261 PMID: 27895957

Stahl, B. C. (2021). *Artificial intelligence for a better future: an ecosystem perspective on the ethics of AI and emerging digital technologies*. Springer Nature. DOI: 10.1007/978-3-030-69978-9

Stephens, P. A., Pettorelli, N., Barlow, J., Whittingham, M. J., & Cadotte, M. W. (2015). Management by proxy? The use of indices in applied ecology. *Journal of Applied Ecology, 52*(1), 1–6. DOI: 10.1111/1365-2664.12383 PMID: 25985647

Strubell, E., Ganesh, A., & McCallum, A. (2019). Energy and policy considerations for deep learning in NLP. *Proceedings of the 57th Annual Meeting of the Association for Computational Linguistics*, 3645–3650. https://doi.org/DOI: 10.18653/v1/P19-1355

Sullivan, B., Wood, C., Iliff, M., Bonney, R., Fink, D., & Kelling, S. (2009). eBird: A citizen-based bird observation network in the biological sciences. *Biological Conservation -. Biological Conservation, 142*(10), 2282–2292. DOI: 10.1016/j.biocon.2009.05.006

Sun, C., Sun, Y., Xu, K., He, Z., Li, H., Li, Y., . . . Chen, H. (2025). A comprehensive benchmark and guide for sequence-function interpretable deep learning models in genomics. bioRxiv, 2025-01.

Sundaram, L., Gao, H., Padigepati, S. R., McRae, J. F., Li, Y., Kosmicki, J. A., & Birnbaum, D. P. (2018). Predicting the clinical impact of human mutation with deep neural networks. *Nature Genetics, 50*(8), 1161–1170. DOI: 10.1038/s41588-018-0167-z PMID: 30038395

Sung, H., & Hopper, J. L. (2023). Artificial intelligence in epidemiology: Predictive modeling and the future of public health surveillance. *Epidemiologic Reviews, 45*(1), 12–25. DOI: 10.1093/epirev/mxad001

Sutton, R. T., Pincock, D., Baumgart, D. C., Sadowski, D. C., Fedorak, R. N., & Kroeker, K. I. (2020). An overview of clinical decision support systems: Benefits, risks, and strategies for success. *NPJ Digital Medicine, 3*(1), 17. DOI: 10.1038/s41746-020-0221-y PMID: 32047862

Szegedy, C., Liu, W., Jia, Y., Sermanet, P., Reed, S., & Anguelov, D.. (2015). Going deeper with convolutions. In *Proceedings of the IEEE Conference on Computer Vision and Pattern Recognition* (pp. 1–9). IEEE. DOI: 10.1109/CVPR.2015.7298594

Szikszai, M., Wang, T. Y., Krueger, R., Mathews, D. H., Ward, M., & Aviran, S. (2025). Deep Learning for RNA Secondary Structure Determination: Gauging Generalizability and Broadening the Scope of Traditional Methods. bioRxiv, 2025-11.

Tabane, E., Mnkandla, E., & Wang, Z. (2025). Optimizing DNA Sequence Classification via a Deep Learning Hybrid of LSTM and CNN Architecture. *Applied Sciences (Basel, Switzerland)*, *15*(15), 8225. DOI: 10.3390/app15158225

Tang, Z., Somia, N., Yu, Y., & Koo, P. K. (2025). Evaluating the representational power of pre-trained DNA language models for regulatory genomics. *Genome Biology*, *26*(1), 203. DOI: 10.1186/s13059-025-03674-8 PMID: 40660356

Tan, M., Tan, Y., Jiang, H., Xue, J., Wu, Q., Zheng, Y., Liu, G., Xiao, Y., Lv, M., Liao, M., Zhang, L., Qu, S., & Liang, W. (2025). Explainable artificial intelligence in forensic DNA analysis: Alleles identification in challenging electropherograms using supervised machine learning methods. *Forensic Science International. Genetics*, *78*, 103289. DOI: 10.1016/j.fsigen.2025.103289 PMID: 40288204

Terrado-Ortuño, N., & May, P. (2025). Forensic DNA phenotyping: A review on SNP panels, genotyping techniques, and prediction models. *Forensic Sciences Research*, *10*(1), owae013. Advance online publication. DOI: 10.1093/fsr/owae013 PMID: 39990695

Testagrose, C., & Boucher, C. (2025). Tokenization and deep learning architectures in genomics: A comprehensive review. *Computational and Structural Biotechnology Journal*, *27*, 3547–3555. DOI: 10.1016/j.csbj.2025.07.038 PMID: 40821715

Thilakarathne, N., Bakar, M. S., Abas, P. E., & Yassin, H. (2022). A Cloud Enabled Crop Recommendation Platform for Machine Learning-Driven Precision Farming. *Sensors (Basel)*, *22*(16), 6299. DOI: 10.3390/s22166299 PMID: 36016060

Tilman, D., Isbell, F., & Cowles, J. M. (2014). Biodiversity and ecosystem functioning. *Annual Review of Ecology, Evolution, and Systematics*, *45*(1), 471–493. DOI: 10.1146/annurev-ecolsys-120213-091917

Tominaga, T.. (2021). Regulatory science as a bridge between science and society. *Clinical Pharmacology and Therapeutics*, *109*(1), 39–48. PMID: 21691273

Topol, E. J. (2019). *Deep medicine: How artificial intelligence can make healthcare human again*. Basic Books.

Topol, E. J. (2019). High-performance medicine: The convergence of human and artificial intelligence. *Nature Medicine*, *25*(1), 44–56. DOI: 10.1038/s41591-018-0300-7 PMID: 30617339

Trafton, A., Yang, L., & Ghassemi, M. (2024). Demographic shortcuts: Why high-performing AI diagnostic models fail to transfer across patient populations. MIT News. https://news.mit.edu/2024/demographic-shortcuts-ai-medical-diagnoses-0628

Treves, A., & Karanth, K. U. (2003). Human–carnivore conflict and perspectives on carnivore management worldwide. *Conservation Biology*, *17*(6), 1491–1499. DOI: 10.1111/j.1523-1739.2003.00059.x

Tripathi, B. M., Stegen, J. C., Kim, M., Dong, K., Adams, J. M., & Lee, Y. K. (2018). Soil pH mediates the balance between stochastic and deterministic assembly of bacteria. *The ISME Journal*, *12*(4), 1072–1083. DOI: 10.1038/s41396-018-0082-4 PMID: 29515169

Trouiller, P., Olliaro, P., Torreele, E., Orbinski, J., Laing, R., & Ford, N. (2002). Drug development for neglected diseases: A deficient market and a public-health policy failure. *Lancet*, *359*(9324), 2188–2194. DOI: 10.1016/S0140-6736(02)09096-7 PMID: 12090998

Truong, D. T., Franzosa, E. A., Tickle, T. L., Scholz, M., Weingart, G., Pasolli, E., & Segata, N. (2015). MetaPhlAn2 for enhanced metagenomic taxonomic profiling. *Nature Methods*, *12*(10), 902–903. DOI: 10.1038/nmeth.3589 PMID: 26418763

Turner, W., Spector, S., Gardiner, N., Fladeland, M., Sterling, E., & Steininger, M. (2003). Remote sensing for biodiversity science and conservation. *Trends in Ecology & Evolution*, *18*(6), 306–314. DOI: 10.1016/S0169-5347(03)00070-3

Tyagi, V. (2017). *Content-based image retrieval*. Springer Nature. DOI: 10.1007/978-981-10-6759-4

U.S. Food and Drug Administration. (2025). Considerations for the use of artificial intelligence to support regulatory decision-making for drug and biological products: Draft guidance for industry. *Federal Register*, *90*(23), 3456–3489.

Ukoba, O. K., & Jen, T. C. (2019, December). Review of atomic layer deposition of nanostructured solar cells. In *Journal of Physics: Conference Series* (Vol. 1378, No. 4, p. 042060). IOP Publishing. https://doi.org/DOI: 10.1088/1742-6596/1378/4/042060

Ullah, Z., Alsubaie, N., Jamjoom, M., Alajmani, S. H., & Saleem, F. (2023). EffiMob-Net: A deep learning-based hybrid model for detection and identification of tomato diseases using leaf images. *Agriculture*, *13*(3), 737. Advance online publication. DOI: 10.3390/agriculture13030737

United Nations Environment Programme. (2024). *Tackling the triple planetary crisis: A roadmap for the digital age*. UNEP Publishing.

United Nations. (2020). *The sustainable development goals report 2020*. https://unstats.un.org/sdgs/report/2020/

Upadhyay, A., Chandel, N. S., Singh, K. P., Chakraborty, S. K., Nandede, B. M., Kumar, M., Subeesh, A., Upendar, K., Salem, A., & Elbeltagi, A. (2025). Deep learning and computer vision in plant disease detection: A comprehensive review of techniques, models, and trends in precision agriculture. *Artificial Intelligence Review*, *58*(3), 92. Advance online publication. DOI: 10.1007/s10462-024-11100-x

Ur Rehman, H., Tufail, H., Arooj, A., Raza, A., Kalsoom, U., Muhammad Abbas Malik, H., Shahid Cholistani, M., Qurban Khan, S., Abdullah Nasir, M., Ali, Z., Irshad, M., & Faheem, H. (2026). The Forensic Science of the Future: A Journey through Omics Technologies. *Journal of Biological and Allied Health Sciences*, *6*(1), 86–99. DOI: 10.56536/jbahs.v6i1.124

Urban, M. C., Bocedi, G., Hendry, A. P., Mihoub, J.-B., Pe'Er, G., Singer, A., Bridle, J. R., Crozier, L. G., De Meester, L., Godsoe, W., Gonzalez, A., Hellmann, J. J., Holt, R. D., Huth, A., Johst, K., Krug, C. B., Leadley, P. W., Palmer, S. C. F., Pantel, J. H., & Travis, J. M. J. (2016). Improving the forecast for biodiversity under climate change. *Science*, *353*(6304), aad8466. DOI: 10.1126/science.aad8466 PMID: 27609898

USPTO. (2024). Inventorship Guidance for AI-Assisted Inventions. United States Patent and Trademark Office.

Valecha, M., & Posada, D. (2022). Somatic variant calling from single-cell DNA sequencing data. *Computational and Structural Biotechnology Journal*, *20*, 2978–2985. DOI: 10.1016/j.csbj.2022.06.013 PMID: 35782734

Vamathevan, J., Clark, D., Czodrowski, P., Dunham, I., Ferran, E., Lee, G., & Bender, A. (2019). Applications of machine learning in drug discovery and development. *Nature Reviews. Drug Discovery*, *18*(6), 463–477. DOI: 10.1038/s41573-019-0024-5 PMID: 30976107

Van der Wal, R., Fischer, A., Selge, S., & Larson, B. M. H. (2015). Neither the public nor experts judge species primarily on their origins. *Environmental Conservation*, *42*(4), 349–355. DOI: 10.1017/S0376892915000053

van Dongen, J., Ehli, E. A., Slieker, R. C., Bartels, M., Weber, Z. M., Davies, G. E., Slagboom, P. E., Heijmans, B. T., & Boomsma, D. I. (2014). Epigenetic variation in monozygotic twins: A genome-wide analysis of DNA methylation in buccal cells. *Genes*, *5*(2), 347–365. DOI: 10.3390/genes5020347 PMID: 24802513

Van Horn, G., Mac Aodha, O., Song, Y., Cui, Y., Sun, C., Shepard, A., Adam, H., Perona, P., & Belongie, S. (2018). The inaturalist species classification and detection dataset. *Proceedings of the IEEE Conference on Computer Vision and Pattern Recognition*, 8769–8778. DOI: 10.1109/CVPR.2018.00914

van Klompenburg, T.. (2021). Machine learning algorithms predict yield and potential issues in crop management. *Field Crops Research, 266*, 108103. DOI: 10.1016/j.fcr.2021.108103

Van Klompenburg, T., Kassahun, A., & Catal, C. (2020). Crop yield prediction using machine learning: A systematic literature review. *Computers and Electronics in Agriculture, 177*, 105709. DOI: 10.1016/j.compag.2020.105709

Vayena, E., Blasimme, A., & Cohen, I. G. (2018). Machine learning in medicine: Addressing ethical challenges. *PLoS Medicine, 15*(11), e1002689. DOI: 10.1371/journal.pmed.1002689 PMID: 30399149

Veale, M., & Borgesius, F. Z. (2021). Demystifying the Draft EU Artificial Intelligence Act. *Computer Law Review International, 22*(4), 97–112. DOI: 10.9785/cri-2021-220402

Vermesan, O., Bahr, R., Ottella, M., Serrano, M., Karlsen, T., Wahlstrøm, T., Sand, H. E., Ashwathnarayan, M., & Gamba, M. T. (2020). Internet of robotic things intelligent connectivity and platforms. *Frontiers in Robotics and AI, 7*, 104. DOI: 10.3389/frobt.2020.00104 PMID: 33501271

Vertinsky, L., & Rice, T. M. (2002). Thinking about thinking machines: Implications of machine inventors for patent law. *Boston University Journal of Science & Technology Law, 8*, 574.

Vesmas, D. (2024). The Ethics of E-Governance. Safeguarding Data Confidentiality and Human Security In Public Administration. *StudiaSecuritatis, 18*(2), 222–232.

Villaronga, E. F., Kieseberg, P., & Li, T. (2018). Humans forget, machines remember: Artificial intelligence and the right to be forgotten. *Computer Law & Security Review, 34*(2), 304–313. DOI: 10.1016/j.clsr.2017.08.007

Villikudathil, A. T., Jayachandran, K., & Radhakrishnan, E. K. (2024). k-Nearest neighbour machine method for predicting resistance gene against Magnaporthe oryzae in rice using proteomic markers. *Journal of Proteins and Proteomics, 15*(4), 601–610. DOI: 10.1007/s42485-024-00159-3

Vinod, R., Chen, P. Y., & Das, P. (2025). Reprogramming pretrained language models for protein sequence representation learning. *Digital Discovery, 4*(6), 1591–1601. DOI: 10.1039/D4DD00195H

Vinuesa, R., & Brunton, S. L. (2022). Enhancing computational fluid dynamics with machine learning. *Nature Computational Science*, *2*(6), 358–366. DOI: 10.1038/s43588-022-00264-7 PMID: 38177587

Vinyals, O., Blundell, C., Lillicrap, T., & Kavukcuoglu, K. (2016). Matching Networks for One-Shot Learning. In *Proceedings of the 30th International Conference on Neural Information Processing Systems (NIPS'16)* (pp. 3637–3645). Curran Associates Inc., https://dl.acm.org/doi/10.5555/3157382.3157504

Wah, C., Branson, S., Welinder, P., Perona, P., & Belongie, S. (2011). Caltech-UCSD Birds-200-2011 Dataset. *The Caltech Institute Archives (California Institute of Technology)*. DOI: DOI: 10.57702/kvkwb8bo

Wahl, B., Cossy-Gantner, A., Germann, S., & Schwalbe, N. R. (2018). Artificial Intelligence (AI) and global health: How can AI contribute to health in resource-poor settings? *BMJ Global Health*, *3*(4), e000798. DOI: 10.1136/bmjgh-2018-000798 PMID: 30233828

Wallach, I., et al. (2015). AtomNet: a deep convolutional neural network for bioactivity prediction in structure-based drug discovery. arXiv preprint.

Wall, J., Wittemyer, G., Klinkenberg, B., & Douglas-Hamilton, I. (2014). Novel opportunities for wildlife conservation and research with real-time monitoring. *Ecological Applications*, *24*(4), 593–601. DOI: 10.1890/13-1971.1 PMID: 24988762

Wan, Z., Hazel, J. W., Clayton, E. W., Vorobeychik, Y., Kantarcioglu, M., & Malin, B. A. (2022). Sociotechnical safeguards for genomic data privacy. *Nature Reviews Genetics 2022 23:7, 23*(7), 429–445. DOI: 10.1038/s41576-022-00455-y

Wang, H., Ding, J., He, S., Feng, C., Zhang, C., Fan, G., Wu, Y., & Zhang, Y. (2023). MFBP-UNet: A network for pear leaf disease segmentation in natural agricultural environments. *Plants*, *12*(18), 3209. Advance online publication. DOI: 10.3390/plants12183209 PMID: 37765373

Wang, J., Fan, Y., Hong, L., Hu, Z., & Li, Y. (2025). Deep learning for RNA structure prediction. *Current Opinion in Structural Biology*, *91*, 102991. DOI: 10.1016/j.sbi.2025.102991 PMID: 39933218

Wang, K., Abid, M. A., Rasheed, A., Crossa, J., Hearne, S., & Li, H. (2023). DNNGP, a deep neural network-based method for genomic prediction using multi-omics data in plants. *Molecular Plant*, *16*(1), 279–293. DOI: 10.1016/j.molp.2022.11.004 PMID: 36366781

Wang, Y., Kong, S., Zhou, C., Wang, Y., Zhang, Y., Fang, Y., & Li, G. (2024). A review of deep learning models for the prediction of chromatin interactions with DNA and epigenomic profiles. *Briefings in Bioinformatics*, *26*(1), bbae651. Advance online publication. DOI: 10.1093/bib/bbae651 PMID: 39708837

Wang, Y., Shen, Y., Li, J., Wang, T., Peng, J., & Shang, X. (2025). Enhanced RNA secondary structure prediction through integrative deep learning and structural context analysis. *Nucleic Acids Research*, *53*(11), gkaf533. DOI: 10.1093/nar/gkaf533 PMID: 40530692

Wang, Y., Yao, Q., Kwok, J. T., & Ni, L. M. (2020). Generalizing from a Few Examples. *ACM Computing Surveys*, *53*(3), 1–34. DOI: 10.1145/3386252

Wang, Z., Yang, Z., Shi, H., Han, F., Liu, Q., Qi, J., & Lu, Y. (2020). Ecosystem health assessment of world natural heritage sites based on remote sensing and field sampling verification: Bayanbulak as case study. *Sustainability (Basel)*, *12*(7), 2610. DOI: 10.3390/su12072610

Warnat-Herresthal, S., Schultze, H., Shastry, K. L., Manamohan, S., Mukherjee, S., Garg, V., Sarveswara, R., Händler, K., Pickkers, P., Aziz, N. A., Ktena, S., Tran, F., Bitzer, M., Ossowski, S., Casadei, N., Herr, C., Petersheim, D., Behrends, U., Kern, F., & Schultze, J. L. (2021). Swarm Learning for decentralized and confidential clinical machine learning. *Nature*, *594*(7862), 265–270. DOI: 10.1038/s41586-021-03583-3 PMID: 34040261

Webb, S. (2018). Deep learning for biology. *Nature*, *554*(7693), 555–557. DOI: 10.1038/d41586-018-02174-z

Wegmann, M., Leutner, B., & Dech, S. (2016). *Remote sensing and GIS for ecologists: using open source software*. Pelagic Publishing Ltd. DOI: 10.2307/jj.28833743

Wei, Y., Han, C., & Yu, Z. (2023). An environment safety monitoring system for agricultural production based on artificial intelligence, cloud computing and big data networks. *Journal of Cloud Computing (Heidelberg, Germany)*, *12*(1), 83. DOI: 10.1186/s13677-023-00463-1

WHO. (2021). *Ethics and governance of artificial intelligence for health*. World Health Organization.

Wildlife conservation in the digital age: How AI powers World Wildlife Day. (n.d.). Retrieved November 28, 2024, from https://indiaai.gov.in/article/wildlife-conservation-in-the-digital-age-how-ai-powers-world-wildlife-day

Withers, C. A., Rufai, A. M., Venkatesan, A., Tirunagari, S., Lobentanzer, S., Harrison, M., & Zdrazil, B. (2025). Natural language processing in drug discovery: Bridging the gap between text and therapeutics with artificial intelligence. *Expert Opinion on Drug Discovery, 20*(6), 765–783. DOI: 10.1080/17460441.2025.2490835 PMID: 40298230

Wolfert, S., Ge, L., Verdouw, C., & Bogaardt, M. J. (2017). Big data in smart farming: A review. *Agricultural Systems, 153*, 69–80. DOI: 10.1016/j.agsy.2017.01.023

Wong, F., de la Fuente-Nunez, C., & Collins, J. J. (2023). Leveraging artificial intelligence in the fight against infectious diseases. *Science, 381*(6654), 164–170. DOI: 10.1126/science.adh1114 PMID: 37440620

World Economic Forum. (2022). *The global AI agenda: AI as an enabler for the sustainable development goals*. https://www.weforum.org/reports

World Health Organization Regional Office for Europe. (n.d.). Targets of Sustainable Development Goal 3. https://www.who.int/europe/about-us/our-work/sustainable-development-goals/targets-of-sustainable-development-goal-3

World Health Organization. (2020). Global strategy on digital health 2020–2025. https://www.who.int/docs/default-source/documents/gs4dhdaa2a9f352b0445bafbc79ca7d1d9f14.pdf

World Health Organization. (2022). *Global digital health initiative*. https://www.who.int/health-topics/digital-health

World Health Organization. (2025). *Ethics and governance of artificial intelligence for health: Guidance for the post-pandemic era*. WHO Academy Press.

Wu, D., Karhade, D. S., Pillai, M., Jiang, M. Z., Huang, L., Li, G., Cho, H., Roach, J., Li, Y., & Divaris, K. (2021). Machine Learning and Deep Learning in Genetics and Genomics. *Machine Learning in Dentistry*, 163–181. DOI: 10.1007/978-3-030-71881-7_13

Wu, B., Tian, F., Zhang, M., Zeng, H., & Zeng, Y. (2020). Cloud services with big data provide a solution for monitoring and tracking sustainable development goals. *Geography and Sustainability, 1*(1), 25–32. DOI: 10.1016/j.geosus.2020.03.006

Wu, Q., & LI, X. Y. (2015). Remote sensing dynamic monitoring of mining geo-environment based on computational geometry and information TUPU. *Meitan Xuebao*, (1), 26–34.

Wu, S., Chen, N., Xiao, A., Zhang, P., Jiang, C., & Zhang, W. (2025). AI-Empowered Virtual Network Embedding: A Comprehensive Survey. *IEEE Communications Surveys and Tutorials*, *27*(2), 1395–1426. DOI: 10.1109/COMST.2024.3424533

Xia, Q., Luo, L., Yerzati, Y., Ahmed, M. M., Chen, Y., Wang, S., Qin, J., Chen, L., Jin, Q., Guo, Z., & Zhang, R. (2026). Correction: Near-infrared prediction of tannin content in walnut kernels using wavelet transform combined with interpretable machine learning models. *Frontiers in Plant Science*, *17*, 1806058. DOI: 10.3389/fpls.2026.1806058 PMID: 41835266

Xu, Y.. (2021). Artificial intelligence in healthcare: A narrative review. *Stroke and Vascular Neurology*, *6*(3), 340–350. PMID: 33468639

Yang, K. K., Wu, Z., & Arnold, F. H. (2019). Machine-learning-guided directed evolution for protein engineering. *Nature Methods*, *16*(8), 687–694. DOI: 10.1038/s41592-019-0496-6 PMID: 31308553

Yaseen, Z. M. (2022). The next generation of soil and water bodies heavy metals prediction and detection: New expert system based edge cloud server and federated learning technology. *Environmental Pollution*, *313*, 120081. DOI: 10.1016/j.envpol.2022.120081 PMID: 36075340

Yeshey, K., Keenan, R. J., Ford, R. M., & Nitschke, C. R. (2023). Sustainable development implications of human–wildlife conflict: An analysis of subsistence farmers in Bhutan. *International Journal of Sustainable Development and World Ecology*, *30*(5), 548–563. DOI: 10.1080/13504509.2023.2167242

You, N., Liu, C., Lin, H., Wu, S., Chen, G., & Shen, N. (2025). Benchmarking pre-trained genomic language models for RNA sequence-related predictive applications. *Nature Communications*, *17*(1), 223. DOI: 10.1038/s41467-025-66899-y PMID: 41354659

Yuan, Y., Xu, Z., & Lu, G. (2021). SPEDCCNN: Spatial pyramid oriented encoder decoder cascade convolution neural network for crop disease leaf segmentation. *IEEE Access : Practical Innovations, Open Solutions*, *9*, 14849–14866. DOI: 10.1109/ACCESS.2021.3052769

Zaid, A. A., Arqawi, S. M., Mwais, R. M. A., Al Shobaki, M. J., & Abu-Naser, S. S. (2020). The impact of Total quality management and perceived service quality on patient satisfaction and behavior intention in Palestinian healthcare organisations. *Technology Reports of Kansai University*, *62*(03), 221–232.

Zeevi, D., Korem, T., Zmora, N., Israeli, D., Rothschild, D., Weinberger, A., & Segal, E. (2015). Personalized nutrition by prediction of glycemic responses. *Cell*, *163*(5), 1079–1094. DOI: 10.1016/j.cell.2015.11.001 PMID: 26590418

Zeiler, M. D., & Fergus, R. (2014). Visualizing and understanding convolutional networks. In D. Fleet, T. Pajdla, B. Schiele, & T. Tuytelaars (Eds.), Computer Vision – ECCV 2014 (pp. 818–833). Springer. DOI: 10.1007/978-3-319-10590-1_53

Zhang, D., & Li, W. (2021). Artificial intelligence in environmental monitoring: A review. *Environmental Science & Technology*, *55*(8), 4249–4263.

Zhang, K., & Aslan, A. B. (2021). AI technologies for education: Recent research & future directions. *Computers and Education: Artificial Intelligence*, *2*, 100025. DOI: 10.1016/j.caeai.2021.100025

Zhang, Q. C., Petrey, D., Garzón, J. I., Deng, L., & Honig, B. (2012). Structure-based prediction of protein–protein interactions on a genome-wide scale. *Science*, *338*(6110), 1107–1112. DOI: 10.1126/science.1224912 PMID: 23086998

Zhang, X., & Li, H. (2022). The role of AI in renewable energy optimization: A systematic review. *Renewable & Sustainable Energy Reviews*, *156*, 111890.

Zhang, Y., Jiang, B., Zhang, Z., & Li, X. (2019). Machine learning in infectious disease epidemiology and outbreak prediction. *Clinical Microbiology Reviews*, *32*(4), e00014–e00019. DOI: 10.1128/CMR.00014-19

Zhang, Y., Wang, J., & Xu, P. (2021). AI-based distributed energy resource management in residential microgrids: A review. *Applied Energy*, *298*, 117199.

Zhao, Y. (2025). Evolving Paradigms in Low-Template DNA Analysis: From Fixed Analytical Thresholds to Probabilistic Modeling. *Transactions on Materials. Biotechnology and Life Sciences*, *8*, 423–431. DOI: 10.62051/mjbq7s89

Zhavoronkov, A., Ivanenkov, Y. A., Aliper, A., Veselov, M. S., Aladinskiy, V. A., Aladinskaya, A. V., & Aspuru-Guzik, A. (2019). Deep learning enables rapid identification of potent DDR1 kinase inhibitors. *Nature Biotechnology*, *37*(9), 1038–1040. DOI: 10.1038/s41587-019-0224-x PMID: 31477924

Zheng, A., Lamkin, M., Zhao, H., Wu, C., Su, H., & Gymrek, M. (2021). Deep neural networks identify sequence context features predictive of transcription factor binding. *Nature Machine Intelligence 2021 3:2, 3*(2), 172–180. DOI: 10.1038/s42256-020-00282-y

Zhou, J., & Troyanskaya, O. G. (2015). Predicting effects of noncoding variants with deep learning-based sequence model. *Nature Methods*, *12*(10), 931–934. DOI: 10.1038/nmeth.3547 PMID: 26301843

Zhou, X., Zhang, Y., & Wei, Q. (2022). Few-Shot Fine-Grained Image classification via GNN. *Sensors (Basel)*, *22*(19), 7640. DOI: 10.3390/s22197640 PMID: 36236743

Zhou, Z., Riley, R., Kautsar, S., Wu, W., Egan, R., Hofmeyr, S., & Wang, Z. (2025). GenomeOcean: An Efficient Genome Foundation Model Trained on Large-Scale Metagenomic Assemblies. *bioRxiv*.

Zitnik, M., Nguyen, F., Wang, B., Leskovec, J., Goldenberg, A., & Hoffman, M. M. (2019). Machine learning for integrating data in biology and medicine: Principles, practice, and opportunities. *Information Fusion*, *50*, 71–91. DOI: 10.1016/j.inffus.2018.09.012 PMID: 30467459

Zubair, M., Khan, A. H., Bilal, S. F., & Li, J. (2025). Deep learning approaches for resolving genomic discrepancies in cancer: A systematic review and clinical perspective. *Briefings in Bioinformatics*, *26*(6), bbaf541. DOI: 10.1093/bib/bbaf541 PMID: 41176790

About the Contributors

Kyla Tennin is a United Nations Global Ambassador for Peace (for the African Union 6th Region, African Diaspora) for women's empowerment, economic development, financial investment, and sustainable development goals achievement. Including an industry practitioner and DM (Doctor) in Management and Organizational Leadership. She focuses on Change Management and mixed methods empirical research as a Management Consultant, for continuous improvement, business transformation, and recovery during and after internal/external crisis to nations, institutions, and individual's lives. Her career began in December 2002 in banking and financial services at Wells Fargo Bank, N.A. (North America), Wells Fargo Financial, LLC, US Bank, and SunTrust Banks, Inc. with affiliate relationships at Wells Fargo Advisors, Wells Fargo Investments, Wells Fargo Business Banking (Commercial), Wells Fargo Student Loans, Wells Fargo Home Mortgage, and Wells Fargo Card Services. With roles in/as banker, underwriting, financial crime compliance (OFAC, MTR, CTR, SEC), loan administration, for example, and leadership roles in Credit Management & Analysis, Credit Risk, Risk Management, Cash Management, Front Office (in the Branch) and Back Office (in the operations center), Bank Products and Services (cross-selling), Lending, Suppliers/Vendor Management and Procurement, Operations, and Corporate Banking (like for hospitals and clinics). She has media and press appearance in Forbes, ABC, Fox News, Yahoo! Finance, U.S. News & World Reports, Financial Times, CNBC Arabia on Middle East and US Economic Outlook. She is an on-call Grant Reviewer for organizations like the United States Department of Agriculture (e.g. responsible for awarding funding for programs up to $65M in funding) and other institutions. Inclusive, a part-time PhD Faculty of Corporate Governance at Woolf University and WDC Leadership School UK/USA (see here http://wdc.woolf.university/faculty/93d5d684-4a7a-4f5b-afd3-44645e9ecc90), and Fellow-in-Residence and Lead Researcher at University of Phoenix College of Doctoral Studies, Center for Leadership Studies and Organizational Research (CLSOR). Also, she is a leader in various roles at investment forums, conferences, and summits where she works with Presidents of Nations, Heads of Banks, Prime Ministers, High Commissioners, and Chambers of Commerce and other business leaders to strategize and make business decisions to help businesses start-up, advance, and/or scale to improve economic development, promote social justice, and create jobs in various nations. Including at the World Business Angels Investment Forum (WBAF). An affiliated partner of the G20 Global Partnership for Financial Inclusion (GPFI), where the vision is to promote social justice, create jobs, financial inclusion, and innovative financial instruments for innovators, startups, and SMEs, helping them to scale. Dr. Tennin is also affiliated with the International Monetary Fund, World Bank Group, and Parliamentary Network to enhance economies and various forms of businesses, and has spoken on stages with Ms. Rula Ghani, First Lady of Afghanistan, Princess Dina Mired of Jordan, Ms. Intissar El-Sisi, spouse of Egyptian President, Dr. Mariem Mint Dah, First Lady Of Mauritania, Dr. Madeeha bint Ahmed al Shibaniyah, Minister of Education (Oman), and Ms. Amina J. Mohammed, Deputy Secretary-General of the UN & Chair of the UN Sustainable Development Group. She is an OB/GYN MD track turned DM (doctor) with background in biology, biotechnology, and organic and biochemistry, with major emphasis in women and gender studies, has project management and lean six sigma certifications, a Master of Business Administration (MBA), and is a Ph.D. Fellow at the Global Peace Institute, headquartered in the United Kingdom. She contributed to the accreditation process at the University of Phoenix College of Doctoral Studies and completed additional studies (to launch and scale corporations) at Harvard, Harvard Law (e.g. contracts and negotiation), Stanford, Yale, University of Cape Town, Cornell University - The Bank of America Institute for Women's Entrepreneurship at Cornell (Certificate in Women's Entrepreneurship – sponsored by Bank of America), and University of Glasgow to name a few. She has personally written approximately 18 books, 3 book chapters with colleagues, 3 patents, several other intellectual properties, and nearly 85 research reports, later conceptualized into products and services for global businesses. Along with additional major book deals with approximately 8 books in the pipeline with forthcoming publishing. In areas of Financial Crimes, Fraud, Banking, Artificial Intelligence, Case Studies, Sustainable Development (ESG), and forthcoming of Teaching Notes and Textbooks (see a few here https://www.igi-global.com/affiliate/kyla-latrice-tennin/426258). Pipeline book deals and current book deals with IGI Global Publishing, NOVA Publishers, Emerald Publishing, Springer, Springer Nature, Palgrave MacMillian, Wiley, Routledge, and Taylor & Francis Group (CRC Press). Dr. Tennin has presented research and posters in practitioner and academic conferences, workshops, and international webinars. She has professional experience in the financial services and investments, hospitality, restaurant, consumer products/beauty, and healthcare industries, with 95% of her experience spent in financial services and higher education. Dr. Tennin has 23 years of experience in finance, financial services, and banking industries and 17 years owning her own global conglomerate corporation, which she built while homeless, to recover for selfsufficiency, to recover financially, and create jobs for people. Operating in 31 countries, with 24 locations, on 6 continents, and 2 world headquarters. She was awarded over 30 corporate awards, primarily in the financial services industry, mentored entrepreneurs for 22 years with global organizations and world-renowned corporations and Development Banks and Commercial and Retail Banks as sponsors, and currently owns a business academy, women's entrepreneurship academy, and women's entrepreneurship bootcamp program where entrepreneurship, resilience, finance, pricing and tax strategy, research methodology and data analysis for consultants, personal and business credit, corporate governance and risk management, digital transformation, property ownership (real estate (e.g. commercial and residential)), and other courses are taught. Development Banks like Jamaica Development Bank, Qatar Development Bank, Asian Development Bank (ADB), and European Bank for Reconstruction Development Bank for corporations and SMEs. Her areas of expertise and research interests are financial and revenue strategy, entrepreneurship (SME), cosmeceuticals, economic development, emerging markets, multinationals, corporate governance, well-being/burnout, and women's empowerment.

Rohit Kumar Pal is a dynamic academic, researcher, and educator specializing in biological, environmental, and management sciences. Currently serving as a professional doctorate in Education Management & Leadership from European International University, Paris, France & an MBA, he also holds a leadership role as Head of the Department at R.I.C.E Group, Kolkata, where he has been teaching since 2017. With a Professional Doctorate in Education Management and Leadership and an MBA in Human Resource Management, Dr. Pal brings an interdisciplinary perspective to education, sustainability, and organizational studies. He is a certified educator with the National Geographic Society and actively contributes to academia as a reviewer and editorial board member for several international journals. Dr. Pal has an extensive publication record, with numerous research articles, book chapters, and books focusing on leadership, sustainability, environmental awareness, and education systems. His work often bridges the gap between theory and practice, particularly in the context of modern education and emerging technologies like AI. A UGC NET qualifier and WBSET achiever, he is also recognized for excellence in science teaching. His commitment to research, teaching, and global academic collaboration underscores his vision of fostering sustainable and innovative education systems.

Mitali Talukdar is a distinguished academic leader, researcher, and management educator with over 24 years of academic and industry experience. She has served as Officiating Director and Head of Institute at Amity University Kolkata and as Head of the MBA Program at Symbiosis University, Pune. Her expertise spans organizational leadership, curriculum development, accreditation, FDP, Researcher, Author, Editor, patent holder and corporate engagement. A recipient of the Kadambini Award and Nari Shakti Award, she has published extensively in Scopus-indexed journals, presented research at premier institutions including IIT Roorkee and IIM Indore, and serves as reviewer and editor for leading international publishers.

***

Hayitov Abdulla is working as a Lecturer at Department of Transports Systems, Urgench State University, Uzbekistan

Anmol Anand is associated with the Department of Computer Science and Engineering at Jaypee University of Engineering and Technology (JUET), Guna, India. His research interests include Artificial Intelligence, Machine Learning, and Software Development. He has contributed to research in machine learning and has experience in backend software development, with a focus on designing intelligent and scalable software solutions.

Anorgul Ashirova is working as an Associate Professor at the Department of General Professional Sciences, Mamun University, Khiva, Uzbekistan

MonishaB. is a Research Scholar at Christ (Who is deemed to be a university student) in Bengaluru. She graduated from The Tamil Nadu Dr. Ambedkar Law University, Chennai, with an LL.M. in Intellectual Property Law in the year 2021. She has been working as an Assistant Professor of Law with various esteemed Institutions, such as KLE Society's Law College, Bengaluru, and SRM University, Chennai, in the faculty of law. She has an avid interest in research on patents and access to healthcare.

Tatayya Bommali is a distinguished educator, researcher, and academic leader in Finance, currently serving as Prof & Head, Department of Management Studies, and Controller of Exam (COE), Sanketika Vidya Parishad Engineering College,Visakhapatnam, AP, India. He holds a Ph.D. in Finance, along with M.Com, MBA, and B.Ed., and brings 20 years of extensive teaching and research experience. A Cambridge-trained lecturer and IBDP Business Management Facilitator, Dr. Bommali has taught MBA, MCA, B.Tech, B.Com, and CA students, fostering academic excellence across diverse educational frameworks. His global academic exposure, including roles in India, Malaysia, and Dhaka, highlights his contribution to educational leadership. A prolific researcher, he has published extensively in reputed journals and contributed to books in Finance, Accounting, Banking, and Economics. His research interests focus on financial inclusion, digital finance, and sustainable investment strategies, bridging theory with practical applications through mentorship, curriculum innovation, and academic leadership.

Sanjukta Chakraborty is an Assistant Professor and Head of the Department of Computer Science and Engineering at Seacom Skills University, Santiniketan, West Bengal, India With over 12 years of teaching experience, she holds a Ph.D. in CSE, specializing in Artificial Intelligence and Machine Learning. She has completed M.Tech.(CSE), MCA, and B.Sc. (Physics Honours) from reputed Indian universities. Her research interests include AI, ML, Deep Learning, Cyber Security, Data Science, Sentiment Analysis, Big Data Analytics, Software Engineering, and Business Intelligence. Dr. Chakraborty has published 20 research articles and 7 book chapters, and has guided numerous students in their final year dissertations.

Deep Debnath is attached with Dept. of F&A at Ghani Khan Choudhury Institute of Engineering and Technology (GKCIET) which is located at Malda, West Bengal, India. He has obtained BCA, and MSc in Computer and Information Science from Raiganj University, West Bengal, India. He secured 1st position in his MSc in Computer and Information Science exam. His areas of interest include Cyber Security, Business Information Technology, ICT Applications, etc. He has involved in many National and International Conferences of repute and contributed. He has some of the publications in peer reviewed journals and edited volume.

Kiranbhai Ramabhai Dodiya is an experienced Cyber Forensic Expert and Assistant Professor specializing in cyber security and digital forensics. He has taught at prestigious institutions and is currently pursuing a PhD in Biochemistry and Forensic Science. With a strong background in ethical hacking and malware analysis, Kiranbhai is passionate about educating students and contributing to the field of cyber forensics. His expertise includes malware forensic, computer forensics, and data recovery software. Kiranbhai's commitment to professional growth is evident through his participation in workshops, seminars, and research publications. He is dedicated to shaping the next generation of professionals in the field and combatting cybercrimes.

Divyansh Jain is associated with the Department of Computer Science and Engineering at Jaypee University of Engineering and Technology (JUET), Guna, India. His research interests include Artificial Intelligence, Machine Learning, Computer Vision, and intelligent software systems. He has experience in developing

AI-enabled digital solutions using Flutter and backend technologies, with a focus on creating scalable, user-centric applications. His work emphasizes the integration of AI/ML techniques with modern software engineering practices to address real-world challenges.

Naga Jogayya Kothakota is a distinguished molecular biologist and forensic scientist specializing in genetics, genomics, and wildlife forensics. With over 15 years of experience, he has contributed significantly to the fields of conservation genetics, molecular ecology, and environmental biology. Currently an Associate Professor at Centurion University of Technology and Management, Dr. Kothakota combines his academic expertise with practical experience in next-generation sequencing technologies and DNA profiling to address challenges in biodiversity conservation and forensic investigations. His interdisciplinary approach bridges molecular biology with applied forensic science, aiming to enhance both environmental protection and legal enforcement.

Kesavardhini K., PG and Research Department of Microbiology, Sacred Heart College (Autonomous), Tirupattur, Tamil Nadu, India

Sandhiya K., PG and Research Department of Microbiology, Sacred Heart College (Autonomous), Tirupattur, Tamil Nadu, India

Khushi Rahul Kamble is a fourth-year Computer Science Engineering student at Vishwakarma Institute of Information Technology (VIIT), Pune, specializing in Artificial Intelligence and Machine Learning. As a K-Katalyst scholar, she is deeply passionate about AI and problem-solving, consistently striving to develop practical AI-based solutions. Khushi has contributed to several research projects, including the paper "Movie Recommendation System Using Hybrid-Based Method" (K. Kamble, 2025, International Conference on Automation and Computation (AUTOCOM), pp. 119–126) and "Software for Corrosion Protection Methods for Various Metals in Different pH Environments" (K.R. Kamble, Cureus Journals 2(1), 2025). Her work often involves techniques such as clustering and vectorization to deliver effective solutions. Alongside her academic research, she has actively participated in competitions, including the SIH Hackathon 2024, where her team won first place with "Virtual Traffic Police", an AI-based traffic optimization system using dynamic signal management. Khushi continues to leverage her skills and experience to address real-world challenges through AI-driven innovation. She can be contacted at kamblekhushi33@gmail.com .

Mustafa Kayyali is an ardent advocate for excellence in higher education, driven by a relentless pursuit of quality, recognition, and innovation. My journey in academia started with a Master's in Quality Management and Evaluation in Higher Education from Universitat Oberta de Catalunya, followed by a Ph.D. in Quality Management from Azteca University. These academic pursuits have laid a strong foundation for my expertise in Accreditation, Quality Assurance, and Higher Education Rankings. As an entrepreneur, researcher, translator, and publisher. My diverse interests encompass Management, Translation, Interpretation, and Academic consulting, contributing to a well-rounded understanding of the industry. Throughout my career, I have made significant contributions to academic literature, with more than 20 published papers and 7 book chapters to my name. Additionally, I take pride in having translated 5 books, bridging language gaps, and promoting knowledge exchange on a global scale.

Kunj Bihari Meena is a faculty member and researcher in Computer Science and Engineering at Jaypee University and Engineering Technology Guna. Dr. Meena has published multiple papers in reputed international journals and conferences, including SCI and Scopus-indexed publications. He has received Best Paper Awards for his research contributions and serves as a reviewer for prestigious SCI/Scopus journals and conferences. His areas of research include image processing, digital image forensics, deep learning, and blockchain.

Nilamadhab Mishra is an Associate Professor in the School of Computer Science and Engineering, AI-ML Division, at VIT Bhopal University, India. He received his Doctor of Philosophy (Ph.D.) in Computer Science and Information Engineering with a specialization in AI-data science and machine learning from Chang Gung University, Taiwan. He has over 23 years of national and international involvement in academic teaching and research at recognized Indian, Taiwanese, and African universities. He has over 150 publications in SCI/SCIE and SCOPUS-indexed journals, ISBN books and chapters, Indian and Australian patents, IEEE conference proceedings, etc. He has been a reviewer, associate editor, and editorial board member for SCI/SCIE-indexed journals and conferences. He has been involved with several professional bodies: “Fellow Member of “ISROSET”; Fellow Member of “ISRD”; Senior Member of “ASR” (Hong Kong); Senior Member of “IEDRC” (Hong Kong); and Member of “IEEE Collabratec”. He is currently an Associate Editor for two SCOPUS-indexed journals, and his research interests span the areas of AI, Data Science, Machine Learning, and Cognitive Analytics Policy & Applications.

Anastasius Moumtzoglou is a thought leader pioneering a post-Deming, human-centric paradigm for healthcare quality. His research merges quality theory with e-health and complexity science to create new scientific disciplines. He proposes frameworks that view healthcare as a complex adaptive system, where patients are co-producers and technology enables holistic care. His work aims to reshape healthcare for the digital age, moving beyond obsolete industrial models.

R. Nisha, PG and Research Department of Microbiology, Sacred Heart College (Autonomous), Tirupattur, Tamil Nadu, India

Ragavi. P, PG and Research Department of Microbiology, Sacred Heart College (Autonomous), Tirupattur, Tamil Nadu, India

SaranrajP. holds a Ph.D. in Microbiology from Annamalai University, Chidambaram, India, and is NET (ASRB–ICAR) qualified in Microbiology. He is currently serving as Assistant Professor and Head, Department of Microbiology, at Sacred Heart College (Autonomous), Tirupattur, Tamil Nadu, India. In addition, he is the Proprietor and Managing Director of JPS Scientific Publications, contributing actively to academic publishing and research dissemination. Dr. Saranraj has over 15 years of research experience and 10 years of teaching experience in the fields of Environmental Microbiology, Agricultural Microbiology, Industrial Microbiology, Bioremediation, Microbial Biotechnology, and Sustainable Ecosystem Management. His research work has focused on Microbial degradation of pollutants, Effluent treatment, Soil health improvement, Plant–microbe interactions, and Applied environmental sustainability. He has an extensive publication record, with more than 240 international research articles, over 90 books (international and national), and 70+ book chapters published with reputed academic publishers.

Sagnika Pal is an aspiring psychologist from Howrah, West Bengal, India, dedicated to contributing to the field of mental health through academic excellence and practical service. She is currently pursuing her M.Sc. in Applied Psychology at the University of Calcutta and has successfully qualified for both GATE and UGC-NET (Assistant Professor and PhD) as of 2025. Beyond her academic pursuits, she is the founder and private tutor at Pookie Didi, where she mentors students and aspirants preparing for academic and competitive psychology examinations. Her experience also includes serving as a trainee volunteer and research coordinator at Parindey Foundation, along with actively participating in fundraising initiatives for social causes. She is deeply passionate about leveraging her research and communication skills to promote psychological awareness and contribute meaningfully to the field.

Prantosh K. Paul holds prestigious and unique D.Sc. (Doctor of Science) in Information Science and Technology (IST) & D.Tech. (Doctor of Technology) in Agricultural Informatics & Management degree from Azteca University, Mexico under Dr. Agustin Yoh International Postdoctoral Fellowship (DAYIPF) Program.

Dr. Paul perhaps the first person globally who holds DSc in Infornation Science & Technology and DTech in Agricultural Informatics. He also hold PhD in IST from IIEST Shibpur, Post Doctoral Fellow Certificate (PDFC) Program in Environmental Informatics. He holds degrees like M.Tech. (by Research), MS (by Research) in addition to M.Sc. (Double). He is working as Asst. Professor (IST) & Executive Director (MCIS Program), at Dept. of Computer & Information Sciences, and additionally Information Scientist Position at Raiganj University, West Bengal, India. Significantly He is also holds many National and International designations and position such as Honorary Professor, Logos University International, Louisiana, USA, Chief Advisor (Innovative Program & Research Planning), Srinivas University,Mangalore, India. He has credited over 35 Books in diverse areas of interdisciplinary Information Science. He holds 400+ research/review and policy papers, Members in more than 150+ Journals and 250+ Int. Conf. Membership, organization and holds more than a dozen of Awards

Ujan Pradhan is a BTech-CSE pursuing candidate at Department of Computing Technologies, SRM Institute of Science and Technology, Chennai, TN, India. He completed his internship at leading universities and institutions viz. National Institute of Technology Surathkal, Karnataka, India, Indian Institute of Technology Roorkee, Uttarakhand, India, Samsung R&D Institute Bangaluru, India. In addition to educational programs he has also been part of various National and International Conference of repute, papers in leading journals and edited volume, etc.

Sabyasachi Pramanik is a professional IEEE member. He obtained a PhD in Computer Science and Engineering from Sri Satya Sai University of Technology and Medical Sciences, Bhopal, India. Presently, he is an Associate Professor, Department of Computer Science and Engineering, Haldia Institute of Technology, India. He has many publications in various reputed international conferences, journals, and book chapters (Indexed by SCIE, Scopus, ESCI, etc). He is doing research in the fields of Artificial Intelligence, Data Privacy, Cybersecurity, Network Security, and Machine Learning. He also serves on the editorial boards of several international journals. He is a reviewer of journal articles from IEEE, Springer, Elsevier, Inderscience, IET and IGI Global. He has reviewed many conference papers, has been a keynote speaker, session chair, and technical program committee member at many international conferences. He has authored a book on Wireless Sensor Network. He has edited 8 books from IGI Global, CRC Press, Springer and Wiley Publications.

ValarmathiR. is a Distinguished Professor of Law at Christ (Deemed to be University), Bengaluru. She specialises in commercial law and has worked with many premium institutions such as CMR University and Bengaluru. She is a recipient of the Global Excellence Award for her distinguished service in Legal Service. She has presented over 10 research Papers and is currently guiding five research scholars in the field of Intellectual Property Law.

Vijayalakshmi R., PG and Research Department of Microbiology, Sacred Heart College (Autonomous), Tirupattur, Tamil Nadu, India

Pranati Rakshit is the Principal of Tarakeswar Degree College, Tarakeswar, Hooghly, West Bengal, India. She belongs to Department of Computer Science. She previously served as Professor and Head of the Department of Computer Science and Engineering at JIS College of Engineering, Kalyani, West Bengal, India. She has 25+ years of teaching experience. She has completed her Ph.D. degree from Jadavpur University, Kolkata, West Bengal, India, in the field of Pattern Recognition, Machine Learning, Medical image analysis. She has completed her Master degree and B.E. Degree in Computer Science & Engineering. She has worked in the field of Pattern Recognition, Machine Learning, Deep Learning, Medical image analysis, Sentiment Analysis etc. She has supervised more than 22 M.Tech. Projects and presently supervising 1 Ph.D. student. She has 70 numbers of research publications in International journals and conferences including SCI and Scopus indexing. She has 43 patents (Granted-6, Published-37), 2 copyrights and 1 design IPR in her account. She is a life member of Indian Society of Technical Education.

Kim Sanders is a Senior Research Strategist specializing in participatory planning processes that bridge the gap between inclusive engagement and operational execution. She brings extensive expertise in establishing robust data security protocols and accountability frameworks, which has resulted in improved compliance across organizational, regulatory, and contractual requirements. Dr. Sanders has supervised and mentored investigative teams, successfully embedding harmonized indicators and consistent protocols that foster a culture of transparency and data integrity. She excels in building strong partnerships with diverse stakeholders, including federal agencies and community groups, to advance data sharing agreements and policies rooted in social justice and equity. Dr. Sanders professional experience includes directing AI and equity-centered research projects aligned with a sector-wide data flow architecture, translating complex findings into practical frameworks, KPIs, and recommendations for senior leaders to support decision-making and program improvement. She also has experience in designing and leading strategic table talks for intergenerational learning experiences, integrating experiential learning opportunities to bridge theory and practice.

Sushil Sharma is a Vice Provost Dean, Colege of CBT at Texas A&M University, Texarkana. In this role, Dr. Sharma has the unique distinction of earning two doctoral degrees. Dr. Sharma is responsible for overseeing and leading; the Office of Research and Sponsored Projects (ORSP), the Office of Graduate Studies (OGS), the Office of International Studies (ISO), the Office of Institutional Effectiveness and Research (OIER), Extended Education and Community Development (EECD), Honors Program, and Library. Before joining Texas, A&M, Dr. Sharma was a tenured faculty member (2001-2021) and served as the Associate Dean (2012-2021) of the AACSB-accredited Miller College of Business at Ball State University (R2 research university) in Indiana. Before his Associate Deanship, Dr. Sharma has been the Executive Director for MBA and other Certificate Programs (2011-2013) and Department Chair for Information Systems and Operations Management (2008 – 2012) at Ball State University, besides being Chair of the Quantitative Systems Group, Chair of the IT Systems group (1992- 2000), Placement Chair, Chair of Student Affairs and Alumni at world-class premier top tier research R1 business school; the Indian Institute of Management (IIM) India. Dr. Sharma has a uniquely international academic and professional background, having taught as full-time or visiting faculty in premier institutions of international acclaim and repute at IIM (India) (1990-2000) University of Waterloo (2000 - 2001), Canada (known to be the MIT of Canada), Turku School of Economics and Victoria University of Wellington, New Zealand (2011). Dr. Sharma has directed executive development programs and consulted with companies/organizations in the USA and India.

Gitanjali R. Shinde has 15 years of experience teaching. Presently, she is working as Associate Professor of Computer Science and Engineering (AI and ML), at Vishwakarma Institute of Technology, Pune. She earned a Ph.D. in Wireless Communication from Aalborg University, Denmark, with a focus area of "A Cluster Framework for Internet of People, Things, and Services." Dr. Shinde took research funding for IoT authentication, more than 50 papers published, and 10+ books, such as Data Analytics for Pandemics: A COVID-19 Case Study, which won the award.

Vipin Tyagi is a distinguished academician, researcher and author, currently serving as Dean (Academics and Research) at Jaypee University of Engineering and Technology in Guna, Madhya Pradesh. With a rich background in Computer Science, his expertise spans across Image Processing, Cyber Forensics and Speech Recognition. Prof. Tyagi has a prolific publication record, contributing significantly to his fields of interest. His scholarly work includes numerous journal articles, conference papers, and books. Some of his notable publications are in the areas of image forgery detection, image denoising and content-based image retrieval. He has authored and edited several books and papers in the field of Computer Science, such as "Digital Image Forgery Detection: Techniques, Challenges and Applications", "Predictive Computing and Information Security" and "Content-Based Image

Retrieval: Ideas, Influences, and Current Trends" published by Springer Singapore. His book "Understanding Digital Image Processing" using MATLAB, published by Taylor and Francis is highlighted by Mathworks at: https://in.mathworks.com/academia/books/understanding-digital-image-processing-tyagi.html His scholarly work has earned him a respectable h-index and numerous citations in the academic community.

Uttkarsh is associated with the Department of Computer Science and Engineering at Jaypee University of Engineering and Technology (JUET), Guna, India. His research interests include Artificial Intelligence, Machine Learning, Computer Vision, and medical image analysis. He has contributed to research on brain hemorrhage detection using CT scan images and has experience in deep learning and data-driven approaches for healthcare applications. His broader interests encompass intelligent systems and AI-enabled solutions for real-world challenges.

Pravallika Vataparthi is an Assistant Professor in the School of Forensic Science with teaching and research interests spanning forensic genetics, molecular biology, and applied forensic methodologies. Her work focuses on practical improvements in DNA profiling workflows, quality control in forensic laboratories, and the translation of new genomic methods into casework. She actively supervises students on projects involving STR/SNP analysis, evidence handling, and laboratory accreditation practices, and contributes to interdisciplinary efforts that integrate computational tools into routine forensic workflows.

Jennifer Washington, PhD, is an independent health services researcher whose work focuses on health equity, mental health access, and the social and structural determinants of care utilization in the United States. Her doctoral research examined racial, socioeconomic, and geographic disparities in mental health service use during the COVID-19 pandemic using national survey data and systems-based theory. Her current scholarship extends this foundation by integrating equity, governance, and responsible artificial intelligence frameworks to advance more just, sustainable, and population-centered health systems.

Index

A

B

C

D

E

F

G

H

I

L

M

N

P

R

S

T

U

V

W